Nonlinear Optimization

Nonlinear Optimization

Andrzej Ruszczyński

PRINCETON UNIVERSITY PRESS
PRINCETON AND OXFORD

Published by Princeton University Press, 41 William Street, Princeton, New Jersey 08540
In the United Kingdom: Princeton University Press, 3 Market Place, Woodstock, Oxfordshire OX20 1SY

Library of Congress Cataloging-in-Publication Data

Ruszczyński, Andrzej P.
Nonlinear optimization / Andrzej Ruszczyński.
p. cm.
Includes bibliographical references and index.
ISBN-13: 978-0-691-11915-1 (acid-free paper)
ISBN-10: 0-691-11915-5 (acid-free paper)
1. Mathematical optimization. 2. Nonlinear theories. I. Title.

QA402.5.R87 2006
519.6–dc22 2005049614

British Library Cataloging-in-Publication Data is available

The publisher would like to acknowledge the author of this volume
for providing the camera-ready copy from which this book was printed.

This book has been composed in *Times Roman* and *MathTime* using LaTeX.

Printed on acid-free paper. ∞

pup.princeton.edu

Printed in the United States of America

10 9 8 7 6 5 4 3 2 1

to Darinka

Wieleż lat czekać trzeba, nim się przedmiot świeży
Jak figa ucukruje, jak tytuń uleży?

– Adam Mickiewicz, *Dziady, Część III*

In how few years are fresh themes matured,
As figs grow candied or tobacco cured?

– Adam Mickiewicz, *Forefathers' Eve, Part III*

Contents

Preface

Optimization is a mature and very successful area of applied mathematics. It deals with the analysis and solution of problems of finding the best element in a set. The structure of the set and the way the term "best" is understood determine different fields of the optimization theory: linear, integer, stochastic, nonsmooth optimization, optimal control, semi-infinite programming, etc. Within these fields, and across them, several directions of research are pursued: optimality conditions, duality theory, sensitivity analysis, and numerical methods.

Applications of optimization are very broad and diverse. The most notable areas of application include engineering, statistics, economics, management sciences, computer science, and mathematics itself. Optimization problems arise in approximation theory, probability theory, structure design, chemical process control, routing in telecommunication networks, image reconstruction, experiment design, radiation therapy, asset valuation, portfolio management, supply chain management, facility location, among others.

Many books discuss specific topics within the area of optimization, different groups of numerical methods, and particular application areas. My intention is to provide a book that will help to enter the area of optimization and to understand its modern ideas, principles, and methods within a reasonable time, but without sacrificing mathematical precision. Optimization problems arising in applications usually require additional analysis. It is my firm conviction, supported by many years of experience, that understanding of the mathematical foundations of optimization is the key to successful analysis of applied problems and to the choice or construction of an appropriate numerical method.

Our way to achieve these objectives is to present both the theory and the methods in a unified and coherent way. Such an approach requires a careful selection of the material, in order to focus attention on the most important principles.

The first part of the book is devoted to the theory of optimization. We present theoretical foundations of convex analysis, with the focus on the elements that are necessary for development of the optimization theory. Then we develop necessary and sufficient conditions of optimality and the duality theory for nonlinear optimization problems. In the second part of the book, we apply the theory to construct numerical methods of optimization, and to analyze their applicability and efficiency. While we focus on the most important ideas, we do not restrict our presentation to basic topics only. We cover rather modern concepts and techniques, such as methods for solving optimization problems involving nonsmooth functions. This is possible because of the unified treatment of the theory and methods, which clarifies and simplifies the description of the algorithmic ideas.

This book grew out of my lecture notes developed during 25 years of teaching optimization at Warsaw University of Technology, Princeton University, University of Wisconsin-Madison, and Rutgers University, for students of engineering, applied mathematics, and management sciences. A one-semester course can be taught on the basis of the chapters on convex analysis (without the subdifferential calculus), optimality conditions and duality for smooth problems, and numerical methods for smooth problems. It is well-suited for graduate students in engineering, applied mathematics, or management sciences, who have had no prior classes on optimization. The inclusion of the material on nondifferentiable optimization is possible in a one-year course, or in a course for students who have already had contact with optimization on a basic level and who have a good background in multivariate calculus and linear algebra.

The book can also be used for independent study. It is self-contained, it has many examples and figures illustrating the relevance and applications of the theoretical concepts, and it has a collection of carefully selected exercises. I hope that it will serve as a stepping stone for all those who want to pursue further studies of advanced topics in the area of optimization.

I would like to thank Jacek Szymanowski and Andrzej Wierzbicki for introducing me to the area of nonlinear optimization. Many thanks also go to Robert Vanderbei for encouraging me to undertake this project and for his support. Ekkehard Sachs provided me with invaluable feedback, which led to significant improvements in the chapters on numerical methods. And special thanks are due to my wife and closest collaborator, Darinka Dentcheva, whose continuous encouragement, creative suggestions, and infinite patience helped me see this project to its conclusion.

Princeton, New Jersey, July 2005 *Andrzej Ruszczyński*

Nonlinear Optimization

Chapter One

Introduction

A general optimization problem can be stated very simply as follows. We have a certain set X and a function f which assigns to every element of X a real number. The problem is to find a point $\hat{x} \in X$ such that

$$f(\hat{x}) \leq f(x) \quad \text{for all} \quad x \in X.$$

The set X is usually called the *feasible set*, and the function f is called the *objective function.*

At this level of generality, very little can be said about optimization problems. In this book, we are interested in problems in which X is a subset of a finite dimensional real space $\mathbb{R}^n$, and the function f is sufficiently regular, for example, convex or differentiable. Frequently, the definition of X involves systems of equations and inequalities, which we call *constraints.*

The simplest case is when $X = \mathbb{R}^n$. Such a problem is called the *unconstrained optimization problem.* If X is a strict subset of the space $\mathbb{R}^n$, we speak about the *constrained problem.* The most popular constrained optimization problem is the *linear programming* problem, in which f is a linear function and the set X is defined by finitely many linear equations and inequalities.

If the objective function f or some of the equations or inequalities defining the feasible set X are nonlinear, the optimization problem is called the *nonlinear optimization* (or nonlinear programming) *problem.* In this case, the specific techniques and theoretical results of linear programming cannot be directly applied, and a more general approach is needed.

Nonlinear optimization problems have attracted the attention of science since ancient times. We remember classical geometrical problems like the problem of finding the largest area rectangle inscribed in a given circle, or the problem of finding the point that has the smallest sum of distances to the vertices of a triangle. We know that optimization occurs in nature: many laws of physics are formulated as principles of minimum or maximum of some scalar characteristics of observed objects or systems, like energy or entropy. Interestingly, bee cells have the optimal shape that minimizes the average amount of wall material per cell.

But the most important questions leading to optimization problems are associated with human activities. What is the cheapest way of reaching a certain goal? What is the maximum effect that can be achieved, given limited resources? Which model reproduces the results of observations in the best way? These, and many other similar questions arise in very different areas of science, and motivate the development of nonlinear optimization models, nonlinear optimization theory, and numerical methods of optimization.

Let us examine several typical application problems. Of course, the versions of problems that we present here are drastically simplified for the purpose of illustrating their main features.

Image Reconstruction

One of the diagnostic methods used in modern medicine is *computer tomography*. Its idea is to send a beam through a part of the body at many angles, and to reconstruct a three-dimensional image of the internal organs from the results. Formally, this procedure can be described as follows. A part of the body (for example, the head) is represented as a set $S \subset \mathbb{R}^3$. For the purpose of analysis we assume that S is composed of very small cells S_i, $i = 1, \ldots, n$ (for example, cubes). We further assume that the density of the matter is constant within each cell, similar to the darkness of each pixel in a digital picture. We denote by x_i the density within cell S_i. These quantities are unknown and the purpose of the analysis is to establish their values.

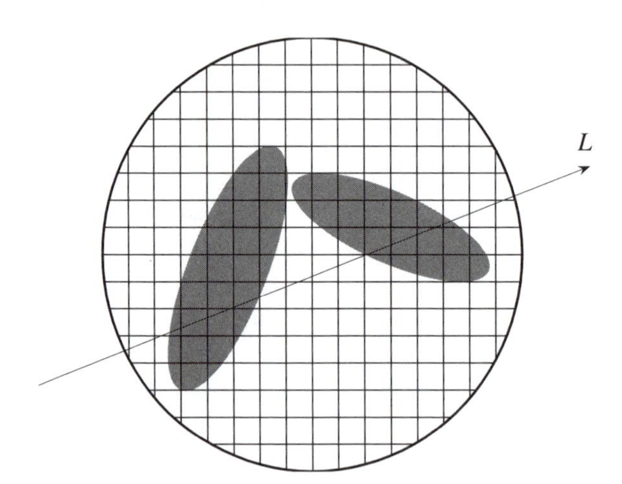

Figure 1.1. Beam L intersects a subset of cells.

A beam L passing through the set S intersects some of the cells (see Figure 1.1). Let $I(L)$ be this subset of cells. For every $i \in I(L)$ we know the length $a_i(L)$ of the path of the beam within cell i. The attenuation of the beam's energy depends on the total amount of matter on its way, $\sum_{i \in I(L)} a_i(L)x_i$. For every beam, given the measurements of its energy at the exit point, we can estimate this sum by some quantity $b(L)$. Thus, we obtain the linear relation

$$\sum_{i \in I(L)} a_i(L)x_i = b(L).$$

If this analysis is repeated for a collection of beams L_j, $j = 1, \dots, m$, we obtain the system of linear equations

$$\sum_{i \in I(L_j)} a_i(L_j)x_i = b(L_j), \quad j = 1, \dots, m. \tag{1.1}$$

In addition, the physical nature of the problem dictates the inequalities:

$$x_i \geq 0, \quad i = 1, \dots, n. \tag{1.2}$$

Ideally, if all quantities were measured perfectly and the set S was actually built of small, uniform cells, the unknown density vector x would be a solution of this system. But several difficulties immediately arise. The number of beams, m, is not usually equal to the number of cells, n. The system of equations (1.1) may have too few equations to determine x in a unique way, or it may have too many or inconsistent equations. Its solution may violate inequalities (1.2), or it may have no solution at all. Moreover, the cell model is an approximation, the measurements have errors of their own, and it may be unreasonable to attempt to perfectly match their results.

For all these reasons, it is better to abandon the idea of perfectly satisfying equations (1.1) and to use instead some measure of fit. It may be the sum of the squares

$$f(x) = \sum_{j=1}^{m} \Big(b(L_j) - \sum_{i \in I(L_j)} a_i(L_j)x_i \Big)^2.$$

The smaller the values of f, the better, and therefore we formulate the following problem:

$$\begin{aligned} &\text{minimize} \ \sum_{j=1}^{m} \Big(b(L_j) - \sum_{i \in I(L_j)} a_i(L_j)x_i \Big)^2 \\ &\text{subject to} \ \ x \geq 0. \end{aligned} \tag{1.3}$$

The above notation means that we want to find a point $\hat{x} \geq 0$ such that

$$f(\hat{x}) \leq f(x) \quad \text{for all} \quad x \geq 0.$$

This problem will always have an optimal solution, because the function $f(x) \to \infty$ if any coordinate $x_i \to \infty$, and $f(\cdot)$ is a continuous function.

Our optimization problem (1.3) may still have some flaws. If the number of cells is larger than the number of beams used, the problem may have many optimal solutions. We would prefer a model that selects one of them in some reasonable way. One possibility is to modify the function $f(x)$ by adding to it the total amount of matter:

$$f(x) = \sum_{j=1}^{m} \Big(b(L_j) - \sum_{i \in I(L_j)} a_i(L_j) x_i \Big)^2 + \delta \sum_{j=1}^{n} x_j,$$

where $\delta > 0$ is a model parameter. The objective function of problem (1.3) and of the last modification is quadratic, and the constraints are linear. Such problems are called *quadratic programming problems*.

There are numerous other ways to formulate a meaningful objective function in this problem. They may involve nonlinear terms representing our knowledge about the properties of the image. We may formulate a stochastic model of attenuation and an appropriate maximum likelihood estimator. There is no need to discuss all these forms here. Most important for us is that the optimization model provides the flexibility that the system of equations did not have.

Portfolio Optimization

We want to invest our capital K in n assets. If we invest an amount z_i in asset i, after a fiscal quarter the value of this investment will become $(1 + R_i)z_i$, where R_i is the quarterly return rate of asset i. The return rate R_i is a random variable. Assume that all capital K is invested: $z_1 + z_2 + \cdots + z_n = K$. Introducing the variables $x_i = z_i/K$, $i = 1, \ldots, n$, we can express the quarterly return rate of the entire portfolio as

$$\begin{aligned} R(x) &= \frac{1}{K}\Big((1 + R_1)z_1 + (1 + R_2)z_2 + \cdots + (1 + R_n)z_n \Big) - 1 \\ &= R_1 x_1 + R_2 x_2 + \cdots + R_n x_n. \end{aligned}$$

It is a random quantity depending on our asset allocations x_i, $i = 1, \ldots, n$. By the definition of the variables x_i, they satisfy the equation

$$x_1 + x_2 + \cdots + x_n = 1.$$

We cannot sell assets that we do not own and thus we require that

$$x_i \geq 0, \quad i = 1, \ldots, n.$$

We denote by X the set of asset allocations defined by the last two relations. Our idea is to decide about the asset allocations by an optimization problem.

The expected return rate of the portfolio has the form

$$\mathbb{E}\big[R(x)\big] = \mathbb{E}\big[R_1\big]x_1 + \mathbb{E}\big[R_2\big]x_2 + \cdots + \mathbb{E}\big[R_n\big]x_n.$$

However, the problem

$$\underset{x \in X}{\text{maximize}} \ \mathbb{E}\big[R(x)\big]$$

has a trivial and meaningless solution: invest everything in the asset j^* having the highest expected return rate $\mathbb{E}\big[R_{j^*}\big]$. In order to correctly model our preferences, we have to modify the problem. We have to introduce the concept of *risk* into the model and to express the aversion to risk in the objective function. One way to accomplish this is to consider the variance of the portfolio return rate, $\mathbb{V}\big[R(x)\big]$, as the measure of risk. Then we can formulate a much more meaningful optimization problem

$$\underset{x \in X}{\text{maximize}} \ \mathbb{E}\big[R(x)\big] - \delta\mathbb{V}\big[R(x)\big].$$

The parameter $\delta > 0$ is fixed here. It represents the trade-off between the mean return rate and the risk. It is up to the modeler to select its value.

It is convenient to introduce a notation for the mean return rates,

$$\mu_i = \mathbb{E}\big[R_i\big], \quad i = 1, \ldots, n,$$

for their covariances

$$c_{ij} = \operatorname{cov}\big[R_i, R_j\big] = \mathbb{E}\big[(R_i - \mu_i)(R_j - \mu_j)\big], \quad i, j = 1, \ldots, n,$$

and for the covariance matrix

$$C = \begin{bmatrix} c_{11} & c_{12} & \cdots & c_{1n} \\ c_{21} & c_{22} & \cdots & c_{2n} \\ \vdots & \vdots & & \vdots \\ c_{n1} & c_{n2} & \cdots & c_{nn} \end{bmatrix}.$$

Then we can express the mean return rate and the variance of the return rate as follows:

$$\mathbb{E}\big[R(x)\big] = \sum_{i=1}^{n} \mu_i x_i = \langle \mu, x \rangle,$$

$$\mathbb{V}\big[R(x)\big] = \sum_{i=1}^{n} \sum_{j=1}^{n} c_{ij} x_i x_j = \langle x, Cx \rangle.$$

The optimization problem assumes the lucid form

$$\underset{x \in X}{\text{maximize}} \ \langle \mu, x \rangle - \delta \langle x, Cx \rangle. \tag{1.4}$$

The objective function of this problem is quadratic and the constraints are linear. Such an optimization problem is called a *quadratic optimization problem*.

Our model can be modified in various ways. Most importantly, the variance is not a good measure of risk, as it penalizes the surplus return (over the mean) equally as the shortfall. One possibility is to measure risk by the central semideviation of order p:

$$\sigma_p\big[R(x)\big] = \Big(\mathbb{E}\big[\max\big(0, \mathbb{E}[R(x)] - R(x)\big)\big]^p\Big)^{1/p},$$

with $p \geq 1$. Then the optimization problem

$$\underset{x \in X}{\text{maximize}}\ \langle \mu, x \rangle - \delta \sigma_p\big[R(x)\big]$$

becomes more difficult, because it involves a nonsmooth objective function.

We can also introduce additional constraints on asset allocations. And for all these models, we can change the trade-off coefficient $\delta > 0$ to analyze its influence on the solution to the problem.

Signal Processing

A transmitter sends in short time intervals I_k harmonic signals

$$s_k(t) = a_k \cos(\omega t), \quad t \in I_k, \quad k = 1, 2, \ldots.$$

The pulsation ω is fixed. The signals are received by n receivers. Because of their different locations, each of the receivers sees the signal with a different amplitude, and with a different phase shift. The incoming signals are equal to

$$u_{kj} = c_j a_k \cos(\omega t - \psi_j), \quad j = 1, \ldots, n, \quad k = 1, 2, \ldots,$$

where c_j and ψ_j denote the amplitude multiplier and the phase shift associated with receiver j.

It is convenient to represent the signal s_k as a complex number S_k with the real and imaginary parts $\Re(S_k) = a_k$ and $\Im(S_k) = 0$. Similarly, each amplitude and phase shift are represented by the complex number

$$H_j = c_j e^{-i\psi_j}, \quad j = 1, \ldots, n.$$

Then the incoming signals can be represented in the complex domain as follows:

$$U_{kj} = H_j S_k, \quad j = 1, \ldots, n, \quad k = 1, 2, \ldots.$$

Because of various noises and interferences, the receivers actually receive the signals

$$X_{kj} = U_{kj} + Z_{kj}, \quad j = 1, \dots, n, \quad k = 1, 2, \dots.$$

Here Z_{kj} denotes the complex representation of the noise at receiver j in the kth interval. To facilitate the reconstruction of the transmitted signal $\{S_k\}$, we want to combine the received signals with some complex weights W_1, W_2, ..., W_n to obtain the output signals:

$$Y_k = \sum_{j=1}^{n} W_j^* X_{kj}, \quad k = 1, 2, \dots.$$

In the formula above, W_j^* denotes the complex conjugate of W_j.

Introduce the notation:

$$H = \begin{bmatrix} H_1 \\ \vdots \\ H_n \end{bmatrix}, \qquad W = \begin{bmatrix} W_1 \\ \vdots \\ W_n \end{bmatrix}, \qquad Z_k = \begin{bmatrix} Z_{k1} \\ \vdots \\ Z_{kn} \end{bmatrix}.$$

We have

$$Y_k = \sum_{j=1}^{n} W_j^* \big(H_j S_k + Z_{kj} \big) = (W^* H) S_k + W^* Z_k, \quad k = 1, 2, \dots,$$

where $W^* = \big[W_1^* \; W_2^* \; \dots \; W_n^* \big]$.

Our objective is to find the weights W_j to maximize the useful part of the signal Y_k, $k = 1, 2, \dots$, relative to the part due to the noise and interferences. To formalize this objective we consider the sequences $\{S_k\}$ and $\{Z_k\}$ as discrete-time stochastic processes with values in $\mathbb{C}$ and $\mathbb{C}^n$, respectively.† The noise process $\{Z_k\}$ is assumed to be uncorrelated with the signal $\{S_k\}$ and to have the covariance matrix

$$\Theta = \mathbb{E}\big[Z_k Z_k^* \big].$$

The energy of the transmitted signal is denoted by

$$\sigma^2 = \mathbb{E}\big[S_k S_k^* \big].$$

The energy of the output signal $\{Y_k\}$ can be calculated as follows:

$$\mathbb{E}\big[Y_k Y_k^* \big] = \sigma^2 \| W^* H \|^2 + W^* \Theta W. \tag{1.5}$$

†The symbol $\mathbb{C}$ denotes the space of complex numbers.

Here the symbol $\|A\|$ denotes the norm of the complex vector $A \in \mathbb{C}^n$,

$$\|A\| = \Big(\sum_{j=1}^{n} |A_j|^2\Big)^{1/2} \quad \text{with} \quad |A_j|^2 = A_j A_j^*.$$

The first component on the right hand side of (1.5) represents the part of the output signal energy due to the transmitted signal $\{S_k\}$, while the second component is the energy resulting from the noise. The basic problem of minimizing the noise-to-signal ratio can now be formalized as follows:

$$\begin{aligned} &\text{minimize} \;\; W^* \Theta W \\ &\text{subject to} \;\; \|W^* H\| \geq 1. \end{aligned} \tag{1.6}$$

Unfortunately, in practice the vector H is not exactly known. We would like to formulate a problem that allows H to take any value of form $H_0 + \Delta$, where H_0 is some nominal value and $\|\Delta\| \leq \delta$, with the uncertainty radius $\delta > 0$ given. Problem (1.6) becomes:

$$\begin{aligned} &\text{minimize} \;\; W^* \Theta W \\ &\text{subject to} \;\; \|W^*(H_0 + \Delta)\| \geq 1 \quad \text{for all} \quad \|\Delta\| \leq \delta. \end{aligned}$$

It is a rather complex nonlinear optimization problem with a quadratic objective and with the feasible set defined by infinitely many nonlinear constraints. We can transform it to a simpler form as follows.

First, we notice that the worst value of Δ for each W is $\Delta = -\delta W/\|W\|$. Thus all constraints are equivalent to the inequality

$$\|W^* H\| \geq 1 + \delta \|W\|.$$

Also, adding a phase shift to all components of W does not change the objective and the constraints, and therefore we can assume that the imaginary part of $W^* H$ is zero. This yields the problem

$$\begin{aligned} \text{minimize} \;\; & W^* \Theta W \\ \text{subject to} \;\; & \Re(W^* H) \geq 1 + \delta \|W\|, \\ & \Im(W^* H) = 0. \end{aligned} \tag{1.7}$$

It has a convex objective function, one convex nonlinear constraint, and one linear constraint.

Classification

We have two sets: $A = \{a_1, \ldots, a_k\}$, and $B = \{b_1, \ldots, b_m\}$ in $\mathbb{R}^n$. Each vector a_i may be, for example, the set of results of various measurements or tests performed on patients for whom a particular treatment was effective, while the vectors b_j may represent results of the same tests for patients who did not respond to this treatment.

Our goal is to find a *classifier*: ideally it would be a function $\varphi : \mathbb{R}^n \to \mathbb{R}$ such that $\varphi(a_i) > 0$ for all $i = 1, \dots, k$, and $\varphi(b_j) < 0$ for all $j = 1, \dots, m$. When a new vector of results $c \in \mathbb{R}^n$ appears, we can assign it to sets A or B, depending on the sign of $\varphi(c)$.

This problem is usually modeled as follows. We first determine a family of functions $\varphi(\cdot)$ in which we look for a suitable classifier. Suppose we decided to use linear classifiers of the form

$$\varphi(a) = \langle v, a \rangle - \gamma .$$

In this formula $v \in \mathbb{R}^n$ and $\gamma \in \mathbb{R}$ are unknown. To find the "best" classifier in this family means to determine the values of v and γ.

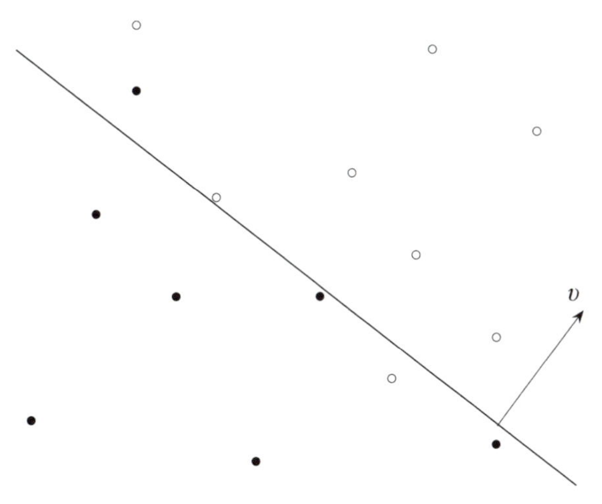

Figure 1.2. Linear classifier. The white points belong to the set A and the dark points belong to the set B.

When a fixed family of classifiers is used, we cannot expect to be able to separate the sets A and B with a classifier from this family. There is always a possibility of misclassification, and we have to determine the way in which we shall decide which classifier is "better." Figure 1.2 illustrates this issue for two sets and for a linear classifier. One way to measure the quality of classification is to consider for each point a_i the classification error:

$$e_-(a_i; v, \gamma) = \begin{cases} 0 & \text{if } \langle v, a_i \rangle - \gamma \geq 0, \\ \gamma - \langle v, a_i \rangle & \text{if } \langle v, a_i \rangle - \gamma < 0. \end{cases}$$

Similarly, for each point b_j the classification error is

$$e_+(b_j; v, \gamma) = \begin{cases} 0 & \text{if } \langle v, b_j \rangle - \gamma \le 0, \\ \langle v, b_j \rangle - \gamma & \text{if } \langle v, b_j \rangle - \gamma > 0. \end{cases}$$

Then our problem can be formulated as an optimization problem

$$\operatorname*{minimize}_{v,\gamma} \sum_{i=1}^{k} e_-(a_i; v, \gamma) + \sum_{j=1}^{m} e_+(b_j; v, \gamma). \tag{1.8}$$

We immediately notice a flaw in this model: setting $\gamma = 0$ and $v = 0$ we can make all classification errors zero, but the resulting classifier is useless. We need to restrict the set of parameters to ensure a discriminating power of the classifier. One possible condition is

$$\|v\| = 1. \tag{1.9}$$

Problem (1.8)–(1.9) is a nonlinear optimization problem with a piecewise linear objective function and a nonlinear constraint. By introducing *slack variables* $s \in \mathbb{R}^k$ and $z \in \mathbb{R}^m$, we can rewrite this problem as follows:

$$\begin{aligned} \operatorname*{minimize}_{v,\gamma,s,z} \quad & \sum_{i=1}^{k} s_i + \sum_{j=1}^{m} z_j \\ \text{subject to} \quad & \langle v, a_i \rangle - \gamma + s_i \ge 0, \quad i = 1, \dots, k, \\ & \langle v, b_j \rangle - \gamma - z_j \le 0, \quad j = 1, \dots, m, \\ & s \ge 0, \ z \ge 0, \\ & \|v\| = 1. \end{aligned} \tag{1.10}$$

Another way to guarantee a discriminating power of the classifier is to enforce a "buffer zone" between the two sets, by requiring that

$$\begin{aligned} & \langle v, a_i \rangle - \gamma + s_i \ge 1, \quad i = 1, \dots, k, \\ & \langle v, b_j \rangle - \gamma - z_j \le -1, \quad j = 1, \dots, m. \end{aligned}$$

The width of the buffer zone is equal to $2/\|v\|$. The condition $\|v\| = 1$ is no longer needed, because $v = 0$ and $\gamma = 0$ are no longer attractive. However, we have to prevent $\|v\|$ from becoming too large (otherwise the buffer zone has little meaning). One way to accomplish this is to add to the objective function a term which quickly grows with $\|v\|$, as in the problem below:

$$
\begin{aligned}
\underset{v,\gamma,s,z}{\text{minimize}} \quad & \sum_{i=1}^{k} s_i + \sum_{j=1}^{m} z_j + \delta\|v\|^2 \\
\text{subject to} \quad & \langle v, a_i \rangle - \gamma + s_i \geq 1, \quad i = 1, \dots, k, \\
& \langle v, b_j \rangle - \gamma - z_j \leq -1, \quad j = 1, \dots, m, \\
& s \geq 0, \; z \geq 0.
\end{aligned}
\tag{1.11}
$$

Here $\delta > 0$ is a fixed parameter of the problem. This quadratic optimization problem, under the name *support vector machine*, is commonly used in computer science for data classification.

Optimal Control

A robot has to move an object of mass M from position x^{ini} to position x^{fin} within time $[0, T]$ avoiding an obstacle (see Figure 1.3).

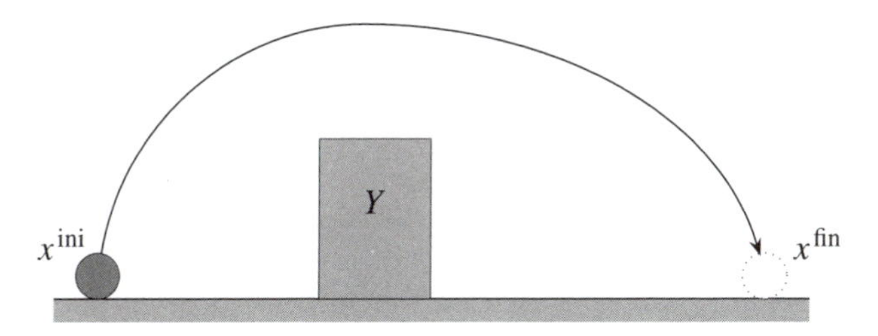

Figure 1.3. The object has to be moved from x^{ini} to x^{fin} avoiding the obstacle Y.

One way to approach this problem is to split the time interval $[0, T]$ into n equal pieces

$$
I_k = \left[\frac{kT}{n}, \frac{(k+1)T}{n}\right), \quad k = 0, \dots, n-1,
$$

and to apply a constant force to the object within each interval. Denote by F_k the force used in interval k, and by v_k the velocity of the object at the beginning of the interval k. Both F_k and v_k are three-dimensional vectors. The velocity at the end of interval k equals

$$
v_{k+1} = v_k + \frac{T}{nM}(F_k + G), \quad k = 1, \dots, n. \tag{1.12}
$$

Here G denotes the gravitation force. We ignore, for simplicity, the mass of the robot arm and the mechanics of the arm.

Denoting by x_k the three-dimensional vector representing the location of the object at the beginning of interval k, we can write the equation for the position at the end of interval k:

$$x_{k+1} = x_k + \frac{T}{n} v_k + \frac{T^2}{2n^2 M}(F_k + G), \quad k = 1, \dots, n. \tag{1.13}$$

Equations (1.12)–(1.13) are an example of state equations for a discrete-time dynamical system. The six-dimensional vector (x_k, v_k) represents the *state* of the system at the beginning of the kth interval. The three-dimensional vector F_k is the *control* applied in the kth interval.

The condition that the object cannot hit the obstacle can be modeled as follows. We describe the obstacle as a certain closed set $Y \subset \mathbb{R}^3$ and we introduce the distance function†

$$d(x, Y) \triangleq \min_{y \in Y} \|x - y\|.$$

Our requirement can now be formulated as follows:

$$d(x_k, Y) \geq \delta, \quad k = 0, 1, \dots, n,$$

where $\delta > 0$ is some minimum distance to the obstacle allowed. For the purpose of practical tractability we may replace this condition with a slightly weaker one, formulated for a sufficiently dense net of points $y^j \in Y$, $j = 1, \dots, J$:

$$\|x^k - y^j\| \geq \delta, \quad k = 0, 1, \dots, n, \quad j = 1, \dots, J. \tag{1.14}$$

Other constraints are obvious from the problem formulation:

$$\begin{gathered} (x_0, v_0) = (x^{\text{ini}}, 0), \quad (x_n, v_n) = (x^{\text{fin}}, 0), \\ x_{k3} \geq 0, \; k = 1, \dots, n-1. \end{gathered} \tag{1.15}$$

Finally, to formulate an optimization problem, we need to define a suitable objective function. Many reasonable formulations are possible. We can be concerned with the comfort of the travel by minimizing the variation of the force:

$$f_1(F) = \|F_0 + G\| + \sum_{k=1}^{n-1} \|F_k - F_{k-1}\| + \|F_{k-1} + G\|.$$

We can minimize the total work:

$$f_2(x, F) = \sum_{k=1}^{n-1} \langle F_k, x_{k+1} - x_k \rangle,$$

†The symbol $\triangleq$ means equal by definition.

a combination of the two functions, or some other expression representing our preferences. The decision variables in the model are (x_k, v_k, F_k), $k = 1, \dots, n$, and the state equations (1.12)–(1.13), as well as additional conditions (1.14)–(1.15), are constraints of the problem.

Approximation of Functions

We have a certain space of functions $\mathcal{S}$ defined on a domain $D \subset \mathbb{R}^m$, and a function $\psi \in \mathcal{S}$. The space $\mathcal{S}$ may be, for example, the space $\mathcal{C}(D)$ of continuous functions on a compact set D, or the space of integrable functions $\mathcal{L}_p(D, \mathcal{B}, \mu)$, where $p \in [1, \infty]$, D is a Lebesgue measurable set in $\mathbb{R}^m$, μ is the Lebesgue measure, and $\mathcal{B}$ is a σ-algebra of measurable subsets of D. The space $\mathcal{S}$ is equipped with a metric $\operatorname{dist}(\cdot, \cdot)$.

We are also given a mapping $A : \mathbb{R}^n \to \mathcal{S}$. For example, it may have the form

$$A(x) = \sum_{j=1}^{n} x_j \varphi_j,$$

where φ_j, $j = 1, \dots, n$, are given functions in $\mathcal{S}$. The last relation is understood as follows: $A(x)$ is a function with values

$$\big[A(x)\big](u) = \sum_{j=1}^{n} x_j \varphi_j(u), \quad u \in D.$$

The *approximation problem* is simply the optimization problem:

$$\underset{x \in \mathbb{R}^n}{\text{minimize}} \; \operatorname{dist}(\psi, A(x)).$$

Several special cases are of interest. If $\mathcal{S} = \mathcal{C}(D)$ and the distance between two continuous functions is measured by the maximum norm, we obtain the *Chebyshev approximation* problem

$$\underset{x \in \mathbb{R}^n}{\text{minimize}} \; \max_{u \in D} \big|\psi(u) - \big[A(x)\big](u)\big|.$$

If $\mathcal{S} = \mathcal{L}_p(D, \mathcal{B}, \mu)$, with some $p \in [1, \infty)$, then our problem has the form

$$\underset{x \in \mathbb{R}^n}{\text{minimize}} \int_D \big|\psi(u) - \big[A(x)\big](u)\big|^p \, d\mu.$$

A special case of practical importance arises when the domain D is finite: $D = \{u_1, u_2, \dots, u_m\}$. Then the approximated function ψ can be viewed as a finite collection of data $y_i = \psi(u_i)$, $i = 1, \dots, m$. The approximation

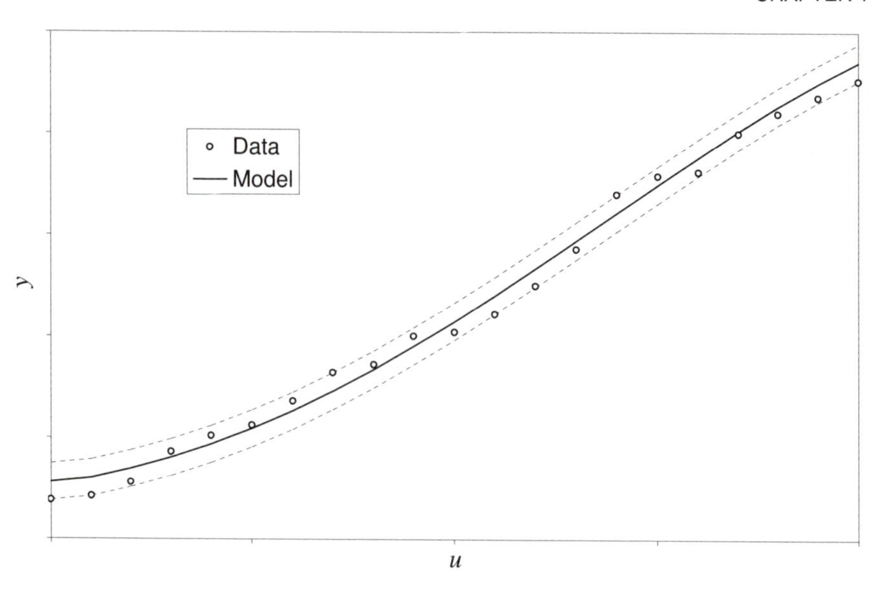

Figure 1.4. Chebyshev approximation.

problem is then the problem of constructing a model which is close to the data in the sense of the selected metric $\text{dist}(\cdot, \cdot)$.

Figure 1.4 illustrates the solution of a Chebyshev approximation problem with a finite set of data.

Optimization problems arise in engineering, economics, statistics, business, medicine, physics, chemistry, biology, and in mathematics itself. New optimization problems appear all the time, and scientists analyze their properties and look for their solutions. Frequently, as in the examples discussed above, the models have to be adjusted or modified, to reflect the intentions of their authors and the specificity of the application. Because of these reasons, almost no applied nonlinear optimization problem is exactly the same as models discussed in textbooks. Therefore, it is our intention to provide readers with tools that will help them to analyze their models, to choose the best solution methods, and to improve the models if the results are not appropriate. Such tools necessarily involve solid mathematical foundations.

PART 1

Theory

Chapter Two

Elements of Convex Analysis

2.1 CONVEX SETS

2.1.1 Basic Properties

The notion of a convex set is central to optimization theory. A convex set is such that, for any two of its points, the entire segment joining these points is contained in the set (see Figure 2.1).

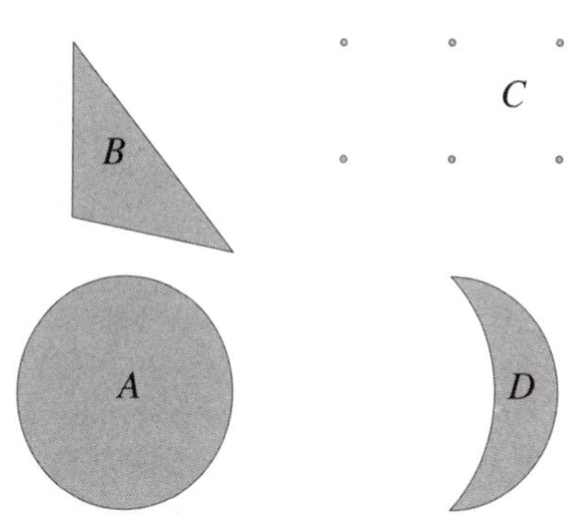

Figure 2.1. Convex sets (A, B) and nonconvex sets (C, D).

DEFINITION 2.1. A set $X \subset \mathbb{R}^n$ is called *convex* if for all $x^1 \in X$ and $x^2 \in X$ it contains all points

$$\alpha x^1 + (1 - \alpha)x^2, \quad 0 < \alpha < 1.$$

Convexity is preserved by the operation of intersection.

LEMMA 2.2. *Let I be an arbitrary index set. If the sets $X_i \subset \mathbb{R}^n$, $i \in I$, are convex, then the set $X = \bigcap_{i \in I} X_i$ is convex.*

Proof. If two points x^1 and x^2 are elements of X, they belong to each X_i. Since X_i's are convex, the segment joining these points is included in X_i, for all i. It is, therefore, included in their intersection. □

Sets can be subjected to algebraic operations similarly to points in $\mathbb{R}^n$.

We can multiply a set X by a scalar c to get

$$cX \triangleq \{y \in \mathbb{R}^n : y = cx,\ x \in X\}.$$

The *Minkowski sum* of two sets is defined as follows:

$$X + Y \triangleq \{z \in \mathbb{R}^n : z = x + y,\ x \in X,\ y \in Y\}.$$

These operations preserve convexity.

LEMMA 2.3. *Let X and Y be convex sets in $\mathbb{R}^n$ and let c and d be real numbers. Then the set $Z = cX + dY$ is convex.*

Proof. If $z^1 \in Z$ then $z^1 = cx^1 + dy^1$ with $x^1 \in X$ and $y^1 \in Y$. Similarly, $z^2 \in Z$ has the form $z^2 = cx^2 + dy^2$ with $x^2 \in X$ and $y^2 \in Y$. Then, for every $\alpha \in [0, 1]$,

$$\alpha z^1 + (1 - \alpha)z^2 = c(\alpha x^1 + (1 - \alpha)x^2) + d(\alpha y^1 + (1 - \alpha)y^2) \in Z,$$

as required. □

A point $\alpha x^1 + (1 - \alpha)x^2$, where $\alpha \in [0, 1]$, appearing in Definition 2.1, belongs to the segment joining x^1 and x^2. We can "join" more points by constructing their convex hull.

DEFINITION 2.4. A point x is called a *convex combination* of points $x^1, \dots, x^m$ if there exist $\alpha_1 \geq 0, \dots, \alpha_m \geq 0$ such that

$$x = \alpha_1 x^1 + \alpha_2 x^2 + \cdots + \alpha_m x^m$$

and

$$\alpha_1 + \alpha_2 + \cdots + \alpha_m = 1.$$

DEFINITION 2.5. The *convex hull* of the set X (denoted by $\operatorname{conv} X$) is the intersection of all convex sets containing X.

The relation between these two concepts is the subject of the next lemma.

LEMMA 2.6. *The set* $\operatorname{conv} X$ *is the set of all convex combinations of points of X.*

Proof. Let us consider the set Y of all convex combinations of elements of X. If $y^1 \in Y$ and $y^2 \in Y$, then

$$y^1 = \alpha_1 x^1 + \alpha_2 x^2 + \cdots + \alpha_m x^m,$$
$$y^2 = \beta_1 z^1 + \beta_2 z^2 + \cdots + \beta_l z^l,$$

where $x^1 \dots, x^m, z^1, \dots, z^l \in X$, all α's and β's are nonnegative, and

$$\sum_{i=1}^{m} \alpha_i = 1, \qquad \sum_{i=1}^{l} \beta_i = 1.$$

Therefore, for every $\lambda \in (0, 1)$, the point

$$\lambda y^1 + (1 - \lambda) y^2 = \sum_{i=1}^{m} \lambda \alpha_i x^i + \sum_{i=1}^{l} (1 - \lambda) \beta_i z^i$$

is a convex combination of the points $x^1, \dots, x^m, z^1, \dots, z^l$. Consequently, the set Y is convex. Obviously, $Y \supset X$. Thus

$$\operatorname{conv} X \subset Y.$$

On the other hand, if $y \in Y$ then y, as a convex combination of points of X, must be contained in every convex set containing X. Hence, $\operatorname{conv} X \supset Y$, which completes the proof. □

For example, the convex hull of three points in $\mathbb{R}^2$ is the triangle having these points as vertices.

In the above result we considered all convex combinations, that is, for all numbers m of points and for arbitrary selections of these points. Both m and the points selected can be restricted. We now show that the number of points m need not be larger than $n + 1$. In Section 2.3 we further show that only some special points need to be considered, if the set X is compact.

Lemma 2.7. *If $X \subset \mathbb{R}^n$, then every element of* $\operatorname{conv} X$ *is a convex combination of at most $n + 1$ points of X.*

Proof. Let x be a convex combination of $m > n + 1$ points of X. We shall show that m can be reduced by one. If $\alpha_j = 0$ for some j, then we can delete the jth point and we are done. So, let all $\alpha_i > 0$. Since $m > n + 1$, one can find $\gamma_1, \gamma_2, \dots, \gamma_m$, not all equal 0, so that

$$\gamma_1 \begin{bmatrix} x^1 \\ 1 \end{bmatrix} + \gamma_2 \begin{bmatrix} x^2 \\ 1 \end{bmatrix} + \dots + \gamma_m \begin{bmatrix} x^m \\ 1 \end{bmatrix} = 0. \tag{2.1}$$

Let $\tau = \min \left\{ \frac{\alpha_i}{\gamma_i} : \gamma_i > 0 \right\}$. Note that τ is well defined, because some $\gamma_j > 0$, if their sum is zero. Let $\bar{\alpha}_i = \alpha_i - \tau \gamma_i, i = 1, 2, \dots, m$. By (2.1) we still have $\sum_{i=1}^{m} \bar{\alpha}_i = 1$ and $\sum_{i=1}^{m} \bar{\alpha}_i x^i = x$. By the definition of τ, at least one $\bar{\alpha}_j = 0$ and we can delete the jth point. Continuing in this way, we can reduce the number of points to $n + 1$. □

The last result is known as *Carathéodory's Lemma.*

We end this section with some basic relations of convexity and topological properties of a set.

LEMMA 2.8. *If X is convex, then its interior* int X *and its closure $\overline{X}$ are convex.*

Proof. Let B denote the unit ball. If $x^1 \in \text{int}\, X, x^2 \in \text{int}\, X$, then one can find $\varepsilon > 0$ such that $x^1 + \varepsilon B \subset X$ and $x^2 + \varepsilon B \subset X$. Thus $\alpha x^1 + (1-\alpha)x^2 + \varepsilon B \subset X$ for $0 < \alpha < 1$. Therefore $\alpha x^1 + (1-\alpha)x^2 \in \text{int}\, X$. To prove the second part of the lemma, let $x^k \to x$ and $y^k \to y$ with $x^k \in X$ and $y^k \in X$. Then the sequence of points $\alpha x^k + (1-\alpha)y^k$ is contained in X and converges to $\alpha x + (1-\alpha)y \in \overline{X}$. □

LEMMA 2.9. *Assume that the set $X \subset \mathbb{R}^n$ is convex. Then* int $X = \emptyset$ *if and only if X is contained in a linear manifold of dimension smaller than n.*

Proof. Let $x_0 \in X$. Consider the collection of vectors $x - x_0$ for all $x \in X$. Let m be the maximum number of linearly independent vectors in this collection. Then all vectors $x - x_0$, where $x \in X$, can be expressed as linear combinations of m selected vectors $v_1, v_2, \dots, v_m$. Denoting by $\text{lin}\{v_1, v_2, \dots, v_m\}$ the subspace of all linear combinations of the vectors $v_1, \dots, v_m$ we can write

$$X \subset x_0 + \text{lin}\{v_1, v_2, \dots, v_m\}.$$

If the set X has a nonempty interior, we can choose $x_0 \in \text{int}\, X$. Then a small ball about x_0 is included in X and we can choose exactly n independent vectors v_i in the construction above. Therefore, $m = n$ in this case. Conversely, suppose the set $\{x - x_0 : x \in X\}$ contains n linearly independent vectors $v_1, \dots, v_n$. By the convexity of X we have

$$X \supset \Big\{x_0 + \sum_{i=1}^{n} \gamma_i v_i : \sum_{i=1}^{n} \gamma_i \le 1,\ \gamma_i \ge 0,\ i = 1, \dots, n\Big\}.$$

Due to linear independence of the vectors v_i, the set on the right hand side is an n-dimensional simplex having a nonempty interior. □

2.1.2 Projection

Let us consider a convex closed set $V \subset \mathbb{R}^n$ and a point $x \in \mathbb{R}^n$. We call the point in V that is closest to x the *projection* of x on V and we denote it by $\Pi_V(x)$. Obviously, if $x \in V$ then $\Pi_V(x) = x$, but the projection is always well defined, as the following result shows.

THEOREM 2.10. *If the set $V \subset \mathbb{R}^n$ is nonempty, convex and closed, then for every $x \in \mathbb{R}^n$ there exists exactly one point $z \in V$ that is closest to x.*

Proof. Let $\mu = \inf\{\|z - x\| : z \in V\}$. Since V is nonempty, μ is finite. Let us consider a sequence of points $z^k \in V$ such that $\|z^k - x\| \to \mu$, as $k \to \infty$. It is bounded, so it must have a convergent subsequence, $\{z^k\}$, $k \in \mathcal{K}$. Denote the limit of this subsequence by z. We have

$$\|z - x\| = \lim_{\substack{k\to\infty \\ k\in\mathcal{K}}} \|z^k - x\| = \mu.$$

Since V is closed, $z \in V$. This proves the existence.

Suppose two different points $z^1 \in V$ and $z^2 \in V$ have distance μ from x. Consider the point $z = (z^1 + z^2)/2$. It belongs to V, by convexity. Its distance to x can be calculated by the Pythagorean theorem:

$$\|z - x\|^2 = \mu^2 - \frac{1}{4}\|z^1 - z^2\|^2 < \mu^2,$$

a contradiction. Thus the projection is unique. □

Lemma 2.11. *Assume that $V \subset \mathbb{R}^n$ is a closed convex set and let $x \in \mathbb{R}^n$. Then $z = \Pi_V(x)$ if and only if $z \in V$ and*

$$\langle v - z, x - z\rangle \le 0 \quad \textit{for all} \quad v \in V. \tag{2.2}$$

Proof. Let $z = \Pi_V(x)$ and $v \in V$ (see Figure 2.2). Consider points of the form

$$w(\alpha) = \alpha v + (1 - \alpha)z, \quad 0 \le \alpha \le 1.$$

By convexity, they are all in V and their distance to x cannot be smaller than $\|z - x\|$. We have

$$\begin{aligned}\|w(\alpha) - x\|^2 &= \langle z + \alpha(v - z) - x, z + \alpha(v - z) - x\rangle \\ &= \|z - x\|^2 + 2\alpha\langle z - x, v - z\rangle + \alpha^2\|v - z\|^2.\end{aligned}$$

Consider the above expression as a function of $\alpha \in [0, 1]$. It is bounded below by $\|z - x\|^2$ if and only if the linear term has a nonnegative coefficient. The last statement is identical to (2.2).

Suppose (2.2) for some $z \in V$. Setting $v = \Pi_V(x)$ in (2.2) we get

$$\langle \Pi_V(x) - z, x - z\rangle \le 0.$$

By the first part of our proof, as $z \in V$, inequality (2.2) has to be satisfied:

$$\langle z - \Pi_V(x), x - \Pi_V(x)\rangle \le 0.$$

Adding the last two inequalities we get

$$\langle \Pi_V(x) - z, \Pi_V(x) - z\rangle \le 0,$$

which is possible only if $\Pi_V(x) = z$. □

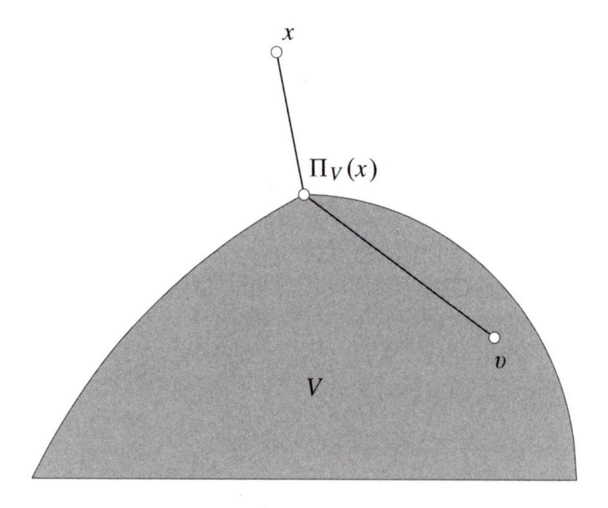

Figure 2.2. Projection.

In particular, if the set V is a linear manifold, for every $v \in V$ we have

$$w = 2\Pi_V(x) - v \in V$$

as well. Therefore the inequalities

$$\langle v - \Pi_V(x), x - \Pi_V(x) \rangle \le 0,$$
$$\langle w - \Pi_V(x), x - \Pi_V(x) \rangle \le 0,$$

are equivalent to

$$\langle v - \Pi_V(x), x - \Pi_V(x) \rangle = 0.$$

Consequently, $x - \Pi_V(x) \perp V$.

Example 2.12. A model represents the output variable, $y \in \mathbb{R}$, as a linear function,

$$y = \sum_{i=1}^{n} x_i u_i,$$

of input variables $u_1, \dots, u_n$. The quantities $x_1, \dots, x_n$ are unknown model coefficients. We have N observations of input and output variables: (u^j, y^j), $j = 1, \dots, N$. One way to determine the values of the coefficients $x_1, \dots, x_n$ is to minimize the sum of the squared errors:

$$f(x) = \sum_{j=1}^{N} \Big(y^j - \sum_{i=1}^{n} x_i u_i^j \Big)^2.$$

Defining the matrix of input data

$$U = \begin{bmatrix} u_1^1 & u_2^1 & \dots & u_n^1 \\ u_1^2 & u_2^2 & \dots & u_n^2 \\ \vdots & \vdots & & \vdots \\ u_1^N & u_2^N & \dots & u_n^N \end{bmatrix}$$

and the set of all possible model outcomes

$$V = \{Ux : x \in \mathbb{R}^n\},$$

we see that this step of the estimation is equivalent to finding the orthogonal projection of the vector $y = (y^1, \dots, y^N)$ on V:

$$\hat{y} = \Pi_V(y).$$

If there exist more than one x such that $Ux = \hat{y}$, the final step is to find in the set

$$X = \{x \in \mathbb{R}^n : Ux = \hat{y}\}$$

the minimum norm element $\hat{x}$, that is,

$$\hat{x} = \Pi_X(0).$$

The projection operator is nonexpansive.

THEOREM 2.13. *Assume that $V \subset \mathbb{R}^n$ is a closed convex set. Then for all $x \in \mathbb{R}^n$ and $y \in \mathbb{R}^n$ we have*

$$\|\Pi_V(x) - \Pi_V(y)\| \le \|x - y\|.$$

Proof. By Lemma 2.11,

$$\langle \Pi_V(y) - \Pi_V(x), x - \Pi_V(x) \rangle \le 0,$$
$$\langle \Pi_V(x) - \Pi_V(y), y - \Pi_V(y) \rangle \le 0.$$

Adding both sides we get

$$\|\Pi_V(x) - \Pi_V(y)\|^2 + \langle \Pi_V(x) - \Pi_V(y), y - x \rangle \le 0.$$

We obviously have that

$$0 \le \frac{1}{2}\|\Pi_V(x) - \Pi_V(y) + (y - x)\|^2$$
$$= \frac{1}{2}\|\Pi_V(x) - \Pi_V(y)\|^2 + \langle \Pi_V(x) - \Pi_V(y), y - x \rangle + \frac{1}{2}\|y - x\|^2.$$

Adding the last two inequalities we conclude that

$$\frac{1}{2}\|\Pi_V(x) - \Pi_V(y)\|^2 \le \frac{1}{2}\|y - x\|^2,$$

which was what we set out to prove. □

2.1.3 Separation Theorems

A convex closed set and a point outside of it can be separated by a plane.

THEOREM 2.14. *Let $X \subset \mathbb{R}^n$ be a closed convex set and let $x \notin X$. Then there exist a nonzero $y \in \mathbb{R}^n$ and $\varepsilon > 0$ such that*

$$\langle y, v \rangle \le \langle y, x \rangle - \varepsilon \quad \textit{for all} \quad v \in X.$$

Proof. Let z be the point of X which is closest to x. It exists, since X is closed. By Lemma 2.11,

$$\langle x - z, v - z \rangle \le 0 \quad \text{for all} \quad v \in X.$$

Define $y = x - z$. Note that y is nonzero, because $x \notin X$. For each $v \in X$ the last inequality implies that

$$\langle y, v \rangle \le \langle y, z \rangle = \langle y, x \rangle + \langle y, z - x \rangle = \langle y, x \rangle - \|y\|^2.$$

Setting $\varepsilon = \|y\|^2$ we obtain the assertion of the theorem. □

See Figure 2.3 for a geometrical interpretation of this property.

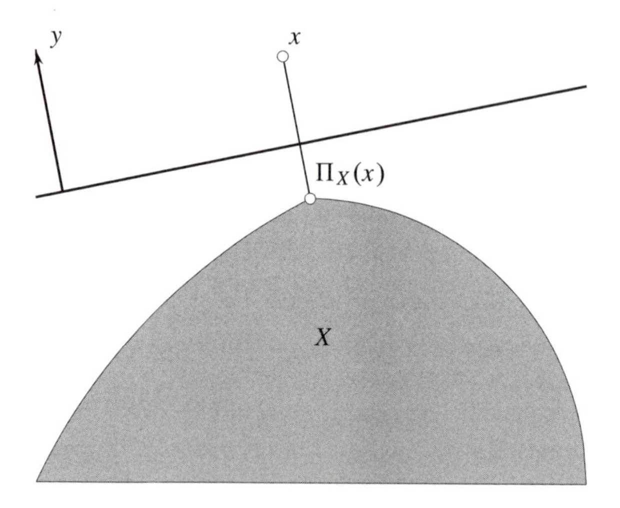

Figure 2.3. The main separation theorem.

If the set X is not closed, the point $x \notin X$ may be a boundary point and we have a weaker version of the above result.

THEOREM 2.15. *Let $X \subset \mathbb{R}^n$ be a convex set and let $x \notin X$. Then there exists a nonzero $y \in \mathbb{R}^n$ such that*

$$\langle y, v \rangle \le \langle y, x \rangle \quad \textit{for all} \quad v \in X.$$

Proof. Let $x^k \to x$, $x^k \notin \overline{X}$. By Theorem 2.14 we can find $y^k \neq 0$ such that

$$\langle y^k, v \rangle \leq \langle y^k, x^k \rangle \quad \text{for all} \quad v \in X. \tag{2.3}$$

Since $y^k \neq 0$, with no loss of generality we may assume that $\|y^k\| = 1$, because we can divide both sides of (2.3) by $\|y^k\|$. Let y be an accumulation point of the sequence $\{y^k\}$. Passing to the limit (over a subsequence, if necessary) in (2.3) we obtain the required result. □

Separation is possible also when both objects being separated are convex sets.

THEOREM 2.16. *Let X_1 and X_2 be convex sets in $\mathbb{R}^n$. If $X_1 \cap X_2 = \emptyset$, then there exists a nonzero $y \in \mathbb{R}^n$ such that*

$$\langle y, x^1 \rangle \leq \langle y, x^2 \rangle \quad \textit{for all} \quad x^1 \in X_1, \; x^2 \in X_2.$$

Proof. Define $X = X_1 - X_2$. Since $x = 0$ does not belong to X, we can use Theorem 2.15 to find $y \neq 0$ separating 0 from X. This yields $\langle y, v \rangle \leq 0$ for all v of the form $x^1 - x^2$ with $x^1 \in X_1$ and $x^2 \in X_2$. □

If one of the sets is bounded, strict separation is possible.

THEOREM 2.17. *Let X_1 and X_2 be closed convex sets in $\mathbb{R}^n$ and let X_1 be bounded. If $X_1 \cap X_2 = \emptyset$, then there exist a nonzero $y \in \mathbb{R}^n$ and $\varepsilon > 0$ such that*

$$\langle y, x^1 \rangle \leq \langle y, x^2 \rangle - \varepsilon$$

for all $x^1 \in X_1$ and all $x^2 \in X_2$.

Proof. By the closedness of X_1 and X_2 and boundedness of X_1, the set $X = X_1 - X_2$ is closed. Since $0 \notin X$, we can apply Theorem 2.14, similar to the proof of Theorem 2.16. □

The boundedness assumption in Theorem 2.16 is essential. The closed sets $X_1 = \{x \in \mathbb{R}^2 : x_2 \leq 0\}$ and $X_2 = \{x \in \mathbb{R}^2 : x_2 \geq e^{-x_1}\}$ cannot be strictly separated.

These geometrical facts have profound implications, as we shall see later in this chapter.

2.2 CONES

2.2.1 Basic Concepts

A particular class of convex sets, convex cones, play a significant role in optimization theory.

DEFINITION 2.18. A set $K \subset \mathbb{R}^n$ is called a *cone* if for every $x \in K$ and all $\alpha > 0$ one has $\alpha x \in K$. A *convex cone* is a cone that is a convex set.

We use only $\alpha > 0$ in the cone definition to allow easy formulation of the cone separation theorem (Theorem 2.32), later in this chapter.

A simple example of a convex cone in $\mathbb{R}^n$ is the nonnegative orthant:

$$\mathbb{R}^n_+ = \{x \in \mathbb{R}^n : x_j \geq 0, \ j = 1, \ldots, n\}.$$

For convex cones, positive combinations remain in the set, similar to convex combinations for convex sets.

LEMMA 2.19. *Let K be a convex cone. If $x^1 \in K$, $x^2 \in K$, ..., $x^m \in K$ and $\alpha_1 > 0$, $\alpha_2 > 0$,..., $\alpha_m > 0$, then $\alpha_1 x^1 + \alpha_2 x^2 + \cdots + \alpha_m x^m \in K$.*

Proof. By convexity,

$$\frac{\alpha_1 x^1 + \alpha_2 x^2 + \cdots + \alpha_m x^m}{\alpha_1 + \alpha_2 + \cdots + \alpha_m} \in K.$$

As K is a cone, we can multiply the point on the left hand side by $\alpha_1 + \alpha_2 + \ldots + \alpha_m$ and stay in K. □

Let us consider two important examples of convex cones.

LEMMA 2.20. *Assume that X is a convex set. Then the set*

$$\operatorname{cone}(X) = \{\gamma x : x \in X, \ \gamma \geq 0\}$$

is a convex cone.

Proof. The set $\operatorname{cone}(X)$ is a cone, because for all its elements $d = \gamma x$ and all $\beta \geq 0$, we also have $\beta d = (\beta\gamma)x \in \operatorname{cone}(X)$. To prove that it is convex, let us consider

$$\begin{aligned} d^1 &= \gamma_1 x^1, \quad x^1 \in X, \\ d^2 &= \gamma_2 x^2, \quad x^2 \in X, \\ d &= \alpha d^1 + (1-\alpha) d^2, \end{aligned}$$

with $\alpha \in [0, 1]$. If $\gamma_1 = 0$ or $\gamma_2 = 0$, then d is an element of $\operatorname{cone}(X)$ in a trivial way. We need to consider only the case with $\gamma_1 > 0$ and $\gamma_2 > 0$. We have

$$d = \alpha\gamma_1 x^1 + (1-\alpha)\gamma_2 x^2 = (\alpha\gamma_1 + (1-\alpha)\gamma_2) \frac{\alpha\gamma_1 x^1 + (1-\alpha)\gamma_2 x^2}{\alpha\gamma_1 + (1-\alpha)\gamma_2}.$$

The fraction on the right hand side, which we denote by x, is a convex combination of x^1 and x^2 and is therefore an element of X. Thus $d = \gamma x$ with $\gamma = \alpha\gamma_1 + (1-\alpha)\gamma_2$, and therefore $d \in \operatorname{cone}(X)$. □

The set cone(X) is called the *cone generated by the set* X. For a convex set X and a point $x \in X$ the set

$$K_X(x) \triangleq \text{cone}(X - x)$$

is called the *cone of feasible directions* of X at x (or the *radial* cone). It follows directly from the definition that it is a convex cone (see Figure 2.4).

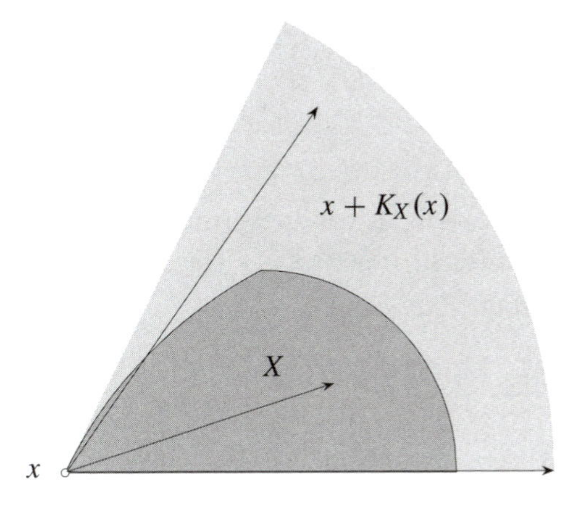

Figure 2.4. The cone of feasible directions (translated to the point x).

Example 2.21. Assume that the set $X \subset \mathbb{R}^n$ is a closed convex cone itself, and $x \in X$. Let us calculate the cone of feasible directions for X at x:

$$\begin{aligned} K_X(x) &= \{d \in \mathbb{R}^n : d = \tau(y - x),\ y \in X,\ \tau \geq 0\} \\ &= \{d \in \mathbb{R}^n : d = h - \tau x,\ h \in X,\ \tau \geq 0\} \\ &= X - \{\tau x : \tau \geq 0\} = X + \{\tau x : \tau \in \mathbb{R}\}. \end{aligned}$$

In the last two equations we used the fact that X is a cone.

Definition 2.22. Let $X \subset \mathbb{R}^m$ be a convex set. The set

$$X_\infty \triangleq \{d : X + d \subset X\}$$

is called the *recession cone* of X.

We shall show that X_∞ is a convex cone. We first note that for each $d \in X_\infty$ and for every m:

$$X + md \subset X + (m-1)d \subset \cdots \subset X + d \subset X.$$

Using convexity of X we infer that $X + \tau d \subset X$ for all $\tau \geq 0$. Hence $\tau d \in X_\infty$ for all $\tau \geq 0$. This means that X_∞ is a cone.

The fact that X_∞ is convex can be verified directly from the definition. Indeed, if $d^1 \in X_\infty$ and $d^2 \in X_\infty$, then

$$x + (\alpha d^1 + (1-\alpha)d^2) = \alpha(x + d^1) + (1-\alpha)(x + d^2) \in X,$$

for all $x \in X$ and all $\alpha \in (0, 1)$.

DEFINITION 2.23. Let K be a cone in $\mathbb{R}^n$. The set

$$K^\circ \triangleq \{y \in \mathbb{R}^n : \langle y, x\rangle \leq 0 \text{ for all } x \in K\}$$

is called the *polar cone* of K.

For example, the set $K = \{x \in \mathbb{R}^n : x_i \geq 0,\ i = 1, \ldots, n\}$ is a convex closed cone. Its polar is $-K$, as can be verified directly. Another example of a polar cone is shown in Figure 2.5. The negative of the polar cone, $K^* \triangleq -K^\circ$, is called the *dual cone*.

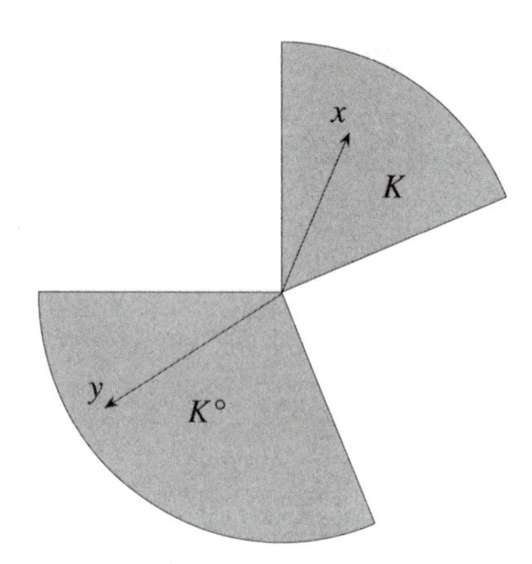

Figure 2.5. The polar cone.

Example 2.24. Let $K_1, \ldots, K_m$ be cones in $\mathbb{R}^n$ and let $K = K_1 + \cdots + K_m$. Clearly, K is a cone. We shall calculate its polar cone. If $z \in K^\circ$ then for every $x^1 \in K_1, \ldots, x^m \in K_m$ we have

$$\langle z, x^1\rangle + \cdots + \langle z, x^m\rangle \leq 0. \tag{2.4}$$

Let us choose $j \in \{1, \ldots, n\}$. Setting all $x^i = 0$ in (2.4), except for $i = j$, we conclude that

$$\langle z, x^j\rangle \leq 0 \quad \text{for all} \quad x^j \in K_j.$$

Consequently, $z \in K_j^\circ$. As j was arbitrary,

$$K^\circ \subset K_1^\circ \cap \cdots \cap K_m^\circ.$$

On the other hand, for every element z of $K_1^\circ \cap \cdots \cap K_m^\circ$ inequality (2.4) is satisfied, and thus $z \in K^\circ$. Therefore,

$$(K_1 + \cdots + K_m)^\circ = K_1^\circ \cap \cdots \cap K_m^\circ.$$

Directly from the definition we obtain the following properties.

LEMMA 2.25. *For every convex cone $K \subset \mathbb{R}^n$:*

(i) *The polar cone K° is convex and closed;*

(ii) $K^\circ = (\overline{K})^\circ$.

We also have the following useful property.

LEMMA 2.26. *Let K be a cone in $\mathbb{R}^n$ and let $y \in \mathbb{R}^n$ be such that the scalar product $\langle y, x\rangle$ is bounded from above for all $x \in K$. Then $y \in K^\circ$.*

Proof. Suppose $y \notin K^\circ$. Then we can find $x \in K$ such that $\langle y, x\rangle > 0$. Since K is a cone, $\alpha x \in K$ for all $\alpha > 0$. But then $\langle y, \alpha x\rangle$ is not bounded from above, which contradicts our assumption. □

We use the notation $K^{\circ\circ}$ for the polar cone of the polar cone of K. The following result is known as the *Bipolar Theorem.*

THEOREM 2.27. *If $K \subset \mathbb{R}^n$ is a closed convex cone, then*

$$K^{\circ\circ} = K.$$

Proof. If $x \in K$ then $\langle y, x\rangle \le 0$ for all $y \in K^\circ$, so $x \in K^{\circ\circ}$. Hence $K \subset K^{\circ\circ}$. We shall prove equality here. Suppose one can find $z \in K^{\circ\circ} \setminus K$. Since K is closed, we can use Theorem 2.14 to strictly separate z and K. So, for some $y \neq 0$ and $\varepsilon > 0$ we have

$$\langle y, x\rangle \le \langle y, z\rangle - \varepsilon \quad \text{for all} \quad x \in K.$$

Thus the left hand side is bounded from above and Lemma 2.26 yields $y \in K^\circ$. On the other hand, setting $x = 0$ we get $\langle y, z\rangle \ge \varepsilon$, which contradicts our assumption that $z \in K^{\circ\circ}$. □

Polar cones to pre-images of cones under linear transformations can be calculated in an explicit form.

THEOREM 2.28. *Assume that C is a closed convex cone in $\mathbb{R}^m$ and A is an $m \times n$ matrix. Let*

$$K = \{x \in \mathbb{R}^n : Ax \in C\}.$$

Then K is a closed convex cone and

$$K^\circ = \{A^T\lambda : \lambda \in C^\circ\}.$$

Proof. K is a closed convex cone by construction. If $z = A^T\lambda$ with $\lambda \in C^\circ$, then for every $x \in K$ we have

$$\langle x, z\rangle = \langle x, A^T\lambda\rangle = \langle Ax, \lambda\rangle \le 0,$$

because $Ax \in C$. Hence

$$\{A^T\lambda : \lambda \in C^\circ\} \subset K^\circ.$$

The set on the left hand side is convex and closed. Suppose $h \in K^\circ \setminus \{A^T\lambda : \lambda \in C^\circ\}$. Then we can strictly separate h and the set in question: there exist $y \neq 0$ and $\varepsilon > 0$ such that

$$\langle y, h\rangle \ge \langle y, A^T\lambda\rangle + \varepsilon \quad \text{for all} \quad \lambda \in C^\circ. \tag{2.5}$$

Thus the product $\langle y, A^T\lambda\rangle = \langle Ay, \lambda\rangle$ is bounded from above for all $\lambda \in C^\circ$. By Lemma 2.26, $Ay \in C^{\circ\circ}$. Since C is closed, Theorem 2.27 implies that $Ay \in C$ and therefore $y \in K$. Since $h \in K^\circ$, we obtain $\langle y, h\rangle \le 0$. But then inequality (2.5) with $\lambda = 0$ yields a contradiction: $0 \ge \varepsilon$. □

Theorem 2.28 allows us to explicitly calculate polar cones of cones defined by systems of linear inequalities. Setting

$$C = \{y \in \mathbb{R}^m : y_i \le 0,\ i = 1, \dots, m\}$$

we obtain the following statement (known as *Farkas Lemma*).

Corollary 2.29. *Let A be an $m \times n$ matrix and let*

$$K = \{x \in \mathbb{R}^n : Ax \le 0\}.$$

Then

$$K^\circ = \{y \in \mathbb{R}^n : y = A^T\lambda,\ \lambda \in \mathbb{R}^m,\ \lambda \ge 0\}. \tag{2.6}$$

The above fact is frequently formulated as an alternative: *exactly one of the following two systems has a solution, either*

(i) $Ax \le 0$ and $\langle c, x\rangle > 0$; *or*

(ii) $c = A^T\lambda$, $\lambda \ge 0$.

Indeed, system (i) is not solvable if and only if $c \in K^\circ$, which is equivalent to (ii), by virtue of Corollary 2.29.

Example 2.30. Suppose we have n securities with present prices $c_1, \dots, c_n$. The securities can be either bought or sold *short*, that is, borrowed and sold for cash. Any amounts can be traded.

At some future time one of m states may occur. The price of security j in state i will be equal to a_{ij}, $i = 1, \dots, m$, $j = 1, \dots, n$. At this time we will liquidate our holding, that is, the securities held will be sold, and the short positions will be covered by purchasing and returning the amounts borrowed.

An *arbitrage*, in the simplest version, is the existence of $\bar{x} \in \mathbb{R}^n$ such that $\langle c, \bar{x} \rangle < 0$ and $A\bar{x} \geq 0$. Such a portfolio $\bar{x}$ can be "purchased" with an immediate profit and then liquidated at a future point in time with no additional loss, whatever the future state. If arbitrage is present, considering portfolios $M\bar{x}$, with $M \to \infty$, we can increase the profits without any limits.

Defining the convex cone

$$K = \{x \in \mathbb{R}^n : Ax \geq 0\},$$

we see that the absence of arbitrage is equivalent to the fact that $\langle c, x \rangle \geq 0$ for all $x \in K$. The latter is nothing else but

$$-c \in K^\circ.$$

Corollary 2.29 implies that there exist *state prices* $p \geq 0$ such that

$$c = A^T p. \tag{2.7}$$

The converse is obvious: if such state prices exist, then

$$\langle c, x \rangle = \langle A^T p, x \rangle = \langle p, Ax \rangle \geq 0 \quad \text{for all} \quad x \in \mathbb{R}^n.$$

Suppose the first security (cash) has a constant unit price of 1 for all states: $c_1 = 1$ and $a_{i1} = 1, i = 1, \dots, m$. Then it follows from the first relation in (2.7) that

$$\sum_{i=1}^{m} p_i = 1.$$

The vector p may be interpreted as a vector of implied probabilities of states $1, \dots, m$.

Similar considerations can be carried out for other definitions of arbitrage (see Example 2.33 below).

Example 2.31. Consider the space $\mathbb{S}^n$ of symmetric matrices of dimension $n \times n$. The set $\mathbb{S}^n_+$ of all positive semidefinite matrices is a cone in this space, as can be verified directly from the definition.

In order to speak about the polar cone, we need to define the scalar product $\langle A, B \rangle_{\mathbb{S}}$ for symmetric matrices A and B. We set

$$\langle A, B \rangle_{\mathbb{S}} = \operatorname{tr}(AB) = \sum_{i=1}^{n} \sum_{j=1}^{n} a_{ij} b_{ij}.$$

This function is called the Frobenius inner product of matrices. We then have

$$\left[\mathbb{S}_+^n\right]^\circ = \{B \in \mathbb{S}^n : \operatorname{tr}(BA) \le 0, \text{ for all } A \in \mathbb{S}_+^n\}.$$

We shall prove that

$$\left[\mathbb{S}_+^n\right]^\circ = -\mathbb{S}_+^n. \tag{2.8}$$

Suppose $B \in -\mathbb{S}_+^n$ and $A \in \mathbb{S}_+^n$. Let $z_1, \dots, z_n$ be eigenvectors of A having length 1, and let $\lambda_1, \dots, \lambda_n$ be the corresponding eigenvalues. We have the equation

$$\begin{aligned}\operatorname{tr}(BA) &= \operatorname{tr}\Big(B \sum_{j=1}^n \lambda_j z_j z_j^T\Big) = \sum_{j=1}^n \lambda_j \operatorname{tr}\big(B z_j z_j^T\big) \\ &= \sum_{j=1}^n \lambda_j \langle z_j, B z_j\rangle.\end{aligned}$$

Since the eigenvalues of A are nonnegative and B is negative semidefinite, we conclude that $\operatorname{tr}(BA) \le 0$ for all $A \in \mathbb{S}_+^n$. Thus

$$\left[\mathbb{S}_+^n\right]^\circ \supset -\mathbb{S}_+^n.$$

To prove the converse inclusion, consider any $B \in \left[\mathbb{S}_+^n\right]^\circ$. For an arbitrary $z \in \mathbb{R}^n$ we can set $A = zz^T$ and get

$$0 \ge \operatorname{tr}(BA) = \operatorname{tr}\big(Bzz^T\big) = \langle z, Bz\rangle.$$

Hence B is negative semidefinite and formula (2.8) holds true.

2.2.2 Separation of Cones

In Section 2.1.3 we considered separation of two disjoint convex sets. Clearly, convex cones can also be separated by the same principle. It follows from Theorem 2.16 that if K_1 and K_2 are convex cones, and $K_1 \cap K_2 = \emptyset$, then there exists $y \ne 0$ such that

$$\langle y, x^1\rangle \le \langle y, x^2\rangle \quad \text{for all} \quad x^1 \in K_1,\ x^2 \in K_2.$$

Using Lemma 2.26 we deduce that $y \in K_1^\circ$ and $-y \in K_2^\circ$. Therefore Theorem 2.16 can be rephrased as follows: *there exist* $y^1 \in K_1^\circ$ *and* $y^2 \in K_2^\circ$ *such that*

$$y^1 + y^2 = 0.$$

In this algebraic formulation, the separation theorem for cones has a very useful generalization.

THEOREM 2.32. *Let $K_1, K_2 \dots, K_m$ be convex cones in $\mathbb{R}^n$. If $K_1 \cap K_2 \cap \dots \cap K_m = \emptyset$, then there exist $y^i \in K_i^\circ$, $i = 1, 2, \dots, m$, not all equal 0, such that*

$$y^1 + y^2 + \dots + y^m = 0.$$

Proof. Let us define two cones in $\mathbb{R}^n \times \mathbb{R}^n \times \dots \times \mathbb{R}^n = \mathbb{R}^{mn}$:

$$C_1 = \{z = (z^1, \dots, z^m) : z^i \in K_i,\ i = 1, 2, \dots, m\},$$
$$C_2 = \{w = (x, \dots, x) : x \in \mathbb{R}^n\}.$$

Since $\bigcap_{i=1}^m K_i = \emptyset$, we have $C_1 \cap C_2 = \emptyset$. Then by Theorem 2.16 we can find a nonzero $y \in \mathbb{R}^{mn}$ such that

$$\langle y, w \rangle \geq \langle y, z \rangle$$

for all $w = (x, \dots, x)$ and all $z \in C_1$. Writing $y = (y^1, y^2, \dots, y^m)$ we obtain

$$\langle y^1 + y^2 + \dots + y^m, x \rangle \geq \langle y^1, z^1 \rangle + \langle y^2, z^2 \rangle + \dots + \langle y^m, z^m \rangle \qquad (2.9)$$

for all $x \in \mathbb{R}^n$ and all $z^i \in K_i, i = 1, 2, \dots, m$. Setting $x = 0$, we see that the right hand side is bounded from above for all $z^i \in K_i, i = 1, 2, \dots, m$, which implies that each $\langle y^i, z^i \rangle$ is bounded from above for all $z^i \in K_i$. By Lemma 2.26, $y^i \in K_i^\circ$. The left hand side of (2.9) is bounded from below for all $x \in \mathbb{R}^n$, which is possible only when $y^1 + y^2 + \dots + y^m = 0$. □

Example 2.33. Let us return to Example 2.30 and let us define a *weak arbitrage* as the existence of $\bar{x} \in \mathbb{R}^n$ such that $\langle c, \bar{x} \rangle \leq 0$, and $A\bar{x} \geq 0$, $A\bar{x} \neq 0$. In other words, portfolio $\bar{x}$ can be purchased without any additional cash, and at a future point in time it can be sold with no loss, and with some profit for at least one state.

Defining the convex cones

$$K_1 = \{x \in \mathbb{R}^n : \langle c, x \rangle \leq 0\},$$
$$K_2 = \{x \in \mathbb{R}^n : Ax \geq 0,\ Ax \neq 0\},$$

we see that the absence of weak arbitrage is equivalent to the fact that

$$K_1 \cap K_2 = \emptyset.$$

It follows that there exists $y \neq 0$ such that $y \in K_1^\circ$ and $-y \in K_2^\circ$. The first polar cone is obvious:

$$K_1^\circ = \{\alpha c : \alpha \geq 0\}.$$

To calculate the second polar, we assume that $K_2 \neq \emptyset$. Then

$$\overline{K_2} = \{x \in \mathbb{R}^n : Ax \geq 0\}.$$

By Lemma 2.25, we have $K_2^\circ = (\overline{K}_2)^\circ$. Theorem 2.32 yields:

$$(\overline{K}_2)^\circ = \{-A^T\lambda : \lambda \geq 0\}.$$

Consequently, for some $\alpha \geq 0$ and a nonnegative vector λ we have

$$y = \alpha c = A^T\lambda.$$

Since y is nonzero, $\alpha > 0$. Dividing by α and setting $p = \lambda/\alpha$ we conclude that there exists a vector of state prices $p \in \mathbb{R}^m$, $p \geq 0$, such that

$$c = A^T p.$$

The converse statement can be proved as in Example 2.30. Again, if one of the securities is cash, then p may be interpreted as a vector of probabilities.

Note that if $K_2 = \emptyset$ the existence of state prices is not guaranteed.

To use Theorem 2.32 we need to be able to calculate polar cones. The following technical lemma is useful.

LEMMA 2.34. *If $x \in \operatorname{int} K$, then $\langle y, x\rangle < 0$ for all nonzero $y \in K^\circ$.*

Proof. Suppose $\langle y, x\rangle = 0$ for some nonzero $y \in K^\circ$. Define $z = x + \varepsilon y$. Since $x \in \operatorname{int} K$, for sufficiently small $\varepsilon > 0$ we have $z \in K$ and $\langle y, z\rangle > 0$, a contradiction with $y \in K^\circ$. □

We are now in a position to present the general formula for a polar cone of an intersection.

THEOREM 2.35. *Let $K_1, \ldots, K_m$ be convex cones in $\mathbb{R}^n$ and let $K = K_1 \cap \cdots \cap K_m$. If $K_1 \cap \operatorname{int} K_2 \cap \cdots \cap \operatorname{int} K_m \neq \emptyset$, then*

$$K^\circ = K_1^\circ + K_2^\circ + \cdots + K_m^\circ.$$

Proof. If $y^1 \in K_1^\circ$, $y^2 \in K_2^\circ$, ..., $y^m \in K_m^\circ$, then for all $x \in K$ we get

$$\langle y^1 + y^2 + \cdots + y^m, x\rangle \leq 0,$$

so

$$K_1^\circ + K_2^\circ + \cdots + K_m^\circ \subset K^\circ.$$

We shall prove the inverse inclusion. Choose $y \in K^\circ$ and define the cone

$$C = \{x : \langle x, y\rangle > 0\}.$$

Clearly, $C \cap K = \emptyset$, because $\langle x, y\rangle \leq 0$ for all $x \in K$. So, $C \cap K_1 \cap \cdots \cap K_m = \emptyset$ and by Theorem 2.32 we can find $d \in C^\circ$, $y^1 \in K_1^\circ, \ldots, y^m \in K_m^\circ$, not all zero, such that

$$d + y^1 + \cdots + y^m = 0. \tag{2.10}$$

Directly from the definition of C we see that $d = -\alpha y$ for some $\alpha \geq 0$.

If $\alpha = 0$, then $d = 0$. By assumption, there exists $x \in K_1 \cap \operatorname{int} K_2 \cap \cdots \cap \operatorname{int} K_m$. Taking the scalar product of x and both sides of (2.10) we get

$$\langle x, y^1 \rangle + \cdots + \langle x, y^m \rangle = 0.$$

All components on the left hand side are nonpositive, so $\langle x, y^i \rangle = 0$, $i = 1, \ldots, m$. Since $x \in \operatorname{int} K_i$, $i = 2, \ldots, m$, Lemma 2.34 implies that $y^i = 0$, $i = 2, \ldots, m$. Equation (2.10) then yields $y^1 = 0$, a contradiction. Thus $\alpha > 0$.

Dividing both sides of (2.10) by α and rearranging terms we get

$$y = -\frac{1}{\alpha}d = \frac{1}{\alpha}y^1 + \frac{1}{\alpha}y^2 + \cdots + \frac{1}{\alpha}y^m \in K_1^\circ + K_2^\circ + \cdots + K_m^\circ.$$

Recall that y was an arbitrary element of K°. Therefore

$$K^\circ \subset K_1^\circ + K_2^\circ + \cdots + K_m^\circ,$$

which completes the proof. □

If the cones $K_1, \ldots, K_m$ are polyhedral, the regularity assumption $K_1 \cap \operatorname{int} K_2 \cap \cdots \cap \operatorname{int} K_m \neq \emptyset$ is not needed; the result follows directly from Corollary 2.29 (see Exercise 2.7).

Another application of Theorem 2.32 is the calculation of the polar to the cone

$$K = \{x \in K_1 : Ax \in K_2\}, \tag{2.11}$$

which arises in a natural way in the analysis of optimality conditions (see Chapter 3). In (2.11) the sets $K_1 \subset \mathbb{R}^n$ and $K_2 \subset \mathbb{R}^m$ are closed convex cones and A is an $m \times n$ matrix.

THEOREM 2.36. *Assume that K_1 and K_2 are closed convex cones, and K is defined by* (2.11). *If*

$$0 \in \operatorname{int}\{Ax - y : x \in K_1,\ y \in K_2\}, \tag{2.12}$$

then

$$K^\circ = K_1^\circ + \{A^T\lambda : \lambda \in K_2^\circ\}.$$

Proof. Define $D = \{x \in \mathbb{R}^n : Ax \in K_2\}$. Clearly,

$$K = K_1 \cap D.$$

Hence

$$K^\circ \supset K_1^\circ + D^\circ.$$

We shall show the equality here. Choose any $v \in K^\circ$ and define the cone

$$C = \{x : \langle x, v \rangle > 0\}.$$

We have $C \cap K = \emptyset$, because $\langle x, v \rangle \le 0$ for all $x \in K$. Thus,

$$C \cap K_1 \cap D = \emptyset.$$

By Theorem 2.32, we can find $d \in C^\circ$, $w \in K_1^\circ$, $z \in D^\circ$, not all zero, such that

$$d + w + z = 0. \tag{2.13}$$

Directly from the definition of C we see that $d = -\alpha v$ for some $\alpha \ge 0$. Suppose $\alpha = 0$. Then (2.13) yields $w = -z$. Using Theorem 2.28 we see that

$$z = A^T \lambda, \quad \lambda \in K_2^\circ.$$

It follows that

$$-A^T \lambda \in K_1^\circ,$$

where $\lambda \in K_2^\circ$. Thus, for every $x \in K_1$ we obtain

$$-\langle x, A^T \lambda \rangle \le 0,$$

that is,

$$\langle Ax, \lambda \rangle \ge 0.$$

As $\lambda \in K_2^\circ$, we also have $\langle y, \lambda \rangle \le 0$ for all $y \in K_2$. Hence

$$\langle Ax - y, \lambda \rangle \ge 0 \quad \text{for all} \quad x \in K_1 \quad \text{and all} \quad y \in K_2.$$

Since the set of all $Ax - y$, where $x \in K_1$ and $y \in K_2$, contains a neighborhood of 0 by assumption, we get $\lambda = 0$. This yields $z = 0$ and $w = 0$, a contradiction. Consequently, $\alpha > 0$.

Dividing (2.13) by α and rearranging terms we obtain

$$v = \frac{1}{\alpha} w + \frac{1}{\alpha} A^T \lambda, \quad w \in K_1^\circ, \quad \lambda \in K_2^\circ.$$

Thus $v \in K_1^\circ + D^\circ$, as required. □

Again, when all cones are polyhedral, the regularity assumption (2.12) is not needed.

2.2.3 Normal Cones

DEFINITION 2.37. Consider a convex closed set $X \subset \mathbb{R}^n$ and a point $x \in X$. The set

$$N_X(x) \triangleq [\text{cone}(X - x)]^\circ$$

is called the *normal cone* to X at x.

As a polar cone, the normal cone is closed and convex. It follows from the definition that $v \in N_X(x)$ if and only if

$$\langle v, y - x \rangle \leq 0 \quad \text{for all} \quad y \in X. \tag{2.14}$$

Lemma 2.11 therefore implies the following characterization of the normal cone.

LEMMA 2.38. *Let X be a closed convex set and let $x \in X$. Then*

$$N_X(x) = \{v \in \mathbb{R}^n : \Pi_X(x + v) = x\}.$$

This is illustrated in Figure 2.6.

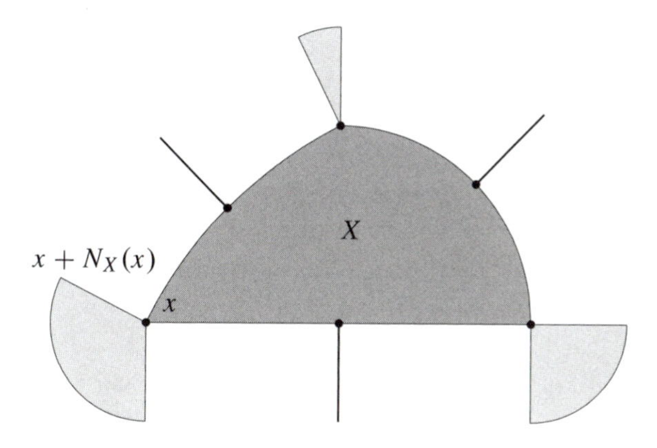

Figure 2.6. Normal cones (translated to the points at which they are calculated).

Example 2.39. Suppose C is a closed convex cone in $\mathbb{R}^n$ and that $z \in \mathbb{R}^n$. Consider the projection

$$x = \Pi_C(z).$$

Define $y = z - x$. It follows from Lemma 2.38 that

$$y \in N_C(x) = \left[K_C(x)\right]^\circ.$$

By Example 2.21, $K_C(x) = C + \{\tau x : \tau \in \mathbb{R}\}$. Example 2.24 then yields

$$N_C(x) = C^\circ \cap \{y \in \mathbb{R}^n : y \perp x\}.$$

We conclude that

$$z = x + y, \qquad x \in C, \qquad y \in C^\circ, \qquad y \perp x.$$

By symmetry, $y = \Pi_{C^\circ}(z)$.

Normal cones to intersections can be calculated as sums of normal cones, but under an additional regularity assumption.

Lemma 2.40. *Assume that $X = X_1 \cap \cdots \cap X_m$, where X_i are closed convex sets, $i = 1, \ldots, m$, and let $x \in X$. If $X_1 \cap \operatorname{int} X_2 \cap \cdots \cap \operatorname{int} X_m \neq \emptyset$, then*

$$N_X(x) = N_{X_1}(x) + \cdots + N_{X_m}(x).$$

Proof. We have

$$\operatorname{cone}(X - x) = \operatorname{cone}(X_1 - x) \cap \cdots \cap \operatorname{cone}(X_m - x).$$

By assumption,

$$\operatorname{cone}(X_1 - x) \cap \operatorname{int}\big(\operatorname{cone}(X_2 - x)\big) \cap \cdots \cap \operatorname{int}\big(\operatorname{cone}(X_m - x)\big) \neq \emptyset.$$

Applying Theorem 2.35, we obtain the required result. □

The regularity assumption cannot be dropped, in general, as the following example shows.

Example 2.41. Let

$$X_1 = \{x \in \mathbb{R}^2 : \|x\| \le 1\}, \quad X_2 = \{x \in \mathbb{R}^2 : x_1 = 1\},$$

and let $x = (1, 0)$. We have

$$N_{X_1}(x) = \{v \in \mathbb{R}^2 : v_1 \ge 0,\ v_2 = 0\}, \quad N_{X_2}(x) = \{v \in \mathbb{R}^2 : v_2 = 0\}.$$

On the other hand $X = X_1 \cap X_2$ contains just the point x, and

$$N_X(x) = \mathbb{R}^2.$$

The operation $x \mapsto N_X(x)$ is upper semicontinuous in the following sense.

Lemma 2.42. *Assume that $X \subset \mathbb{R}^n$ is a closed convex set and that a sequence $\{x^k\}$ of elements of X is convergent to a point $\hat{x}$. Then for every convergent sequence of elements $v^k \in N_X(x^k)$ its limit $\hat{v}$ is an element of $N_X(\hat{x})$.*

Proof. Take any $y \in X$. It follows from (2.14) at the point x^k that

$$\langle v^k, y - x^k \rangle \le 0, \quad k = 1, 2, \ldots.$$

Passing to the limit in the above inequality we obtain (2.14) at $\hat{x}$. □

2.3 EXTREME POINTS

Definition 2.43. A point x of a convex set X is called an *extreme point* of X if no other points $x^1 \in X$ and $x^2 \in X$ exist such that

$$x = \frac{1}{2}x^1 + \frac{1}{2}x^2.$$

The notion of an extreme point is illustrated in Figure 2.7.

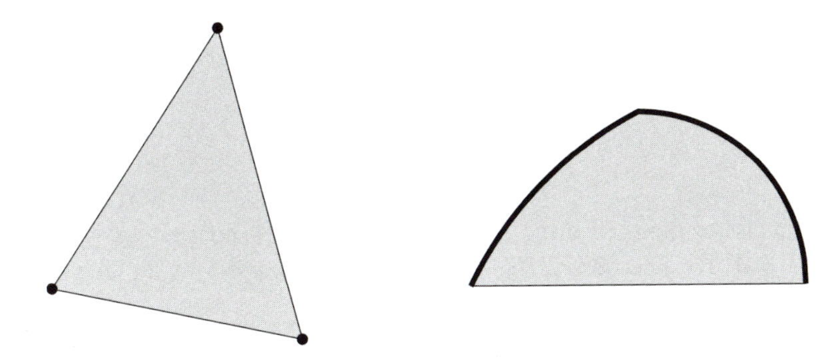

Figure 2.7. Extreme points of convex compact sets.

Extreme points of a compact convex set fully characterize the set.[†]

Theorem 2.44. *A convex and compact set in $\mathbb{R}^n$ is equal to the convex hull of the set of its extreme points.*

Proof. The proof is by induction on the dimension of the space n. For $n = 1$ the theorem is true, because all compact convex sets in $\mathbb{R}$ are bounded intervals $[a, b]$ and their extreme points are the ends: a and b. Supposing that the theorem holds true for all $k \leq n$ we shall prove it for $n + 1$.

Let $X \subset \mathbb{R}^{n+1}$ be a convex, closed and bounded set. We may assume that $\text{int}\, X \neq \emptyset$, because otherwise, by Lemma 2.9, we would be able to define a space of a smaller dimension containing X and our result would follow immediately.

So, let p be an arbitrary point of X and let us take some $x_0 \in \text{int}\, X$ which is different from p. Define $d = p - x_0$ and consider the line $x_0 + \tau d$ for $\tau \in \mathbb{R}$. Since X is convex, closed and bounded, the line must cross the boundary of X at its two points: $x^1 = x_0 + \tau_1 d$ and $x^2 = x_0 + \tau_2 d$ with $\tau_1 > 0$ and $\tau_2 < 0$. Obviously, p is a convex combination of x^1 and x^2. It suffices to show that both x^1 and x^2 are convex combinations of extreme points of X (see Figure 2.8).

[†]Theorem 2.44 is known as *Minkowski's Theorem.*

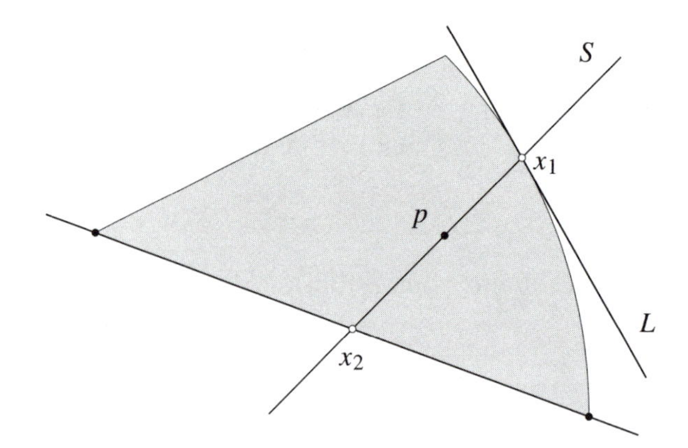

Figure 2.8. Construction of a convex combination of extreme points.

Let us focus on x^1. Define the set $S = \{x = x_0 + \tau d : \tau > \tau_1\}$. Clearly, $X \cap S = \emptyset$. By Theorem 2.15, we can separate X and S by a plane defined by some $y \neq 0$ such that

$$\langle y, x \rangle \leq \langle y, x_0 + \tau d \rangle$$

for all $x \in X$ and all $\tau > \tau_1$. Letting $\tau \to \tau_1$ we obtain

$$\langle y, x \rangle \leq \langle y, x^1 \rangle \tag{2.15}$$

for all $x \in X$.

Let us define the hyperplane

$$L = \{x : \langle y, x \rangle = \langle y, x^1 \rangle\}.$$

Consider the set $X_1 = X \cap L$. It is a convex, closed and bounded set contained in a linear manifold L of dimension n and $x^1 \in X_1$. So, by our assumption, x^1 is a convex combination of extreme points of X_1.

We shall show that extreme points of X_1 are also extreme points of X. Suppose it is not true and an extreme point v of X_1 is not an extreme point of X. Then v is a center of an interval with both ends in X. Since X lies on one side of L, the ends must be in L, hence in X_1, a contradiction.

Summing up, x^1 is a convex combination of extreme points of X. We treat x^2 in a similar way and conclude that our assertion is true for sets in $\mathbb{R}^{n+1}$. By induction, the result holds true for all n. □

The above result can be extended to unbounded convex sets with the use of the concept of the recession cone.

Lemma 2.45. *A closed convex set X is bounded if and only if $X_\infty = \{0\}$.*

Proof. If X is bounded, it cannot contain any rays, so $X_\infty = \{0\}$. It remains to prove the converse. Suppose $X_\infty = \{0\}$. If the set X is unbounded, there exists a sequence $x^1, x^2, \ldots$ of elements of X whose norms diverge. As the vectors $y^i = x^i/\|x^i\|$ have norm 1, they have an accumulation point y. Consider the point $z = x + \lambda y$ for some $x \in X$ and some positive number λ. We construct the points

$$z^i = \Big(1 - \frac{\lambda}{\|x^i\|}\Big)x + \Big(\frac{\lambda}{\|x^i\|}\Big)x^i.$$

For i large enough, we have $\lambda/\|x^i\| \in (0, 1)$ and thus $z^i \in X$. On the other hand, the points z^i have an accumulation point z and the set X is closed. Thus, $z \in X$. This implies that $y \in X_\infty$, since the point $x \in X$ and the number $\lambda > 0$ were arbitrary. We obtain a contradiction, which proves the statement. □

We are now ready to state our main result. We denote by $E(X)$ the set of all extreme points of a set X.

THEOREM 2.46. *A closed convex set X, which has at least one extreme point, can be represented as follows:*

$$X = \operatorname{conv} E(X) + X_\infty.$$

Proof. We first observe that the set X does not contain any line. Indeed, if it contains a line defined by some vector d, then d and $-d$ belong to X_∞. Then no point x can be extreme, because $x = (x + d)/2 + (x - d)/2$.

We can now proceed as in the proof of Theorem 2.44 with the following modification. Taking a line through the point x_0, we may obtain either two boundary points x^1 and x^2, as before, or one boundary point (say x^1) and a vector (e.g., $-d$) from the recession cone X_∞. □

This result can be further refined with the use of the concept of an *extreme ray* of the recession cone (see Exercise 2.3).

Our results are important for linear programming.

THEOREM 2.47. *Let A be an $m \times n$ matrix and let the set $X \subset \mathbb{R}^n$ be defined as*

$$X = \{x \in \mathbb{R}^n : Ax = b,\ x \geq 0\}.$$

A point x is an extreme point of X if and only if the columns of A that correspond to positive components of x are linearly independent.

Proof. Let $J = \{j : x_j > 0\}$ and let the columns a^j, $j \in J$, be linearly dependent. Then one can find γ_j, $j \in J$, not all equal to 0, such that $\sum_{j\in J} \gamma_j a^j = 0$. Let us define two points x^1 and x^2 by setting $x_j^1 = x_j + \tau\gamma_j$ and $x_j^2 = x_j - \tau\gamma_j$ for $j \in J$, and $x_j^1 = x_j^2 = x_j$ for $j \notin J$. Then for sufficiently small $\tau > 0$ both x^1 and x^2 belong to X and $x = \frac{1}{2}x^1 + \frac{1}{2}x^2$. Therefore x cannot be an extreme point of X.

To prove the inverse implication, consider a point x which is not extreme. Then we can find two different points x^1 and x^2 in X such that $x = \frac{1}{2}x^1 + \frac{1}{2}x^2$. Let J be the set of indices for which $x_j^1 \neq x_j^2$. Clearly, J is nonempty and $x_j > 0$ for $j \in J$. Since $Ax^1 = Ax^2 = b$, we have

$$\sum_{j\in J}(x_j^1 - x_j^2)a^j = 0,$$

which proves that the columns a^j, $j \in J$ are linearly dependent. □

Solutions of the system $Ax = b$ whose nonzero components correspond to linearly independent columns of A, are called in linear programming *basic solutions*. If they are nonnegative, they are called *basic feasible solutions*. Their role is now evident.

THEOREM 2.48. *If the set $X = \{x \in \mathbb{R}^n : Ax = b,\ x \geq 0\}$ is bounded, it is the convex hull of the set of basic feasible solutions.*

Proof. The result follows immediately from Theorems 2.44 and 2.47. □

We use this observation in Section 2.4.

Extreme points play a crucial role in the analysis of the *problem of moments* in the theory of probability. We present a simplified version of this problem.

Example 2.49. Consider the set $\mathcal{S}$ of all random variables with realizations in the set $T = \{t_1, t_2, \ldots, t_n\} \subset \mathbb{R}$. We know the first m moments of a certain random variable $X \in \mathcal{S}$:

$$\mu_k = \mathbb{E}(X^k), \quad k = 1, \ldots, m,$$

where $m < n$. We shall prove that there exists a random variable $Y \in \mathcal{S}$ with identical values of its first m moments and such that no more than $m+1$ realizations from the set T have positive probability.

Define the set of possible probability distributions on T:

$$P = \{p \in \mathbb{R}^n : \sum_{j=1}^{n} p_j = 1,\ p_j \geq 0,\ j = 1, \ldots, n\}.$$

Let

$$A = \begin{bmatrix} t_1 & t_2 & \dots & t_n \\ t_1^2 & t_2^2 & \dots & t_n^2 \\ \vdots & \vdots & & \vdots \\ t_1^m & t_2^m & \dots & t_n^m \end{bmatrix}.$$

The set

$$M = \{Ap : p \in P\}$$

is the collection of all m-tuples of moments of random variables having realizations in T. The set M is convex and compact, and $\mu = (\mu_1, \dots, \mu_m)$ is its element. Therefore there exist $m + 1$ extreme points z^i of M such that

$$\mu = \sum_{i=1}^{m+1} \alpha_i z^i, \quad \sum_{i=1}^{m+1} \alpha_i = 1, \quad \alpha_i \geq 0, \quad i = 1, \dots, m+1.$$

We shall show that each z^i can be represented as $z^i = Ae^{j(i)}$, where e^j is the jth unit vector in $\mathbb{R}^n$.

Consider a fixed i. We have $z^i = Ap^i$, where $p^i \in P$. Writing

$$p^i = \sum_{j=1}^{n} p_j^i e^j$$

we see that

$$z^i = \sum_{j=1}^{n} p_j^i A e^j, \quad \sum_{j=1}^{n} p_j^i = 1, \quad p_j^i \geq 0, \quad j = 1, \dots, n.$$

It follows that z^i is a convex combination of points Ae^j, $j = 1, \dots, m$. Since z^i is an extreme point of M, this may happen only if $z^i = Ae^j$ whenever $p_j^i > 0$. We can, therefore, select one index $j(i)$ with $p_{j(i)}^i > 0$, and conclude that $z^i = Ae^{j(i)}$. This can be done for all i and thus

$$\mu = \sum_{i=1}^{m+1} \alpha_i A e^{j(i)} = A \sum_{i=1}^{m+1} \alpha_i e^{j(i)}.$$

It follows that $\mu = A\hat{p}$, where the vector of probabilities $\hat{p}$ is given by

$$\hat{p}_k = \sum_{\{i:j(i)=k\}} \alpha_i, \quad k = 1, \dots, n.$$

There can be no more than $m + 1$ different indices $j(i)$, when $i = 1, \dots, m+1$. Thus no more than $m + 1$ of probabilities $\hat{p}_k$ are positive.

Another way to obtain this conclusion is to consider the set S of all vectors $p \in \mathbb{R}^n$ such that

$$
\begin{aligned}
Ap &= \mu, \\
\sum_{j=1}^{n} p_j &= 1, \\
p &\geq 0.
\end{aligned}
$$

By Theorem 2.48 its extreme points have no more than $m + 1$ positive components.

2.4 CONVEX FUNCTIONS

2.4.1 Basic Concepts

In our analysis it is convenient to consider functions which may take, in addition to real values, two special values: $-\infty$ and $+\infty$. The real line augmented with these special values will be denoted by $\overline{\mathbb{R}}$.

With every function $f : \mathbb{R}^n \to \overline{\mathbb{R}}$ we can associate two sets: the *domain*

$$\operatorname{dom} f \triangleq \{x : f(x) < +\infty\}$$

and the *epigraph*

$$\operatorname{epi} f \triangleq \{(x, v) \in \mathbb{R}^n \times \mathbb{R} : v \geq f(x)\}.$$

Definition 2.50. A function f is called *convex* if $\operatorname{epi} f$ is a convex set.

An example of a convex function,

$$
f(x) = \begin{cases} x \ln(x) - x & \text{if } x > 0, \\ 0 & \text{if } x = 0, \\ +\infty & \text{if } x < 0. \end{cases}
$$

is shown in Figure 2.9.

Definition 2.51. A function f is called *concave* if $-f$ is convex.

Definition 2.52. A function f is called *proper* if $f(x) > -\infty$ for all x and $f(x) < +\infty$ for at least one x.

From now on we will always deal with proper convex functions.

Lemma 2.53. *A function f is convex if and only if for all x^1 and x^2 and for all $0 \leq \alpha \leq 1$ we have*

$$f(\alpha x^1 + (1 - \alpha)x^2) \leq \alpha f(x^1) + (1 - \alpha) f(x^2). \tag{2.16}$$

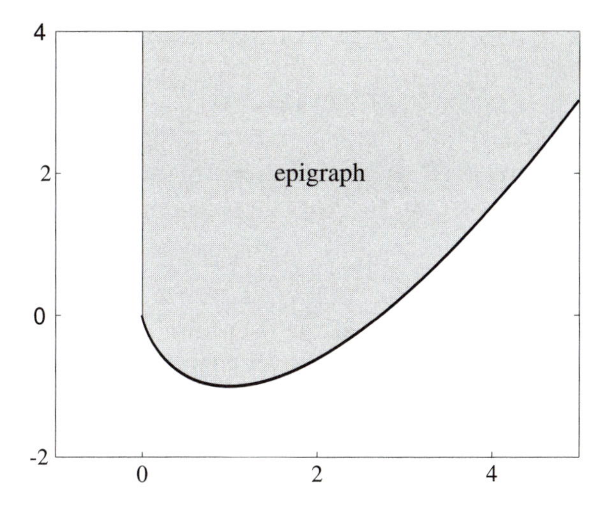

Figure 2.9. Example of a convex function.

Proof. If $x^1 \notin \operatorname{dom} f$ or $x^2 \notin \operatorname{dom} f$, the above inequality is satisfied trivially. It remains to consider the case when $x^1 \in \operatorname{dom} f$ and $x^2 \in \operatorname{dom} f$. Then the points

$$\begin{bmatrix} x^1 \\ f(x^1) \end{bmatrix} \quad \text{and} \quad \begin{bmatrix} x^2 \\ f(x^2) \end{bmatrix}$$

belong to $\operatorname{epi} f$. If f is convex, then their convex combination is in the epigraph:

$$\begin{bmatrix} \alpha x^1 + (1-\alpha)x^2 \\ \alpha f(x^1) + (1-\alpha) f(x^2) \end{bmatrix} \in \operatorname{epi} f, \tag{2.17}$$

which implies (2.16). On the other hand, (2.16) entails (2.17), which in turn yields the convexity of the epigraph. □

Inequality (2.16) can be used as an alternative definition of proper convex functions.

Example 2.54. The function

$$f(x) = \|x\|_\diamond,$$

where $\|\cdot\|_\diamond$ is a norm in $\mathbb{R}^n$, is a convex proper function. Indeed, using the triangle inequality for the norm, for all x, y and $\alpha \in [0, 1]$ we obtain

$$\|\alpha x + (1-\alpha)y\|_\diamond \le \|\alpha x\|_\diamond + \|(1-\alpha)y\|_\diamond = \alpha\|x\|_\diamond + (1-\alpha)\|y\|_\diamond.$$

In the last equation we also used the fact that a norm is positively homogeneous.

Example 2.55. Assume Z is a closed convex set in $\mathbb{R}^n$. The distance to Z,

$$f(x) = \min_{z \in Z} \|x - z\|_{\Diamond},$$

where $\|\cdot\|_{\Diamond}$ is a norm in $\mathbb{R}^n$, is a convex proper function. Indeed, consider two points, x and y, and any $\alpha \in (0, 1)$. Define the points $v \in Z$ and $w \in Z$ such that

$$f(x) = \|x - v\|_{\Diamond}, \qquad f(y) = \|y - w\|_{\Diamond}.$$

They exist, because Z is a closed set. In the particular case when $\|\cdot\|_{\Diamond}$ is the Euclidean norm, v and w are the orthogonal projections of x and y on Z. By the convexity of Z, the convex combination of these points, $\alpha v + (1 - \alpha)w$, with $\alpha \in (0, 1)$, is an element of Z as well. Therefore,

$$\begin{aligned} f(\alpha x + (1-\alpha)y) &= \min_{z \in Z} \|\alpha x + (1-\alpha)y - z\|_{\Diamond} \\ &\le \|\alpha x + (1-\alpha)y - [\alpha v + (1-\alpha)w]\|_{\Diamond} \\ &\le \alpha \|x - v\|_{\Diamond} + (1-\alpha)\|y - w\|_{\Diamond} \\ &= \alpha f(x) + (1-\alpha) f(y). \end{aligned}$$

In the last inequality we used the properties of the norm.

The convexity of the set Z is relevant here; the distance to a nonconvex set is not a convex function.

Definition 2.56. A function f is called *strictly convex* if inequality (2.16) is strict for all $x^1 \neq x^2$ and for all $0 < \alpha < 1$.

Lemma 2.57. *If f is convex, then* dom f *is a convex set.*

Proof. If $x^1 \in \operatorname{dom} f$ and $x^2 \in \operatorname{dom} f$, then, by Lemma 2.53, $f(\alpha x^1 + (1 - \alpha)x^2) < +\infty$. □

Lemma 2.58. *If f_i, $i \in I$, is a family of convex functions, then*

$$f(x) = \sup_{i \in I} f_i(x)$$

is convex.

Proof. Noting that

$$\operatorname{epi} f = \bigcap_{i \in I} \operatorname{epi} f_i$$

we obtain the required result from Lemma 2.2. □

Example 2.59. For a symmetric matrix A, its maximum eigenvalue $\lambda_{\max}(A)$ is well defined. We can consider, therefore, the function

$$f(A) = \lambda_{\max}(A),$$

defined on the space $\mathbb{S}^n$ of symmetric matrices of dimension $n \times n$. As

$$\lambda_{\max}(A) = \max_{\|y\|=1} \langle y, Ay \rangle,$$

and each function

$$f_y(A) = \langle y, Ay \rangle$$

is linear, the function $\lambda_{\max}(\cdot)$ is convex.

Lemma 2.60. *If f is a convex function, then for all $x^1, x^2, \dots, x^m$ and all $\alpha_1 \geq 0, \alpha_2 \geq 0, \dots, \alpha_m \geq 0$ such that $\alpha_1 + \alpha_2 + \cdots + \alpha_m = 1$, one has*

$$f(\alpha_1 x^1 + \alpha_2 x^2 + \cdots + \alpha_m x^m) \leq \alpha_1 f(x^1) + \alpha_2 f(x^2) + \cdots + \alpha_m f(x^m).$$

Proof. The result follows from the convexity of epi f: the points

$$\begin{bmatrix} x^i \\ f(x^i) \end{bmatrix}, \quad i = 1, 2, \dots, m,$$

belong to the epigraph, so their convex combination is in epi f, too. □

Lemma 2.61. *If the functions f_i, $i = 1, 2, \dots, m$, are convex, then for all $c_1 \geq 0, c_2 \geq 0, \dots, c_m \geq 0$ the function*

$$f(x) = c_1 f_1(x) + c_2 f_2(x) + \cdots + c_m f_m(x)$$

is convex.

Proof. Since (2.16) holds true for each f_i, we can multiply these inequalities by $c_i \geq 0$ and sum up to get (2.16) for f. □

The epigraph of a function $f : \mathbb{R}^n \to \overline{\mathbb{R}}$ can be used to characterize continuity properties of the function, irrespective of its convexity. A function $f : \mathbb{R}^n \to \overline{\mathbb{R}}$ is called *lower semicontinuous*, if for every convergent sequence of points $\{x^k\}$ we have

$$f\Big(\lim_{k\to\infty} x^k\Big) \leq \liminf_{k\to\infty} f(x^k).$$

Lemma 2.62. *A function $f : \mathbb{R}^n \to \overline{\mathbb{R}}$ is lower semicontinuous if and only if its epigraph is a closed set.*

Proof. Consider a sequence $\{(x^k, \alpha^k)\}$ of points of epi f, and suppose $x^k \to x$ and $\alpha^k \to \alpha$, as $k \to \infty$. If f is lower semicontinuous, then

$$f(x) \le \liminf_{k\to\infty} f(x^k) \le \lim_{k\to\infty} \alpha^k = \alpha,$$

which proves that $(x, \alpha) \in \operatorname{epi} f$.

Suppose the epigraph is closed, but f is not lower semicontinuous. Then there exists a sequence $\{x^k\} \subset \mathbb{R}^n$ convergent to some point $x \in \mathbb{R}^n$ such that

$$f(x) > \lim_{k\to\infty} f(x^k),$$

where the limit on the right hand side may be $-\infty$, in general. Then there exists $\varepsilon > 0$ such that $f(x^k) < f(x) - \varepsilon$ for all sufficiently large k. Hence

$$(x^k, f(x) - \varepsilon) \in \operatorname{epi} f$$

for all sufficiently large k. Since the epigraph is closed, the limit of these points, $(x, f(x) - \varepsilon)$, is an element of the epigraph of f as well. This means that $f(x) - \varepsilon \ge f(x)$, a contradiction. Thus f must be lower semicontinuous. □

Lemma 2.63. *If $f : \mathbb{R}^n \to \overline{\mathbb{R}}$ is convex, then for each $\beta \in \mathbb{R}$ the set*

$$M_\beta = \{x : f(x) \le \beta\} \tag{2.18}$$

is convex. If, in addition, f is lower semicontinuous, then the set M_β is closed for all β.

Proof. If $x \in M_\beta$ and $y \in M_\beta$, then, by Lemma 2.53,

$$f(\alpha x + (1-\alpha)y) \le \alpha f(x) + (1-\alpha) f(y) \le \beta,$$

so $\alpha x + (1-\alpha)y \in M_\beta$.

If f is lower semicontinuous, its epigraph is closed (Lemma 2.62). Consider the set in $\mathbb{R}^n \times \mathbb{R}$:

$$M_\beta \times \{\beta\} = \operatorname{epi} f \cap \{(x, \beta) : x \in \mathbb{R}^n\}.$$

It is closed, because epi f is closed. Thus M_β is closed. □

The set M_β in the last lemma is called the *level set* of f.

Not every function that has convex level sets is convex, as the example of $f(x) = \sqrt{|x|}$ clearly demonstrates.

Lemma 2.64. *Let $X \subset \mathbb{R}^n$ be a convex set and let $f : \mathbb{R}^n \to \overline{\mathbb{R}}$ be a convex function. Then the set $\hat{X}$ of solutions of the optimization problem*

$$\underset{x \in X}{\text{minimize}}\ f(x) \tag{2.19}$$

is convex.

Proof. If (2.19) has no solutions, $\hat{X}$ is convex, because it is empty. Let $\hat{X} \neq \emptyset$ and let $\hat{x} \in \hat{X}$, $\beta = f(\hat{x})$. Then

$$\hat{X} = X \cap M_\beta$$

with M_β defined by (2.18). By Lemma 2.2, $\hat{X}$ is convex. □

The maxima of convex functions in convex sets can be characterized as well.

Theorem 2.65. *Let $f : \mathbb{R}^n \to \overline{\mathbb{R}}$ be a convex function and let $X \subset \operatorname{dom} f$ be a convex, closed and bounded set. Then the set of solutions of the problem*

$$\underset{x \in X}{\text{maximize}}\ f(x) \tag{2.20}$$

contains at least one extreme point of X. If, in addition, the function $f(\cdot)$ is affine, then the set of solutions of (2.20) is the convex hull of the set of the extreme points of X that are solutions of (2.20).

Proof. Let $\hat{x}$ be a solution of (2.20). By Theorem 2.44 we can find extreme points $x^1, x^2, \dots, x^m$ of X such that

$$\hat{x} = \alpha_1 x^1 + \alpha_2 x^2 + \cdots + \alpha_m x^m$$

with some $\alpha_1 > 0, \alpha_2 > 0, \dots, \alpha_m > 0$, $\alpha_1 + \alpha_2 + \cdots + \alpha_m = 1$. By Lemma 2.60,

$$f(\hat{x}) \le \sum_{i=1}^{m} \alpha_i f(x^i).$$

This can be rewritten as

$$\sum_{i=1}^{m} \alpha_i \big[f(\hat{x}) - f(x^i) \big] = 0. \tag{2.21}$$

Because $\hat{x}$ is optimal, $f(\hat{x}) \ge f(x^i)$, $i = 1, \dots, m$. Then it follows from (2.21) that $f(\hat{x}) = f(x^i)$, $i = 1, \dots, m$. All points x^i are optimal as well. We see that the set of solutions of (2.20) is included in the convex hull of the extreme points that are solutions of (2.20).

If the function $f(\cdot)$ is affine then $-f(\cdot)$ is convex. The set of solutions of problem (2.20) is the same as the set of minima of $-f(x)$ over $x \in X$. It is convex, by virtue of Lemma 2.64. Therefore the convex hull of the the extreme points that are solutions of (2.20) is included in the set of all solutions of (2.20). □

We can now return to the linear programming problem, which we analyzed at the end of Section 2.3.

THEOREM 2.66. *If the feasible set of the linear programming problem*

$$\begin{aligned} \text{minimize} \quad & \langle c, x \rangle \\ \text{subject to} \quad & Ax = b, \\ & x \geq 0, \end{aligned} \tag{2.22}$$

is bounded, then the set of optimal solutions is the convex hull of the set of optimal basic feasible solutions.

Proof. Let us define $f(x) = -\langle c, x \rangle$. It is an affine function and each solution of (2.22) solves (2.20) with

$$X = \{x : Ax = b, \ x \geq 0\}.$$

The result follows directly from Theorem 2.65. □

The above fact is used by the simplex method for solving linear programming problems. It moves from one basic feasible solution to a better one, as long as progress is possible. The best basic feasible solution is guaranteed to be optimal. It can be found after finitely many steps if a solution exists. If the set is unbounded, we may discover a ray from the recession cone, along which the objective can be decreased without limits. In this case no optimal solution exists.

2.4.2 Smooth Convex Functions

We now formulate convexity criteria for smooth functions. We denote by $\nabla f(x)$ the *gradient* of the function f at x,

$$\nabla f(x) \triangleq \begin{bmatrix} \frac{\partial f(x)}{\partial x_1} \\ \frac{\partial f(x)}{\partial x_2} \\ \vdots \\ \frac{\partial f(x)}{\partial x_n} \end{bmatrix}.$$

Here $x_1, x_2, \ldots, x_n$ denote the coordinates of the vector x.

If f is twice continuously differentiable, $\nabla^2 f(x)$ denotes the *Hessian* of f at x,

$$\nabla^2 f(x) \triangleq \begin{bmatrix} \frac{\partial^2 f(x)}{\partial x_1^2} & \frac{\partial^2 f(x)}{\partial x_1 \partial x_2} & \cdots & \frac{\partial^2 f(x)}{\partial x_1 \partial x_n} \\ \frac{\partial^2 f(x)}{\partial x_2 \partial x_1} & \frac{\partial^2 f(x)}{\partial x_2^2} & \cdots & \frac{\partial^2 f(x)}{\partial x_2 \partial x_n} \\ \vdots & \vdots & \ddots & \vdots \\ \frac{\partial^2 f(x)}{\partial x_n \partial x_1} & \frac{\partial^2 f(x)}{\partial x_n \partial x_2} & \cdots & \frac{\partial^2 f(x)}{\partial x_n^2} \end{bmatrix}.$$

We use the symbol $\mathcal{C}^2$ to denote the set of twice continuously differentiable real valued functions. For $f \in \mathcal{C}^2$ the Hessian is a symmetric matrix, because the values of the mixed derivatives do not depend on the order of differentiation.

THEOREM 2.67. *Assume that a function f is continuously differentiable. Then*

(i) *f is convex if and only if for all x and y*

$$f(y) \geq f(x) + \langle \nabla f(x), y - x \rangle; \tag{2.23}$$

(ii) *f is strictly convex if and only if for all $x \neq y$*

$$f(y) > f(x) + \langle \nabla f(x), y - x \rangle. \tag{2.24}$$

Proof. (i) Suppose f is convex, but for some x and y

$$f(y) \leq f(x) + \langle \nabla f(x), y - x \rangle - \varepsilon$$

with $\varepsilon > 0$. Let us consider $z = \alpha y + (1 - \alpha)x$ for $0 < \alpha < 1$. By Lemma 2.53,

$$f(z) \leq \alpha f(y) + (1 - \alpha) f(x) \leq f(x) + \alpha \langle \nabla f(x), y - x \rangle - \alpha \varepsilon.$$

Rearranging terms and dividing by α we get

$$\frac{f(z) - f(x)}{\alpha} \leq \langle \nabla f(x), y - x \rangle - \varepsilon. \tag{2.25}$$

Let $\alpha \downarrow 0$. Since $z = x + \alpha d$ with $d = y - x$, the left side of (2.25) converges to the directional derivative at x in the direction d

$$f'(x; d) = \langle \nabla f(x), d \rangle$$

and we obtain a contradiction in (2.25).

To prove the opposite implication, let us assume (2.23). Let y and z be arbitrary points, $y \neq z$, and let $x = \alpha y + (1-\alpha)z$ with $\alpha \in (0,1)$. Then, by (2.23) we have

$$f(y) \geq f(x) + \langle \nabla f(x), y - x \rangle,$$
$$f(z) \geq f(x) + \langle \nabla f(x), z - x \rangle.$$

Multiplying these inequalities by α and $1-\alpha$, respectively, and adding them one gets

$$\alpha f(y) + (1-\alpha) f(z) \geq f(x),$$

which was what we set out to prove.

(ii) If f is strictly convex, then f is convex and (2.23) is true. We shall prove that the inequality in (2.23) is strict, if $y \neq x$ and $\alpha \in (0,1)$. Suppose equality in (2.23). Let $z = \frac{1}{2}x + \frac{1}{2}y$. Then by strict convexity of f and by equality in (2.23)

$$f(z) < \frac{1}{2} f(x) + \frac{1}{2} f(y) = f(x) + \frac{1}{2} \langle \nabla f(x), y - x \rangle. \tag{2.26}$$

Let $v = \beta x + (1-\beta)z$ with $0 < \beta < 1$. Then, by convexity and by (2.26)

$$f(v) < \beta f(x) + (1-\beta) f(z) < f(x) + \frac{1}{2}(1-\beta)\langle \nabla f(x), y - x \rangle.$$

Since $v - x = (1-\beta)(z-x) = \frac{1}{2}(1-\beta)(y-x)$, the last inequality reads

$$f(v) < f(x) + \langle \nabla f(x), v - x \rangle,$$

which contradicts (2.23).

To prove that (2.24) implies strict convexity, let us simply note that the argument used in case (i) applies again, but with sharp inequalities. $\square$

It follows from the proof of the above theorem that if $f : \mathbb{R}^n \to \mathbb{R}$ is convex and differentiable at the point x then (2.23) holds true for all $y \in \mathbb{R}^n$. If f is strictly convex, (2.24) remains valid as well, for all $y \in \mathbb{R}^n$.

Example 2.68. Consider the function $f : \mathbb{R}^n \to \mathbb{R}$ defined as the quadratic form,

$$f(x) = \langle x, Ax \rangle,$$

where A is a symmetric matrix. The function f is convex if and only if A is a positive semidefinite matrix, and f is strictly convex if and only if A is a positive definite matrix. Indeed,

$$\nabla f(x) = 2Ax,$$

and for all x and all y we have the equation

$$\begin{aligned} f(y) - f(x) - \langle \nabla f(x), y - x \rangle &= \langle y, Ay \rangle - \langle x, Ax \rangle - 2\langle Ax, y - x \rangle \\ &= \langle y, Ay \rangle + \langle x, Ax \rangle - 2\langle Ax, y \rangle \\ &= \langle y - x, A(y - x) \rangle. \end{aligned}$$

The expression on the right hand side is nonnegative for all x and y if and only if A is positive semidefinite. This expression is positive for all $y \neq x$ if and only if A is positive definite.

Examples of quadratic functions in $\mathbb{R}^2$ corresponding to a positive definite matrix and to an indefinite matrix are illustrated in Figure 2.10.

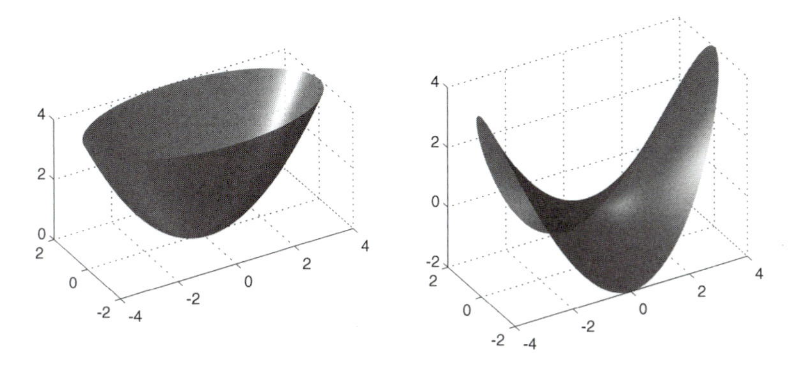

Figure 2.10. Convex and nonconvex quadratic functions.

The theorem below uses the observation of Example 2.68 in a more general setting.

THEOREM 2.69. *Assume that $f : \mathbb{R}^n \to \mathbb{R}$ is twice continuously differentiable. Then*

(i) *f is convex if and only if its Hessian $\nabla^2 f(x)$ is positive semidefinite for all $x \in \mathbb{R}^n$;*

(ii) *if the Hessian $\nabla^2 f(x)$ is positive definite for all $x \in \mathbb{R}^n$ then f is strictly convex.*

Proof. If $f \in \mathcal{C}^2$, then for all x and all y we have

$$f(y) = f(x) + \langle \nabla f(x), y - x \rangle + \frac{1}{2}\langle y - x, \nabla^2 f(x_\theta)(y - x) \rangle, \quad (2.27)$$

where $x_\theta = x + \theta(y - x)$ with some $0 \leq \theta \leq 1$. If the Hessian $\nabla^2 f(x)$ is positive semidefinite for all x, then the quadratic term in (2.27) is nonnegative and we obtain (2.23). If the Hessian is positive definite, then for $y \neq x$ the quadratic term is positive and we get (2.24).

Suppose the Hessian is not positive semidefinite for some x. Then there exists d such that

$$\langle d, \nabla^2 f(x) d\rangle < 0.$$

Let $y = x + \varepsilon d$ for some $\varepsilon > 0$. If ε is small enough, then y and x_θ are so close to x that

$$\langle d, \nabla^2 f(x_\theta) d\rangle < 0,$$

because the Hessian is continuous. But then the quadratic term in (2.27) is negative and we obtain a contradiction with (2.23). □

Note that statement (ii) of the theorem does not have the "only if" part, which appeared in the quadratic case. For example, the function $f(x) = x^4$ is strictly convex, but its second derivative vanishes at 0.

2.4.3 Directional Derivatives

So far we have discussed convex functions that are at least once continuously differentiable. In many applications one has to deal with functions that are nonsmooth. For example the Euclidean norm

$$\|x\| = \Big(\sum_{j=1}^{n} x_j^2\Big)^{1/2}$$

is not differentiable at 0. In fact, no norm is differentiable at 0. Some, like

$$\|x\|_1 = \sum_{j=1}^{n} |x_j| \quad \text{or} \quad \|x\|_\infty = \max_{1\le j\le n} |x_j|,$$

are nondifferentiable at many points (see Figure 2.11). Nonsmooth functions

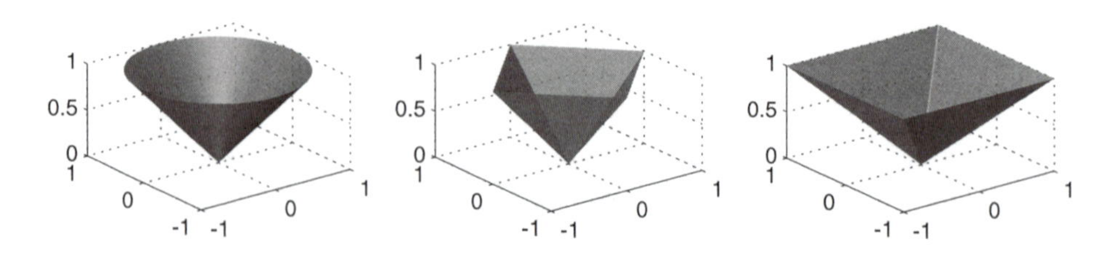

Figure 2.11. The norms $\|\cdot\|$, $\|\cdot\|_1$ and $\|\cdot\|_\infty$.

are very common in optimization models.

The concept of the gradient of a smooth function can be generalized to the case of nonsmooth functions, and in particular: convex nonsmooth functions. To understand this construction we must first prove several important properties of convex functions.

LEMMA 2.70. *Let $f : \mathbb{R}^n \to \overline{\mathbb{R}}$ be a convex function. For every $x \in \operatorname{int} \operatorname{dom} f$ there exist $\delta > 0$ and L such that*

$$|f(y) - f(x)| \le L\|y - x\| \quad \textit{whenever} \quad \|y - x\| < \delta.$$

Proof. Since $x \in \operatorname{int} \operatorname{dom} f$, we can choose a small simplex S about x such that it is included in the domain of f. Every point y of the simplex is a convex combination of its vertices $v^1, \dots, v^{n+1}$ with some nonnegative coefficients $\alpha_1, \dots, \alpha_{n+1}$ totaling 1. By the convexity of f,

$$f(y) \le \sum_{i=1}^{n+1} \alpha_i f(v^i) \le \max_{1 \le i \le n+1} f(v^i) = M.$$

Let $\varepsilon > 0$ be the lower bound on the distance from x to the boundary of S. Suppose $\|y - x\| < \varepsilon/2$. We can represent y as a convex combination of x and a point w at which the line passing through x and y intersects the boundary of S (we choose the intersection point that is closer to y). We have

$$y = (1 - \alpha)x + \alpha w,$$

with

$$\alpha = \frac{\|y - x\|}{\|w - x\|} \le \frac{\|y - x\|}{\varepsilon}.$$

By the convexity of f

$$\begin{aligned} f(y) - f(x) &\le \alpha[f(w) - f(x)] \\ &\le \alpha[M - f(x)] \le \frac{M - f(x)}{\varepsilon}\|y - x\|. \end{aligned} \tag{2.28}$$

We can also represent x as a convex combination of y and the point z at which the same line crosses the boundary of S on the other side:

$$x = (1 - \beta)y + \beta z$$

with

$$\beta = \frac{\|y - x\|}{\|y - z\|} \le \frac{\|y - x\|}{\varepsilon}.$$

By the convexity of f,

$$f(x) \le (1-\beta)f(y) + \beta f(z) \le (1-\beta)f(y) + \beta M.$$

Hence

$$f(y) - f(x) \ge \frac{2(f(x) - M)}{\varepsilon}\|y - x\|.$$

Combining the last inequality with (2.28) we conclude that

$$|f(y) - f(x)| \le \frac{2(M - f(x))}{\varepsilon}\|y - x\| \quad \text{whenever} \quad \|y - x\| < \varepsilon/2.$$

The required inequality is satisfied with $L = 2(M - f(x))/\varepsilon$. □

Let $f : \mathbb{R}^n \to \overline{\mathbb{R}}$ be a convex function and let $x \in \operatorname{dom} f$. Then for every $d \in \mathbb{R}^n$ the quantity

$$f'(x; d) = \lim_{\tau \downarrow 0} \frac{f(x + \tau d) - f(x)}{\tau}, \tag{2.29}$$

is called the *directional derivative* of f at x in the direction d.

Lemma 2.71. *For every $x \in \operatorname{dom} f$ and every $d \in \mathbb{R}^n$ the limit in (2.29) exists (finite or infinite). If $x \in \operatorname{int} \operatorname{dom} f$, then $f'(x; d)$ is finite for all d.*

Proof. Consider the differential quotient

$$Q(\tau) = \frac{f(x + \tau d) - f(x)}{\tau}.$$

If $f(x + \tau d) = +\infty$ for all $\tau > 0$, we have $Q(\tau) = +\infty$ for all $\tau > 0$ and $f'(x; d) = +\infty$. If $f(x + \tau_0 d) < +\infty$ for some $\tau_0 > 0$, the convexity of f implies that $f(x + \tau d) < +\infty$ for all $0 < \tau < \tau_0$, and $Q(\tau)$ is well defined for these τ. Let $0 < \tau_1 < \tau_2 < \tau_0$. We have

$$x + \tau_1 d = \left(1 - \frac{\tau_1}{\tau_2}\right)x + \frac{\tau_1}{\tau_2}(x + \tau_2 d).$$

The convexity of $f(\cdot)$ renders

$$f(x + \tau_1 d) \le \left(1 - \frac{\tau_1}{\tau_2}\right)f(x) + \frac{\tau_1}{\tau_2}f(x + \tau_2 d),$$

which can be rewritten as

$$f(x + \tau_1 d) - f(x) \le \frac{\tau_1}{\tau_2}[f(x + \tau_2 d) - f(x)].$$

Dividing by τ_1 we see that the differential quotients are monotone:

$$Q(\tau_1) \le Q(\tau_2) \quad \text{for all} \quad 0 < \tau_1 \le \tau_2. \tag{2.30}$$

Therefore the limit in (2.29) exists (finite or equal to $-\infty$). The monotonicity of $Q(\cdot)$ implies that

$$f'(x; d) \le Q(\tau) \quad \text{for all} \quad \tau > 0. \tag{2.31}$$

If $x \in \operatorname{int} \operatorname{dom} f$, Lemma 2.70 implies that for all sufficiently small τ

$$|Q(\tau)| = \frac{|f(x + \tau d) - f(x)|}{\tau} \le L\|d\|,$$

and therefore the limit of $Q(\tau)$ for $\tau \downarrow 0$ must be finite. □

2.5 SUBDIFFERENTIAL CALCULUS

2.5.1 Subgradients and Subdifferentials

For general convex functions the concept of the gradient is generalized to the notion of a *subgradient*. It is defined by an inequality similar to inequality (2.23) which characterizes smooth convex functions.

Definition 2.72. Let $f : \mathbb{R}^n \to \overline{\mathbb{R}}$ be a proper convex function and let $x \in \operatorname{dom} f$. A vector $g \in \mathbb{R}^n$ such that

$$f(y) \ge f(x) + \langle g, y - x \rangle \quad \text{for all} \quad y \in \mathbb{R}^n \tag{2.32}$$

is called a *subgradient* of f at x.

The set of all subgradients of f at x is called the *subdifferential* of f at x and is denoted by $\partial f(x)$.

The notion of the subgradient has a useful geometrical interpretation. Suppose $g \in \partial f(x)$. Inequality (2.32) means that the epigraph of f is located on or above the graph of the affine function $l(y) = f(x) + \langle g, y - x \rangle$. This is illustrated in Figure 2.12.

For every point $(y, v) \in \operatorname{epi} f$ we have

$$v \ge f(y) \ge f(x) + \langle g, y - x \rangle,$$

which can be rewritten as

$$\langle g, y - x \rangle + (-1)(v - f(x)) \le 0.$$

Consequently, $(g, -1)$ is an element of the normal cone $N_{\operatorname{epi} f}(x, f(x))$.

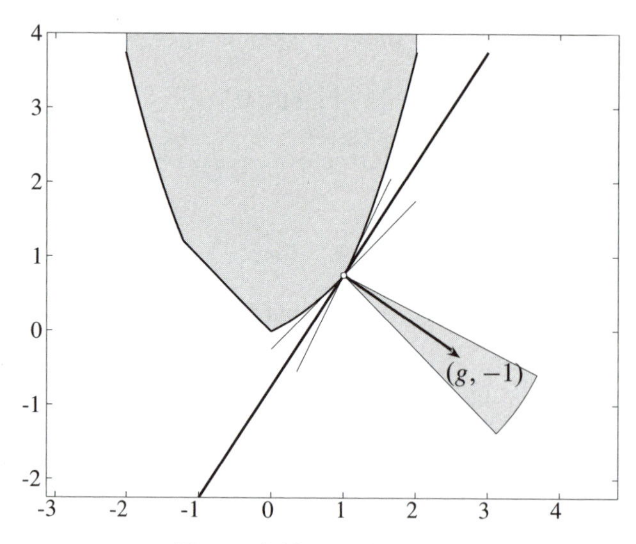

Figure 2.12. A subgradient.

The reverse is also true, under some conditions. If a vector $(u, \gamma) \in N_{\operatorname{epi} f}(x, f(x))$ and $\gamma \neq 0$, then $g = -u/\gamma$ is a subgradient of f at x. The condition that $\gamma \neq 0$ is essential here. Consider the convex function shown in Figure 2.9 on page 45. At the point (0,0) all normals to the epigraph have the form $(u, 0)$, where $u < 0$, and f has no subgradient at $x = 0$.

Subgradients, if they exist, provide lower bounds for directional derivatives.

Lemma 2.73. *Assume that $f : \mathbb{R}^n \to \overline{\mathbb{R}}$ is a proper convex function and let $x \in \operatorname{dom} f$. A vector g is a subgradient of f at x if and only if*

$$f'(x; d) \geq \langle g, d \rangle \quad \text{for every} \quad d \in \mathbb{R}^n. \tag{2.33}$$

Proof. Suppose (2.33) is true. Then for every y, by (2.31),

$$f(y) \geq f(x) + f'(x; y - x) \geq f(x) + \langle g, y - x \rangle,$$

which shows that g is a subgradient. To prove the converse, suppose $g \in \partial f(x)$. Then for every d and every $\tau > 0$ the subgradient inequality (2.32) yields

$$\frac{f(x + \tau d) - f(x)}{\tau} \geq \frac{\langle g, \tau d \rangle}{\tau} = \langle g, d \rangle.$$

Passing to the limit with $\tau \downarrow 0$ we get (2.33). □

We are now in a position to resolve the question of the existence of subgradients and to fully characterize the directional derivatives.

THEOREM 2.74. *Let $f : \mathbb{R}^n \to \overline{\mathbb{R}}$ be a convex function. Assume that $x \in \operatorname{int}\operatorname{dom} f$. Then $\partial f(x)$ is a nonempty, convex, closed and bounded set. Furthermore, for every direction $d \in \mathbb{R}^n$ one has*

$$f'(x; d) = \max_{g \in \partial f(x)} \langle g, d \rangle.$$

Proof. In view of Lemma 2.73, it is sufficient to show that for every d there exists $g \in \partial f(x)$ such that

$$f'(x; d) = \langle g, d \rangle. \tag{2.34}$$

Consider two sets in $\mathbb{R}^{n+1}$:

$$E = \{(y, v) : v > f(y)\},$$

and

$$L = \{(x + \tau d, f(x) + \tau f'(x; d)) : \tau \in \mathbb{R}\}.$$

Since

$$f(x + \tau d) \geq f(x) + \tau f'(x; d) \quad \text{for all} \quad \tau \in \mathbb{R},$$

these sets have no common points. They are convex and therefore can be separated by a plane (see Figure 2.13).

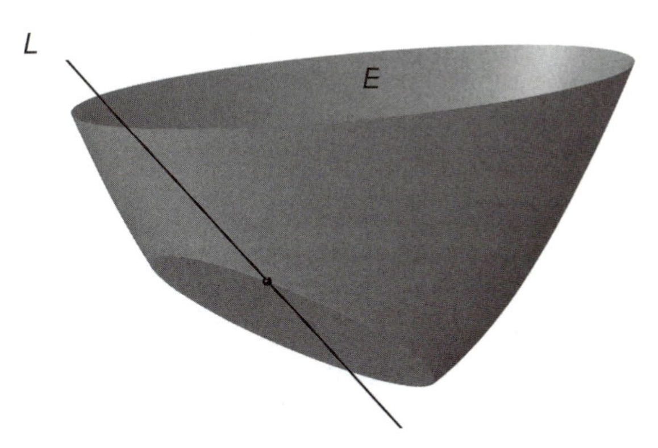

Figure 2.13. The set E and line L can be separated.

By Theorem 2.15, there exists a nonzero

$$z = \begin{bmatrix} u \\ \gamma \end{bmatrix}$$

such that for every point $(y, v) \in E$ and every $\tau \in \mathbb{R}$ we have

$$\langle u, y\rangle + \gamma v \geq \langle u, x + \tau d\rangle + \gamma\,[f(x) + \tau f'(x; d)]. \tag{2.35}$$

We deduce that $\gamma \geq 0$, for otherwise letting $v \to \infty$ leads to a contradiction.

Suppose $\gamma = 0$. As x is an interior point of the domain, we can choose y from a small ball B about x in such a way that there exists $v > f(y)$. Setting $\tau = 0$ in (2.35), we get $\langle u, y\rangle \geq \langle u, x\rangle$ for all $y \in B$, which is possible only when $u = 0$ and contradicts the condition that $z \neq 0$. Thus $\gamma > 0$.

Dividing both sides of (2.35) by γ, setting $g = -u/\gamma$ and letting $v \downarrow f(y)$ we obtain:

$$f(y) - \langle g, y\rangle \geq f(x) + \tau f'(x; d) - \langle g, x + \tau d\rangle$$

for all $y \in \operatorname{int}\operatorname{dom} f$ and all $\tau \in \mathbb{R}$. It follows from this inequality that

$$\tau\big[f'(x; d) - \langle g, d\rangle\big] \leq f(y) - f(x) - \langle g, y - x\rangle,$$

for all $\tau \in \mathbb{R}$. This is possible only if the coefficient multiplied by τ is zero, and thus (2.34) holds true. It remains to show that g is a subgradient.

Setting $\tau = 0$ we conclude that

$$f(y) \geq f(x) + \langle g, y - x\rangle, \quad \text{for all} \quad y \in \operatorname{int}\operatorname{dom} f. \tag{2.36}$$

By convexity, for every $y \in \operatorname{dom} f$,

$$f(y) - f(x) \geq 2[f((x + y)/2) - f(x)],$$

and $(x + y)/2 \in \operatorname{int}\operatorname{dom} f$. Applying (2.36) to $f((x + y)/2)$ we see that (2.32) holds true for all $y \in \operatorname{dom} f$, and thus for all $y \in \mathbb{R}^n$. Therefore g is a subgradient of f at x and the subdifferential is nonempty.

Suppose $g^1 \in \partial f(x)$ and $g^2 \in \partial f(x)$. Then for all y

$$f(y) \geq f(x) + \langle g^1, y - x\rangle,$$
$$f(y) \geq f(x) + \langle g^2, y - x\rangle.$$

Adding these inequalities multiplied by α and by $1-\alpha$, where $\alpha \in (0, 1)$, we conclude that $\alpha g^1 + (1 - \alpha)g^2 \in \partial f(x)$. Thus the subdifferential is convex.

If $g^k \in \partial f(x)$ and $g^k \to g$, then we can pass to the limit in the inequality

$$f(y) \geq f(x) + \langle g^k, y - x\rangle,$$

to conclude that $g \in \partial f(x)$. Therefore the subdifferential is closed.

Let us prove the boundedness. Let $g \in \partial f(x)$. By Lemma 2.36 for $x + \tau d$ sufficiently close to x we have

$$f(x + \tau d) - f(x) \leq \tau L\|d\|.$$

Hence

$$f'(x; d) \le L\|d\|, \quad \text{for all} \quad d.$$

Using Lemma 2.73 we obtain:

$$\langle g, d\rangle \le f'(x; d) \le L\|d\|, \quad \text{for all} \quad d,$$

which implies $\|g\| \le L$. □

It follows from the proof that the subdifferential $\partial f(x)$ is a convex and closed set at every point x at which $f(\cdot)$ has at least one subgradient (is subdifferentiable).

The formula for the directional derivative can be generalized to boundary points at which f is subdifferentiable.

LEMMA 2.75. *If a convex function $f : \mathbb{R}^n \to \overline{\mathbb{R}}$ is subdifferentiable at a point x, then for every d*

$$f'(x; d) = \sup_{g \in \partial f(x)} \langle g, d\rangle.$$

Moreover, if $f'(x; d) < \infty$ then the supremum above is attained.

Proof. By Lemma 2.71, the directional derivative $f'(x; d)$ exists (finite or infinite). Also, from the definition of the subdifferential it follows that for each $g \in \partial f(x)$ we have

$$f'(x; d) = \lim_{\tau \downarrow 0} \frac{f(x + \tau d) - f(x)}{\tau} \ge \langle g, d\rangle.$$

Suppose $d \in \mathbb{R}^n$ and $\varepsilon > 0$ exist such that $f'(x; d) > \langle g, d\rangle + \varepsilon$ for all $g \in \partial f(x)$. Then

$$\mu = \sup_{g \in \partial f(x)} \langle g, d\rangle + \epsilon < \infty.$$

We can define the line

$$L = \{(x + \tau d, f(x) + \tau\mu) : \tau \in \mathbb{R}\}.$$

Proceeding exactly as in the proof of Theorem 2.74 we can separate the line L and the set

$$E = \{(y, v) \in \mathbb{R}^{n+1} : v > f(y)\}$$

and find a new subgradient $\bar{g}$ such that $\mu = \langle \bar{g}, d\rangle$. This leads to a contradiction and proves the result. □

From Lemma 2.73 and Theorem 2.74 we can immediately draw the following conclusion.

LEMMA 2.76. *A convex function $f : \mathbb{R}^n \to \overline{\mathbb{R}}$ is differentiable at x if and only if the subdifferential $\partial f(x)$ has only one element, in which case it is the gradient of f at x.*

Proof. A function f is differentiable at x if and only if its directional derivative $f'(x; d)$ is linear in d. Then

$$f'(x; d) = \langle \nabla f(x), d \rangle \quad \text{for all} \quad d.$$

It follows from Lemma 2.73 that $\nabla f(x) \in \partial f(x)$. If there exists another subgradient $g \neq \nabla f(x)$, then we can set $d = g - \nabla f(x)$ to obtain from Lemma 2.73

$$f'(x; d) \geq \langle g, d \rangle = \|g\|^2 - \langle g, \nabla f(x) \rangle.$$

By differentiability,

$$f'(x; d) = \langle \nabla f(x), d \rangle = -\|\nabla f(x)\|^2 + \langle g, \nabla f(x) \rangle.$$

Subtracting the last two relations we conclude that

$$\|g\|^2 - 2\langle g, \nabla f(x) \rangle + \|\nabla f(x)\|^2 \leq 0,$$

which is possible only when $g = \nabla f(x)$.

To prove the opposite implication, suppose the subdifferential has only one element, g. Theorem 2.74 implies that

$$f'(x; d) = \langle g, d \rangle \quad \text{for all} \quad d.$$

Thus the directional derivative is linear in the direction, which is equivalent to the differentiability of f at x and to $g = \nabla f(x)$. □

For a convex function $f : \mathbb{R}^n \to \overline{R}$, which is subdifferentiable at a point x, we define the *direction of steepest descent* as the vector $d \in \mathbb{R}^n$ of length not exceeding 1 for which the directional derivative $f'(x; d)$ is minimal.

LEMMA 2.77. *If f is subdifferentiable at x and $0 \notin \partial f(x)$, then the direction of steepest descent of f at x has the form*

$$\hat{d} = -\frac{g}{\|g\|},$$

where g is the minimum norm element of $\partial f(x)$.

Proof. Consider the minimum norm element $g \in \partial f(x)$. It is nonzero, because $0 \notin \partial f(x)$ and the subdifferential is closed. As g is the projection of 0 on $\partial f(x)$, Lemma 2.11 yields the inequality

$$\langle 0 - g, s - g \rangle \le 0 \quad \text{for all} \quad s \in \partial f(x).$$

This can be rewritten as follows:

$$\max_{s \in \partial f(x)} \langle -g, s \rangle = -\|g\|^2.$$

Dividing by $\|g\|$ and using Lemma 2.75 we obtain

$$f'(x; \hat{d}) = \max_{s \in \partial f(x)} \langle \hat{d}, s \rangle = \max_{s \in \partial f(x)} \Big\langle -\frac{g}{\|g\|}, s \Big\rangle = -\|g\|.$$

For every other d of length at most 1 we have

$$f'(x; d) = \max_{s \in \partial f(x)} \langle d, s \rangle \ge \langle d, g \rangle \ge -\|g\|.$$

This proves that $f'(x; \hat{d})$ is the smallest possible. □

Let us consider four examples of convex functions and their subdifferentials.

Example 2.78. Let $\|\cdot\|$ be the Euclidean norm in $\mathbb{R}^n$, and let

$$f(x) = \|x\|.$$

Then

$$\partial f(0) = \{g \in \mathbb{R}^n : \|g\| \le 1\}.$$

Indeed, for every direction d,

$$f'(0; d) = \lim_{\tau \downarrow 0} \frac{\|\tau d\|}{\tau} = \|d\|.$$

By Lemma 2.73, a vector $g \in \partial f(0)$ if and only if

$$\langle g, d \rangle \le \|d\| \quad \text{for all} \quad d,$$

which is equivalent to $\|g\| \le 1$.

Our result can be extended to any norm $\|\cdot\|_\diamond$ in $\mathbb{R}^n$. By an identical argument we see that its subdifferential at 0 is the set

$$\partial \|0\|_\diamond = \{g \in \mathbb{R}^n : \langle g, d \rangle \le \|d\|_\diamond \text{ for all } d\}.$$

Defining the *dual norm*

$$\|g\|_* = \sup_{d \ne 0} \frac{\langle g, d \rangle}{\|d\|_\diamond} = \sup_{\|d\|_\diamond = 1} \langle g, d \rangle$$

we find that

$$\partial\|0\|_\diamond = \{g \in \mathbb{R}^n : \|g\|_* \le 1\}.$$

In particular, the subdifferential of $\|\cdot\|_1$ at 0 is the set $\{g : \|g\|_\infty \le 1\}$, and the subdifferential of $\|\cdot\|_\infty$ at 0 is the set $\{g : \|g\|_1 \le 1\}$.

Let us now consider an arbitrary nonzero point x. For every vector $g \in \partial\|x\|_\diamond$ we must have

$$\begin{aligned} 2\|x\|_\diamond &= \|x + x\|_\diamond \ge \|x\|_\diamond + \langle g, x\rangle, \\ 0 &= \|x - x\|_\diamond \ge \|x\|_\diamond - \langle g, x\rangle, \end{aligned}$$

which implies that

$$\langle g, x\rangle = \|x\|_\diamond.$$

Dividing both sides by $\|x\|_\diamond$ (which is nonzero), we conclude that

$$\|g\|_* \ge \langle g, \frac{x}{\|x\|_\diamond}\rangle = 1.$$

Suppose $\|g\|_* > 1$. Then there exists d such that $\|d\|_\diamond = 1$ and $\langle g, d\rangle > 1$. We obtain

$$\|x\|_\diamond + 1 = \|x\|_\diamond + \|d\|_\diamond \ge \|x + d\|_\diamond \ge \|x\|_\diamond + \langle g, d\rangle > \|x\|_\diamond + 1,$$

a contradiction. We claim, therefore, that

$$\partial\|x\|_\diamond = \{g \in \mathbb{R}^n : \|g\|_* \le 1,\ \langle g, x\rangle = \|x\|_\diamond\}. \tag{2.37}$$

Since we have already established that every subgradient is an element of the set on the right hand side, it remains to prove the converse. Let g be an element of the set on the right hand side. Since $\|g\|_* \le 1$, for every y we have

$$\|y\|_\diamond \ge \langle g, y\rangle = \langle g, x\rangle + \langle g, y - x\rangle = \|x\|_\diamond + \langle g, y - x\rangle,$$

so g is a subgradient of $\|\cdot\|_\diamond$ at x.

Our formula for the subdifferential of the norm is now valid for $y = 0$ and for $y \ne 0$.

Example 2.79. Let Z be a closed convex set in $\mathbb{R}^n$ and $\|\cdot\|_\diamond$ be a norm. Consider the function

$$f(x) = \min_{z \in Z} \|x - z\|_\diamond.$$

From Example 2.55 we know that f is a convex function. Let us calculate its subdifferential at a point x. Choose $\hat{z} \in Z$ such that $\|x - \hat{z}\|_\diamond = f(x)$. Such a point exists because Z is a closed set, but it does not have to be unique, if the norm $\|\cdot\|_\diamond$ is not strictly convex. For every y we have

$$f(y) \le \|y - \hat{z}\|_\diamond,$$

and at $y = x$ the above inequality becomes an equation. Thus, for every $g \in \partial f(x)$ we obtain

$$\|y - \hat{z}\|_\diamond \geq f(y) \geq \langle g, y - x \rangle.$$

This implies that $g \in \partial \|x - \hat{z}\|_\diamond$. Therefore,

$$\partial f(x) \subset \partial \|x - \hat{z}\|_\diamond.$$

For every $g \in \partial f(x)$ and for every $z \in Z$ we must have

$$0 = f(z) \geq f(x) + \langle g, z - x \rangle = \|x - \hat{z}\|_\diamond + \langle g, \hat{z} - x \rangle + \langle g, z - \hat{z} \rangle.$$

We already know that $g \in \partial \|x - \hat{z}\|_\diamond$, so we can use the results of the calculation in Example 2.78. We have two cases. Either $x = \hat{z}$, in which case the first two terms on the right hand side vanish, or $x \neq \hat{z}$. In the latter case, formula (2.37) yields $\|x - \hat{z}\|_\diamond = \langle g, x - \hat{z} \rangle$. In both cases the sum of the first two terms on the right hand side is zero. The last displayed inequality simplifies to

$$\langle g, z - \hat{z} \rangle \leq 0 \quad \text{for all} \quad z \in Z.$$

Hence

$$g \in N_Z(\hat{z}).$$

We claim that

$$\partial f(x) = N_Z(\hat{z}) \cap \partial \|x - \hat{z}\|_\diamond. \tag{2.38}$$

Note that the point $\hat{z}$ need not be unique, in general, so we also claim that the set on the right hand side is the same for all possible projections $\hat{z}$.

We have already proved that for every $\hat{z}$ each subgradient of f at x is an element of the set at the right hand side of (2.38). Let us prove the converse. Let g be an element of the set at the right hand side of (2.38). Since $\|g\|_* \leq 1$, for every y we have

$$\begin{aligned} f(y) &= \inf_{z \in Z} \|y - z\|_\diamond \geq \inf_{z \in Z} \langle g, y - z \rangle \\ &= \langle g, y - x \rangle + \langle g, x - \hat{z} \rangle + \inf_{z \in Z} \langle g, \hat{z} - z \rangle. \end{aligned}$$

As $g \in \partial \|x - \hat{z}\|_\diamond$ we have $\langle g, x - \hat{z} \rangle = \|x - \hat{z}\|_\diamond = f(x)$. We also have in (2.38) that $g \in N_Z(\hat{z})$, so $\langle g, \hat{z} - z \rangle \geq 0$ for all $z \in Z$. Consequently, the last displayed inequality yields

$$f(y) \geq f(x) + \langle g, y - x \rangle \quad \text{for all} \quad y,$$

and g is a subgradient.

In particular, if $\|\cdot\|_\diamond$ is the Euclidean norm and $x \notin Z$, then the point $\hat{z}$ is the orthogonal projection of x on Z. Therefore

$$\partial f(x) = N_Z(\Pi_Z(x)) \cap \partial \|x - \Pi_Z(x)\|.$$

But the subdifferential of the norm contains only one element:

$$\partial \|x - \Pi_Z(x)\| = \{x - \Pi_Z(x)\},$$

and

$$x - \Pi_Z(x) \in N_Z(\Pi_Z(x)).$$

Therefore the distance function is differentiable at every $x \notin Z$ and its gradient is given by

$$\nabla f(x) = x - \Pi_Z(x).$$

Example 2.80. Let

$$f(x) = \max_{i \in I} f_i(x),$$

where I is a finite set, and f_i are convex and differentiable functions for each $i \in I$. Define

$$\hat{I}(x) = \{i \in I : f_i(x) = f(x)\}.$$

Then

$$\partial f(x) = \operatorname{conv}\{\nabla f_i(x) : i \in \hat{I}(x)\}.$$

Indeed, if $s \in \hat{I}(x)$ we have

$$f(y) \geq f_s(y) \geq f_s(x) + \langle \nabla f_s(x), y - x \rangle = f(x) + \langle \nabla f_s(x), y - x \rangle.$$

Thus $\nabla f_s(x) \in \partial f(x)$. Since the subdifferential is convex, we have therefore

$$\operatorname{conv}\{\nabla f_i(x) : i \in \hat{I}(x)\} \subset \partial f(x).$$

To prove the equality of these sets, suppose $g \in \partial f(x)$ exists such that

$$g \notin \operatorname{conv}\{\nabla f_i(x) : i \in \hat{I}(x)\}.$$

By Theorem 2.14 we can strictly separate g and the convex hull above: there exist $d \neq 0$ and $\varepsilon > 0$ such that

$$\langle g, d \rangle \geq \max_{i \in \hat{I}(x)} \langle \nabla f_i(x), d \rangle + \varepsilon.$$

Thus for all $\tau > 0$

$$f(x + \tau d) \geq f(x) + \tau \langle g, d \rangle \geq f(x) + \tau \max_{i \in \hat{I}(x)} \langle \nabla f_i(x), d \rangle + \tau\varepsilon. \tag{2.39}$$

For a sufficiently small $\tau > 0$ we can always find $r \in \hat{I}(x)$ such that $f(x + \tau d) = f_r(x + \tau d)$. By the differentiability of the f_i's there exist functions $o_i(\tau)$ such that

$$f_i(x + \tau d) = f_i(x) + \tau \langle \nabla f_i(x), d \rangle + o_i(\tau)$$

and $o_i(\tau)/\tau \to 0$ as $\tau \downarrow 0$. Thus, for small $\tau > 0$

$$f(x + \tau d) \leq f(x) + \tau \max_{i \in \hat{I}(x)} \langle \nabla f_i(x), d \rangle + \max_{i \in \hat{I}(x)} o_i(\tau).$$

Combining this inequality with (2.39) we obtain

$$f(x) + \tau \max_{i \in \hat{I}(x)} \langle \nabla f_i(x), d \rangle + \max_{i \in \hat{I}(x)} o_i(\tau) \geq f(x) + \tau \max_{i \in \hat{I}(x)} \langle \nabla f_i(x), d \rangle + \tau \varepsilon.$$

Simplifying, dividing by τ and letting $\tau \downarrow 0$ we obtain $0 \geq \varepsilon$, a contradiction.

The function in Figure 2.12 on page 58 is the maximum of four smooth functions. Two of them are "active" at the selected point, and all convex combinations of their gradients are subgradients of f at this point.

Example 2.81. Let C be a closed convex set in $\mathbb{R}^n$ and let us consider the *indicator function of C*

$$\delta_C(x) = \begin{cases} 0 & \text{if } x \in C \\ +\infty & \text{otherwise.} \end{cases}$$

It is convex. Let us calculate its subdifferential at a point $x \in C$. We have $g \in \partial \delta(x)$ if and only if $\delta_C(y) - \delta_C(x) \geq \langle g, y - x \rangle$ for all y. This is trivially satisfied for $y \notin C$, so the only meaningful case is when $y \in C$. Thus, for g to be a subgradient of $\delta_C(\cdot)$ at x it is necessary and sufficient that

$$\langle g, y - x \rangle \leq 0, \quad \text{for all} \quad y \in C.$$

Hence

$$\langle g, d \rangle \leq 0 \quad \text{for all} \quad d \in \text{cone}(C - x).$$

We have obtained, therefore, the representation of the subdifferential as the normal cone:

$$\partial \delta_C(x) = [\text{cone}(C - x)]^\circ = N_C(x).$$

In particular, if K is a closed convex cone, then

$$\partial \delta_K(0) = K^\circ.$$

2.5.2 Chain Rules of Subdifferential Calculus

Many simple rules of calculus extend to subdifferentials.

Lemma 2.82. *Assume that $f : \mathbb{R}^n \to \overline{\mathbb{R}}$ is a convex function, $\alpha > 0$, and $h(x) = \alpha f(x)$. Then h is convex, and $\partial h(x) = \alpha \partial f(x)$, for every x.*

Proof. The relation follows directly from the definition. We have $g \in \partial f(x)$ if and only if for all y

$$h(y) = \alpha f(y) \geq \alpha[f(x) + \langle g, y - x \rangle] = h(x) + \langle \alpha g, y - x \rangle,$$

which is equivalent to $\alpha g \in \partial h(x)$. □

Lemma 2.83. *Assume that $f : \mathbb{R}^m \to \overline{\mathbb{R}}$ is a convex function, A is an $m \times n$ matrix, and $h(x) = f(Ax)$. Then $\partial h(x) = A^T \partial f(Ax)$, for every x.*

Proof. The relation follows directly from the definition. We have $g \in \partial f(Ax)$ if and only if

$$h(y) = f(Ay) \geq f(Ax) + \langle g, Ay - Ax \rangle = h(x) + \langle A^T g, y - x \rangle,$$

which is equivalent to $A^T g \in \partial h(x)$. □

Example 2.84. Consider the function

$$h(x) = \sqrt{\langle x, Ax \rangle},$$

where A is a symmetric positive definite matrix. Factorizing $A = U^T U$, with some nonsingular U, we see that

$$h(x) = \|Ux\|.$$

Therefore we can use Example 2.78 and Lemma 2.83 to conclude that

$$\begin{aligned}\partial h(0) &= \{U^T g : \|g\| \leq 1\} \\ &= \{v : \|(U^T)^{-1} v\| \leq 1\} = \{v : \langle v, A^{-1} v \rangle \leq 1\}.\end{aligned}$$

In fact, $h(x)$ is a norm, whose dual norm is $\|v\|_{A^{-1}} = \sqrt{\langle v, A^{-1} v \rangle}$. Our calculation re-establishes in this special case the general result about subdifferentials of norms presented in Example 2.78.

THEOREM 2.85. *Assume that* $f = f_1 + f_2$, *where* $f_1 : \mathbb{R}^n \to \overline{\mathbb{R}}$ *and* $f_2 : \mathbb{R}^n \to \overline{\mathbb{R}}$ *are convex proper functions. If there exists a point* $x_0 \in \operatorname{dom} f$ *such that* f_1 *is continuous at* x_0, *then*

$$\partial f(x) = \partial f_1(x) + \partial f_2(x), \quad \textit{for all} \quad x \in \operatorname{dom} f.$$

Proof. Suppose $g^1 \in \partial f_1(x)$ and $g^2 \in \partial f_2(x)$. Then, for all y

$$\begin{aligned}f(y) = f_1(y) + f_2(y) &\geq f_1(x) + \langle g^1, y - x \rangle + f_2(x) + \langle g^2, y - x \rangle \\ &= f(x) + \langle g^1 + g^2, y - x \rangle.\end{aligned}$$

This proves that

$$\partial f(x) \supset \partial f_1(x) + \partial f_2(x).$$

We shall show the equality here.

Suppose a subgradient $g \in \partial f(x)$ exists such that $g \notin \partial f_1(x) + \partial f_2(x)$. To apply the separation theorem (Theorem 2.14) we need to verify that $\partial f_1(x) + \partial f_2(x)$ is a closed convex set. Its convexity follows from the convexity of the subdifferentials $\partial f_1(x)$ and $\partial f_2(x)$ (see Theorem 2.74 and the remark after its proof) and from Lemma 2.3.

Both subdifferentials are closed and the sum $\partial f_1(x) + \partial f_2(x)$ is closed as well. We prove this technical property after the main line of the proof.

If $\partial f_1(x) + \partial f_2(x)$ is closed, we can invoke Theorem 2.14 and separate g from the sum $\partial f_1(x) + \partial f_2(x)$. There exists a direction $d \in \mathbb{R}^n$ and $\varepsilon > 0$ such that

$$\langle g, d \rangle \geq \langle g^1 + g^2, d \rangle + \varepsilon, \tag{2.40}$$

for all $g^1 \in \partial f_1(x)$ and all $g^2 \in \partial f_2(x)$. It follows that $\langle g^1, d \rangle$ is bounded from above for all $g^1 \in \partial f_1(x)$. Thus, by Lemma 2.75,

$$f_1'(x; d) = \max_{g^1 \in \partial f_1(x)} \langle g^1, d \rangle < \infty.$$

Similarly,

$$f_2'(x; d) = \max_{g^2 \in \partial f_2(x)} \langle g^2, d \rangle < \infty.$$

Taking the supremum of the right hand side of (2.40) over $g^1 \in \partial f_1(x)$ and $g^2 \in \partial f_2(x)$ and using the last two relations, we obtain

$$\langle g, d \rangle \geq \max_{g^1 \in \partial f_1(x)} \langle g^1, d \rangle + \max_{g^2 \in \partial f_2(x)} \langle g^2, d \rangle + \varepsilon = f_1'(x; d) + f_2'(x; d) + \varepsilon.$$

On the other hand, $\langle g, d \rangle \leq f'(x; d)$ and the last inequality yields

$$f'(x; d) \geq f_1'(x; d) + f_2'(x; d) + \varepsilon,$$

a contradiction. Therefore a subgradient $g \in \partial f(x) \setminus \big(\partial f_1(x) + \partial f_2(x)\big)$ cannot exist.

Let us now return to the issue of the closedness of the sum $\partial f_1(x) + \partial f_2(x)$. If one of the functions f_1 or f_2 is continuous at x, then its subdifferential is compact, and the sum $\partial f_1(x) + \partial f_2(x)$ is closed, too. The only remaining case is when both subdifferentials are unbounded. Consider two sequences $g_1^k \in \partial f_1(x)$ and $g_2^k \in \partial f_2(x)$ such that

$$g_1^k + g_2^k \to s, \quad \text{as} \quad k \to \infty.$$

Suppose $s \notin \partial f_1(x) + \partial f_2(x)$. As mentioned above, $\|g_1^k\| \to \infty$ and $\|g_2^k\| \to \infty$, because all accumulation points of these sequences must be elements of the corresponding subdifferentials. Consider the sequence

$$z^k = \frac{g_1^k}{\|g_1^k\|}.$$

By choosing a subsequence, if necessary, we may assume that $\{z^k\}$ has limit z. We have

$$f_1(x_0) - f_1(x) \geq \langle g_1^k, x_0 - x \rangle.$$

Dividing by $\|g_1^k\|$ and passing to the limit with $k \to \infty$ we conclude that $\langle z, x_0 - x\rangle \le 0$. We also have $g_2^k = s - g_1^k$, and thus

$$f_2(x_0) - f_2(x) \ge \langle s - g_1^k, x_0 - x\rangle.$$

Dividing by $\|g_1^k\|$ and passing to the limit we obtain $\langle z, x_0 - x\rangle \ge 0$. Therefore

$$\langle z, x_0 - x\rangle = 0.$$

Since f_1 is continuous at x_0, we can find $\varepsilon > 0$ such that $f_1(x_0 + \varepsilon z) < \infty$. We obtain

$$f_1(x_0 + \varepsilon z) - f_1(x) \ge \langle g_1^k, x_0 + \varepsilon z - x\rangle.$$

Dividing by $\|g_1^k\|$ and passing to the limit we conclude that

$$0 \ge \langle z, x_0 - x + \varepsilon z\rangle = \varepsilon\|z\|^2 = \varepsilon,$$

a contradiction. Therefore the sum $\partial f_1(x) + \partial f_2(x)$ is closed. □

The last theorem is called the *Moreau–Rockafellar Theorem.*

Example 2.86. Consider the set

$$K = K_1 \cap K_2 \cap \cdots \cap K_m,$$

where each K_i, $i = 1, \ldots, m$ is a convex cone. From Example 2.81 we know that

$$\partial\delta_K(0) = K^\circ.$$

On the other hand,

$$\delta_K(x) = \delta_{K_1}(x) + \delta_{K_2}(x) + \cdots + \delta_{K_m}(x).$$

If $K_1 \cap \operatorname{int} K_2 \cap \cdots \cap \operatorname{int} K_m \ne \emptyset$, we can employ Theorem 2.85 to get

$$\partial\delta_K(0) = \partial\delta_{K_1}(0) + \partial\delta_{K_2}(0) + \cdots + \partial\delta_{K_m}(0) = K_1^\circ + K_2^\circ + \cdots + K_m^\circ.$$

We have thus established again the result of Theorem 2.35.

More generally, let X_i, $i = 1, \ldots, m$ be closed convex sets, and let

$$X = X_1 \cap X_2 \cap \cdots \cap X_m.$$

We have

$$\partial\delta_X(x) = N_X(x).$$

If $X_1 \cap \operatorname{int} X_2 \cap \cdots \cap \operatorname{int} X_m \ne \emptyset$, Theorem 2.85 yields again

$$N_X(x) = N_{X_1}(x) + N_{X_2}(x) + \cdots + N_{X_m}(x),$$

as previously established in Lemma 2.40.

2.5.3 The Subdifferential of the Maximum Function

Consider the function

$$F(x) = \sup_{y \in Y} f(x, y).$$

We assume that $f : \mathbb{R}^n \times Y \to \overline{\mathbb{R}}$ satisfies the following conditions:

(i) $f(\cdot, y)$ is convex for all $y \in Y$;

(ii) $f(x, \cdot)$ is upper semicontinuous for all x in a certain neighborhood of a point x_0;

(iii) The set $Y \subset \mathbb{R}^m$ is compact.

The maximum function F is convex by virtue of Lemma 2.58. It is proper, due to (ii). Our intention is to describe the subdifferential of F at the point x_0. We have already considered a special case of this problem in Example 2.80, and we shall generalize the formula developed there. Our analysis follows similar lines, but with more technical complications.

We denote by $\hat{Y}(x)$ the set of $y \in Y$ at which $f(x, y) = F(x)$. Since $f(x, \cdot)$ is upper semicontinuous and Y is compact, the set $\hat{Y}(x)$ is nonempty and compact, for every x in a certain neighborhood of x_0.

We use the symbol $\partial_x f(x_0, y)$ to denote the subdifferential of the function $f(\cdot, y)$ at x_0.

THEOREM 2.87. *Assume conditions* (i)–(iii). *Then*

$$\partial F(x_0) \supset \operatorname{conv}\Big(\bigcup_{y \in \hat{Y}(x_0)} \partial_x f(x_0, y) \Big).$$

If, in addition, the function $f(\cdot, y)$ is continuous at x_0 for all $y \in Y$, then

$$\partial F(x_0) = \operatorname{conv}\Big(\bigcup_{y \in \hat{Y}(x_0)} \partial_x f(x_0, y) \Big). \tag{2.41}$$

Proof. Suppose $g \in \partial_x f(x_0, y_0)$ for some $y_0 \in \hat{Y}(x_0)$. Then, for every x we have

$$F(x) = \sup_{y \in Y} f(x, y) \geq f(x, y_0) \geq f(x_0, y_0) + \langle g, x - x_0 \rangle.$$

Therefore $g \in \partial F(x_0)$. Since the subdifferential is convex, our first assertion holds true.

We now prove that the set on the right hand side of (2.41) is closed. Consider a convergent sequence of vectors $s_k \in \partial_x f(x_0, y_k)$, with $y_k \in$

$\hat{Y}(x_0)$, and let $s^* = \lim_{k\to\infty} s_k$. As the set $\hat{Y}(x_0)$ is compact, by passing to a subsequence, if necessary, we may also assume that the sequence $\{y_k\}$ is convergent. Its limit, y^*, is an element of $\hat{Y}(x_0)$. For every x in a small neighborhood of x_0, the upper semicontinuity of $f(x,\cdot)$ and the fact that s_k is a subgradient imply

$$f(x,y^*) \geq \limsup_{k\to\infty} f(x,y_k) \geq \limsup_{k\to\infty} \big[f(x_0,y_k) + \langle s_k, x - x_0\rangle \big]$$
$$= f(x_0,y^*) + \langle s^*, x - x_0\rangle.$$

In the last transformation we use the fact that $f(x_0,y_k) = F(x_0) = f(x_0,y^*)$. This proves that $s^* \in \partial f(x_0,y^*)$.

Suppose our second assertion is false, that is, there exists $g \in \partial F(x_0)$ such that

$$g \notin \operatorname{conv}\Big(\bigcup_{y\in\hat{Y}(x_0)} \partial_x f(x_0,y)\Big).$$

Since the set on the right hand side is convex and closed, Theorem 2.14 implies that there exist $d \neq 0$ and $\varepsilon > 0$ such that

$$\langle g, d\rangle \geq \langle s, d\rangle + \varepsilon \tag{2.42}$$

for all $s \in \partial_x f(x_0,y)$ and all $y \in \hat{Y}(x_0)$. Consider points of the form $x_0 + \tau_k d$, where $\tau_k \downarrow 0$. By the convexity of F we have

$$\frac{F(x_0+\tau_k d) - F(x_0)}{\tau_k} \geq \langle g, d\rangle.$$

Define the sets

$$Y_k = \Big\{ y \in Y : \frac{f(x_0+\tau_k d, y) - F(x_0)}{\tau_k} \geq \langle g, d\rangle \Big\}, \quad k = 1,2,\ldots.$$

They are closed, because $f(x,\cdot)$ is upper semicontinuous. They are nonempty, because $\hat{Y}(x_0+\tau_k d) \subset Y_k$. Moreover, for every $y \in Y$ the expression

$$\frac{f(x_0+\tau d,y) - F(x_0)}{\tau} = \frac{f(x_0+\tau d,y) - f(x_0,y)}{\tau} + \frac{f(x_0,y) - F(x_0)}{\tau}$$

defines an increasing function of τ. Indeed, the first fraction is the difference quotient of a convex function (see formula (2.30) on page 57), and the second fraction has a fixed nonpositive numerator. This implies that

$$Y_1 \supset Y_2 \supset Y_3 \supset \cdots.$$

The sets Y_k are compact and nonempty, and thus there exists a point $\bar{y}$ which belongs to all of them. Then

$$\frac{f(x_0 + \tau_k d, \bar{y}) - F(x_0)}{\tau_k} \geq \langle g, d \rangle, \quad k = 1, 2, \ldots.$$

Because $f(x_0 + \tau_k d, \bar{y}) \to f(x_0, \bar{y})$ as $k \to \infty$, we must have $f(x_0, \bar{y}) = F(x_0)$, i.e., $\bar{y} \in \hat{Y}(x_0)$. Passing to the limit in the last inequality with $k \to \infty$ and employing Theorem 2.74 we conclude that

$$\lim_{k\to\infty} \frac{f(x_0 + \tau_k d, \bar{y}) - F(x_0)}{\tau_k} = \langle s, d \rangle \geq \langle g, d \rangle,$$

with some $s \in \partial_x f(x_0, \bar{y})$. This contradicts (2.42). □

Example 2.88. We have n continuous functions $\varphi_i(t)$, $i = 1, \ldots, n$, where $t \in [a, b] \subset \mathbb{R}$. We call them *basic functions*. Our objective is to approximate a continuous function $\psi(t)$ by a linear combination of basic functions:

$$\sum_{i=1}^{n} x_i \varphi_i(t) \approx \psi(t), \quad t \in [a, b].$$

In the last formula x_i, $i = 1, \ldots, n$, are unknown coefficients. If we are interested in a uniform approximation in the interval $[a, b]$, it is natural to measure the approximation error as follows:

$$F(x) = \max_{a \leq t \leq b} \Big| \sum_{i=1}^{n} x_i \varphi_i(t) - \psi(t) \Big|.$$

Approximation based on this criterion is frequently referred to as *Chebyshev approximation*.

Theorem 2.87 allows us to calculate the subdifferential of the error function $F(x)$ at any point x. Define the set

$$\hat{T}(x) = \Big\{ t \in [a, b] : \Big| \sum_{i=1}^{n} x_i \varphi_i(t) - \psi(t) \Big| = F(x) \Big\}.$$

In the typical case, when $F(x) > 0$, we obtain

$$\partial F(x) = \operatorname{conv} \Big\{ \operatorname{sgn}\Big(\sum_{i=1}^{n} x_i \varphi_i(t) - \psi(t) \Big) \begin{bmatrix} \varphi_1(t) \\ \varphi_2(t) \\ \vdots \\ \varphi_n(t) \end{bmatrix} : t \in \hat{T}(x) \Big\}.$$

The symbol $\operatorname{sgn}(\cdot)$ denotes the sign function.

Example 2.89. Consider the maximum eigenvalue function $\lambda_{\max}(\cdot)$ defined on the space $\mathbb{S}^n$ of symmetric matrices of dimension $n \times n$. It is convex (cf. Example 2.59). Since

$$\lambda_{\max}(A) = \max_{\|y\|=1} \langle y, Ay \rangle,$$

we can calculate its subdifferential by Theorem 2.87. The set

$$\hat{Y}(A) = \{y \in \mathbb{R}^n : \langle y, Ay \rangle = \lambda_{\max}(A), \ \|y\| = 1\},$$

is the set of all eigenvectors of A corresponding to the maximum eigenvalue and having length 1.

As in Example 2.31, we use the Frobenius inner product,

$$\langle A, H \rangle_{\mathbb{S}} = \operatorname{tr}(AH) = \sum_{i=1}^{n} \sum_{j=1}^{n} a_{ij} h_{ij},$$

and we rewrite the function as follows:

$$\lambda_{\max}(A) = \max_{\|y\|=1} \langle yy^T, A \rangle_{\mathbb{S}}.$$

For every $y \in \mathbb{R}^n$ the function $f_y(A) = \langle yy^T, A \rangle_{\mathbb{S}}$ is linear and its gradient equals $\nabla f_y(A) = yy^T$. Consequently

$$\partial \lambda_{\max}(A) = \operatorname{conv}\{yy^T : Ay = \lambda_{\max}(A)y, \ \|y\| = 1\}.$$

Our final observation is that the set of matrices of the form $W = yy^T$ can be equivalently characterized as follows: W has rank one and $W \in \mathbb{S}^n_+$. Then

$$\operatorname{tr}(W) = \|y\|^2 \quad \text{and} \quad \langle A, W \rangle_{\mathbb{S}} = \langle y, Ay \rangle.$$

Therefore the requirement that y is an eigenvector of length 1 corresponding to the maximum eigenvalue is equivalent to the conditions

$$\operatorname{tr}(W) = 1 \quad \text{and} \quad \langle A, W \rangle_{\mathbb{S}} = \lambda_{\max}(A).$$

We obtain

$$\partial \lambda_{\max}(A) = \operatorname{conv}\{W \in \mathbb{S}^n_+ : \langle A, W \rangle_{\mathbb{S}} = \lambda_{\max}(A), \ \operatorname{tr}(W) = 1, \ \operatorname{rank}(W) = 1\}.$$

We can omit the convex hull and the rank restriction in the last representation of the subdifferential:

$$\partial \lambda_{\max}(A) = \{W \in \mathbb{S}^n_+ : \langle A, W \rangle_{\mathbb{S}} = \lambda_{\max}(A), \ \operatorname{tr}(W) = 1\}.$$

Indeed, every element of this set has the form $W = \sum_{j=1}^n \mu_j y_j y_j^T$, with orthogonal eigenvectors y_j of W having norm 1, and with the corresponding eigenvalues $\mu_j \geq 0$. The condition $\operatorname{tr}(W) = 1$ implies $\sum_{j=1}^n \mu_j = 1$. Hence

$$\langle A, W \rangle_{\mathbb{S}} = \sum_{j=1}^{n} \mu_j \langle y_j, Ay_j \rangle \leq \sum_{j=1}^{n} \mu_j \lambda_{\max}(A) = \lambda_{\max}(A).$$

Equality occurs if and only if y_j's corresponding to positive μ_j's are eigenvectors of A corresponding to its maximum eigenvalue. Then W is a convex combination of rank one matrices $y_j y_j^T$ formed of these eigenvectors, with the weights μ_j.

2.6 CONJUGATE DUALITY

2.6.1 The Conjugate Function

The domain of a proper convex function is a convex set. If it has a nonempty interior, then the function is subdifferentiable at all points of the interior (Theorem 2.74). If the domain has no interior, we may restrict our considerations to the linear manifold of the smallest dimension containing the domain. The function restricted to this manifold (after an appropriate change of variables) will be subdifferentiable in the corresponding space at interior points of its domain.

It follows that for every proper convex function f there exists a point $x_0 \in \operatorname{dom} f$ such that f is subdifferentiable at x_0. Then for every $x \in \mathbb{R}^n$ inequality (2.32) holds true:

$$f(x) \geq f(x_0) + \langle s_0, x - x_0 \rangle,$$

where $s_0 \in \partial f(x_0)$. Let us consider the set of $\alpha \in \mathbb{R}$, for which the affine function

$$l_\alpha(x) = \langle s_0, x \rangle - \alpha \tag{2.43}$$

is a *minorant* of f, that is, it satisfies the inequality

$$l_\alpha(x) \leq f(x) \quad \text{for all} \quad x \in \mathbb{R}^n.$$

We call s_0 the *slope* of the affine minorant $l_\alpha(\cdot)$. Since $s_0 \in \partial f(x_0)$, the set of affine minorants having slope s_0 is nonempty and has the maximum element, corresponding to

$$\alpha^* = \langle s_0, x_0 \rangle - f(x_0).$$

This motivates the following two questions:

1. For which vectors $s \in \mathbb{R}^n$ do affine minorants of the form (2.43) exist, and do they have a finite supremum?
2. Is it possible to represent f as a supremum of all possible affine minorants, for all $s \in \mathbb{R}^n$?

In order to analyze these issues, we introduce the following concept.

DEFINITION 2.90. Let $f : \mathbb{R}^n \to \overline{\mathbb{R}}$ be a proper convex function. The function $f^* : \mathbb{R}^n \to \overline{\mathbb{R}}$ defined by

$$f^*(s) \triangleq \sup_x \left\{ \langle s, x \rangle - f(x) \right\}$$

is called the *conjugate function* to f.

It follows directly from the definition that for every $x \in \mathbb{R}^n$ and every $s \in \mathbb{R}^n$ we have:

$$f(x) + f^*(s) \geq \langle s, x \rangle. \tag{2.44}$$

If $f^*(s) < +\infty$ our first question has a positive answer: f has affine minorants with slope s. If $f^*(s) = +\infty$, no affine minorants with slope s exist.

In fact, we can define the conjugate function f^* for *every* extended real valued function $f : \mathbb{R}^n \to \overline{\mathbb{R}}$.

LEMMA 2.91. *If the function $f : \mathbb{R}^n \to \overline{\mathbb{R}}$ is proper and has an affine minorant, then the conjugate function is proper, convex, and lower semicontinuous.*

Proof. By Definition 2.90, the epigraph of the function f^* has the form

$$\operatorname{epi} f^* = \bigcap_{x \in \operatorname{dom} f} \left\{ (s, v) : v \geq \langle s, x \rangle - f(x) \right\}.$$

As an intersection of closed convex sets, it is a closed convex set (Lemma 2.2). Thus f^* is a convex function. Moreover, by Lemma 2.62, f^* is a lower semicontinuous function.

By assumption, an affine minorant of the form (2.43) exists. Because f is proper, there exists a point $\bar{x}$ such that $-\infty < f(\bar{x}) < +\infty$. We obtain

$$f^*(s_0) = \sup_x \left\{ \langle s_0, x \rangle - f(x) \right\} \geq \langle s_0, \bar{x} \rangle - f(\bar{x}).$$

On the other hand, (2.43) is a minorant of f, and thus

$$f^*(s_0) = \sup_x \left\{ l_\alpha(x) + \alpha - f(x) \right\} \leq \alpha.$$

Hence $f^*(s_0)$ is finite.

For every $s \in \mathbb{R}^n$ we have

$$f^*(s) \geq \langle s, x_0 \rangle - f(x_0) = f^*(s_0) + \langle s - s_0, x_0 \rangle.$$

Therefore f^* is proper. □

If f is convex and proper then it always has an affine minorant, and thus f^* is convex, proper, and lower semicontinuous.

Example 2.92. Consider a convex cone $K \subset \mathbb{R}^n$ and its indicator function

$$\delta_K(x) = \begin{cases} 0 & \text{if } x \in K \\ +\infty & \text{otherwise.} \end{cases}$$

Let us calculate its conjugate δ_K^*. We have

$$\delta_K^*(s) = \sup_{x \in \mathbb{R}^n} \left\{ \langle s, x \rangle - \delta_K(x) \right\} = \sup_{x \in K} \langle s, x \rangle.$$

By Lemma 2.26, $\langle s, x \rangle$ has finite supremum over $x \in K$ if and only if $s \in K^\circ$. Therefore

$$\delta_K^*(s) = \begin{cases} 0 & \text{if } s \in K^\circ \\ +\infty & \text{otherwise,} \end{cases}$$

that is, $\delta_K^* = \delta_{K^\circ}$.

Let us observe that the conjugate of δ_{K° is $\delta_{K^{\circ\circ}}$, which is equal to δ_K again, if K is closed (Theorem 2.27). We discuss this in the next section.

Example 2.93. Let $\|\cdot\|_\Diamond$ be a norm in $\mathbb{R}^n$. Let us consider the unit ball,

$$B = \{x \in \mathbb{R}^n : \|x\|_\Diamond \le 1\},$$

and its indicator function,

$$\delta_B(x) = \begin{cases} 0 & \text{if } \|x\|_\Diamond \le 1 \\ +\infty & \text{otherwise.} \end{cases}$$

The conjugate function has the form

$$(\delta_B)^*(s) = \sup_{x \in \mathbb{R}^n} \left\{ \langle s, x \rangle - \delta_B(x) \right\} = \sup_{\|x\|_\Diamond \le 1} \langle s, x \rangle = \|s\|_*,$$

where $\|\cdot\|_*$ is the dual norm to $\|\cdot\|_\Diamond$.

Example 2.94. Let $\|\cdot\|_\Diamond$ be a norm in $\mathbb{R}^n$. The conjugate function to $f(x) = \|x\|_\Diamond$ has the form

$$f^*(s) = \sup_{x \in \mathbb{R}^n} \left\{ \langle s, x \rangle - \|x\|_\Diamond \right\}.$$

Let $\|\cdot\|_*$ be the dual norm. If $\|s\|_* > 1$ we can find $\bar{x}$ such that $\|\bar{x}\|_\Diamond = 1$ and $\langle s, \bar{x} \rangle > 1$. Multiplying $\bar{x}$ by an arbitrary $\beta > 0$ we see that the supremum on the right hand side is $+\infty$.

On the other hand, if $\|s\|_* \le 1$, then $\langle s, x \rangle \le \|x\|_\Diamond$. Thus the supremum on the right hand side is 0. Consequently,

$$f^*(s) = \delta_{B_*}(s),$$

where

$$B_* = \{s \in \mathbb{R}^n : \|s\|_* \le 1\}$$

is the unit ball with respect to the dual norm.

Let us combine this result with Example 2.93 and note that $\| \cdot \|_\diamond$ is the dual norm to $\| \cdot \|_*$. It follows that $\| \cdot \|_\diamond$ and $\delta_{B_*}(\cdot)$ are mutually conjugate. We explain this in the next section.

2.6.2 The Biconjugate Function

Let us now consider the conjugate function to the conjugate function f^*:

$$f^{**}(x) \triangleq \sup_s \big\{\langle s, x\rangle - f^*(s)\big\}. \tag{2.45}$$

We call it the *biconjugate* function.

By Definition 2.90, each function $l_s : \mathbb{R}^n \to \mathbb{R}$ given by

$$l_s(x) = \langle s, x\rangle - f^*(s)$$

is the pointwise supremum of all affine minorants of f having slope s. Thus $f^{**}(\cdot)$ is the pointwise supremum of *all* affine minorants of f. The following result is known as the *Fenchel–Moreau Theorem.*

THEOREM 2.95. *If an extended real valued function $f : \mathbb{R}^n \to \overline{\mathbb{R}}$ has at least one affine minorant, then*

$$\operatorname{epi} f^{**} = \overline{\operatorname{conv}}(\operatorname{epi} f).$$

In particular, if f is a proper, lower semicontinuous and convex function, then

$$f^{**} = f.$$

Proof. As we saw before stating the theorem, $f^{**} \le f$, and thus

$$\operatorname{epi} f \subset \operatorname{epi} f^{**}.$$

It follows from (2.45) that the epigraph of f^{**} is an intersection of closed halfspaces, and is therefore a closed convex set. Consequently,

$$\overline{\operatorname{conv}}(\operatorname{epi} f) \subset \operatorname{epi} f^{**}.$$

Suppose a strict inclusion above. Then there exists a point

$$(y, \beta) \in \big[\operatorname{epi} f^{**}\big] \setminus \big[\overline{\operatorname{conv}}(\operatorname{epi} f)\big].$$

Since the set $\overline{\text{conv}}(\text{epi}\, f)$ is closed and convex, by virtue of Theorem 2.14 we can strictly separate (y, β) from this set. There exist $s \in \mathbb{R}^n$, $\gamma \in \mathbb{R}$, and $\varepsilon > 0$ such that

$$\langle s, y\rangle + \gamma\beta \le \langle s, x\rangle + \gamma\,\alpha - \varepsilon,$$

for all $(x, \alpha) \in \overline{\text{conv}}(\text{epi}\, f)$. Letting $\alpha \to \infty$ we deduce that $\gamma \ge 0$.

If $\gamma > 0$ we can divide both sides of the last inequality by γ. Setting $\alpha = f(x)$ we conclude that

$$l(x) = \Big\langle \frac{s}{\alpha}, y - x\Big\rangle + \beta + \frac{\varepsilon}{\gamma} \le f(x),$$

for all $x \in \text{dom}\, f$. Thus $l(\cdot)$ is an affine minorant of f. It follows that $f^{**}(y) \ge l(y) = \beta + \frac{\varepsilon}{\gamma} > \beta$, which contradicts the inclusion $(y, \beta) \in \text{epi}\, f^{**}$.

If $\gamma = 0$, the separating plane is vertical:

$$\langle s, y\rangle \le \langle s, x\rangle - \varepsilon \quad \text{for all} \quad x \in \text{dom}\, f. \tag{2.46}$$

By assumption, f has an affine minorant:

$$\langle z, x\rangle + c \le f(x) \quad \text{for all} \quad x \in \mathbb{R}^n.$$

Adding (2.46) multiplied by $M > 0$ to the last displayed inequality and rearranging terms we obtain

$$\langle z - Ms, x\rangle + M\langle s, y\rangle + c + M\varepsilon \le f(x) \quad \text{for all} \quad x \in \mathbb{R}^n.$$

Consequently, the function on the left hand side is an affine minorant of f, and its value at y must not be greater than $f^{**}(y)$. It follows that

$$\langle z, y\rangle + c + M\varepsilon \le f^{**}(y) \quad \text{for all} \quad M > 0.$$

Letting $M \to \infty$ we conclude that $f^{**}(y) = +\infty$. This contradicts the inclusion $(y, \beta) \in \text{epi}\, f^{**}$. □

If f has no affine minorants, then $f^*(s) = +\infty$ for all $s \in \mathbb{R}^n$, and thus $f^{**}(x) = -\infty$ for all x.

Theorem 2.95 explains the phenomenon that is evident in Examples 2.92, 2.93, and 2.94. We can also use it to calculate conjugate functions.

Example 2.96. Let Z be a set in $\mathbb{R}^n$. Let us define the *support function* of Z as follows:

$$\sigma_Z(s) = \sup_{x \in Z} \langle s, x\rangle.$$

We notice that $\sigma_Z(\cdot)$ is the conjugate of the indicator function

$$\delta_Z(x) = \begin{cases} 0 & \text{if } x \in Z \\ +\infty & \text{otherwise.} \end{cases}$$

Therefore $\sigma_Z^* = \delta_Z^{**}$. By Theorem 2.95,

$$\text{epi}\delta_Z^{**} = \overline{\text{conv}}(\text{epi}\delta_Z) = \text{epi}\delta_{\overline{\text{conv}}(Z)}.$$

It follows that

$$\sigma_Z^*(\cdot) = \delta_{\overline{\text{conv}}(Z)}(\cdot). \tag{2.47}$$

If Z is convex and closed, the functions $\sigma_Z(\cdot)$ and $\delta_Z(\cdot)$ are mutually conjugate. If Z is not convex and closed, the biconjugate $\delta_Z^{**}(\cdot)$ is the indicator function of the closed convex hull of Z.

In particular, a norm $\|\cdot\|_\diamond$ is the support function of the dual ball B_*, and Examples 2.93 and 2.94 illustrate the general formula (2.47).

Example 2.97. Assume that an uncertain outcome may take one of n values: x_1, $\ldots, x_n$. Define $\mathbb{1}$ as the vector in $\mathbb{R}^n$ having all components equal to 1. A *convex risk function* is defined as a convex function $f : \mathbb{R}^n \to \mathbb{R}$ satisfying the following conditions:

(i) If $x \leq y$ then $f(x) \leq f(y)$;

(ii) For all $x \in \mathbb{R}^n$ and $a \in \mathbb{R}$ we have $f(x + a\mathbb{1}) = f(x) + a$;

(iii) $f(\cdot)$ is positively homogeneous.

We use Theorem 2.95 to derive an alternative representation of a risk function. As $f(\cdot)$ is convex and finite-valued, it is continuous. Therefore $f = f^{**}$. More specifically, for all $x \in \mathbb{R}^n$,

$$f(x) = \sup_{p \in \mathbb{R}^n} \big\{\langle p, x\rangle - f^*(p)\big\}. \tag{2.48}$$

We use the symbol p for the dual variable, in view of the following characterization of the domain of the conjugate function.

We first show that dom $f^* \subset \mathbb{R}^n_+$. Suppose we can find $\bar{p} \in \text{dom } f^*$ which is not an element of the cone $\mathbb{R}^n_+$. Then it can be strictly separated from $\mathbb{R}^n_+$. There exist $z \in \mathbb{R}^n$ and $\varepsilon > 0$ such that:

$$\langle z, p\rangle \leq \langle z, \bar{p}\rangle - \varepsilon \quad \text{for all} \quad p \in \mathbb{R}^n_+.$$

It follows from Lemma 2.26 that $z \leq 0$. Setting $p = 0$ we also infer that $\langle z, \bar{p}\rangle \geq \varepsilon > 0$. Consider an arbitrary point $x \in \mathbb{R}^n$ and points $y = x + tz$, for $t \geq 0$. As $y \leq x$, condition (i) implies that $f(y) \leq f(x)$. From the definition of the conjugate function we deduce that

$$\begin{aligned} f^*(\bar{p}) &= \sup_{x \in \mathbb{R}^n} \big\{\langle \bar{p}, x\rangle - f(x)\big\} = \sup_{x \in \mathbb{R}^n} \big\{\langle \bar{p}, x + tz\rangle - f(x + tz)\big\} \\ &\geq \sup_{x \in \mathbb{R}^n} \big\{\langle \bar{p}, x\rangle - f(x)\big\} + t\langle \bar{p}, z\rangle \geq f^*(\bar{p}) + t\varepsilon. \end{aligned}$$

As t may be an arbitrary positive number, we conclude that $f^*(\bar{p}) = \infty$, a contradiction. Consequently, dom $f^* \subset \mathbb{R}^n_+$. Condition (ii) implies that for all $a \in \mathbb{R}$:

$$f^*(p) = \sup_{x\in\mathbb{R}^n} \{\langle p, x\rangle - f(x)\} = \sup_{x\in\mathbb{R}^n} \{\langle p, x + a\mathbb{1}\rangle - f(x + a\mathbb{1})\}$$
$$= \sup_{x\in\mathbb{R}^n} \{\langle p, x\rangle - f(x)\} + a(\langle p, \mathbb{1}\rangle - 1) = f^*(p) + a(\langle p, \mathbb{1}\rangle - 1).$$

This means that $f^*(p) < \infty$ only if $\langle p, \mathbb{1}\rangle = 1$. We conclude that the domain of the conjugate function is included in the set of probability vectors

$$P = \Big\{p \in \mathbb{R}^n_+ : \sum_{j=1}^{n} p_j = 1\Big\}.$$

If the function $f(\cdot)$ satisfies condition (iii), for every $t > 0$ and every $p \in$ dom f^* we have

$$f^*(p) = \sup_{x\in\mathbb{R}^n} \{\langle p, x\rangle - f(x)\} = \sup_{x\in\mathbb{R}^n} \{\langle p, tx\rangle - f(tx)\}$$
$$= \sup_{x\in\mathbb{R}^n} \{t\langle p, x\rangle - tf(x)\} = tf^*(p).$$

This implies that $f^*(p) = 0$. Consequently, $f^*(\cdot)$ is the indicator function of the certain set $\mathcal{A} \subset P$. Formula (2.48) simplifies to

$$f(x) = \sup_{p\in\mathcal{A}} \langle p, x\rangle,$$

which can be interpreted as the largest expected value of x, with respect to probability vectors from the set $\mathcal{A}$.

2.6.3 Subgradients of Conjugate Functions

Let us recall that if a convex function $f : \mathbb{R}^n \to \overline{\mathbb{R}}$ is subdifferentiable at x and if $s \in \partial f(x)$, then for all $y \in \mathbb{R}^n$ the following inequality holds true:

$$f(y) \geq f(x) + \langle s, y - x\rangle.$$

At $y = x$ the above inequality becomes an equation, and thus the right hand side is the largest affine minorant of f having slope s. Consequently, inequality (2.44) becomes an equation. In fact, this equation can be used as a characterization of subgradients of f and f^*.

Theorem 2.98. *Assume that $f : \mathbb{R}^n \to \overline{\mathbb{R}}$ is a proper convex function. Then the following two statements are equivalent:*

(i) $s \in \partial f(x)$;

(ii) $f(x) + f^*(s) = \langle s, x\rangle$.

If, in addition, f is lower semicontinuous then both statements are equivalent to

(iii) $x \in \partial f^*(s)$.

Proof. Suppose assertion (i) is true. If $s \in \partial f(x)$, then, as discussed before the theorem, inequality (2.44) becomes an equation. Thus (ii) holds true.

By the definition of the conjugate function, for every $y \in \mathbb{R}^n$ we have

$$f^*(s) \geq \langle s, y \rangle - f(y).$$

Suppose equation (ii) holds true. Manipulating the above inequality and using (ii) we obtain:

$$f(y) \geq \langle s, y \rangle - f^*(s) = f(x) + \langle s, y - x \rangle.$$

Thus assertion (i) is true.

Consider the function $g(x) = f^*(x)$. Since f is lower semicontinuous, Theorem 2.95 implies that $g^* = f$. Therefore assertion (ii) can be rewritten as follows:

$$g^*(x) + g(s) = \langle s, x \rangle.$$

We have proved that this is equivalent to (i) (with the roles of s and x reversed), which now reads

$$x \in \partial g(s).$$

This is (iii) for f, as required. $\square$

Example 2.99. Let C be a closed convex set in $\mathbb{R}^n$ and let us consider the indicator function of C,

$$\delta_C(x) = \begin{cases} 0 & \text{if } x \in C \\ +\infty & \text{otherwise.} \end{cases}$$

Let us calculate its subdifferential at a point $x \in C$. From Example 2.96 we know that

$$\delta_C^*(s) = \sigma_C(s),$$

where $\sigma_C(\cdot)$ is the support function of S. Both functions are convex and lower semicontinuous. We can, therefore, use Theorem 2.98 to conclude that for $x \in C$

$$\begin{aligned} \partial\delta_C(x) &= \big\{s \in \mathbb{R}^n : \sigma_C(s) + \delta_C(x) = \langle s, x \rangle\big\} \\ &= \big\{s \in \mathbb{R}^n : \sup_{y \in C} \langle s, y \rangle = \langle s, x \rangle\big\}. \end{aligned}$$

It follows that $s \in \partial\delta_C(x)$ if and only if

$$\langle s, y - x \rangle \leq 0 \quad \text{for all} \quad y \in C.$$

This is equivalent to $s \in N_C(x)$, as already proved in Example 2.81.

Example 2.100. Let $a_1, \dots, a_m$ be arbitrary vectors in $\mathbb{R}^n$. Consider the piecewise linear function:

$$f(x) = \max_{1 \le i \le m} \langle a_i, x \rangle.$$

To calculate its conjugate function we define the set $A = \{a_1, \dots, a_m\}$ and we notice that f is the support function of A:

$$f(x) = \max_{a \in A} \langle a, x \rangle = \sigma_A(x).$$

By Theorem 2.95 and Example 2.96,

$$f^*(s) = \sigma_A^*(s) = \delta_A^{**}(s) = \delta_{\mathrm{conv}(A)}(s),$$

that is, f^* is the indicator function of the convex hull of the vectors $a_1, \dots, a_m$.

The subdifferential of f at 0 is, by Theorem 2.98, equal to the set of s such that

$$f(0) + \delta_{\mathrm{conv}(A)}(s) = \langle s, 0 \rangle.$$

This simplifies to $\delta_{\mathrm{conv}(A)}(s) = 0$ and yields

$$\partial f(0) = \mathrm{conv}\{a_1, \dots, a_m\}.$$

This is the same as obtained in Example 2.80 on page 66 and in Section 2.5.3; however, our techniques there were more specific, and the results deeper.

Example 2.101. Subdifferentials of norms and distances are very easy to calculate with the application of Theorem 2.98. Any norm $\|\cdot\|_\Diamond$ in $\mathbb{R}^n$ is the support function of the dual ball:

$$\|x\|_\Diamond = \sup_{s \in B_*} \langle s, x \rangle = \sigma_{B_*}(x).$$

Its conjugate function, by Example 2.94, is the indicator function $\delta_{B_*}(\cdot)$ of the dual ball. Therefore

$$\begin{aligned} \partial \|x\|_\Diamond &= \big\{ s \in \mathbb{R}^n : \langle s, x \rangle = \|x\|_\Diamond + \delta_{B_*}(s) \big\} \\ &= \big\{ s \in B_* : \langle s, x \rangle = \|x\|_\Diamond \big\}. \end{aligned}$$

We have obtained this formula directly in Example 2.78.

Example 2.102. Consider a discrete random variable Y attaining values $y_1 < y_2 < \cdots < y_m$ with probabilities $p_1, p_2, \dots, p_m$. For each target level $x \in \mathbb{R}$ we can calculate the *expected shortfall* below x:

$$f(x) = \mathbb{E}\big[\max(0, x - Y)\big] = \sum_{\{k : y_k < x\}} p_k (x - y_k).$$

The function f is convex and nondecreasing. It is also piecewise linear with break points at $y_1, \dots, y_m$ and with slopes

$$s_i = \sum_{k=1}^{i} p_k, \quad \text{if} \quad y_i < x < y_{i+1}, \quad i = 1, \dots, m-1.$$

To the left of y_1 the function f has value 0 and slope $s_0 = 0$, and to the right of y_m it has slope $s_m = 1$.

Let us derive its conjugate function. For every $i = 0, 1, \ldots, m-1$ we have

$$\big[s \in \partial f(y_{i+1})\big] \Leftrightarrow \big[s_i \le s \le s_{i+1}\big].$$

It follows from Theorem 2.98 that for $s \in [s_i, s_{i+1}]$ the conjugate function has the form:

$$\begin{aligned} f^*(s) = s y_{i+1} - f(y_{i+1}) &= s y_{i+1} - \sum_{k=1}^{i} p_k (y_{i+1} - y_k) \\ &= (s - s_i) y_{i+1} + \sum_{k=1}^{i} p_k y_k. \end{aligned}$$

Clearly, $f^*(0) = 0$ and $f^*(1) = \sum_{k=1}^{m} p_k y_k = \mathbb{E}[Y]$. If $s < 0$ or $s > 1$, then $f^*(s) = +\infty$.

The function $f^*(s)$ is convex, piecewise linear, and has break points at cumulative probabilities s_i, $i = 0, 1, \ldots, m$. Its value at the point s_i is the contribution of the smallest i realizations of Y to the expected value of Y. In economics it is called the *absolute Lorenz curve*.

EXERCISES

2.1. There are n mutual funds and m asset categories. Let a_{ij} be the fraction of the capital of fund j invested in category i. We have initial capital C and we want to invest all or part of it in these funds. We denote by x_j the amount invested in fund $j = 1, \ldots, n$. No short selling is allowed, so x_j has to be nonnegative.

(a) Describe the set $X \subset \mathbb{R}^n$ of all possible amounts invested in these funds (fund portfolios). What are its extreme points?

(b) Describe the set $Y \subset \mathbb{R}^m$ of all possible amounts invested in this way in the m asset categories (asset portfolios).

(c) Show that if a point y is an extreme point of Y, it has the form $y = Ax$, where x is an extreme point of X.

(d) Suppose $y \in Y$ is an asset portfolio obtained by investing in some of the available funds. Prove that you can construct it by investing in no more than $m + 1$ funds.

2.2. Prove that x is an extreme point of a convex set X if and only if $X \setminus \{x\}$ is convex.

2.3. Let K be a closed convex cone in $\mathbb{R}^n$. For a point $x \in K$ the set $\{\alpha x : \alpha \ge 0\}$ is called a *ray*. An *extreme ray* of K is a ray which cannot be expressed as a convex combination of two other rays.

(a) Prove that a ray T is extreme if and only if $K \setminus T$ is convex.

(b) Prove that a closed convex cone which does not contain any line is the convex hull of the set of its extreme rays.

2.4. Let S be the set of all $n \times n$ matrices A such that $a_{ij} \geq 0$ for $i, j = 1, \dots, n$ and $\sum_{i=1}^{n} a_{ij} = 1$ for $j = 1, \dots, n$, $\sum_{j=1}^{n} a_{ij} = 1$ for $i = 1, \dots, n$. Prove that S is convex and that its extreme points are permutation matrices, that is, matrices $P \in S$ such that $p_{ij} \in \{0, 1\}$, $i, j = 1, \dots, n$.

2.5. The cone K in $\mathbb{R}^n$ is defined as follows:

$$K = \{x \in \mathbb{R}^n : 0 \leq x_1 \leq x_2 \leq \cdots \leq x_n\}.$$

Find the polar cone K°.

2.6. C is a closed convex cone in $\mathbb{R}^n$. The cone $K \in \mathbb{R}^{mn}$ is defined by

$$K = \{y = (y^1, y^2, \dots, y^m) : y^i \in \mathbb{R}^n,\ i = 1, \dots, m,\\ y^1 + y^2 + \cdots + y^m \in C\}.$$

Find the polar cone K°.

2.7. Assume that the cones $K_1, \dots, K_m$ in $\mathbb{R}^n$ are polyhedral, that is,

$$K_i = \{x \in \mathbb{R}^n : A_i x \leq 0\},$$

with some matrices A_i of dimension $l_i \times n$, $i = 1, \dots, m$. Prove that

$$K^\circ = K_1^\circ + K_2^\circ + \cdots + K_m^\circ.$$

2.8. Let K be a closed convex cone in $\mathbb{R}^n$. Prove that every $x \in \mathbb{R}^n$ can be represented as

$$x = \Pi_K(x) + \Pi_{K^\circ}(x)$$

and

$$\langle \Pi_K(x), \Pi_{K^\circ}(x) \rangle = 0.$$

2.9. Let K be a closed convex cone. Find the normal cone to K at $x \in K$.

2.10. Prove the *generalized geometric–arithmetic mean inequality*: for all $x_1 > 0, x_2 > 0, \dots, x_n > 0$ and all $\alpha_1 \geq 0, \alpha_2 \geq 0, \dots, \alpha_n \geq 0$ such that $\alpha_1 + \alpha_2 + \cdots + \alpha_n = 1$ we have

$$x_1^{\alpha_1} x_2^{\alpha_2} \cdots x_n^{\alpha_n} \leq \alpha_1 x_1 + \alpha_2 x_2 + \cdots + \alpha_n x_n.$$

2.11. Let a_i be vectors in $\mathbb{R}^n$, $i = 1, \dots, m$. Prove that the function

$$f(x) = \ln\Big(\sum_{i=1}^{m} e^{\langle a_i, x \rangle}\Big)$$

is convex.

2.12. Prove that if a convex function f has a nonempty and bounded level set $M_\beta = \{x : f(x) \le \beta\}$ for some $\beta \in \mathbb{R}$, then all level sets of f are bounded.

2.13. Prove that for a norm $\|\cdot\|_\diamond$ in $\mathbb{R}^n$ its epigraph K is a cone in $\mathbb{R}^{n+1}$. Prove that its polar cone has the form

$$K^\circ = \big\{(\lambda, \alpha) \in \mathbb{R}^n \times \mathbb{R} : \alpha \le -\|\lambda\|_*\big\},$$

where $\|\cdot\|_*$ is the dual norm. Use Example 2.94.

2.14. Prove that for a convex function $f : \mathbb{R}^n \to \overline{\mathbb{R}}$ its subdifferential is a *monotone mapping*, that is,

$$\langle g^2 - g^1, x^2 - x^1 \rangle \ge 0,$$

for all $x^1, x^2 \in \operatorname{dom} f$ and all $g^1 \in \partial f(x^1)$, $g^2 \in \partial f(x^2)$.

2.15. A model represents the output variable, $y \in \mathbb{R}$, as a linear function,

$$y = \sum_{i=1}^{n} x_i u_i,$$

of input variables $u_1, \dots, u_n$. The quantities $x_1, \dots, x_n$ are unknown model coefficients. We have N observations of input and output variables: (u^j, y^j), $j = 1, \dots, N$. One way to determine the values of the coefficients $x_1, \dots, x_n$ is to minimize the sum of the absolute errors:

$$f(x) = \sum_{j=1}^{N} \Big| y^j - \sum_{i=1}^{n} x_i u_i^j \Big|.$$

Calculate the subdifferential of the function f at a point x.

2.16. For a vector $x \in \mathbb{R}^n$ we define $x_{[j]}$, $j = 1, \dots, n$, as its ordered coordinates:

$$x_{[1]} \ge x_{[2]} \ge \cdots \ge x_{[n]}.$$

Prove that for any $1 \le k \le n$ the function

$$f_k(x) = \sum_{j=1}^{k} x_{[j]}$$

is convex and calculate its subdifferential.

2.17. Let X be a convex compact polyhedron in $\mathbb{R}^n$ and let A be an $m \times n$ matrix. Prove that the function $F : \mathbb{R}^m \to \mathbb{R}$,

$$F(\lambda) = \max_{x \in X} \langle \lambda, Ax \rangle$$

is convex. Calculate its subdifferential at a point λ.

2.18. Assume that $f : \mathbb{R}^m \to \overline{\mathbb{R}}$ and let A be an $m \times n$ matrix. Derive the formula for the conjugate of the function $g : \mathbb{R}^n \to \overline{\mathbb{R}}$ given by $g(x) = f(Ax)$.

2.19. Let A be a positive definite symmetric matrix of dimension $n \times n$. Calculate the conjugate function to

$$f_A(x) = \frac{1}{2}\langle x, Ax\rangle.$$

Use this result to prove that if A and B are positive definite, and $A - B$ is positive semidefinite, then $B^{-1} - A^{-1}$ is positive semidefinite.

2.20. Let Z be a convex closed set in $\mathbb{R}^n$ and let

$$\text{dist}(x, Z) = \min_{z \in Z} \|x - z\|,$$

where $\|x - z\|$ is the Euclidean norm in $\mathbb{R}^n$. Prove that the function

$$f(x) = \|x\|^2 - \big[\,\text{dist}(x, Z)\big]^2$$

is convex.
Hint: Represent $f(x)$ using a conjugate of a convex function.

2.21. Derive Theorem 2.27 (the bipolar theorem) as a special case of the Fenchel–Moreau Theorem (Theorem 2.95).
Hint: Use Example 2.96.

2.22. Let Z be a convex closed set in $\mathbb{R}^n$ and let

$$f(x) = \text{dist}(x, Z) = \min_{z \in Z} \|x - z\|_\diamond,$$

where $\|x - z\|_\diamond$ is a norm in $\mathbb{R}^n$. Prove that the conjugate function f^* has the form

$$f^*(s) = \sigma_Z(s) + \delta_{B_*}(s),$$

where $\sigma_Z(\cdot)$ is the support function of Z, and $\delta_{B_*}(\cdot)$ is the indicator function of the dual ball $B_* = \{s \in \mathbb{R}^n : \|s\|_* \le 1\}$.

Use this result to calculate the subdifferential of f.

Chapter Three

Optimality Conditions

3.1 UNCONSTRAINED MINIMA OF DIFFERENTIABLE FUNCTIONS

In this chapter we analyze conditions that have to be satisfied at points which are local minima of optimization problems. Let $f : \mathbb{R}^n \to \mathbb{R}$, $X \subset \mathbb{R}^n$, and let us consider the problem

$$\underset{x \in X}{\text{minimize}}\ f(x).$$

A point $\hat{x} \in X$ is called a *local minimum* of this problem if there exists $\varepsilon > 0$ such that

$$f(y) \geq f(\hat{x}) \quad \text{for all} \quad y \in X \quad \text{such that} \quad \|y - \hat{x}\| \leq \varepsilon.$$

If $f(y) \geq f(\hat{x})$ for all $y \in X$, the point $\hat{x}$ is called the *global minimum*. When $X = \mathbb{R}^n$ we discuss *unconstrained* local and global minima.

From the classical multivariate calculus we know the following theorem.

THEOREM 3.1. *Assume that $f : \mathbb{R}^n \to \mathbb{R}$ is differentiable at a point $\hat{x}$.*

(i) *If $f(\cdot)$ attains its unconstrained local minimum at $\hat{x}$, then*

$$\nabla f(\hat{x}) = 0. \tag{3.1}$$

(ii) *If $f(\cdot)$ is convex and (3.1) is satisfied, then $\hat{x}$ is an unconstrained global minimum of $f(\cdot)$.*

Proof. (i) From the definition of the gradient, for all $y \in \mathbb{R}^n$

$$f(y) = f(\hat{x}) + \langle \nabla f(\hat{x}), y - x \rangle + r(\hat{x}, y)$$

where

$$\lim_{y \to \hat{x}} \frac{r(\hat{x}, y)}{\|y - \hat{x}\|} = 0.$$

If $\nabla f(\hat{x}) \neq 0$, then we consider the points

$$y(\tau) = \hat{x} - \tau \nabla f(\hat{x}), \quad \tau > 0.$$

We obtain

$$\begin{aligned} f(y(\tau)) - f(\hat{x}) &= \langle \nabla f(\hat{x}), y(\tau) - \hat{x} \rangle + r(\hat{x}, y(\tau)) \\ &= -\tau \|\nabla f(\hat{x})\|^2 + r(\hat{x}, y(\tau)). \end{aligned} \tag{3.2}$$

Since the rest $r(\hat{x}, y(\tau))$ is infinitely smaller than $\|y(\tau) - \hat{x}\|$, when $\tau \to 0$, we can find $\bar{\tau} > 0$ such that for all $\tau \in (0, \bar{\tau})$

$$\frac{r(\hat{x}, y(\tau))}{\|y(\tau) - \hat{x}\|} \le \frac{1}{2}\|\nabla f(\hat{x})\|.$$

This can be equivalently expressed as

$$r(\hat{x}, y(\tau)) \le \frac{1}{2}\tau \|\nabla f(\hat{x})\|^2.$$

Substituting the last inequality into (3.2) we conclude that for all $\tau \in (0, \bar{\tau})$

$$f(y(\tau)) - f(\hat{x}) \le -\frac{1}{2}\tau \|\nabla f(\hat{x})\|^2 < 0.$$

Therefore $\hat{x}$ cannot be a local minimum.

(ii) If $f(\cdot)$ is convex, then for all $y \in \mathbb{R}^n$ we have

$$f(y) \ge f(\hat{x}) + \langle \nabla f(\hat{x}), y - x \rangle = f(\hat{x}).$$

Therefore, $\hat{x}$ is a global minimum. □

Points satisfying condition (3.1) are called *stationary*.

Theorem 3.1 has many applications.

Example 3.2. A real random variable X has the probability density function $f(x, \theta)$, where $\theta \in \mathbb{R}^m$ is a vector of distribution parameters. We have collected n independent observations $x_1, \dots, x_n$ of X. The *maximum likelihood estimator* $\hat{\theta}$ of θ is obtained by maximizing (with respect to θ) the function

$$L(\theta) = f(x_1, \theta) f(x_2, \theta) \cdots f(x_n, \theta).$$

Assume that the set of θ, for which all the factors above are positive, is nonempty. Since the function $\ln(\cdot)$ is strictly increasing, $\hat{\theta}$ is also the maximum of

$$\ln(L(\theta)) = \sum_{j=1}^{n} \ln(f(x_j, \theta)).$$

If the density function is differentiable with respect to the parameters θ at $\hat{\theta}$, we conclude that

$$\sum_{j=1}^{n} \frac{\nabla_\theta f(x_j, \hat{\theta})}{f(x_j, \hat{\theta})} = 0.$$

For example, if $f(\cdot)$ is the density function of the normal distribution,

$$f(x) = \frac{1}{\sigma\sqrt{2\pi}} e^{-\frac{(x-\mu)^2}{2\sigma^2}},$$

and $\theta = (\mu, \sigma)$, we obtain

$$-\ln(L(\theta)) = \frac{n}{2}\ln(2\pi) + n\ln(\sigma) + \frac{1}{2\sigma^2}\sum_{j=1}^{n}(x_j - \mu)^2.$$

The minimum of this function is obtained at a point at which

$$\frac{\partial\big[-\ln(L(\theta))\big]}{\partial\mu} = -\frac{1}{\sigma^2}\sum_{j=1}^{n}(x_j - \mu) = 0,$$

and

$$\frac{\partial\big[-\ln(L(\theta))\big]}{\partial\sigma} = \frac{n}{\sigma} - \frac{1}{\sigma^3}\sum_{j=1}^{n}(x_j - \mu)^2 = 0.$$

These equations render

$$\hat{\mu} = \frac{1}{n}\sum_{j=1}^{n} x_j, \quad \hat{\sigma} = \sqrt{\frac{1}{n}\sum_{j=1}^{n}(x_j - \hat{\mu})^2}.$$

Example 3.3. A model represents the output variable, $y \in \mathbb{R}$, as a linear function,

$$y = \sum_{i=1}^{n} x_i u_i,$$

of input variables $u_1, \dots, u_n$. The quantities $x_1, \dots, x_n$ are unknown model coefficients. We have N observations of the values of input and output variables: (u^j, y^j), $j = 1, \dots, N$. One way to determine the values of the coefficients $x_1, \dots, x_n$ is to minimize the sum of the squared errors:

$$f(x) = \sum_{j=1}^{N}\Big(y^j - \sum_{i=1}^{n} x_i u_i^j\Big)^2.$$

Defining the matrix of input data

$$U = \begin{bmatrix} u_1^1 & u_2^1 & \cdots & u_n^1 \\ u_1^2 & u_2^2 & \cdots & u_n^2 \\ \vdots & \vdots & & \vdots \\ u_1^N & u_2^N & \cdots & u_n^N \end{bmatrix},$$

we obtain

$$f(x) = \|y - Ux\|^2.$$

The gradient of $f(\cdot)$ has the form

$$\nabla f(x) = 2U^T(Ux - y),$$

and the best values of the coefficients can be calculated by solving the system of equations

$$U^T Ux = U^T y.$$

If U has rank n, we obtain

$$\hat{x} = (U^T U)^{-1} U^T y.$$

If the rank of U is smaller than n, many minima exist.

Example 3.4. Two points, a and b, are connected by a chain consisting of n mass points joined by extensible links (see Figure 3.1). Each point has mass m, and the spring constant of the links equals K.

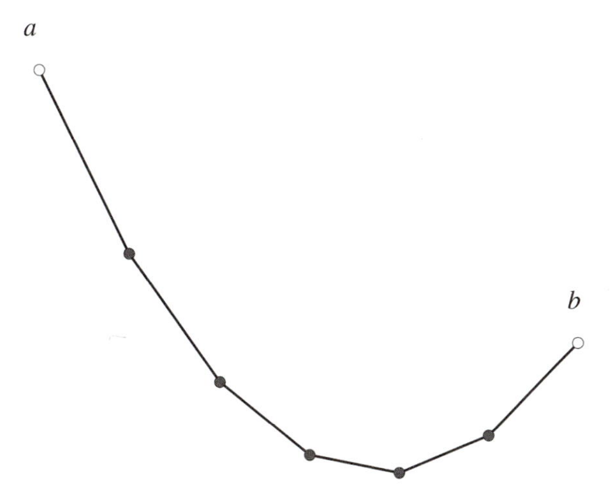

Figure 3.1. The chain with elastic links.

To calculate the location of the points at equilibrium we use the principle of minimum energy. Let us denote by x^i the location of the ith point, $i = 1, \dots, n$. Also, for the uniformity of the notation, we write $x^0 = a$, $x^{n+1} = b$.

The potential energy of the link between the points i and $i+1$ equals $\frac{K}{2}\|x^{i+1} - x^i\|^2$. The potential energy of the ith mass point is equal to $mg\langle x^i, e\rangle$, where $e = (0, 1)$ and g is the gravitation constant. The total energy, therefore, has the form

$$f(x^1, \dots, x^n) = \frac{K}{2}\sum_{i=1}^{n} \|x^{i+1} - x^i\|^2 + mg\sum_{i=1}^{n}\langle e, x^i\rangle.$$

Differentiating with respect to x^i we obtain:

$$\nabla_{x^i} f(x) = K(2x^i - x^{i-1} - x^{i+1}) + mge = 0, \quad i = 1, \ldots, n.$$

This equation represents the equilibrium of forces. Setting $z^i = x^{i+1} - x^i$, $i = 0, 1, \ldots, n$, we get

$$z^i = z^0 + \frac{imge}{K}, \quad i = 1, \ldots, n.$$

Therefore

$$x^i = a + iz^0 + \frac{i(i-1)mge}{2K}, \quad i = 1, \ldots, n+1,$$

and the value of z^0 can be easily found from the equation $x^{n+1} = b$. Finally,

$$x^i = \frac{n+1-i}{n+1}a + \frac{i}{n+1}b - \frac{i(n+1-i)mge}{2K}, \quad i = 1, \ldots, m.$$

It follows that the horizontal positions of the mass points are equally spaced. As the second differences, $z^i - z^{i-1}$, are constant, the mass points are located on a parabola.

3.2 UNCONSTRAINED MINIMA OF CONVEX FUNCTIONS

If a function $f : \mathbb{R}^n \to \overline{\mathbb{R}}$ is convex, all its local minima are in fact global minima. Indeed, suppose $\hat{x}$ is a local minimum of $f(\cdot)$, and there exists a point y such that $f(y) < f(\hat{x})$. Consider points on the segment connecting $\hat{x}$ and y:

$$z(\tau) = (1-\tau)\hat{x} + \tau y, \quad \tau \in [0, 1].$$

The convexity of $f(\cdot)$ implies that $f(z(\tau)) < f(\hat{x})$ for $\tau \in (0, 1]$ and therefore $\hat{x}$ cannot be a local minimum.

It follows that conditions for a local minimum and a global minimum are identical. Using the concept of a subdifferential of $f(\cdot)$ we easily obtain a counterpart of Theorem 3.1 on page 88.

THEOREM 3.5. *Let $f : \mathbb{R}^n \to \overline{\mathbb{R}}$ be a proper convex function. A point $\hat{x}$ is a global minimum of $f(\cdot)$ if and only if*

$$0 \in \partial f(\hat{x}).$$

Proof. A point $\hat{x}$ is a global minimum if and only if for all y

$$f(y) \geq f(\hat{x}) + \langle 0, y - x \rangle,$$

which is equivalent to the assertion. □

This simple observation, together with our techniques for calculating subdifferentials, helps to analyze many applied problems involving distances and maxima of families of functions.

Example 3.6. Let us return to Example 3.3, but with the sum of absolute errors as the measure of the quality of the model:

$$f(x) = \sum_{j=1}^{N} \Big| \sum_{i=1}^{n} x_i u_i^j - y^j \Big|.$$

In more abstract terms,

$$f(x) = \|Ux - y\|_1.$$

with

$$U = \begin{bmatrix} u_1^1 & u_2^1 & \cdots & u_n^1 \\ u_1^2 & u_2^2 & \cdots & u_n^2 \\ \vdots & \vdots & & \vdots \\ u_1^N & u_2^N & \cdots & u_n^N \end{bmatrix} = \begin{bmatrix} (u^1)^T \\ (u^2)^T \\ \vdots \\ (u^N)^T \end{bmatrix}.$$

The subdifferential of $f(\cdot)$ can be calculated using Lemma 2.83 and Example 2.78 (see also Exercise 2.15). In the typical case, when $y \neq Ux$,

$$\begin{aligned} \partial f(x) &= \big\{ U^T g : g \in \partial \|z - y\|,\ z = Ux \big\} \\ &= \big\{ U^T g : \|g\|_\infty = 1,\ \langle g, Ux - y \rangle = \|y - Ux\|_1 \big\}. \end{aligned}$$

At the optimal x, zero is an element of the subdifferential. We obtain

$$0 = U^T \hat{g}. \tag{3.3}$$

To express the conditions on $\hat{g}$ introduce the sets

$$J^+ = \Big\{ j : y^j > \sum_{i=1}^{n} \hat{x}_i u_i^j \Big\},$$

$$J^- = \Big\{ j : y^j < \sum_{i=1}^{n} \hat{x}_i u_i^j \Big\},$$

$$J^0 = \Big\{ j : y^j = \sum_{i=1}^{n} \hat{x}_i u_i^j \Big\}.$$

It follows from the equation $\langle g, Ux - y \rangle = \|y - Ux\|_1$ that

$$\hat{g}_j = \begin{cases} -1 & \text{if } j \in J^+, \\ 1 & \text{if } j \in J^-, \\ \beta_j & \text{if } j \in J^0, \end{cases}$$

with $\beta_j \in [-1, 1]$. This result can be transformed as follows. Using u^j to denote the transpose of the jth row of U we can write conditions (3.3) as follows:

$$\sum_{j\in J^+} u^j - \sum_{j\in J^-} u^j = \sum_{j\in J^0} \beta_j u^j.$$

Example 3.7. We have n continuous functions $\varphi_i(t)$, $i = 1, \dots, n$, where $t \in [a, b] \subset \mathbb{R}$. Our objective is to approximate in a uniform way a continuous function $\psi(t)$ by a linear combination of the basic functions,

$$\sum_{i=1}^{n} x_i \varphi_i(t) \approx \psi(t), \quad t \in [a, b],$$

where x_i, $i = 1, \dots, n$, are unknown coefficients. The approximation error is defined as follows:

$$F(x) = \max_{a\le t\le b} \Big| \sum_{i=1}^{n} x_i \varphi_i(t) - \psi(t) \Big|.$$

This approach is frequently referred to as Chebyshev approximation. In Example 2.88 on page 73 we calculated the subdifferential of $F(\cdot)$ at points x at which $F(x) > 0$:

$$\partial F(x) = \overline{\text{conv}} \left\{ \text{sgn}\Big(\sum_{i=1}^{n} x_i \varphi_i(t) - \psi(t) \Big) \begin{bmatrix} \varphi_1(t) \\ \varphi_2(t) \\ \vdots \\ \varphi_n(t) \end{bmatrix}, \ t \in \hat{T}(x) \right\},$$

where

$$\hat{T}(x) = \left\{ t \in [a, b] : \Big| \sum_{i=1}^{n} x_i \varphi_i(t) - \psi(t) \Big| = F(x) \right\}$$

and sgn$(\cdot)$ is the sign function,

$$\text{sgn}(v) = \begin{cases} -1 & v < 0, \\ 0 & v = 0, \\ 1 & v > 0. \end{cases}$$

Suppose $\hat{x}$ is the best value of the coefficients x. Since the set of vectors of the form $(\varphi_1(t), \varphi_2(t), \dots, \varphi_n(t))$, $t \in \hat{T}(\hat{x})$, is closed, Theorem 3.5 implies that

$$0 \in \text{conv} \left\{ \text{sgn}\Big(\sum_{i=1}^{n} \hat{x}_i \varphi_i(t) - \psi(t) \Big) \begin{bmatrix} \varphi_1(t) \\ \varphi_2(t) \\ \vdots \\ \varphi_n(t) \end{bmatrix}, \ t \in \hat{T}(\hat{x}) \right\}.$$

The set on the right hand side is convex and compact, and thus 0 is also a convex combination of at most $n + 1$ of its points (Lemma 2.7). It follows that there exist points $t_1, \ldots, t_m$ belonging to the set $\hat{T}(\hat{x})$, where $m \le n+1$, and nonnegative coefficients α_k, $k = 1, \ldots, m$, totaling one, such that

$$\sum_{k=1}^{m} \alpha_k \operatorname{sgn}\Big(\sum_{i=1}^{n} \hat{x}_i \varphi_i(t_k) - \psi(t_k)\Big) \begin{bmatrix} \varphi_1(t_k) \\ \varphi_2(t_k) \\ \vdots \\ \varphi_n(t_k) \end{bmatrix} = 0.$$

Example 3.8. Let Y be a compact set in $\mathbb{R}^n$. We want to find the smallest ball containing Y. Denoting by x the center of the ball, we see that our objective is to minimize the function

$$f(x) = \max_{y \in Y} \|x - y\|.$$

Let $\hat{x}$ be the best location of the center. Define the set

$$\hat{Y} = \{y \in Y : \|\hat{x} - y\| = f(\hat{x})\}.$$

By Theorem 2.87,

$$\partial f(\hat{x}) = \operatorname{conv}\Big(\bigcup_{y \in \hat{Y}} \partial \|\hat{x} - y\|\Big).$$

We can calculate the subdifferentials of the norm by using Example 2.78 on page 63. In the only interesting case when $f(\hat{x}) > 0$ (the set Y is not a singleton) the norm is differentiable and we obtain

$$\nabla \|\hat{x} - y\| = \frac{\hat{x} - y}{\|\hat{x} - y\|}.$$

Hence

$$0 \in \overline{\operatorname{conv}}\Big(\bigcup_{y \in \hat{Y}} \frac{\hat{x} - y}{\|\hat{x} - y\|}\Big) = \operatorname{conv}\Big(\bigcup_{y \in \hat{Y}} \frac{\hat{x} - y}{\|\hat{x} - y\|}\Big).$$

By Theorem 2.44 and Lemma 2.7, we can choose no more than $n+1$ of its extreme points such that 0 is their convex combination. It follows that there exist $m \le n+1$ points $y^i \in \hat{Y}$ and nonnegative weights α_i totaling 1, such that

$$0 = \sum_{i=1}^{m} \alpha_i \frac{\hat{x} - y^i}{\|\hat{x} - y^i\|}.$$

Because $\|\hat{x} - y^i\| = f(\hat{x}) > 0$, we can multiply both sides by $f(\hat{x})$ and we obtain the equation

$$\hat{x} = \sum_{i=1}^{m} \alpha_i y^i.$$

In other words, the sphere passing through the points y^i encompasses Y, and the center of this sphere is an element of the $m - 1$ dimensional simplex spanned by these points.

Example 3.9. We are given three convex closed sets Y_1, Y_2 and Y_3 in $\mathbb{R}^2$. Our objective is to find a point x whose sum of distances to the given sets is minimal. Formally, we need to find the minimum of the function

$$f(x) = \operatorname{dist}(x, Y_1) + \operatorname{dist}(x, Y_2) + \operatorname{dist}(x, Y_3),$$

where

$$\operatorname{dist}(x, Y) = \min_{y \in Y} \|x - y\|.$$

From Example 2.55 on page 46 we know that $f(\cdot)$ is convex, and Example 2.79 on page 64 provides us with the form of the subdifferential of each of the functions $\operatorname{dist}(x, Y_i)$.

Suppose the optimal point $\hat{x}$ is not an element of any of the given sets. Then each of the functions $\operatorname{dist}(x, Y_i)$ is differentiable with the gradient

$$\nabla\big[\operatorname{dist}(\hat{x}, Y_i)\big] = \frac{\hat{x} - \Pi_{Y_i}(\hat{x})}{\|\hat{x} - \Pi_{Y_i}(\hat{x})\|}.$$

The necessary condition of optimality reads

$$0 = \sum_{i=1}^{3} \frac{\hat{x} - \Pi_{Y_i}(\hat{x})}{\|\hat{x} - \Pi_{Y_i}(\hat{x})\|}.$$

Each of the three vectors in the sum above has length 1, and they sum to 0, which is possible only if they form an equilateral triangle. Consequently, the angles between the vectors

$$\Pi_{Y_i}(\hat{x}) - \hat{x}, \quad i = 1, 2, 3,$$

are all equal to $2\pi/3$.

If such a point does not exist, the optimal solution must be an element of one of the sets. Suppose $\hat{x} \in Y_1$, but $\hat{x} \notin Y_2 \cup Y_3$. Then, according to Example 2.79,

$$\partial \operatorname{dist}(\hat{x}, Y_1) = N_{Y_1}(\hat{x}) \cap \partial\|x - \hat{x}\| = N_{Y_1}(\hat{x}) \cap \{g : \|g\| \le 1\}.$$

We obtain

$$0 \in \partial f(\hat{x}) = \partial \operatorname{dist}(\hat{x}, Y_1) + \frac{\hat{x} - \Pi_{Y_2}(\hat{x})}{\|\hat{x} - \Pi_{Y_2}(\hat{x})\|} + \frac{\hat{x} - \Pi_{Y_3}(\hat{x})}{\|\hat{x} - \Pi_{Y_3}(\hat{x})\|}.$$

Equivalently,

$$\frac{\Pi_{Y_2}(\hat{x}) - \hat{x}}{\|\Pi_{Y_2}(\hat{x}) - \hat{x}\|} + \frac{\Pi_{Y_3}(\hat{x}) - \hat{x}}{\|\Pi_{Y_3}(\hat{x}) - \hat{x}\|} \in N_{Y_1}(\hat{x}),$$

and

$$\left\| \frac{\Pi_{Y_2}(\hat{x}) - \hat{x}}{\|\Pi_{Y_2}(\hat{x}) - \hat{x}\|} + \frac{\Pi_{Y_3}(\hat{x}) - \hat{x}}{\|\Pi_{Y_3}(\hat{x}) - \hat{x}\|} \right\| \le 1.$$

Thus the angle between the vectors $\Pi_{Y_2}(\hat{x}) - \hat{x}$ and $\Pi_{Y_3}(\hat{x}) - \hat{x}$ is at least $2\pi/3$, and its median is normal to Y_1. This is illustrated in Figure 3.2, which also shows

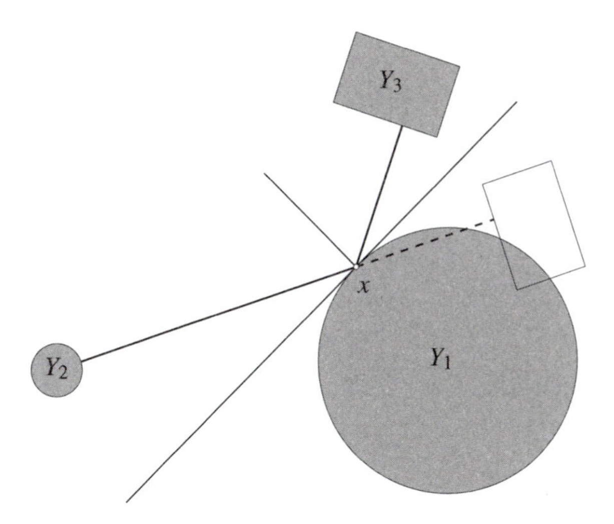

Figure 3.2. The closest point to three sets.

the geometrical interpretation of the optimality condition: x is the reflection point of a ray starting at a normal direction to Y_2, reflecting at Y_1 and hitting Y_3 from its normal cone.

If no set Y_i contains a point satisfying these conditions, the solution must be an element of an intersection of two of them. We leave to the reader the analysis of this case.

Optimality conditions for convex functions allow for an easy proof of a classical geometrical result, known as *Helly's Theorem.*

THEOREM 3.10. *Assume that I is a finite family of closed convex sets in $\mathbb{R}^n$. If every $n+1$ of them have a nonempty intersection, then they all have a nonempty intersection.*

Proof. Let Y_i, $i \in I$, be the given sets and let

$$\operatorname{dist}(x, Y_i) = \min_{y \in Y_i} \|x - y\|.$$

Consider the function

$$f(x) = \max_{i \in I} \big[\operatorname{dist}(x, Y_i) \big].$$

It is convex, finite everywhere, and it converges to $+\infty$, as $\|x\| \to \infty$. Therefore $f(\cdot)$ has a minimal point $\hat{x}$.

Suppose the intersection of all sets is empty and thus $f(\hat{x}) > 0$. Define

$$\hat{I} = \{i \in I : \operatorname{dist}(\hat{x}, Y_i) = f(\hat{x})\}.$$

By Theorem 2.87 we have

$$0 \in \partial f(\hat{x}) = \operatorname{conv}\Big(\bigcup_{i \in \hat{I}} \partial \operatorname{dist}(\hat{x}, Y_i) \Big),$$

Since $f(\hat{x}) > 0$, we obtain

$$0 \in \operatorname{conv}\Big(\bigcup_{i \in \hat{I}} \frac{\hat{x} - \Pi_{Y_i}(\hat{x})}{\|\hat{x} - \Pi_{Y_i}(\hat{x})\|} \Big).$$

It follows from Lemma 2.7 that there exists a subfamily $J \subset \hat{I}$ of no more than $n+1$ sets Y_i such that

$$0 \in \operatorname{conv}\Big(\bigcup_{i \in J} \frac{\hat{x} - \Pi_{Y_i}(\hat{x})}{\|\hat{x} - \Pi_{Y_i}(\hat{x})\|} \Big).$$

The right hand side of this expression is the subdifferential of the function

$$f_J(x) = \max_{i \in J} \big[\operatorname{dist}(x, Y_i) \big].$$

By virtue of Theorem 3.5 the point $\hat{x}$ is its global minimizer. Since the subfamily J has cardinality at most $n+1$, it must have a nonempty intersection, and thus

$$\hat{x} \in \bigcap_{i \in J} Y_i.$$

This contradicts the assumption that $\operatorname{dist}(\hat{x}, Y_i) = f(\hat{x}) > 0$ for $i \in \hat{I}$. □

For application of this theorem, see Exercise 3.18.

3.3 TANGENT CONES

3.3.1 Basic Concepts

If an optimization problem has constraints $x \in X$, the derivation of optimality conditions at a point $\hat{x}$ must involve the analysis of ways to perturb $\hat{x}$ while remaining in X. The fundamental concept in this analysis is that of a *tangent direction*. The most convenient way to define it is to reverse the perturbation operation and to consider sequences of points of X converging to $\hat{x}$.

Definition 3.11. A direction d is called *tangent* to the set $X \subset \mathbb{R}^n$ at the point $x \in X$ if there exist sequences of points $x^k \in X$ and scalars $\tau_k > 0$, $k = 1, 2, \ldots,$ such that $\tau_k \downarrow 0$ and

$$d = \lim_{k \to \infty} \frac{x^k - x}{\tau_k}.$$

It is implicit in this definition that $x^k \to x$, otherwise the limit above does not exists.

LEMMA 3.12. *Let $X \subset \mathbb{R}^n$ and let $x \in X$. The set $T_X(x)$ of all tangent directions for X at x is a closed cone.*

Proof. Suppose $d \in T_X(x)$. For every $\beta > 0$ we have

$$\beta d = \lim_{k\to\infty} \frac{x^k - x}{(\tau_k/\beta)},$$

so the sequences $\{x^k\}$ and $\{\tau_k/\beta\}$ satisfy Definition 3.11 for the direction βd. Hence $T_X(x)$ is a cone.

Let directions d^j be tangent to X at x with the corresponding sequences $\{x^{j,k}\}$ and $\{\tau_{j,k}\}$, $k = 1, 2, \ldots$, satisfying Definition 3.11, and let $\lim_{j\to\infty} d^j = d$. Since directions d^j are tangent, for every j we can find $k(j)$ such that

$$\left\| \frac{x^{j,k(j)} - x}{\tau_{j,k(j)}} - d^j \right\| \le \|d^j - d\|.$$

Therefore

$$\left\| \frac{x^{j,k(j)} - x}{\tau_{j,k(j)}} - d \right\| \le 2\|d^j - d\|.$$

It follows that the sequences $\{x^{j,k(j)}\}$ and $\{\tau_{j,k(j)}\}$, $j = 1, 2, \ldots$, satisfy Definition 3.11 for the direction d. Consequently, the cone $T_X(x)$ is closed. □

The notion of the tangent cone is illustrated in Figure 3.3.

Tangent cones are crucial for developing optimality conditions for nonlinear optimization problems. In Theorem 3.24 on page 113 we establish that a local minimum $\hat{x}$ of problem

$$\operatorname*{minimize}_{x\in X} f(x)$$

satisfies the relation

$$-\nabla f(\hat{x}) \in \left[T_X(\hat{x})\right]^{\circ}.$$

All optimality conditions, in one way or another, decipher this inclusion. This is the main motivation for the considerations of this section.

In general, cones of tangent directions may be nonconvex, which makes the analysis of optimality conditions difficult. But we can still identify some

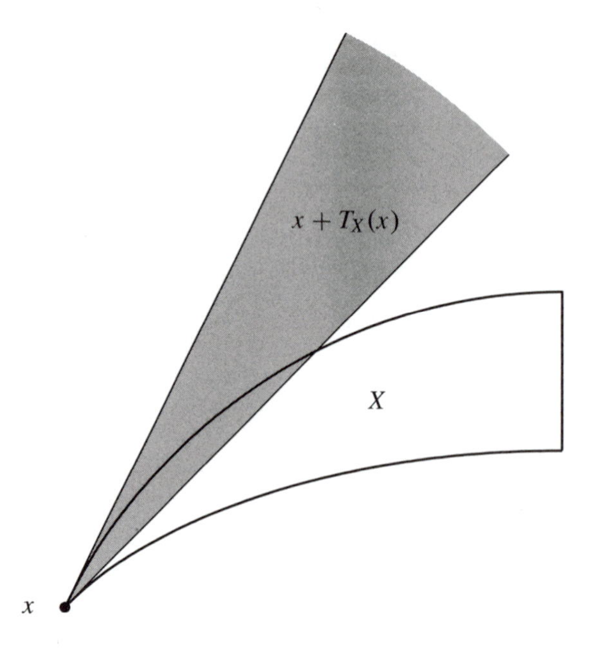

Figure 3.3. The tangent cone (translated to the point at which it is evaluated).

important cases when these cones are convex and we can provide their analytical description. Recall for convenience the definition of the cone of feasible directions at $x \in X$:

$$K_X(x) = \{d \in \mathbb{R}^n : d = \beta(y - x),\ y \in X,\ \beta \geq 0\}.$$

LEMMA 3.13. *Let $X \subset \mathbb{R}^n$ be a convex set and let $x \in X$. Then*

$$T_X(x) = \overline{K_X(x)}.$$

Proof. Each $d \in K_X(x)$ is a tangent direction by definition. Moreover, $K_X(x)$ is a convex cone. Since the tangent cone is closed,

$$\overline{K_X(x)} \subset T_X(x).$$

If these sets are not equal, we can find a direction $h \in T_X(x) \setminus \overline{K_X(x)}$. By the Separation Theorem 2.14 there exists $y \neq 0$ such that $\langle y, h\rangle > 0$ and $\langle y, d\rangle \leq 0$ for all $d \in \overline{K_X(x)}$. The direction h is tangent to X at x. If a sequence $\{x^k\}$ of points of X and a sequence $\tau_k \downarrow 0$ satisfy Definition 3.11 for the direction h, we obtain

$$\langle y, h\rangle = \Big\langle y, \lim_{k\to\infty} \frac{1}{\tau_k}(x^k - x)\Big\rangle = \lim_{k\to\infty} \frac{1}{\tau_k}\langle y, x^k - x\rangle.$$

Each vector $x^k - x$ is an element of $K_X(x)$ and therefore $\langle y, x^k - x\rangle \le 0$. The last displayed equation renders $\langle y, h\rangle \le 0$, which contradicts the postulated properties of the separating y. □

In applications, we encounter sets defined by an intersection:

$$X = X_1 \cap X_2 \cap \cdots \cap X_m.$$

For a point $x \in X$ we always have

$$T_X(x) \subset T_{X_1}(x) \cap T_{X_2}(x) \cap \cdots \cap T_{X_m}(x)$$

but an equality is not guaranteed. The main part of this section is devoted to conditions under which we obtain equality in this relation, for specific forms of the sets X_i. Moreover, we calculate the polar cone $[T_X(x)]^\circ$ in these cases.

3.3.2 Smooth Constraints. Metric Regularity

Feasible sets of nonlinear optimization problems are usually defined by systems of inequalities and equations:

$$\begin{aligned} g_i(x) &\le 0, \quad i = 1, \dots, m, \\ h_i(x) &= 0, \quad i = 1, \dots, p, \end{aligned}$$

with $g_i : \mathbb{R}^n \to \mathbb{R}$, $i = 1, \dots, m$, $h_i : \mathbb{R}^n \to \mathbb{R}$, $i = 1, \dots, p$. Additionally, we may have abstract set constraints of the form $x \in X_0$.

In order to develop algebraic forms of tangent cones to such sets, it is convenient to consider an abstract system

$$\begin{aligned} g(x) &\in Y_0, \\ x &\in X_0. \end{aligned} \tag{3.4}$$

Here $g : \mathbb{R}^n \to \mathbb{R}^m$ is continuously differentiable, Y_0 is a closed convex set in $\mathbb{R}^m$, and X_0 is a closed convex set in $\mathbb{R}^n$. For example, when $Y_0 = \{y \in \mathbb{R}^m : y \le 0\}$, system (3.4) has inequality constraints $g_i(x) \le 0$, $i = 1, \dots, m$. When $Y_0 = \{0\}$ the system has equality constraints $g_i(x) = 0$, $i = 1, \dots, m$. Combinations of equality and inequality constraints can be represented by defining Y_0 as a product of half-lines and zeros.

We use the symbol $g'(x)$ to denote the Jacobian of the function $g(\cdot)$ at the point x:

$$g'(x) = \begin{bmatrix} \frac{\partial g_1(x)}{\partial x_1} & \frac{\partial g_1(x)}{\partial x_2} & \cdots & \frac{\partial g_1(x)}{\partial x_n} \\ \frac{\partial g_2(x)}{\partial x_1} & \frac{\partial g_2(x)}{\partial x_2} & \cdots & \frac{\partial g_2(x)}{\partial x_n} \\ \vdots & \vdots & \vdots & \vdots \\ \frac{\partial g_m(x)}{\partial x_1} & \frac{\partial g_m(x)}{\partial x_2} & \cdots & \frac{\partial g_m(x)}{\partial x_n} \end{bmatrix}.$$

Denote by X the set defined by system (3.4),

$$X = \{x \in X_0 : g(x) \in Y_0\},$$

and consider a point $x_0 \in X$. For a direction d to be tangent to X at x_0 it is necessary that $d \in T_{X_0}(x_0)$. Furthermore, when x_0 is perturbed in the direction d, then $g(x_0)$ is perturbed (in first order terms) in the direction $g'(x_0)d$. Thus it is also necessary that $g'(x_0)d \in T_{Y_0}(y_0)$. Consequently,

$$T_X(x_0) \subset \big\{d \in \mathbb{R}^n : d \in T_{X_0}(x_0),\ g'(x_0)d \in T_{Y_0}(g(x_0))\big\}.$$

We formalize this argument in Theorem 3.15 below.

The last relation becomes an equation, if the sets X_0 and Y_0 are convex closed polyhedra, and if the function $g(\cdot)$ is affine. Indeed, in this case the first order expansions are exact, and the tangent directions are feasible directions.

In general, however, the equality in the relation above is not guaranteed, unless system (3.4) has an additional property, called *metric regularity*. Roughly speaking, it allows compensation of small perturbations in conditions (3.4) with small adjustments of x.

DEFINITION 3.14. System (3.4) is called *metrically regular* at the point $x_0 \in X$ if there exist $\varepsilon > 0$ and C such that for all $\widetilde{x}$ and $\widetilde{u}$ satisfying $\|\widetilde{x} - x_0\| \le \varepsilon$ and $\|\widetilde{u}\| \le \varepsilon$ we can find $x_{\text{R}} \in X_0$ satisfying the inclusion

$$g(x_{\text{R}}) - \widetilde{u} \in Y_0,$$

and such that

$$\|x_{\text{R}} - \widetilde{x}\| \le C\Big(\operatorname{dist}(\widetilde{x}, X_0) + \operatorname{dist}(g(\widetilde{x}) - \widetilde{u}, Y_0)\Big). \tag{3.5}$$

The concept of metric regularity of a system $g(x) = 0, x \in X_0$ is illustrated in Figure 3.4.

We prove in Theorem A.10 in the Appendix that metric regularity is equivalent to the following *Robinson's condition*:

$$\big\{g'(x_0)d - v : d \in K_{X_0}(x_0),\ v \in K_{Y_0}(g(x_0))\big\} = \mathbb{R}^m. \tag{3.6}$$

Observe that the set on the left hand side of (3.6) is a cone, and thus an equivalent expression of Robinson's condition is

$$0 \in \operatorname{int}\big\{g'(x_0)(x - x_0) - (y - g(x_0)) : x \in X_0,\ y \in Y_0\big\}.$$

The condition is clear, but the considerations associated with it are very technical. To avoid a long detour through the area of stability of systems of inclusions, we present these considerations in the Appendix. However,

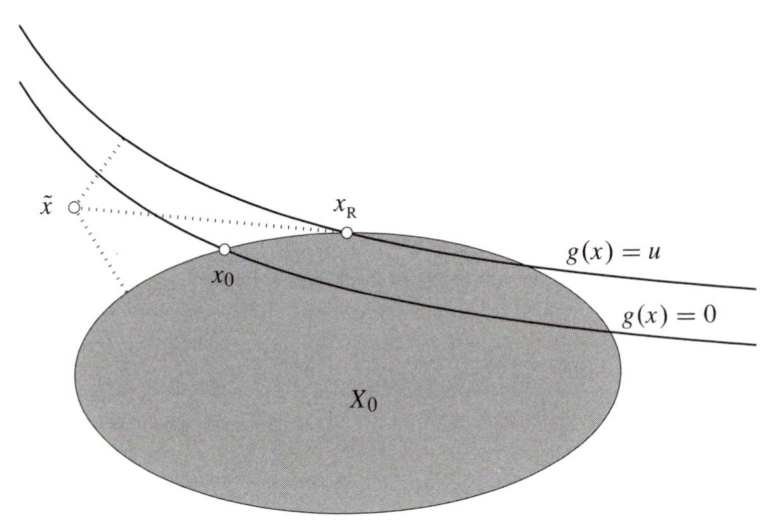

Figure 3.4. Metric regularity of the system. The distance of the perturbed point $\tilde{x}$ to a solution x_R of the perturbed system is proportional to the violation of the two constraints.

it may be useful at this stage to notice that in the special case of equality constraints $g(x) = 0$, when $X_0 = \mathbb{R}^n$ and $Y_0 = \{0\}$, metric regularity is equivalent to the linear independence of the gradients of the constraints

$$\nabla g_i(x_0), \quad i = 1, \dots, m.$$

The relevance of the concept of metric regularity is demonstrated in the following theorem.

THEOREM 3.15. *If system* (3.4) *is metrically regular, then*

$$T_X(x_0) = \left\{ d \in \mathbb{R}^n : d \in T_{X_0}(x_0),\ g'(x_0)d \in T_{Y_0}(g(x_0)) \right\}. \tag{3.7}$$

Proof. Let us prove at first that every tangent direction d is a member of the set on the right hand side of (3.7). Since $X \subset X_0$, the direction d is an element of $T_{X_0}(x_0)$. By Definition 3.11, there exist points $x^k \in X$ and scalars $\tau_k \downarrow 0$ such that

$$d = \lim_{k\to\infty} \frac{x^k - x_0}{\tau_k}.$$

Set $y^k = g(x^k)$, $y_0 = g(x_0)$. We have

$$y^k = y_0 + g'(x_0)(x^k - x_0) + o_k,$$

with $\lim_{k\to\infty}\big(o_k/\tau_k\big) = 0$. Dividing the last displayed equation by τ_k and passing to the limit, we obtain

$$\lim_{k\to\infty} \frac{y_k - y_0}{\tau_k} = g'(x_0)d.$$

As $y^k \in Y_0$, we get $g'(x_0)d \in T_{Y_0}(y_0)$. Summing up, the inclusion "$\subset$" is true in (3.7).

We need to prove the opposite inclusion. Let d be a direction in the set on the right hand side of (3.7). Consider points of the form

$$x(\tau) = x_0 + \tau d, \quad \tau > 0.$$

As $d \in T_{X_0}(x_0)$,

$$\operatorname{dist}(x(\tau), X_0) = o_1(\tau), \tag{3.8}$$

with $o_1(\tau)/\tau \to 0$, when $\tau \downarrow 0$. Also,

$$g(x(\tau)) = g(x_0) + \tau g'(x_0)d + o_2(\tau),$$

with $\|o_2(\tau)\|/\tau \to 0$, when $\tau \downarrow 0$. Since $g'(x_0)d \in T_{Y_0}(g(x_0))$, it follows that

$$\operatorname{dist}(g(x(\tau)), Y_0) \le \|o_2(\tau)\| + \operatorname{dist}(g(x_0) + \tau g'(x_0)d, Y_0) = o_3(\tau), \tag{3.9}$$

with $o_3(\tau)/\tau \to 0$, when $\tau \downarrow 0$. Consequently, the points $x(\tau)$ "almost" belong to X_0, and "almost" satisfy the constraint $g(x) \in Y_0$. The errors are negligible with respect to τ. We can now use the property of metric regularity. Setting $\widetilde{x} = x(\tau)$ and $\widetilde{u} = 0$ in Definition 3.14, we deduce that for sufficiently small $\tau > 0$ we can find points $x_{\mathrm{R}}(\tau) \in X$ such that

$$\|x_{\mathrm{R}}(\tau) - x(\tau)\| \le C\Big(\operatorname{dist}(x(\tau), X_0) + \operatorname{dist}(g(x(\tau)), Y_0)\Big).$$

Using (3.8) and (3.9) we conclude that

$$\lim_{\tau\downarrow 0} \frac{\|x_{\mathrm{R}}(\tau) - x_0\|}{\tau} = d.$$

Hence d is a tangent direction to X at x_0. □

We can now easily develop algebraic forms of tangent cones to systems of equations and inequalities. Consider the system

$$\begin{aligned} g_i(x) &\le 0, \quad i = 1, \dots, m, \\ h_i(x) &= 0, \quad i = 1, \dots, p, \\ x &\in X_0, \end{aligned} \tag{3.10}$$

with continuously differentiable functions $g : \mathbb{R}^n \to \mathbb{R}^m$ and $h : \mathbb{R}^n \to \mathbb{R}^p$ and with a closed convex set X_0. We consider a point x_0 satisfying (3.10) and we define the set of active inequality constraints:

$$I^0(x_0) = \{1 \le i \le m : g_i(x_0) = 0\}.$$

System (3.10) is a special case of system (3.4) with

$$Y_0 = \{(y, 0) \in \mathbb{R}^m \times \mathbb{R}^p : y_i \le 0, \ i \in I^0(x_0)\}.$$

Robinson's condition (3.6) takes on the form:

$$\left\{ \begin{bmatrix} g'(x_0)d - v \\ h'(x_0)d \end{bmatrix} : d \in T_{X_0}(x_0), \ v \in \mathbb{R}^m, \ v_i \le 0, \ i \in I^0(x_0) \right\} = \mathbb{R}^m \times \mathbb{R}^p. \tag{3.11}$$

As the set on the left hand side is a cone, it is equivalent to require that 0 is an interior point of this set. A simpler sufficient condition of metric regularity can be obtained as follows.

LEMMA 3.16. *Assume that there exists a point $x_{\text{MF}} \in \operatorname{int} X_0$ such that*

$$\begin{aligned} \langle \nabla g_i(x_0), x_{\text{MF}} - x_0 \rangle &< 0, \quad i \in I^0(x_0), \\ \langle \nabla h_i(x_0), x_{\text{MF}} - x_0 \rangle &= 0, \quad i = 1, \dots, p, \end{aligned} \tag{3.12}$$

and that the gradients $\nabla h_i(x_0)$, $i = 1, \dots, p$, are linearly independent. Then system (3.10) *is metrically regular at x_0.*

Proof. As x_{MF} is an interior point, there exists $\varepsilon > 0$ such that a ball about x_{MF} of radius ε is also included in the interior of X_0. Define $B = \{s \in \mathbb{R}^n : \|s\| \le \varepsilon\}$. Then

$$x_{\text{MF}} - x_0 + B \subset K_{X_0}(x_0).$$

In view of (3.12) we can choose ε small enough (but positive) and a positive δ, so that

$$\langle \nabla g_i(x_0), x_{\text{MF}} - x_0 + s \rangle < -\delta, \quad \text{for all} \quad i \in I^0(x_0), \quad s \in B. \tag{3.13}$$

Since the gradients $\nabla h_i(x_0)$, $i = 1, \dots, p$, are linearly independent, we have

$$0 \in \operatorname{int}\{h'(x_0)s : s \in B\}. \tag{3.14}$$

By choosing $\delta > 0$ sufficiently small, we can ensure that a ball of radius δ is included in the set on the right hand side of (3.14).

To formally verify Robinson's condition (3.11), choose any $(y, z) \in \mathbb{R}^m \times \mathbb{R}^p$ such that $\|(y, z)\| \le \delta$. By (3.14), we can find $s \in B$ such that $h'(x_0)s = z$. Then the second relation of (3.12) yields

$$h'(x_0)(x_{\text{MF}} - x_0 + s) = z.$$

Relation (3.13) implies that we can find $v \leq 0$ such that

$$g'(x_0)(x_{\text{MF}} - x_0 + s) - v = y.$$

This means that for the direction $d = x_{\text{MF}} - x_0 + s$ and for the selected v the corresponding element of the set at the left hand side of (3.11) equals (y, z). This can be done for all (y, z) such that $\|(y, z)\| \leq \delta$. Consequently, Robinson's condition holds true and the system is metrically regular. □

We have to stress that the conditions of Lemma 3.16 are not equivalent to metric regularity, because we made a strong assumption that x_{MF} is an interior point of X_0.

The assumptions of Lemma 3.16 for $X_0 = \mathbb{R}^n$ are known as the Mangasarian–Fromovitz constraint qualification condition. In this case it is equivalent to metric regularity.

LEMMA 3.17. *System* (3.10) *with* $X_0 = \mathbb{R}^n$ *satisfies the Mangasarian–Fromovitz constraint qualification condition at a point* x_0 *if and only if it is metrically regular at* x_0.

Proof. In view of Lemma 3.10, it remains to prove that metric regularity implies the Mangasarian–Fromovitz condition. With no loss of generality we may assume that all inequality constraints are active at x_0 (otherwise, we could just ignore the inactive constraints). Metric regularity of (3.10) with $X_0 = \mathbb{R}^n$ is equivalent to Robinson's condition:

$$\left\{ \begin{bmatrix} g'(x_0)d - v \\ h'(x_0)d \end{bmatrix} : d \in \mathbb{R}^n,\ v \in \mathbb{R}^m,\ v \leq 0 \right\} = \mathbb{R}^m \times \mathbb{R}^p.$$

In particular, $\{h'(x_0)d : d \in \mathbb{R}^n\} = \mathbb{R}^p$, and thus the gradients $\nabla h_i(x_0)$ are linearly independent. Moreover, for each vector of the form $(y, 0) \in \mathbb{R}^{m+p}$ it follows from Robinson's condition that there exist $d \in \mathbb{R}^n$ and $v \leq 0$ such that

$$\langle \nabla g_i(x_0), d \rangle = y_i + v_i, \quad i = 1, \ldots, m,$$
$$\langle \nabla h_i(x_0), d \rangle = 0, \quad i = 1, \ldots, p.$$

Choosing $y_i < 0$, $i = 1, \ldots, m$, we obtain the Mangasarian–Fromovitz condition. □

We end this section by calculating the polar of the tangent cone described in Theorem 3.15. Theorem 2.36 provides the formula:

$$[T_X(x_0)]^\circ = [T_{X_0}(x_0)]^\circ + \left\{ [g'(x_0)]^T \lambda : \lambda \in [T_{Y_0}(g(x_0))]^\circ \right\}. \tag{3.15}$$

We use it in the analysis of optimality conditions for problems with feasible sets of the form (3.4).

3.3.3 Convex Nonsmooth Constraints

Tangent cones of sets defined by constraints involving nonsmooth convex functions can be described with the use of subdifferentials of the constraint functions. Consider the set

$$X = \{x \in \mathbb{R}^n : g(x) \le 0\},$$

where $g : \mathbb{R}^n \to \overline{\mathbb{R}}$ is a convex function. The key role is played here by *Slater's condition: there exists x_S such that $g(x_S) < 0$.*

THEOREM 3.18. *Assume that $g(x_0) = 0$, the function g is subdifferentiable at x_0, and Slater's condition is satisfied. Then*

$$[K_X(x_0)]^\circ = \operatorname{cone}(\partial g(x_0)).$$

Proof. If $d \in K_X(x_0)$, then there exists $\bar{\tau} > 0$ such that $g(x_0 + \tau d) \le 0$ for all $\tau \in [0, \bar{\tau}]$. Therefore $g'(x_0; d) \le 0$. By Lemma 2.75,

$$g'(x_0; d) = \max_{s \in \partial g(x_0)} \langle s, d \rangle.$$

Hence

$$\langle s, d \rangle \le 0 \quad \text{for all} \quad s \in \operatorname{cone}(\partial g(x_0)).$$

This is equivalent to $d \in [\operatorname{cone}(\partial g(x_0))]^\circ$. Since $d \in K_X(x_0)$ was arbitrary, we have

$$K_X(x_0) \subset [\operatorname{cone}(\partial g(x_0))]^\circ. \tag{3.16}$$

On the other hand, if d is an element of the set on the right hand side, then

$$g'(x_0; d) = \sup_{s \in \partial g(x_0)} \langle s, d \rangle \le 0.$$

If $g'(x_0; d) < 0$, we have $g(x_0 + \tau d) < 0$ for all sufficiently small $\tau > 0$. Thus $x_0 + \tau d \in X$ for these τ and therefore $d \in \operatorname{cone}(X - x_0)$. If $g'(x_0; d) = 0$, we can represent d as the limit of directions

$$d^k = (1 - \alpha_k)d + \alpha_k(x_S - x_0), \quad \alpha_k \downarrow 0,$$

where x_S is the point satisfying Slater's condition. We can estimate the directional derivative as follows:

$$\begin{aligned} g'(x_0; d^k) &= \max_{s \in \partial g(x_0)} \langle s, d^k \rangle \\ &\le (1 - \alpha_k) \max_{s \in \partial g(x_0)} \langle s, d \rangle + \alpha_k \max_{s \in \partial g(x_0)} \langle s, x_S - x_0 \rangle \\ &\le (1 - \alpha_k) g'(x_0; d) + \alpha_k g(x_S) < 0. \end{aligned}$$

It follows that $g(x_0 + \tau d^k) < 0$ for all sufficiently small $\tau > 0$ and thus d^k is an element of $K_X(x_0)$. Consequently, $d \in \overline{K_X(x_0)}$. Since d was an arbitrary element of the set on the right hand side of (3.23), we conclude that

$$[\text{cone}(\partial g(x_0))]^\circ \subset \overline{K_X(x_0)}. \tag{3.17}$$

Combining (3.16) and (3.17), we obtain

$$K_X(x_0) \subset [\text{cone}(\partial g(x_0))]^\circ \subset \overline{K_X(x_0)}.$$

Since the polar cone is always closed, taking the closure of all three sets above we conclude that

$$[\text{cone}(\partial g(x_0))]^\circ = \overline{K_X(x_0)}.$$

We can now calculate the polar cone of both sides of this equation:

$$[\text{cone}(\partial g(x_0))]^{\circ\circ} = [\overline{K_X(x_0)}]^\circ = [K_X(x_0)]^\circ.$$

The set $\text{cone}(\partial g(x_0))$ is closed, and Theorem 2.27 implies that it is equal to the left side of the last equation. This yields the assertion of the theorem. □

Since $\overline{K_X(x_0)} = T_X(x_0)$, we obtain the following conclusion.

COROLLARY 3.19. *Under the assumptions of Theorem* 3.18 *we have*

$$T_X(x_0) = [\text{cone}(\partial g(x_0))]^\circ.$$

These results can be used to analyze the case of finitely many convex inequalities and affine equations. Consider the set

$$X = \{x \in X_0 : g_i(x) \le 0,\ i = 1, \dots, m,\ h_i(x) = 0,\ i = 1, \dots, p\},$$

where $g_i : \mathbb{R}^n \to \mathbb{R}$, $i = 1, \dots, m$, are convex functions, $h_i : \mathbb{R}^n \to \mathbb{R}$, $i = 1, \dots, p$, are affine functions, and X_0 is a closed convex set in $\mathbb{R}^n$.

Slater's condition takes on the form: *there exists a point* $x_s \in X_0$ *such that* $g_i(x_s) < 0$, $i = 1, \dots, m$, $h_i(x_s) = 0$, $i = 1, \dots, p$; *additionally,* $x_s \in \text{int}\, X_0$, *if* $p > 0$.

We define the set of active inequality constraints

$$I^0(x_0) = \{1 \le i \le m : g_i(x_0) = 0\}.$$

In view of applications to optimality conditions, we are mainly interested in the polar to the tangent cone $T_X(x_0)$.

THEOREM 3.20. *Assume that Slater's condition is satisfied. Then*

$$[T_X(x_0)]^\circ = [T_{X_0}(x_0)]^\circ + \sum_{i\in I^0(x_0)} \operatorname{cone}[\partial g_i(x_0)] + \operatorname{lin}\{\nabla h_i(x_0),\ i = 1,\dots,p\}. \tag{3.18}$$

Proof. Suppose $p = 0$ and consider the sets

$$X_i = \{x \in \mathbb{R}^n : g_i(x) \le 0\}, \quad i \in I^0(x_0).$$

As the functions $g_i(\cdot)$ are finite valued, they are continuous (Lemma 2.70 on page 55). Slater's condition implies that $x_S \in \operatorname{int} X_i$, $i = 1,\dots,m$. Hence

$$x_S \in X_0 \cap \operatorname{int} \bigcap_{i\in I^0(x_0)} X_i.$$

All sets X_i, $i = 0, 1,\dots,m$, are closed and convex. We can thus use Lemma 2.40 to obtain

$$N_X(x_0) = N_{X_0}(x_0) + \sum_{i\in I^0(x_0)} N_{X_i}(x_0).$$

By the definition of the normal cone on page 37 and by Lemma 3.13 we can rewrite this relation as

$$[T_X(x_0)]^\circ = [T_{X_0}(x_0)]^\circ + \sum_{i\in I^0(x_0)} [T_{X_i}(x_0)]^\circ.$$

By virtue of Theorem 2.74 on page 59, the functions $g_i(\cdot)$ are subdifferentiable everywhere. Applying Theorem 3.18 and Lemma 2.25(ii) on page 29 we conclude that

$$[T_{X_i}(x_0)]^\circ = [K_{X_i}(x_0)]^\circ = \operatorname{cone}[\partial g_i(x_0)], \quad i \in I^0(x_0),$$

which yields the required result in this case.

If equality constraints are present, we additionally define the set

$$A = \{x \in \mathbb{R}^n : h_i(x) = 0,\ i = 1,\dots,p\}.$$

In this case, Slater's condition implies that

$$x_S \in \operatorname{int} X_0 \cap \operatorname{int}\Big(\bigcap_{i\in I^0(x_0)} X_i\Big) \cap A.$$

We can employ Lemma 2.40 again to conclude that

$$N_X(x_0) = N_{X_0}(x_0) + \sum_{i\in I^0(x_0)} N_{X_i}(x_0) + N_A(x_0).$$

Using Lemma 3.13 again we obtain the assertion of the theorem. □

The tangent cone itself can be now calculated by using Theorem 2.35.

COROLLARY 3.21. *Under the assumptions of Theorem* 3.20,

$$T_X(x_0) = T_{X_0}(x_0) \cap \Bigg(\bigcap_{i \in I^0(x_0)} \big[\operatorname{cone}(\partial g_i(x_0))\big]^{\circ} \Bigg) \cap \Bigg(\bigcap_{1 \le i \le p} \big[\operatorname{lin}(\nabla h_i(x_0))\big]^{\circ} \Bigg).$$

The interior of the set X_0 can be replaced by the relative interior, if necessary.

Example 3.22. Consider the set $\mathbb{S}^n_-$ of symmetric negative semidefinite matrices of dimension n. It is a closed convex cone in the space $\mathbb{S}^n$ of all symmetric matrices of dimension n. We considered it in Example 2.31, where we developed the formula for its polar cone. Now it can be obtained in another way, which is slightly more complicated but quite instructive.

The cone $\mathbb{S}^n_-$ can be expressed algebraically as follows:

$$\mathbb{S}^n_- = \{A \in \mathbb{S}^n : \lambda_{\max}(A) \le 0\},$$

where $\lambda_{\max}(A)$ denotes the maximum eigenvalue of the matrix A.

As discussed in Example 2.59 on page 47, the maximum eigenvalue function $\lambda_{\max}(\cdot)$ is convex. The set $\mathbb{S}^n_-$ satisfies Slater's condition, because $\lambda_{\max}(-I) = -1$.

The subdifferential of the function $\lambda_{\max}(\cdot)$ has been calculated in Example 2.89 on page 74:

$$\partial \lambda_{\max}(A) = \operatorname{conv}\{yy^T : Ay = \lambda_{\max}(A)y,\ \|y\| = 1\}.$$

Suppose A is a boundary point of $\mathbb{S}^n_-$. Then $\lambda_{\max}(A) = 0$, and

$$\partial \lambda_{\max}(A) = \operatorname{conv}\{yy^T : Ay = 0,\ \|y\| = 1\}.$$

We can now apply Theorem 3.18 to obtain

$$[K_{\mathbb{S}^n_-}(A)]^{\circ} = \operatorname{cone}\big[\partial \lambda_{\max}(A)\big] = \operatorname{cone}\big[\operatorname{conv}\{yy^T : Ay = 0,\ \|y\| = 1\}\big].$$

The last set can be written in a more transparent way. By interchanging the operations of taking the convex hull and the conic extension we arrive at the expression

$$\begin{aligned} [K_{\mathbb{S}^n_-}(A)]^{\circ} &= \operatorname{conv}\big[\operatorname{cone}\{yy^T : Ay = 0,\ \|y\| = 1\}\big] \\ &= \operatorname{conv}\{\alpha yy^T : Ay = 0,\ \|y\| = 1,\ \alpha \ge 0\}. \end{aligned}$$

Since $\alpha yy^T = \big(\sqrt{\alpha}y\big)\big(\sqrt{\alpha}y\big)^T$ for $\alpha \ge 0$, we finally obtain

$$N_{\mathbb{S}^n_-}(A) = [K_{\mathbb{S}^n_-}(A)]^{\circ} = \operatorname{conv}\{yy^T : Ay = 0\}. \tag{3.19}$$

At $A = 0$ the normal cone is just the polar cone,

$$N_{\mathbb{S}^n_-}(0) = \big[\mathbb{S}^n_-\big]^{\circ},$$

and thus

$$\left[\mathbb{S}_-^n\right]^\circ = \operatorname{conv}\{yy^T : y \in \mathbb{R}^n\}. \tag{3.20}$$

Each element of the set on the right hand side is a positive semidefinite matrix, because for all $y_i \in \mathbb{R}^n$ and all $\alpha_i \geq 0$, $i = 1, \ldots, k$, we have

$$\left\langle d, \sum_{i=1}^{k} \alpha_i (y_i y_i^T) d \right\rangle = \sum_{i=1}^{k} \alpha_i \langle y_i^T d, y_i^T d \rangle = \sum_{i=1}^{k} \alpha_i \|y_i^T d\|^2 \geq 0,$$

for all $d \in \mathbb{R}^n$. On the other hand, every symmetric matrix Q can be represented as

$$Q = \sum_{j=1}^{n} \lambda_j z_j z_j^T,$$

where $z_1, \ldots, z_n$ are orthogonal eigenvectors of Q of length 1, and $\lambda_1, \ldots, \lambda_n$ are the corresponding eigenvalues. Q is positive semidefinite if and only if its eigenvalues are nonnegative. Therefore every positive semidefinite matrix Q is an element of the set on the right hand side of (3.20) (the fact that the eigenvalues do not total 1 does not matter here, because this set is a cone). Consequently, we obtain the following formula for the polar cone to the cone of negative semidefinite matrices:

$$\left[\mathbb{S}_-^n\right]^\circ = \mathbb{S}_+^n.$$

We have thus re-established the result of Example 2.31.

Suppose A is not necessarily 0, but $\lambda_{\max}(A) = 0$. The normal cone (3.19) can also be derived as follows. We know from Example 2.21 on page 27 that

$$T_{\mathbb{S}_-^n}(A) = \mathbb{S}_-^n - \{\tau A : \tau \geq 0\}.$$

It is a sum of cones, and thus its polar cone is the intersection of the corresponding polar cones (see Example 2.24 on page 28):

$$\begin{aligned} N_{\mathbb{S}_-^n}(A) &= \left[T_{\mathbb{S}_-^n}(A)\right]^\circ = \left[\mathbb{S}_-^n\right]^\circ \cap \left[\{-\tau A : \tau \geq 0\}\right]^\circ \\ &= \mathbb{S}_+^n \cap \{H \in \mathbb{S}^n : \operatorname{tr}(HA) \geq 0\}. \end{aligned}$$

In fact, for every $H \in \mathbb{S}_+^n$ we must have $\operatorname{tr}(HA) \leq 0$, and thus the last equation can also be written as

$$N_{\mathbb{S}_-^n}(A) = \mathbb{S}_+^n \cap \{H \in \mathbb{S}^n : \operatorname{tr}(HA) = 0\}.$$

This is the same as (3.19). By symmetry, at every $A \in \mathbb{S}_+^n$, we also have the formula

$$N_{\mathbb{S}_+^n}(A) = \mathbb{S}_-^n \cap \{H \in \mathbb{S}^n : \operatorname{tr}(HA) = 0\}.$$

This formula is of importance for *semidefinite programming*, which we discuss later in Example 3.36.

3.3.4 Smooth–Convex Constraints

Let us now consider the case when both smooth and nonsmooth (but convex) constraints are present:

$$X = \{x \in X_0 : \ h(x) \in Y_0, \ g(x) \in \mathbb{R}^m_-\}. \tag{3.21}$$

We assume that the function $h : \mathbb{R}^n \to \mathbb{R}^p$ is continuously differentiable, the functions $g_i : \mathbb{R}^n \to \mathbb{R}, i = 1, \dots, m$ are convex, and that X_0 and Y_0 are closed convex sets in $\mathbb{R}^n$ and $\mathbb{R}^p$. The symbol $\mathbb{R}^m_-$ denotes the nonpositive orthant: $\{v \in \mathbb{R}^m : v_j \le 0, \ j = 1, \dots, m\}$.

Our objective is to describe the tangent cone to the set X defined by (3.21) at a point $x_0 \in X$ in terms of the tangent cones to the sets X_0 and Y_0, the derivatives of $h(\cdot)$, and the subdifferentials of $g_i(\cdot)$.

Consider the set defined by the convex constraints,

$$C = \{x \in X_0 : g_i(x) \le 0, \ i = 1, \dots, m\},$$

and let $I^0(\hat{x})$ be the set of active constraints:

$$I^0(\hat{x}) = \{1 \le i \le m : g_i(\hat{x}) = 0\}.$$

Theorem 3.23. *Assume that there exists $x_s \in X_0$ such that $g_i(x_s) < 0$, $i = 1, \dots, m$. Furthermore, let Robinson's condition be satisfied:*

$$h'(\hat{x}) K_C(\hat{x}) - K_{Y_0}(h(\hat{x})) = \mathbb{R}^p.$$

Then

$$\begin{aligned}[T_X(\hat{x})]^\circ = {} & [T_{X_0}(\hat{x})]^\circ + \big\{[h'(\hat{x})]^T \lambda : \lambda \in [T_{Y_0}(h(\hat{x}))]^\circ\big\} \\ & + \sum_{i \in I^0(\hat{x})} \operatorname{cone}(\partial g_i(\hat{x})).\end{aligned} \tag{3.22}$$

Proof. By virtue of Theorem 3.20,

$$T_C(\hat{x}) = T_{X_0}(\hat{x}) \cap \bigcap_{i \in I^0(\hat{x})} [\operatorname{cone}(\partial g_i(\hat{x}))]^\circ.$$

Under Robinson's condition we obtain from Theorem 3.15 that

$$T_X(\hat{x}) = \big\{d \in \mathbb{R}^n : h'(\hat{x}) d \in T_{Y_0}(h(\hat{x})), \ d \in T_C(\hat{x})\big\}.$$

The polar cone can be calculated as follows. First, formula (3.15) yields:

$$[T_X(\hat{x})]^\circ = [T_C(\hat{x})]^\circ + \big\{[h'(\hat{x})]^T \lambda : \lambda \in [T_{Y_0}(h(\hat{x}))]^\circ\big\}.$$

Second, formula (3.18) implies:

$$[T_C(\hat{x})]^\circ = [T_{X_0}(\hat{x})]^\circ + \sum_{i \in I^0(\hat{x})} \operatorname{cone}(\partial g_i(\hat{x})).$$

Combining the last two expressions we get the required expression for the polar of the tangent cone. □

Our system of constraints (3.21) encompasses, as special cases, all systems considered before. The representation of the polar to the tangent cone allows us to formulate necessary conditions of optimality for problems involving such combined constraints. Unfortunately, the verification of the constraint qualification conditions is very difficult in this case.

Another way to derive the representation (3.22) of the polar to the tangent cone is to assume metric regularity of the system of constraints (3.21), as in the analysis of system (3.4). However, the constraint function (g, h) becomes nondifferentiable and nonconvex in this case. The derivation of the tangent cone then requires application of techniques of nonsmooth calculus, which are beyond the scope of this book.

3.4 OPTIMALITY CONDITIONS FOR SMOOTH PROBLEMS

Consider the constrained optimization problem

$$\operatorname*{minimize}_{x \in X} f(x), \tag{3.23}$$

with a differentiable function $f : \mathbb{R}^n \to \mathbb{R}$ and a set $X \subset \mathbb{R}^n$. If its solution $\hat{x}$ is a boundary point of the feasible set X, the necessary condition of optimality formulated in Theorem 3.1 does not have to be satisfied. The main reason is that the perturbations of the point $\hat{x}$ which take it out of the feasible set X are not allowed, and therefore they may correspond to a decrease of the objective function. In order to obtain necessary conditions of optimality, we restrict the set of possible perturbations to tangent directions at $\hat{x}$.

THEOREM 3.24. *Assume that $\hat{x}$ is a local minimum of problem* (3.23) *and that $f(\cdot)$ is differentiable at $\hat{x}$. Let $T_X(\hat{x})$ be the tangent cone to the set X at $\hat{x}$. Then*

$$-\nabla f(\hat{x}) \in \big[T_X(\hat{x})\big]^\circ. \tag{3.24}$$

Conversely, if the function $f(\cdot)$ is convex, the set X is convex, and a point $\hat{x} \in X$ satisfies relation (3.24), *then $\hat{x}$ is a global minimum of problem* (3.23).

Proof. Suppose our assertion is false:

$$-\nabla f(\hat{x}) \notin \big[T_X(\hat{x})\big]^{\circ}.$$

This means that there exists a direction $d \in T_X(\hat{x})$ such that

$$\langle \nabla f(\hat{x}), d \rangle < 0. \tag{3.25}$$

As d is a tangent direction, there exists a sequence of points $x^k \in X$ convergent to $\hat{x}$ and a sequence of scalars $\tau_k \downarrow 0$ such that

$$\lim_{k\to\infty} \frac{x^k - \hat{x}}{\tau_k} = d. \tag{3.26}$$

Since $f(\cdot)$ is differentiable at $\hat{x}$,

$$f(x^k) - f(\hat{x}) = \langle \nabla f(\hat{x}), x^k - \hat{x} \rangle + \alpha_k,$$

where $\alpha_k/\|x^k - \hat{x}\| \to 0$, as $k \to \infty$. Dividing both sides of the last equation by τ_k we obtain

$$\frac{f(x^k) - f(\hat{x})}{\tau_k} = \langle \nabla f(\hat{x}), d \rangle + \Big\langle \nabla f(\hat{x}), \frac{x^k - \hat{x}}{\tau_k} - d \Big\rangle + \frac{\alpha_k}{\tau_k}.$$

Inequality (3.25) implies that $\|d\| \neq 0$ and thus

$$\lim_{k\to\infty} \frac{\alpha_k}{\tau_k} = \lim_{k\to\infty} \frac{\alpha_k \|d\|}{\|x^k - \hat{x}\|} = 0.$$

By virtue of (3.25) and (3.26) we conclude that

$$\lim_{k\to\infty} \frac{f(x^k) - f(\hat{x})}{\tau_k} = \langle \nabla f(\hat{x}), d \rangle < 0.$$

On the other hand, all points x^k are feasible and they approach $\hat{x}$. Since $\hat{x}$ is a local minimum, $f(x^k) \geq f(\hat{x})$ for all sufficiently large k. Hence

$$\liminf_{k\to\infty} \frac{f(x^k) - f(\hat{x})}{\tau_k} \geq 0,$$

a contradiction. Therefore, relation (3.24) is valid.

Assume now that the function $f(\cdot)$ and set X are convex, and that (3.24) is satisfied at a point $\hat{x} \in X$. Since the set X is convex, for every $y \in X$ the direction

$$d = y - \hat{x}$$

is a tangent direction for X at $\hat{x}$ (Lemma 3.13). Thus, condition (3.24) implies that

$$\langle \nabla f(\hat{x}), y - x \rangle \geq 0.$$

The function $f(\cdot)$ is convex and Theorem 2.67 yields

$$f(y) \geq f(\hat{x}) + \langle \nabla f(\hat{x}), y - x \rangle.$$

Therefore $f(y) \geq f(\hat{x})$ for all $y \in X$, as required. □

If the function $f(\cdot)$ has only directional derivatives at $\hat{x}$, then the necessary condition of optimality is

$$f'(\hat{x}; d) \geq 0 \quad \text{for all} \quad d \in T_X(\hat{x}). \tag{3.27}$$

The proof of this condition can easily be obtained by following the argument of the proof of Theorem 3.24, after formula (3.26). We just need to use the directional derivative $f'(\hat{x}, d)$ instead of $\langle \nabla f(\hat{x}), d \rangle$.

The development of necessary conditions of optimality for various classes of nonlinear optimization problems consists mainly in deciphering the fundamental relation (3.24) for different forms of feasible sets X. The formulas for the polar cone of the tangent cone to X at $\hat{x}$ play the key role there.

Let us consider the nonlinear optimization problem

$$\begin{aligned} \text{minimize} \quad & f(x) \\ \text{subject to} \quad & g_i(x) \leq 0, \quad i = 1, \ldots, m, \\ & h_i(x) = 0, \quad i = 1, \ldots, p, \\ & x \in X_0. \end{aligned} \tag{3.28}$$

We assume that the functions $f : \mathbb{R}^n \to \mathbb{R}$, $g_i : \mathbb{R}^n \to \mathbb{R}$, $i = 1, \ldots, m$, and $h_i : \mathbb{R}^n \to \mathbb{R}$, $i = 1, \ldots, p$, are continuously differentiable, and that the set $X_0 \subset \mathbb{R}^n$ is convex and closed. The feasible set of this problem is denoted by X.

The optimality conditions of Theorem 3.24 for problem (3.28) involve the tangent cone to the feasible set X at the optimal point $\hat{x}$. In Theorem 3.15 on page 103 we established that if the system of constraints of problem (3.28) is metrically regular at the point $\hat{x}$, then the tangent cone to the feasible set X has the form

$$\begin{aligned} T_X(\hat{x}) = \big\{ d \in T_{X_0}(\hat{x}) : \; & \langle \nabla g_i(\hat{x}), d \rangle \leq 0, \quad i \in I^0(\hat{x}), \\ & \langle \nabla h_i(\hat{x}), d \rangle = 0, \quad i = 1, \ldots, p \big\}. \end{aligned} \tag{3.29}$$

As before, $I^0(\hat{x})$ is the set of *active* inequality constraints at $\hat{x}$:

$$I^0(\hat{x}) = \{1 \leq i \leq m : g_i(\hat{x}) = 0\}.$$

Formula (3.29) is valid, if all constraint functions $g_i(\cdot)$ and $h_i(\cdot)$ are affine and the set X_0 is a convex polyhedron. In the nonlinear case, Robinson's condition (3.11), which is equivalent to metric regularity, is sufficient for (3.29) to be true. For $X_0 = \mathbb{R}^n$ the Mangasarian–Fromovitz constraint qualification condition is sufficient (see Lemma 3.17 on page 106). Using Corollary 3.21 for smooth convex functions, we see that formula (3.29) is also valid if problem (3.28) satisfies Slater's constraint qualification condition: the functions $g_i(\cdot)$, $i = 1, \dots, m$ are convex, the functions $h_i(\cdot)$, $i = 1, \dots, p$ are affine, there exists a feasible point x_S such that $g_i(x_S) < 0$, $i = 1, \dots, m$, and $x_S \in \operatorname{int} X_0$ if $p > 0$.

If any of the sufficient conditions for (3.29) is satisfied, we say that problem (3.28) satisfies the *constraint qualification condition.*

THEOREM 3.25. *Let $\hat{x}$ be a local minimum of problem* (3.28). *Assume that at $\hat{x}$ the constraint qualification condition is satisfied. Then there exist multipliers $\hat{\lambda}_i \geq 0$, $i = 1, \dots, m$, and $\hat{\mu}_i \in \mathbb{R}$, $i = 1, \dots, p$, such that*

$$0 \in \nabla f(\hat{x}) + \sum_{i=1}^{m} \hat{\lambda}_i \nabla g_i(\hat{x}) + \sum_{i=1}^{p} \hat{\mu}_i \nabla h_i(\hat{x}) + N_{X_0}(\hat{x}), \tag{3.30}$$

and

$$\hat{\lambda}_i g_i(\hat{x}) = 0, \quad i = 1, \dots, m. \tag{3.31}$$

Proof. By the constraint qualification condition, the cone $T_X(\hat{x})$ defined by (3.29) is the tangent cone to the feasible set X at $\hat{x}$. Then, by virtue of Theorem 3.24,

$$-\nabla f(\hat{x}) \in (T_X(\hat{x}))^\circ.$$

It remains to describe the polar to the tangent cone. Assume for simplicity that $I^0(\hat{x}) = \{1, \dots, m_0\}$. Robinson's condition implies that the assumptions of Theorem 2.36 are satisfied with

$$A = \begin{bmatrix} (\nabla g_1(\hat{x}))^T \\ \vdots \\ (\nabla g_{m_0}(\hat{x}))^T \\ (\nabla h_1(\hat{x}))^T \\ \vdots \\ (\nabla h_p(\hat{x}))^T \end{bmatrix}, \qquad K_1 = T_{X_0}(\hat{x}), \qquad K_2 = \mathbb{R}^{m_0}_- \times \{0\}^p.$$

If the constraint functions $g_i(\cdot)$ and $h_i(\cdot)$ are affine and the set X_0 is a convex polyhedron, Theorem 2.36 holds true without the regularity assumption.

Since $K_2^\circ = \mathbb{R}_+^{m_0} \times \mathbb{R}^p$, we deduce that there exist multipliers $\hat{\lambda}_i \geq 0$, $i \in I^0(\hat{x})$, and $\hat{\mu}_i \in \mathbb{R}$, $i = 1, \dots, p$, such that

$$-\nabla f(\hat{x}) \in \sum_{i \in I^0(\hat{x})} \hat{\lambda}_i \nabla g_i(\hat{x}) + \sum_{i=1}^{p} \hat{\mu}_i \nabla h_i(\hat{x}) + [T_{X_0}(\hat{x})]^\circ.$$

For $i \notin I^0(\hat{x})$ we formally define $\hat{\lambda}_i = 0$. As $[T_{X_0}(\hat{x})]^\circ = N_{X_0}(\hat{x})$, the last relation becomes identical with (3.30). The requirement that the multipliers λ_i are 0 for inactive constraints can be algebraically formulated as (3.31). □

Equations (3.30)–(3.31) generalize to the case of inequality constrained problems the classical Lagrange multiplier rule for problems with equality constraints only. The multipliers $\hat{\lambda}_i$, $i = 1, \dots, m$, and $\hat{\mu}_i$, $i = 1, \dots, p$, in (3.30)–(3.31) are called *Lagrange multipliers.* Conditions (3.30)–(3.31) are called *Karush–Kuhn–Tucker (KKT) conditions.*

Lemma 3.26. *Let $\hat{x}$ be a local minimum of problem* (3.28) *and let $\hat{\Lambda}(\hat{x})$ be the set of Lagrange multipliers $\hat{\lambda} \in \mathbb{R}_+^m$ and $\hat{\mu} \in \mathbb{R}^p$ satisfying* (3.30)–(3.31).

(i) *The set $\hat{\Lambda}(\hat{x})$ is convex and closed.*

(ii) *If problem* (3.28) *satisfies Robinson's condition at $\hat{x}$, then the set $\hat{\Lambda}(\hat{x})$ is also bounded.*

Proof. The set $N_{X_0}(\hat{x})$ is convex and closed, and thus the set of $(\hat{\lambda}, \hat{\mu})$ satisfying (3.30) is convex and closed. Conditions (3.31) are linear equations in $\hat{\lambda}$, and thus assertion (i) is true.

Assume that problem (3.28) satisfies Robinson's condition (3.11) at $\hat{x}$ and that $(\hat{\lambda}, \hat{\mu}) \in \hat{\Lambda}(\hat{x})$. Then we can choose $d \in K_{X_0}(\hat{x})$ such that

$$\langle \nabla g_i(\hat{x}), d \rangle \leq -1, \quad i \in I^0(\hat{x}),$$
$$\langle \nabla h_i(\hat{x}), d \rangle = \begin{cases} -1 & \text{if } \hat{\mu}_i \geq 0, \\ 1 & \text{if } \hat{\mu}_i < 0, \end{cases} \quad i = 1, \dots, p.$$

It follows from (3.30) that there exists $z \in N_{X_0}(\hat{x})$ such that

$$0 = \nabla f(\hat{x}) + \sum_{i=1}^{m} \hat{\lambda}_i \nabla g_i(\hat{x}) + \sum_{i=1}^{p} \hat{\mu}_i \nabla h_i(\hat{x}) + z.$$

Multiplying both sides by d we find that

$$0 \leq \langle \nabla f(\hat{x}), d \rangle - \sum_{i \in I^0(\hat{x})} \hat{\lambda}_i - \sum_{i=1}^{p} |\mu_i| + \langle z, d \rangle.$$

As $\langle z, d \rangle \le 0$,

$$\sum_{i \in I^0(\hat{x})} \hat{\lambda}_i + \sum_{i=1}^{p} |\mu_i| \le \langle \nabla f(\hat{x}), d \rangle.$$

This, combined with the nonnegativity condition $\hat{\lambda} \ge 0$, proves the boundedness of the set of multipliers $\hat{\Lambda}(\hat{x})$. □

Let us introduce the *Lagrangian* associated with the constrained problem:

$$L(x, \lambda, \mu) = f(x) + \sum_{i=1}^{m} \lambda_i g_i(x) + \sum_{i=1}^{p} \mu_i h_i(x). \tag{3.32}$$

Using the Lagrangian, condition (3.30) can be written compactly as

$$-\nabla_x L(\hat{x}, \hat{\lambda}, \hat{\mu}) \in N_{X_0}(\hat{x}). \tag{3.33}$$

Thus the necessary condition of Theorem 3.24 for a minimum of the function $L(x, \hat{\lambda}, \hat{\mu})$ over $x \in X_0$ is satisfied at $\hat{x}$. We explore this observation in Theorem 3.27 below and in Chapter 4.

THEOREM 3.27. *Assume that the functions $f(\cdot)$ and $g_i(\cdot)$, $i = 1, \dots, m$, are convex and the functions $h_i(\cdot)$, $i = 1, \dots, p$, are affine. If the point $\hat{x} \in X$ and multipliers $\hat{\lambda}_i \ge 0$, $i = 1, \dots, m$, and $\hat{\mu}_i \in \mathbb{R}$, $i = 1, \dots, p$, satisfy conditions* (3.30)–(3.31), *then $\hat{x}$ is a global minimum of problem* (3.28).

Proof. It follows from the assumptions that the Lagrangian $L(x, \hat{\lambda}, \hat{\mu})$ is convex with respect to x. By (3.33) and the second part of Theorem 3.24,

$$L(\hat{x}, \hat{\lambda}, \hat{\mu}) \le L(x, \hat{\lambda}, \hat{\mu}) \quad \text{for all} \quad x \in X_0.$$

At feasible points of problem (3.28) we have

$$L(x, \hat{\lambda}, \hat{\mu}) \le f(x),$$

and at the point $\hat{x}$ condition (3.31) implies that

$$f(\hat{x}) = L(\hat{x}, \hat{\lambda}, \hat{\mu}).$$

Hence $f(\hat{x}) \le f(x)$ for all $x \in X$. □

Example 3.28. Let A be an $m \times n$ matrix of rank m and let $b \in \mathbb{R}^m$. Consider the set of points in $\mathbb{R}^n$ satisfying the equation

$$Ax = b.$$

It is a linear manifold. Our objective is to calculate the projection of a point $z \in \mathbb{R}^n$ on this manifold (see Section 2.1.2 on page 20). The projection is the solution of the problem

$$\begin{aligned} &\text{minimize } \|x - z\|^2 \\ &\text{subject to } Ax - b = 0. \end{aligned}$$

It has form (3.28) with

$$f(x) = \|x - z\|^2$$

and

$$h_i(x) = \langle a_i, x \rangle - b_i, \quad i = 1, \ldots, m,$$

where a_i denotes the ith row of A. The constraints are affine and therefore the constraint qualification condition is satisfied.

Introducing the vector of Lagrange multipliers $\mu \in \mathbb{R}^m$, we can write the necessary condition of optimality (3.30) as follows:

$$2(x - z) + A^T \mu = 0. \tag{3.34}$$

Let us multiply both sides of this equation by A. Using the fact that $Ax = b$ we obtain

$$AA^T \mu = 2(Az - b).$$

The matrix A has row rank m and thus AA^T is a nonsingular matrix. Therefore

$$\mu = 2(AA^T)^{-1}(Az - b).$$

Substitution into (3.34) renders

$$x = z - \frac{1}{2} A^T \mu = z - A^T (AA^T)^{-1}(Az - b).$$

The reader can verify directly that $Ax = b$ and that the vector $z - x$ is perpendicular to the subspace $\{d : Ad = 0\}$. Thus the necessary and sufficient conditions for the projection, as stated in Lemma 2.11 on page 21, are satisfied.

Example 3.29. Let Q be a symmetric matrix of dimension n. Consider the quadratic form

$$f(x) = \langle x, Qx \rangle$$

on the unit sphere

$$S = \{x \in \mathbb{R}^n : \|x\| = 1\}.$$

To find the minimum value of the quadratic form on the sphere, we formulate the problem

$$\begin{aligned} &\text{minimize } \langle x, Qx \rangle \\ &\text{subject to } 1 - \|x\|^2 = 0. \end{aligned}$$

The gradient of the constraint function is equal to $-2x$. It is never zero for $x \in S$ and therefore Robinson's condition is satisfied. The optimal solution of the problem must satisfy the necessary conditions (3.30)–(3.31).

Introducing the Lagrange multiplier μ associated with the equality constraint, we can write (3.30) as follows:

$$2Qx - 2\mu x = 0,$$

or simply as

$$Qx = \mu x.$$

It follows that x is an eigenvector of Q having length 1, and μ is the corresponding eigenvalue. In order to find the best eigenvector, we calculate the value of the objective function:

$$\langle x, Qx \rangle = \langle x, \mu x \rangle = \mu.$$

Therefore the optimal solution is an eigenvector of length 1 corresponding to the smallest eigenvalue of Q. This is illustrated in Figure 3.5.

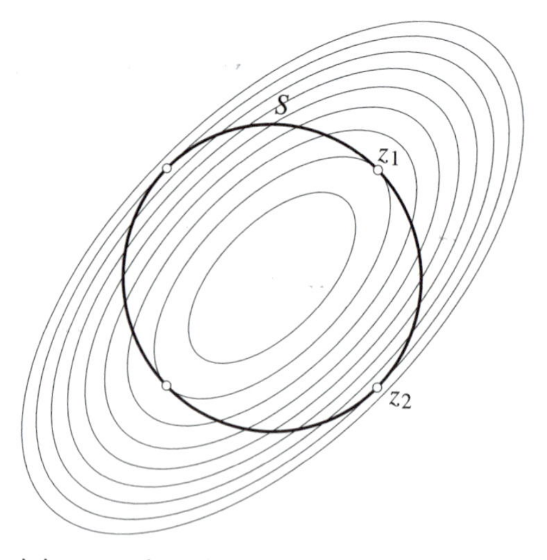

Figure 3.5. The minimum and maximum of a quadratic form on a sphere are attained at eigenvectors $\pm z_1$ and $\pm z_2$.

The largest value of the quadratic form $\langle x, Qx \rangle$ on the sphere can be calculated by minimizing $\langle x, (-Q)x \rangle$ in S. The solution is an eigenvector of Q corresponding to the largest eigenvalue, and the maximum value of the form is equal to the largest eigenvalue.

Example 3.30. Two points, a and b, are connected by a chain consisting of n mass points joined by extensible links. Each point has mass m, and the spring constant of the links equals K. We considered this problem in Example 3.4 on page 91, where we calculated the location of the mass points at equilibrium.

Let us assume now that the point b is located on a horizontal board, which constitutes an obstacle for the chain, and the point a is above the board ($a_2 > b_2$), as illustrated in Figure 3.6. No friction exists between the board and the mass points.

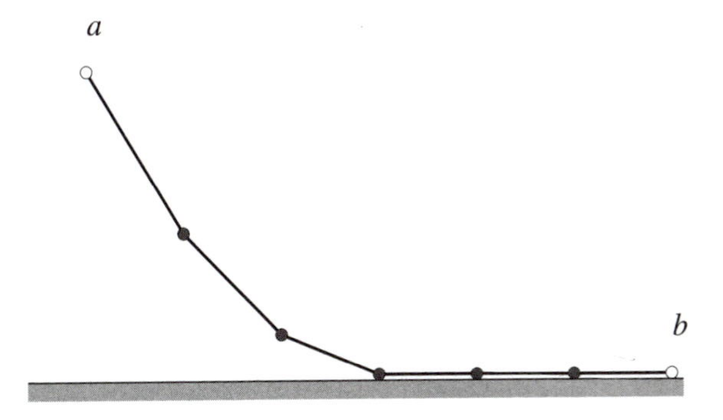

Figure 3.6. The chain with elastic links and a horizontal obstacle.

All mass points have to remain on or above the board, that is, their locations have to satisfy the inequalities

$$x_2^i \geq b_2, \quad i = 1, \dots, n.$$

Our goal is to calculate the location of the mass points at equilibrium.

In Example 3.4 we derived the expression for the potential energy of the mass points:

$$f(x^1, \dots, x^n) = \frac{K}{2} \sum_{i=1}^{n} \|x^{i+1} - x^i\|^2 + mg \sum_{i=1}^{n} \langle e, x^i \rangle,$$

with the convention that $x^0 = a$ and $x^{n+1} = b$, and with $e = (0, 1)$. The problem of minimizing the potential energy in the presence of an obstacle can be formulated as follows:

$$\begin{aligned}
&\text{minimize } \frac{K}{2} \sum_{i=1}^{n} \|x^{i+1} - x^i\|^2 + mg \sum_{i=1}^{n} \langle e, x^i \rangle \\
&\text{subject to } b_2 - \langle e, x^i \rangle \leq 0, \quad i = 1, \dots, n.
\end{aligned}$$

The constraint functions

$$g_i(x) = b_2 - \langle e, x^i \rangle, \quad i = 1, \dots, n,$$

are affine, and therefore the constraint qualification condition is satisfied. Let us introduce Lagrange multipliers $\lambda_i \geq 0, i = 1, \dots, n$. Condition (3.30) has the form

$$\nabla f(x) + \sum_{i=1}^{n} \lambda_i \nabla g_i(x) = 0.$$

Observe that each constraint function $g_i(\cdot)$ depends on x^i only, and thus the last equation can be written for each mass point separately:

$$\nabla_{x^i} f(x) + \lambda_i \nabla_{x^i} g_i(x) = 0, \quad i = 1, \dots, n.$$

Substituting the formulas for $f(\cdot)$ and $g_i(\cdot)$ and differentiating we obtain

$$K(2x^i - x^{i-1} - x^{i+1}) + (mg - \lambda_i)e = 0, \quad i = 1, \dots, n. \tag{3.35}$$

The complementarity conditions (3.31) have the form

$$\lambda_i(b_2 - \langle e, x^i \rangle) = 0, \quad i = 1, \dots, n. \tag{3.36}$$

In general, solving systems like the last one requires considering 2^n cases, for all possible sets I of active constraints. For each set I we require that $\lambda_i = 0$ for $i \notin I$ and that the inequality constraints are satisfied as equations for $i \in I$. In this example we can drastically reduce the number of cases to be considered, by employing our understanding of the physical nature of the problem.

Equation (3.35) for the first coordinate, x_1^i, $i = 1, \dots, n$, reads

$$x_1^{i+1} - x_1^i = x_1^i - x_1^{i-1}, \quad i = 1, \dots, n.$$

That is, the mass points are horizontally equally spaced.

The vertical locations are more difficult to calculate. Common sense suggests that if a mass point j is on the board, then all mass points $j+1, \dots, n$ should remain on the board as well. Formally, we can derive this as follows. Equation (3.35) for the second coordinate has the form:

$$K\left(x_2^{i+1} - x_2^i\right) = K\left(x_2^i - x_2^{i-1}\right) + mg - \lambda_i, \quad i = 1, \dots, n. \tag{3.37}$$

Suppose $x_2^j = b_2$, but $x_2^{j+1} > b_2$. From (3.36) we obtain $\lambda_{j+1} = 0$. The last equation for $i = j + 1$ implies:

$$x_2^{j+2} - x_2^{j+1} = x_2^{j+1} - x_2^j + \frac{mg}{K} > 0.$$

This propagates to the following points and we get $x_2^{i+1} > x_2^i$ for all $i = j, \dots, n$. This contradicts the equation $x_2^{n+1} = b_2$.

Consequently, all points $i = j, \dots, n$ are located on the straight line between x^j and b and are equally spaced.

Let us denote by j^* the first mass point that lies on the board. It follows that $x_2^i > b_2$ for $i = 1, \dots, j^* - 1$, and $x_2^i = b_2$ for $i = j^*, \dots, n$. Equation (3.37) for $i = j^*$ implies that

$$x_2^{j^*} - x_2^{j^*-1} + \frac{mg}{K} = \frac{\lambda_{j^*}}{K} \geq 0. \tag{3.38}$$

Since $\lambda_i = 0$ for $i = 1, \dots, j^* - 1$, equation (3.37) for $i = 1, \dots, j^* - 1$ yields

$$x_2^{i+1} - x_2^i = x_2^1 - x_2^0 + \frac{img}{K}. \tag{3.39}$$

Combining the last relation for $i = j^* - 1$ with (3.38), after simple manipulations we conclude that

$$x_2^1 - x_2^0 \geq -\frac{j^* mg}{K}. \tag{3.40}$$

Summing equations (3.39) for $i = 1, \ldots, j^* - 1$ we obtain

$$b_2 - a_2 = (j^* - 1)\big(x_2^1 - x_2^0\big) + \frac{(j^* - 1) j^* mg}{2K}.$$

Therefore, j^* is the smallest index for which the inequality

$$(j^* - 1)\big(x_2^1 - x_2^0\big) + \frac{j^*(j^* - 1)mg}{2K} \leq b_2 - a_2$$

is satisfied. Dividing by $j^* - 1$ we obtain

$$x_2^1 - x_2^0 \leq \frac{b_2 - a_2}{j^* - 1} - \frac{j^* mg}{2K}.$$

Because of (3.40),

$$\frac{a_2 - b_2}{j^* - 1} \leq \frac{j^* mg}{2K}.$$

The left hand side of the above inequality is a decreasing function of j^*, while the right hand side in an increasing function. The critical index j^* is the smallest integer in $\{1, \ldots, n\}$, for which this inequality is satisfied. If such an integer does not exist, all mass points are above the board. If j^* can be found, the mass points $j^*, \ldots, n$ lie on the board and are equally spaced, and the mass points $1, \ldots, j^* - 1$ can be found from the unconstrained problem with ends at a and x^{j^*}, as in Example 3.4.

The solution, for the same data as in Example 3.4, is displayed in Figure 3.6. Let us observe that the active constraints in the constrained problem cannot be determined by verifying which constraints are violated by the unconstrained solution. The second mass point is below the board level in the unconstrained solution, but the corresponding constraint is not active at the optimal solution of the constrained problem.

If the point b is located above the obstacle, we also have to determine the *last* mass point lying on the board. We leave to the reader the analysis of this case.

If the obstacle is not horizontal, the solution method is similar, but more complicated. The main property, though, that there will be two key mass points – the first and the last one on the obstacle – remains valid.

Example 3.31. Consider the unconstrained optimization problem

$$\operatorname*{minimize}_{x \in \mathbb{R}^n} f(x), \tag{3.41}$$

with the function $f : \mathbb{R}^n \to \mathbb{R}$ defined by

$$f(x) = \max_{i=1,\ldots,N} f_i(x).$$

We assume that each $f_i : \mathbb{R}^n \to \mathbb{R}$, $i = 1, \dots, N$, is continuously differentiable. The necessary condition of optimality for this problem can be developed from optimality conditions for constrained optimization problems. Let us consider the problem

$$\begin{aligned} &\text{minimize } v \\ &\text{subject to } f_i(x) - v \le 0, \quad i \in 1, \dots, N, \end{aligned} \tag{3.42}$$

with decision variables $x \in \mathbb{R}^n$ and $v \in \mathbb{R}$. For every $x \in \mathbb{R}^n$ the minimum value of v is equal to $f(x)$, and therefore this problem is equivalent to (3.41).

Any feasible point (x, v) of problem (3.42) satisfies the Mangasarian–Fromovitz constraint qualification condition with the direction $d = (0, 1) \in \mathbb{R}^n \times \mathbb{R}$.

Denote by $(\hat{x}, \hat{v})$ the optimal solution of (3.42). Optimality conditions (3.30)–(3.31) have the form

$$\begin{bmatrix} 0 \\ 1 \end{bmatrix} + \sum_{i=1}^{N} \lambda_i \begin{bmatrix} \nabla f_i(\hat{x}) \\ -1 \end{bmatrix} = 0, \tag{3.43}$$

$$\lambda_i (f_i(\hat{x}) - \hat{v}) = 0, \quad \lambda_i \ge 0, \quad i = 1, \dots, N. \tag{3.44}$$

At the optimal solution we must have $\hat{v} = f(\hat{x})$ and therefore condition (3.44) means that $\lambda_i = 0$, if $f_i(\hat{x}) < f(\hat{x})$.

The first n rows in equation (3.43) have the form

$$0 = \sum_{i=1}^{N} \lambda_i \nabla f_i(\hat{x}).$$

The last row reads

$$\sum_{i=1}^{N} \lambda_i = 1.$$

Therefore, zero is a convex combination of the gradients $\nabla f_i(\hat{x})$ of these functions $f_i(\cdot)$, for which $f_i(\hat{x}) = f(\hat{x})$ (active pieces).

Example 3.32. We can now apply the results of Example 3.31 to a nonlinear version of the approximation problem of Example 3.3. A model represents the output variable, $y \in \mathbb{R}$, as a function

$$y = \varphi(u, x)$$

of input variables $u_1, \dots, u_m$. Here φ is a function from $\mathbb{R}^m \times \mathbb{R}^n$ to $\mathbb{R}$. The quantities $x_1, \dots, x_n$ are unknown model coefficients.

We have N observations of values of input variables and the corresponding values of output variables: (u^j, y^j), $j = 1, \dots, N$. One way to determine the values of the coefficients $x_1, \dots, x_n$ is to minimize the largest deviation between the values observed and those predicted by the model:

$$f(x) = \max_{1 \le j \le N} \left| y^j - \varphi(u^j, x) \right|.$$

Such uniform approximation (Chebyshev approximation) has already been discussed in Example 3.7 on page 94 (for a model linear in x and for a continuum of "observations"). Assume that for each u^j the function $\varphi(\cdot, u^j)$ is smooth. Then we can represent $f(x)$ as the maximum of $2N$ smooth functions

$$
\begin{aligned}
f_j(x) &= y^j - \varphi(u^j, x), \quad j = 1, \dots, N, \\
f_j(x) &= -f_{j-N}(x), \quad j = N+1, \dots, 2N.
\end{aligned}
$$

We focus on the most interesting case, when the smallest distance is positive (the perfect match is impossible). Then $f(x) > 0$ for all x. It is apparent that if $f_j(x) = f(x)$ then $-f_j(x) \neq f(x)$. Let us define the set

$$
\hat{J}(\hat{x}) = \{1 \le j \le N : \left|y^j - \varphi(u^j, x)\right| = f(\hat{x})\}.
$$

Optimality conditions of Example 3.31 imply that there exist a subset A of no more than $n + 1$ indices from $\hat{J}(\hat{x})$ and nonnegative coefficients α_j, $j \in A$, totaling one and such that

$$
\sum_{j \in A} \alpha_j \operatorname{sgn}\bigl(y^j - \varphi(u^j, x)\bigr) \nabla_x \varphi(u^j, \hat{x}) = 0.
$$

This condition extends to the nonlinear case the conditions obtained in Example 3.7. Of course, here we are comparing the model and the reference outcomes y^j at finitely many points, while in Example 3.7 we considered the maximum over an interval. When the model is nonlinear in parameters x, the maximum function is nonconvex and nonsmooth, and our techniques only allow us to analyze the case of finitely many points.

In Figure 1.4 on page 14 we illustrate the optimal approximation of 21 data points with a model having 5 parameters. As predicted, 6 data points determine the error of the approximation.

3.5 OPTIMALITY CONDITIONS FOR CONVEX PROBLEMS

If the objective function or the constraint functions in a nonlinear optimization problem are convex, but not necessarily differentiable, we use their subgradients in the necessary conditions of optimality. Although the results of this section can be derived as special cases of the results of the following one, we analyze the convex case separately for educational reasons, and because of wide applications.

At first we establish the counterpart of the fundamental relation (3.24) from page 113 and of Theorem 3.5 from page 92. Consider the constrained optimization problem

$$
\underset{x \in X}{\text{minimize}}\ f(x), \tag{3.45}
$$

with a convex function $f : \mathbb{R}^n \to \overline{\mathbb{R}}$ and a convex set $X \subset \mathbb{R}^n$.

THEOREM 3.33. *Assume that $\hat{x}$ is a local minimum of problem* (3.45) *and that there exists a point $x_0 \in X$ such that f is continuous at x_0. Then there exists $s \in \partial f(\hat{x})$ such that*

$$-s \in N_X(\hat{x}). \tag{3.46}$$

Conversely, if a point $\hat{x} \in X$ satisfies relation (3.46) *for some $s \in \partial f(\hat{x})$, then $\hat{x}$ is a global minimum of problem* (3.45).

Proof. Consider the indicator function

$$\delta_X(x) = \begin{cases} 0 & \text{if } x \in X, \\ +\infty & \text{otherwise.} \end{cases}$$

Problem (3.45) is equivalent to the unconstrained optimization problem

$$\underset{x \in \mathbb{R}^n}{\text{minimize}}\ f(x) + \delta_X(x).$$

If $f(\cdot)$ is continuous at $x_0 \in X$ then Theorem 2.85 on page 68 renders

$$\partial\big[f(\hat{x}) + \delta_X(\hat{x})\big] = \partial f(\hat{x}) + \partial\delta_X(\hat{x}).$$

By Theorem 3.5 on page 92, $\hat{x}$ is a minimum if and only if

$$0 \in \partial\big[f(\hat{x}) + \delta_X(\hat{x})\big].$$

Combining the last two relations we see that there must exist a subgradient $s \in \partial f(\hat{x})$ such that

$$-s \in \partial\delta_X(\hat{x}).$$

Using Example 2.81 on page 67 we note that $\partial\delta_X(\hat{x}) = N_X(\hat{x})$, and formula (3.46) holds true.

To prove the opposite implication, suppose (3.46) is satisfied for some $s \in \partial f(\hat{x})$. Then, for every $y \in X$,

$$f(y) \geq f(\hat{x}) + \langle s, y - \hat{x} \rangle.$$

Since $y - \hat{x} \in K_X(\hat{x})$ and $-s \in N_X(\hat{x})$,

$$\langle s, y - \hat{x} \rangle \geq 0.$$

Therefore $f(y) \geq f(\hat{x})$ for all $y \in X$. □

This general fact helps us to obtain optimality conditions for the following problem:

$$
\begin{aligned}
\text{minimize}\ \ & f(x) \\
\text{subject to}\ \ & g_i(x) \le 0, \quad i = 1, \dots, m, \\
& h_i(x) = 0, \quad i = 1, \dots, p, \\
& x \in X_0.
\end{aligned} \tag{3.47}
$$

We assume that the functions $f : \mathbb{R}^n \to \overline{\mathbb{R}}$, and $g_i : \mathbb{R}^n \to \mathbb{R}, i = 1, \dots, m$, are convex, and the functions $h_i : \mathbb{R}^n \to \mathbb{R}$, $i = 1, \dots, p$, are affine. The set X_0 is convex and closed.

Recall from page 109 that Slater's constraint qualification condition for convex constraints takes on the form: There exists a point $x_{\rm s} \in X_0$ such that $g_i(x_{\rm s}) < 0$, $i = 1, \dots, m$, $h_i(x_{\rm s}) = 0$, $i = 1, \dots, p$, and $x_{\rm s} \in \operatorname{int} X_0$, if $p > 0$.

In Theorem 3.20 on page 109 we established that under this condition

$$
N_X(\hat{x}) = [K_{X_0}(\hat{x})]^\circ + \sum_{i \in I^0(\hat{x})} \operatorname{cone}[\partial g_i(\hat{x})] + \sum_{i=1}^{p} \operatorname{lin}[\nabla h_i(\hat{x})]. \tag{3.48}
$$

Here $I^0(\hat{x})$ is the set of $i \in \{1, \dots, m\}$ for which $g_i(\hat{x}) = 0$. This can be combined with Theorem 3.33 to obtain the following optimality conditions for problem (3.47).

THEOREM 3.34. *Assume that $\hat{x}$ is the minimum of problem* (3.47), *the function $f(\cdot)$ is continuous at some feasible point x_0, and Slater's condition is satisfied. Then there exist $\hat{\lambda} \in \mathbb{R}^m_+$ and $\hat{\mu} \in \mathbb{R}^p$ such that*

$$
0 \in \partial f(\hat{x}) + \sum_{i=1}^{m} \hat{\lambda}_i \partial g_i(\hat{x}) + \sum_{i=1}^{p} \hat{\mu}_i \nabla h_i(\hat{x}) + N_{X_0}(\hat{x}) \tag{3.49}
$$

and

$$
\hat{\lambda}_i g_i(\hat{x}) = 0, \quad i = 1, \dots, m. \tag{3.50}
$$

Conversely, if for some feasible point $\hat{x}$ of (3.47) *and some $\hat{\lambda} \in \mathbb{R}^m_+$ and $\hat{\mu} \in \mathbb{R}^p$ conditions* (3.49)–(3.50) *are satisfied, then $\hat{x}$ is the global minimum of problem* (3.47).

Proof. Using Theorem 3.33, the formula (3.48) for the normal cone, and the equation $[K_{X_0}(\hat{x})]^\circ = N_{X_0}(\hat{x})$, we see that there exists $s \in \partial f(\hat{x})$ such that

$$
-s \in N_{X_0}(\hat{x}) + \sum_{i \in I^0(\hat{x})} \operatorname{cone}[\partial g_i(\hat{x})] + \sum_{i=1}^{p} \operatorname{lin}[\nabla h_i(\hat{x})].
$$

This means that there exist $\hat{\lambda}_i \geq 0$, $i \in I^0(\hat{x})$, and $\hat{\mu} \in \mathbb{R}^p$ such that

$$-s \in N_{X_0}(\hat{x}) + \sum_{i \in I^0(\hat{x})} \hat{\lambda}_i \partial g_i(\hat{x}) + \sum_{i=1}^{p} \hat{\mu}_i \nabla h_i(\hat{x}).$$

Setting $\hat{\lambda}_i = 0$ for $i \notin I^0(\hat{x})$ we obtain (3.49)–(3.50).

To prove the converse, assume (3.49)–(3.50). This implies that there exist subgradients $s \in \partial f(\hat{x})$, $s_i \in \partial g_i(\hat{x})$, $i \in I^0(\hat{x})$, and some $v \in N_{X_0}(\hat{x})$ such that

$$0 = s + \sum_{i \in I^0(\hat{x})} \hat{\lambda}_i s_i + \sum_{i=1}^{p} \hat{\mu}_i \nabla h_i(\hat{x}) + v.$$

Suppose y is a feasible point of problem (3.47). Evaluating the scalar product of $y - \hat{x}$ with both sides of the last equation we obtain:

$$0 = \langle s, y - \hat{x} \rangle + \sum_{i \in I^0(\hat{x})} \hat{\lambda}_i \langle s_i, y - \hat{x} \rangle + \sum_{i=1}^{p} \hat{\mu}_i \langle \nabla h_i(\hat{x}), y - \hat{x} \rangle + \langle v, y - \hat{x} \rangle.$$

Since y is feasible, $y - \hat{x} \in K_{X_0}(\hat{x})$. Thus $\langle v, y - \hat{x} \rangle \leq 0$, because $v \in N_{X_0}(\hat{x})$. Moreover, $g_i(y) \leq 0$ and therefore for every $i \in I^0(\hat{x})$ we have

$$\langle s_i, y - \hat{x} \rangle \leq g_i(y) - g_i(\hat{x}) \leq 0.$$

Finally, the feasibility of y implies that

$$\langle \nabla h_i(\hat{x}), y - \hat{x} \rangle = 0, \quad i = 1, \ldots, p.$$

The last three displayed equations yield the inequality

$$\langle s, y - \hat{x} \rangle \geq 0.$$

This means that $f(y) \geq f(\hat{x})$, as postulated. □

Again, using the Lagrangian,

$$L(x, \lambda, \mu) = f(x) + \sum_{i=1}^{m} \lambda_i g_i(x) + \sum_{i=1}^{p} \mu_i h_i(x),$$

we can write the necessary condition of optimality (3.49) as

$$0 \in \partial_x L(\hat{x}, \hat{\lambda}, \hat{\mu}) + N_{X_0}(\hat{x}),$$

which is nothing else but the necessary and sufficient condition of optimality for the problem of minimizing $L(x, \hat{\lambda}, \hat{\mu})$ over $x \in X_0$.

Example 3.35. Given a matrix A of dimension $m \times n$ and vectors $c \in \mathbb{R}^n$ and $b \in \mathbb{R}^m$, the *linear programming problem* is formulated as follows:

$$\begin{aligned} &\text{minimize} \ \langle c, x \rangle \\ &\text{subject to} \ Ax = b, \\ &\qquad\qquad\ \ x \geq 0. \end{aligned}$$

We see that the problem has form (3.47) with

$$\begin{aligned} f(x) &= \langle c, x \rangle \\ h_i(x) &= b_i - \langle a_i, x \rangle, \quad i = 1, \dots, m, \\ X_0 &= \mathbb{R}^n_+. \end{aligned}$$

The constraints are affine, and therefore formula (3.48) for the normal cone to the feasible set remains valid. We can use Theorem 3.34.

Condition (3.49) has (at the optimal solution $\hat{x}$) the form

$$c - \sum_{i=1}^{m} \mu_i a_i \in -[T_{\mathbb{R}^n_+}(\hat{x})]^\circ.$$

We know from Example 2.21 on page 27 that

$$T_{\mathbb{R}^n_+}(\hat{x}) = \mathbb{R}^n_+ + \{\tau \hat{x} : \tau \in \mathbb{R}\}.$$

Thus, by virtue of Example 2.24 on page 28,

$$[T_{\mathbb{R}^n_+}(\hat{x})]^\circ = N_{\mathbb{R}^n_+}(\hat{x}) = \mathbb{R}^n_- \cap \{d \in \mathbb{R}^n : \langle d, \hat{x} \rangle = 0\}.$$

Setting $u = -d$, we conclude that there exist $\mu \in \mathbb{R}^m$ and $u \in \mathbb{R}^n_+$ such that

$$\begin{aligned} c - A^T \mu &= u, \\ \langle u, \hat{x} \rangle &= 0. \end{aligned}$$

These relations can also be derived from direct linear algebraic considerations, but it is instructive to see them in a more general framework.

Example 3.36. Given symmetric matrices $A_1, \dots, A_m$ and C in $\mathbb{S}^n$ the *semidefinite programming problem* is formulated as follows:

$$\begin{aligned} &\text{minimize} \ \operatorname{tr}(CX) \\ &\text{subject to} \ \operatorname{tr}(A_i X) = b_i, \quad i = 1, \dots, m, \\ &\qquad\qquad\ \ X \in \mathbb{S}^n_+. \end{aligned}$$

The last condition means that the decision variable X in this problem is a symmetric positive semidefinite matrix.

We see that the problem has form (3.47) with

$$\begin{aligned} f(X) &= \operatorname{tr}(CX) \\ h_i(X) &= b_i - \operatorname{tr}(A_i X), \quad i = 1, \dots, m, \\ X_0 &= \mathbb{S}^n_+. \end{aligned}$$

The functions $f(\cdot)$ and $h_i(\cdot)$ are affine. Assume Slater's constraint qualification condition (see page 109). In our terms it means that there exists a positive definite matrix X_{S} satisfying the equality constraints. Under this condition we can use Theorem 3.34.

Condition (3.49) has (at the optimal solution $\hat{X}$) the form

$$C - \sum_{i=1}^{m} \mu_i A_i \in -N_{\mathbb{S}^n_+}(\hat{X}).$$

We know from Example 3.22 on page 110 that

$$N_{\mathbb{S}^n_+}(\hat{X}) = \mathbb{S}^n_- \cap \{D \in \mathbb{S}^n : \operatorname{tr}(D\hat{X}) = 0\}.$$

Setting $U = -D$, we conclude that there exists a vector $\mu \in \mathbb{R}^m$ and a quadratic matrix U of dimension n such that

$$\begin{gathered} C - \sum_{i=1}^{m} \mu_i A_i = U, \\ U \in \mathbb{S}^n_+, \\ \operatorname{tr}(U\hat{X}) = 0. \end{gathered} \tag{3.51}$$

These conditions are formally similar to the optimality conditions in linear programming, as presented in Example 3.35. However, their derivation requires application of the formulas for the tangent cone to the positive semidefinite cone $\mathbb{S}^n_+$ and for its polar cone, and the use of the optimality conditions of Theorem 3.34. The optimality conditions cannot be obtained by imitating the linear programming case. In particular, the constraint qualification condition (the existence of a feasible positive definite X_{S}) is crucial in this case, while in linear programming the constraint qualification condition is always satisfied.

Example 3.37. There are n investment opportunities, with random return rates $R_1, \dots, R_n$ in the next year. We assume that the return rates have a joint normal probability distribution. We have a practically unlimited initial capital and our aim is to invest some of it in such a way that the expected value of our investment after a year is maximized, under the condition that the chance of losing more than some fixed amount $b > 0$ is smaller than α, where $\alpha \in (0, 1)$. Such a requirement is called the *Value at Risk* constraint.

Let $x_1, \dots, x_n$ be the amounts invested in the n opportunities. The net increase of the value of our investment after a year is random and equals $G(x, R) = \sum_{i=1}^{n} R_i x_i$.

Its expected value is linear in x:

$$\mathbb{E}\big[G(x, R)\big] = \sum_{i=1}^{n} \bar{r}_i x_i,$$

with $\bar{r}_i = \mathbb{E}\big[R_i\big]$. Our problem takes on the form:

$$\begin{aligned} &\text{maximize} \ \sum_{i=1}^{n} \bar{r}_i x_i \\ &\text{subject to} \ \mathbb{P}\Big\{\sum_{i=1}^{n} R_i x_i \geq -b\Big\} \geq 1 - \alpha, \\ &\qquad\qquad x \geq 0. \end{aligned} \tag{3.52}$$

We do not impose any constraint of the form $x_1 + \ldots + x_n = W_0$, where W_0 is the total invested amount. We also assume that $\bar{r} \neq 0$, otherwise no improvement over $x = 0$ is possible.

Denote by C the covariance matrix of the joint distribution of the return rates. The distribution of the total profit (or loss) is normal, with the expected value $\langle \bar{r}, x\rangle$, and the variance

$$\mathbb{V}\big[G(x, R)\big] = \mathbb{E}\Big(\sum_{i=1}^{n}(R_i - \bar{r}_i)x_i\Big)^2 = \langle x, Cx\rangle.$$

We assume that C is positive definite. If $x \neq 0$ then the random variable

$$\frac{G(x, R) - \langle \bar{r}, x\rangle}{\sqrt{\langle x, Cx\rangle}}$$

has the normal distribution with mean zero and variance one. Our probability constraint is therefore equivalent to the inequality

$$\frac{b + \langle \bar{r}, x\rangle}{\sqrt{\langle x, Cx\rangle}} \geq z_\alpha,$$

where z_α is the $(1 - \alpha)$-quantile of the standard normal random variable. If the risk level $\alpha \leq 1/2$ then $z_\alpha \geq 0$. Therefore the last constraint (after multiplying both sides by $\sqrt{\langle x, Cx\rangle}$) is equivalent to a constraint involving a convex function:

$$z_\alpha \sqrt{\langle x, Cx\rangle} - \langle \bar{r}, x\rangle \leq b.$$

The convexity follows from the positive definiteness of C (see Example 2.84 on page 68). If $x = 0$ the last constraint is satisfied as well, because $b > 0$. Consequently, we obtain the following convex optimization problem equivalent to problem (3.52):

$$\begin{aligned} &\text{minimize} \ -\langle \bar{r}, x\rangle \\ &\text{subject to} \ z_\alpha \sqrt{\langle x, Cx\rangle} - \langle \bar{r}, x\rangle - b \leq 0, \\ &\qquad\qquad x \geq 0. \end{aligned} \tag{3.53}$$

It satisfies Slater's condition with $x_S = 0$.

To formulate the necessary (and sufficient) conditions of optimality, let $\lambda \geq 0$ be the Lagrange multiplier associated with the constraint. Condition (3.49) takes on the form

$$-(1+\lambda)\bar{r} + \frac{\lambda z_\alpha C x}{\sqrt{\langle x, Cx\rangle}} \in -N_{\mathbb{R}^n_+}(\hat{x}).$$

Substituting the explicit form of the normal cone (as in Example 3.35) we conclude that

$$\begin{aligned} -(1+\lambda)\bar{r} + \frac{\lambda z_\alpha C x}{\sqrt{\langle x, Cx\rangle}} &\geq 0, \\ \left\langle x, (1+\lambda)\bar{r} - \frac{\lambda z_\alpha C x}{\sqrt{\langle x, Cx\rangle}} \right\rangle &= 0. \end{aligned} \tag{3.54}$$

If we ignore the nonnegativity constraint on x we can solve this system analytically. The optimality condition becomes

$$(1+\lambda)\bar{r} - \frac{\lambda z_\alpha C x}{\sqrt{\langle x, Cx\rangle}} = 0.$$

Since $\bar{r}$ is nonzero, $\lambda > 0$ and thus the inequality constraint must be satisfied as equality. From the last equation we deduce that the vectors Cx and $\bar{r}$ are colinear: there exists a scalar t such that $Cx = t\bar{r}$. Substitution to the constraint (which is active) yields

$$t = b/\varrho(z_\alpha - \varrho), \quad \lambda = (z_\alpha/\varrho - 1)^{-1}, \quad \text{with} \quad \varrho = \sqrt{\langle \bar{r}, C^{-1}\bar{r}\rangle}.$$

Note that C^{-1} is positive definite and hence $\langle \bar{r}, C^{-1}\bar{r}\rangle$ is positive. If $\varrho = z_\alpha$, no solution to the system exists. If $\varrho > z_\alpha$, the solution has $\lambda < 0$, and thus no optimal solution of the problem exists. In both cases the problem is unbounded.

If $\varrho < z_\alpha$, the vector

$$\hat{x} = \frac{b}{\varrho(z_\alpha - \varrho)} C^{-1}\bar{r}$$

is the solution to the problem without sign restrictions on x. If, in addition, $C^{-1}\bar{r} \geq 0$, then the vector $\hat{x}$ solves our original problem.

If $C^{-1}\bar{r} \ngeq 0$ the sign restrictions on x cannot be ignored. We have to find a subset I of decision variables, such that the problem restricted to this subset (with other variables set to zero) can be solved as above and its solution is nonnegative. Then we need to have

$$C_{(I)}^{-1}\bar{r}_{(I)} \geq 0,$$

with $C_{(I)}$ denoting the quadratic submatrix of C corresponding to I, and $\bar{r}_{(I)}$ is the subvector of $\bar{r}$ with components from I. Moreover, for the remaining variables the first relation of (3.54) must hold, and this requires tedious testing of many subsets I.

Numerical methods of convex optimization, which we discuss in Chapter 6, avoid examining all possible subsets of assets.

3.6 OPTIMALITY CONDITIONS FOR SMOOTH–CONVEX PROBLEMS

Our techniques allow us to develop optimality conditions for problems involving both nondifferentiable and nonconvex smooth functions. We consider the following problem formulation:

$$\begin{aligned} \text{minimize} \ & f(x) \\ \text{subject to} \ & g_i(x) \le 0, \quad i = 1, \dots, m, \\ & h(x) \in Y_0, \\ & x \in X_0. \end{aligned} \tag{3.55}$$

We assume that the functions $g_i : \mathbb{R}^n \to \mathbb{R}$, $i = 1, \dots, m$, are convex, but possibly nonsmooth, and the functions $f : \mathbb{R}^n \to \mathbb{R}$ and $h : \mathbb{R}^n \to \mathbb{R}^p$ are continuously differentiable. The sets $X_0 \subset \mathbb{R}^n$ and $Y_0 \subset \mathbb{R}^p$ are convex and closed.

The assumption that $f(\cdot)$ is smooth is not restrictive, for if $f(\cdot)$ is convex and nonsmooth, we can introduce an additional variable $v \in \mathbb{R}$, an additional constraint $f(x) \le v$, and transform (3.55) to a problem of minimizing v subject to the additional constraint as well as other constraints originally appearing in the problem. Example 3.39 at the end of this section illustrates this transformation.

If the functions $g_i(\cdot)$ are smooth, but nonconvex, we can define

$$h^{\text{new}}(x) = \begin{bmatrix} g_1(x) \\ \dots \\ g_m(x) \\ h(x) \end{bmatrix}, \qquad Y_0^{\text{new}} = \mathbb{R}^m_- \times Y_0,$$

and formally remove the inequality constraints from (3.55). Thus, from the theoretical point of view, it is sufficient to analyze the case when $f(\cdot)$ is smooth and the functions $g_i(\cdot)$ are nonsmooth, but convex.

Consider the set defined by the convex constraints,

$$C = \{x \in X_0 : g_i(x) \le 0, \ i = 1, \dots, m\},$$

and let $I^0(\hat{x})$ be the set of active constraints at the optimal point $\hat{x}$:

$$I^0(\hat{x}) = \{i : g_i(\hat{x}) = 0\}.$$

This renders the formulation

$$\begin{aligned} \text{minimize} \ & f(x) \\ \text{subject to} \ & h(x) \in Y_0, \\ & x \in C. \end{aligned}$$

Recall that Robinson's condition (3.6) for this problem has the form

$$h'(\hat{x})K_C(\hat{x}) - K_{Y_0}(h(\hat{x})) = \mathbb{R}^p. \tag{3.56}$$

The set $K_C(\hat{x})$, in general, is difficult to describe exactly, but under Slater's condition we can find its nonempty subset. Indeed, if there exists $x_S \in X_0$ such that $g_i(x_S) < 0$, $i = 1, \dots, m$, then the set $K_C^0(\hat{x})$ of directions $d \in K_C(\hat{x})$ such that $g_i'(\hat{x}; d) < 0$, $i \in I^0(\hat{x})$, is nonempty. It obviously contains the direction $d = x_S - \hat{x}$. We can, therefore, formulate a more explicit sufficient condition for Robinson's condition by replacing $K_C(\hat{x})$ with $K_C^0(\hat{x})$. In the theorem below it does not constitute any additional restriction.

THEOREM 3.38. *Assume that $\hat{x} \in X_0$ is a local minimum of problem* (3.55). *Assume that the functions $g_i(\cdot)$, $i \in I^0(\hat{x})$, are subdifferentiable at $\hat{x}$ and that there exists $x_S \in X_0$ such that $g_i(x_S) < 0$, $i = 1, \dots, m$. Furthermore, let Robinson's condition be satisfied at $\hat{x}$. Then there exist $\hat{\lambda} \in \mathbb{R}^m_+$ and $\hat{\mu} \in [T_{Y_0}(h(\hat{x}))]^\circ$ such that*

$$0 \in \nabla f(\hat{x}) + \sum_{i=1}^{m} \hat{\lambda}_i \partial g_i(\hat{x}) + [h'(\hat{x})]^T \hat{\mu} + N_{X_0}(\hat{x}) \tag{3.57}$$

and

$$\hat{\lambda}_i g_i(\hat{x}) = 0, \quad i = 1, \dots, m. \tag{3.58}$$

Furthermore, the set $\hat{\Lambda}(\hat{x})$ of multipliers $(\hat{\lambda}, \hat{\mu})$ satisfying the above conditions is convex and compact.

Proof. Let X be the feasible set of problem (3.55). By Theorem 3.24 on page 113,

$$-\nabla f(\hat{x}) \in [T_X(\hat{x})]^\circ.$$

By Theorem 3.23 on page 112, the polar of the tangent cone to X has the form

$$[T_X(\hat{x})]^\circ = N_{X_0}(\hat{x}) + \left\{[h'(\hat{x})]^T \mu : \mu \in [T_{Y_0}(h(\hat{x}))]^\circ\right\} + \sum_{i \in I^0(\hat{x})} \operatorname{cone}(\partial g_i(\hat{x})).$$

It follows that there exist $\hat{\lambda}_i \geq 0$ and $\hat{\mu} \in [T_{Y_0}(h(\hat{x}))]^\circ$ such that

$$0 \in \nabla f(\hat{x}) + \sum_{i \in I^0(\hat{x})} \hat{\lambda}_i \partial g_i(\hat{x}) + [h'(\hat{x})]^T \hat{\mu} + N_{X_0}(\hat{x}).$$

This is equivalent to (3.57)–(3.58).

To prove the compactness of the set of Lagrange multipliers, we can adapt the idea of the proof of Lemma 3.26 in a straightforward manner. We leave the details of this adaptation to the reader. □

If the objective function $f(\cdot)$ is convex and nonsmooth, the standard transformation discussed at the beginning of this section reduces it to the form analyzed in Theorem 3.38. The multiplier associated with the additional constraint $f(x) - v \leq 0$ equals one, and the optimality conditions can be reduced back to (3.57)–(3.58), with the only difference that $\nabla f(\hat{x})$ should be replaced by $\partial f(\hat{x})$ in (3.57).

Theorem 3.38 encompasses as its special cases all formulations of nonlinear optimization problems discussed in the preceding sections. Clearly, the verification of its constraint qualification conditions is very difficult. A formal simplification of the optimality conditions can be achieved by using the concept of metric regularity of the entire system of constraints. Due to the possible nonconvexity and nondifferentiability of the constraints, the analysis of metric regularity requires special techniques of nonsmooth calculus. This leads to other technical complications and exceeds the scope of this book. However, the techniques that we have mastered so far allow us to analyze many important problems involving nonconvex and nonsmooth functions, as in the example below.

Example 3.39. Let $h : \mathbb{R}^n \to \mathbb{R}^m$ be a continuously differentiable function. Consider the unconstrained minimization problem

$$\text{minimize } \|h(x)\|_{\diamond}, \tag{3.59}$$

where $\|\cdot\|_{\diamond}$ is a norm in $\mathbb{R}^m$. The Chebyshev approximation problem of Example 3.32 can be formulated in this way.

Since the objective function is nonsmooth and nonconvex, the optimality conditions for unconstrained optimization of Section 3.1 cannot be directly applied here. However, we can use conditions for constrained problems, after an appropriate transformation.

By introducing the variables $y \in \mathbb{R}^m$ and $v \in \mathbb{R}$ we can transform problem (3.59) to an equivalent constrained problem

$$\begin{aligned} &\text{minimize } v \\ &\text{subject to } \|y\|_{\diamond} - v \leq 0, \\ &\qquad\qquad h(x) - y = 0. \end{aligned} \tag{3.60}$$

Indeed, for every point x the triple: x, $y = h(x)$, and $v = \|y\|_{\diamond}$, is feasible in problem (3.60), and the objective values of both problems coincide. Therefore the optimal value of (3.60) is less than or equal to the optimal value of (3.59). If the triple $\hat{x}$, $\hat{y}$, and $\hat{v}$ are a solution to problem (3.60) then we have $\hat{y} = h(\hat{x})$ and

$\hat{v} = \|\hat{y}\|_\Diamond = \|h(\hat{x})\|_\Diamond$. Thus the optimal value of problem (3.59) is less than or equal to the optimal value of (3.60). This means that both values are the same, and the x-part of the optimal solution of (3.60) is an optimal solution of (3.59).

Let us verify the constraint qualification conditions for problem (3.60). Slater's condition is satisfied, because for each y we can find $v > \|y\|_\Diamond$. Consider the set

$$C = \{(x, y, v) : \|y\|_\Diamond - v \le 0\}.$$

The cone of feasible directions to C at $(\hat{x}, \hat{y}, \hat{v})$ satisfies the inclusion

$$K_C(\hat{x}, \hat{y}, \hat{v}) \supset \{(d_x, d_y, d_v) \in \mathbb{R}^n \times \mathbb{R}^m \times \mathbb{R} : d_v \ge \|d_y\|_\Diamond\}.$$

Indeed, for every $\tau > 0$ and for $d_v \ge \|d_y\|_\Diamond$ we have

$$\|y + \tau d_y\|_\Diamond - (v + \tau d_v) \le \|y\|_\Diamond + \tau\|d_y\|_\Diamond - v - \tau d_v \le \|y\|_\Diamond - v.$$

If the pair (y, v) is feasible for C, the pair $(y + \tau d_y, v + \tau d_v)$ is feasible for all $\tau > 0$, as claimed. Thus each $d_y \in \mathbb{R}^m$ is feasible and

$$\{h'(\hat{x})d_x - d_y : (d_x, d_y, d_v) \in K_C(\hat{x}, \hat{y}, \hat{v})\} = \mathbb{R}^m.$$

Consequently, Robinson's condition is satisfied.

The constraint $\|y\|_\Diamond - v \le 0$ must be active at the solution. It follows from Theorem 3.38 that there exist $\lambda \ge 0$ and $\mu \in \mathbb{R}^m$ such that

$$\begin{bmatrix} 0 \\ 0 \\ 0 \end{bmatrix} \in \begin{bmatrix} 0 \\ 0 \\ 1 \end{bmatrix} + \lambda \begin{bmatrix} 0 \\ \partial\|\hat{y}\|_\Diamond \\ -1 \end{bmatrix} + \begin{bmatrix} [h'(\hat{x})]^T \\ -I \\ 0 \end{bmatrix} \mu.$$

The first block of equations, associated with x, yields

$$[h'(\hat{x})]^T \mu = 0. \tag{3.61}$$

The third one implies that $\lambda = 1$. Then the middle block reads

$$\mu \in \partial\|\hat{y}\|_\Diamond.$$

The subdifferential of the norm has been calculated in Example 2.78 on page 63. We obtain

$$\partial\|\hat{y}\|_\Diamond = \{s \in \mathbb{R}^m : \|s\|_* \le 1,\ \langle s, \hat{y}\rangle = \|\hat{y}\|_\Diamond\}.$$

Here $\|\cdot\|_*$ is the dual norm to $\|\cdot\|_\Diamond$. We conclude that if a point $\hat{x}$ is an optimal solution of problem (3.59) then there exists $\mu \in \mathbb{R}^m$ such that

$$\|\mu\|_* \le 1,$$
$$\langle \mu, h(\hat{x})\rangle = \|h(\hat{x})\|_\Diamond,$$

and equation (3.61) is satisfied. In the typical case, when $\|h(\hat{x})\| > 0$, we actually have $\|\mu\|_* = 1$.

In the special case when $\|\cdot\|_\diamond$ is the Euclidean norm $\|\cdot\|$ this system is equivalent to the equation

$$\left[h'(\hat{x})\right]^T h(\hat{x}) = 0,$$

which can be obtained as the necessary condition of a minimum of the smooth function $\|h(x)\|^2$.

Example 3.40. We are given N points $y^1, \dots, y^N$ on the plane and a convex, bounded and closed set $X \subset \mathbb{R}^2$. Our objective is to find a point $x \in X$ for which the distance to the nearest point y^i is the largest possible. Formally, the problem is

$$\operatorname*{maximize}_{x \in X} \min_{1 \le i \le N} \|x - y^i\|.$$

Defining

$$f_i(x) = -\|x - y^i\|, \quad i = 1, \dots, N,$$

we can equivalently formulate the problem as

$$\operatorname*{minimize}_{x \in X} \max_{1 \le i \le N} f_i(x).$$

The "max" function can be dealt with by introducing an auxiliary variable $v \in \mathbb{R}$ and formulating an equivalent constrained optimization problem

$$\begin{aligned} &\text{minimize } v \\ &\text{subject to } f_i(x) - v \le 0, \quad i = 1, \dots, N, \\ &\qquad\qquad\quad x \in X. \end{aligned}$$

At the optimal solution $(\hat{x}, \hat{v})$ we have $\|\hat{x} - y^i\| > 0$ for all $i = 1, \dots, N$ and $\hat{v} < 0$. Therefore the functions $f_i(\cdot)$ are continuously differentiable in a neighborhood of $\hat{x}$.

Our problem has form (3.55) with

$$h_i(x, v) = f_i(x) - v, \quad i = 1, \dots, N,$$

and $Y_0 = \mathbb{R}^N_-$. Define the set of active constraints

$$\hat{I} = \{1 \le i \le N : \|x - y^i\| = -\hat{v}\},$$

and let m be its cardinality. Robinson's condition at $\hat{x}$ (written for active constraints only) has the form

$$\left\{\left[\langle \nabla f_i(\hat{x}), d_x\rangle - d_v - d_{y_i}\right]_{i \in \hat{I}} : d_x \in K_X(\hat{x}),\ d_v \in \mathbb{R},\ d_y \le 0\right\} = \mathbb{R}^m.$$

It is obviously satisfied, because by setting $d_x = 0$ (which is always tangent to X) we can obtain each vector as a combination of $d_y \le 0$ and a vector with all components equal to d_v (which can be any real number).

The necessary condition (3.57) for a local minimum has the form: there exist multipliers $\mu_i \geq 0, i \in \hat{I}$, such that

$$0 \in \begin{bmatrix} 0 \\ 1 \end{bmatrix} + \sum_{i \in \hat{I}} \mu_i \begin{bmatrix} \frac{y^i - \hat{x}}{\|y^i - \hat{x}\|} \\ -1 \end{bmatrix} + \begin{bmatrix} N_X(\hat{x}) \\ 0 \end{bmatrix}.$$

The first block (corresponding to x) means that there exists a normal vector $s \in N_X(\hat{x})$ such that

$$\sum_{i \in \hat{I}} \mu_i \frac{y^i - \hat{x}}{\|y^i - \hat{x}\|} + s = 0. \tag{3.62}$$

The second block is simply

$$\sum_{i \in \hat{I}} \mu_i = 1.$$

Define

$$\alpha = \sum_{k \in \hat{I}} \frac{\mu_k}{\|y^k - \hat{x}\|},$$

and let

$$\gamma_i = \frac{\mu_i}{\alpha \|y^i - \hat{x}\|}, \quad i \in \hat{I}.$$

By construction, $\sum_{i \in \hat{I}} \gamma_i = 1$. Dividing both sides of (3.62) by α we find that

$$\sum_{i \in \hat{I}} \gamma_i (y^i - \hat{x}) + \frac{s}{\alpha} = 0,$$

which can be rewritten as

$$\hat{x} - \frac{s}{\alpha} = \sum_{i \in \hat{I}} \gamma_i y^i.$$

If the point $\hat{x}$ is an interior point of X, then $s = 0$. Hence $\hat{x}$ is a convex combination of the points y^i, $i \in \hat{I}$. By Lemma 2.7, $\hat{x}$ is a convex combination of at most three points y^i with $i \in \hat{I}$. Since $\hat{x}$ is different than points y^i, the number of points in the convex combination is at least two. Moreover, $\hat{x}$ is the center of a circle passing through these points and not containing strictly any other points y^j.

If the local minimum is a boundary point of X then we may have $s \neq 0$. In this case, $\hat{x} - \bar{s}$ is a convex combination of active y^i, for some normal $\bar{s} = s/\alpha$. Considering points $\hat{x} - \tau\bar{s}$, $\tau \in [0, 1]$, we can choose the smallest τ for which $\hat{x} - \tau\bar{s}$ is a convex combination of active points. Then the number of active points in this combination need not be larger than 2. If there are two such points, the solution $\hat{x}$ is equally distant from both of them. If one point y^i is active, $\hat{x} - y^i$ is normal to X at $\hat{x}$.

All these cases are illustrated in Figure 3.7 for an example of three points y^1, y^2, and y^3, with the set X represented by the shaded area. The point x^1 is a local minimum: it is a convex combination of equally distant y^1, y^2, and y^3. The point

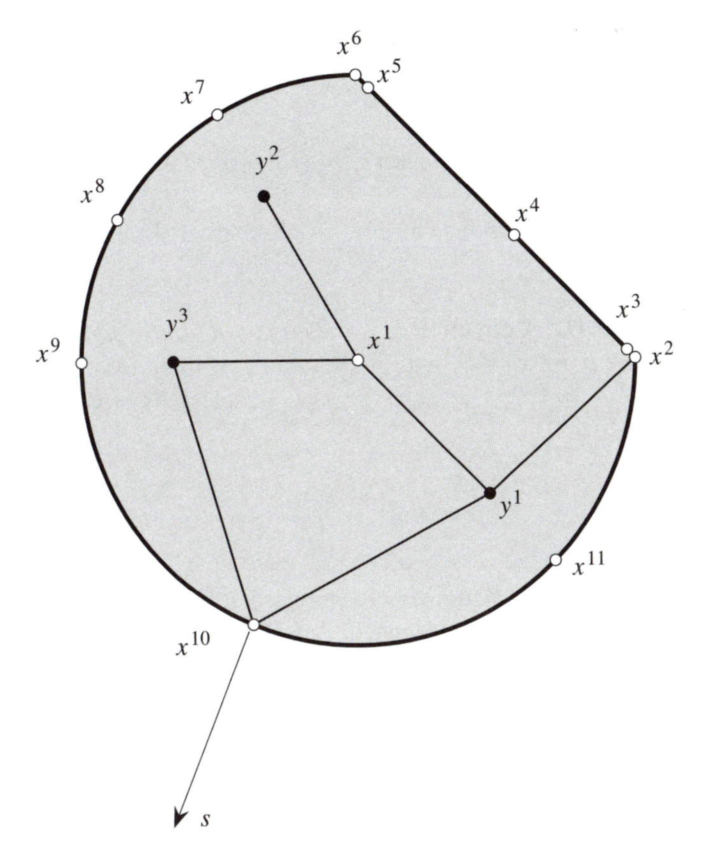

Figure 3.7. Points satisfying necessary conditions of optimality.

x^2 is a local minimum: $x^2 - y^1$ is normal to X at x^2. The point x^{10} is another local minimum: $x^{10} - s$ is a convex combination of y^1 and y^3. In fact, x^{10} is the global minimum. Other local minima are located at x^4, x^6, and x^8. The remaining points: x^3, x^5, x^7, x^9, and x^{11}, satisfy the optimality conditions but are not local minima.

3.7 SECOND ORDER OPTIMALITY CONDITIONS

3.7.1 Second Order Tangent Sets

Consider a set $X \subset \mathbb{R}^n$ and a point $x_0 \in X$. Our development of optimality conditions for problems involving the constraint $x \in X$ was based on the analysis of the tangent set $T_X(x_0)$. This is the set of directions s such that the ray $x_0 + \tau s$, $\tau > 0$, is very close to X:

$$\operatorname{dist}(x_0 + \tau_k s, X) = o(\tau_k),$$

at some points $\tau_k \downarrow 0$. In order to develop second order optimality conditions, we consider *parabolic* trajectories of the form

$$x(\tau) = x_0 + \tau s + \frac{\tau^2}{2} w, \quad \tau \geq 0, \tag{3.63}$$

where s is a tangent direction. Our idea is to have their distance to X infinitely smaller than τ_k^2.

DEFINITION 3.41. A direction w is called *second order tangent* to the set $X \subset \mathbb{R}^n$ at the point $x_0 \in X$ and in direction s, if there exist a sequence of points $x^k \in X, k = 1, 2, \ldots$, and a sequence of scalars $\tau_k > 0, k = 1, 2, \ldots$, such that $\tau_k \downarrow 0$ and

$$w = \lim_{k \to \infty} \frac{x^k - x_0 - \tau_k s}{\frac{1}{2}(\tau_k)^2}.$$

Denoting by o_k the difference between the left hand and right hand sides of the last equation, we can write

$$x^k = x_0 + \tau_k s + \frac{\tau_k^2}{2} w - o_k \frac{\tau_k^2}{2} = x(\tau_k) - o_k \frac{\tau_k^2}{2}.$$

It follows from the definition that the points on the parabolic trajectory (3.63) are, for $\tau = \tau_k$ and $k \to \infty$, very close to the set X:

$$\operatorname{dist}(x(\tau_k), X) = o(\tau_k^2).$$

It is implicit in Definition 3.41 that s is a (first order) tangent direction: $s \in T_X(x_0)$. Indeed, if w is a second order tangent direction, then for the limit in Definition 3.41 to exist, it is necessary that

$$\lim_{k \to \infty} \frac{x^k - x_0 - \tau_k s}{\tau_k} = 0,$$

and thus $(x^k - x_0)/\tau_k \to s$. For $s \notin T_X(x_0)$ no second order tangent direction exists.

The set of all second order tangent directions to X at (x_0, s) is called the *second order tangent set* and denoted by $T_X^2(x_0, s)$.

LEMMA 3.42. *Let $X \subset \mathbb{R}^n$, $x_0 \in X$, $s \in T_X(x_0)$. The set $T_X^2(x_0, s)$ is closed and*

$$T_X^2(x_0, \alpha s) = \alpha^2 T_X^2(x_0, s) \quad \textit{for all} \quad \alpha > 0.$$

Proof. Let $w \in T_X^2(x_0, s)$ and let $\{x^k\}$ and $\{\tau_k\}$ be the sequences satisfying Definition 3.41. For every $\alpha > 0$ we obtain

$$\alpha^2 w = \lim_{k\to\infty} \frac{\alpha^2(x^k - x_0 - \tau_k s)}{\frac{1}{2}(\tau_k)^2}$$
$$= \lim_{k\to\infty} \frac{x^k - x_0 - \frac{\tau_k}{\alpha}(\alpha s)}{\frac{1}{2}\left(\frac{\tau_k}{\alpha}\right)^2}.$$

Thus $\alpha^2 w$ satisfies Definition 3.41 for the direction αs, with the same sequence of points $\{x^k\}$ and with the corresponding sequence of scalars $\{\tau_k/\alpha\}$.

The reverse argument goes the same way.

To prove that the second order tangent set is closed, consider a sequence $\{w^j\}$ converging to some w, where each $w^j \in T_X^2(x_0, s)$. We need to show that w satisfies Definition 3.41 for (x_0, s). Choose any sequence $\varepsilon_j \downarrow 0$. For each w^j, Definition 3.41 is satisfied with some sequences of points $x^{j,k}$ and scalars $\tau_{j,k}$, $k = 1, 2, \ldots$. We can then find $x^{j,k(j)}$ and $\tau_{j,k(j)}$ such that

$$\left\| \frac{x^{j,k(j)} - x_0 - \tau_{j,k(j)} s}{\frac{1}{2}(\tau_{j,k(j)})^2} - w^j \right\| \le \varepsilon_j .$$

Therefore

$$\left\| \frac{x^{j,k(j)} - x_0 - \tau_{j,k(j)} s}{\frac{1}{2}(\tau_{j,k(j)})^2} - w \right\| \le \varepsilon_j + \|w^j - w\|,$$

and w satisfies Definition 3.41 with sequences $x^{j,k(j)}$ and $\tau_{j,k(j)}$, $j = 1, 2, \ldots$

□

In general, the second order tangent set is not a cone, and it may not be convex, even for a convex set X.[†] But for the case of a *polyhedral* set X, the second order tangent set has a nice representation, and is in fact a convex cone.

LEMMA 3.43. *If X is a convex polyhedron then*

$$T_X^2(x_0, s) = T_{T_X(x_0)}(s).$$

Proof. Let us recall that if X is a convex polyhedron, then the tangent cone $T_X(x_0)$ is identical with the cone of feasible directions $K_X(x_0)$, and both cones are polyhedral: they are defined by finitely many linear inequalities.

[†]There exist alternative definitions of second order tangent sets, which guarantee the convexity in this case, but then other complications arise. The reader is referred to the literature cited at the end for further studies of this topic.

Denoting by a_i, $i = 1, \ldots, m$, the vectors defining the inequalities active at x_0, we can write

$$T_X(x_0) = \{s \in \mathbb{R}^n : \langle a_i, s\rangle \le 0,\ i = 1, \ldots, m\}.$$

Consider a direction $s \in T_X(x_0)$ and a sequence

$$x(\tau_k) = x_0 + \tau_k s + \frac{\tau_k^2}{2} w$$

with some $\tau_k \downarrow 0$. The vector w is an element of the second order tangent set if and only if $\operatorname{dist}(x(\tau_k), X) = o(\tau_k^2)$. This is equivalent to

$$\langle a_i, x(\tau_k) - x_0\rangle \le o(\tau_k^2), \quad i = 1, \ldots, m. \tag{3.64}$$

Two cases are possible for each $i = 1, \ldots, m$. If $\langle a_i, s\rangle = 0$ then (3.64) holds true if and only if $\langle a_i, w\rangle \le 0$. If $\langle a_i, s\rangle < 0$ then (3.64) is satisfied for all $w \in \mathbb{R}^n$. Therefore $w \in T_X^2(x_0)$ if and only if

$$\langle a_i, w\rangle \le 0 \quad \text{for all} \quad i \quad \text{such that} \quad \langle a_i, s\rangle = 0.$$

This is equivalent to the inclusion $w \in T_{T_X(x_0)}(s)$. □

Consider now the set X of points $x \in \mathbb{R}^n$ satisfying the system of relations

$$\begin{aligned} g_i(x) &\le 0, \quad i = 1, \ldots, m,\\ h_i(x) &= 0, \quad i = 1, \ldots, p,\\ x &\in X_0. \end{aligned} \tag{3.65}$$

Let $\hat{x} \in X$ and let all functions $g_i(\cdot)$ and $h_i(\cdot)$ be twice continuously differentiable at $\hat{x}$. Furthermore, let X_0 be a convex set.

Under Robinson's condition (3.11), the cone

$$\begin{aligned} T_X(\hat{x}) = \big\{s \in T_{X_0}(\hat{x}) :\ & \\ & \langle \nabla g_i(\hat{x}), s\rangle \le 0,\ i \in I^0(\hat{x}),\\ & \langle \nabla h_i(\hat{x}), s\rangle = 0,\ i = 1, \ldots, p\big\} \end{aligned} \tag{3.66}$$

is the tangent cone to the feasible set at $\hat{x}$. We shall derive an algebraic description of the second order tangent set.

LEMMA 3.44. *Assume that Robinson's condition is satisfied at $\hat{x} \in X$. Then for every $s \in T_X(\hat{x})$,*

$$\begin{aligned} T_X^2(\hat{x}, s) = \big\{w \in T_{X_0}^2(\hat{x}, s) :\ & \\ & \langle \nabla g_i(\hat{x}), w\rangle \le -\langle s, \nabla^2 g_i(\hat{x}) s\rangle,\ i \in I^{00}(\hat{x}, s),\\ & \langle \nabla h_i(\hat{x}), w\rangle = -\langle s, \nabla^2 h_i(\hat{x}) s\rangle,\ i = 1, \ldots, p\big\}, \end{aligned} \tag{3.67}$$

with

$$I^{00}(\hat{x}, s) = \{i \in I^0(\hat{x}) : \langle \nabla g_i(\hat{x}), s\rangle = 0\}.$$

Proof. Let $s \in T_X(\hat{x})$ and let w be an element of the set on the right hand side of (3.67). Consider the parabolic trajectory (3.63). For each $i \in I^0(\hat{x})$ we can expand $g_i(x(\tau))$ as follows:

$$\begin{aligned} g_i(x(\tau)) &= g_i(\hat{x}) + \langle \nabla g_i(\hat{x}), x(\tau) - \hat{x}\rangle \\ &\quad + \frac{\tau^2}{2}\langle x(\tau) - \hat{x}, \nabla^2 g_i(\hat{x})(x(\tau) - \hat{x})\rangle + o(\tau^2) \\ &= \tau\langle \nabla g_i(\hat{x}), s\rangle + \frac{\tau^2}{2}\langle \nabla g_i(\hat{x}), w\rangle + \frac{\tau^2}{2}\langle s, \nabla^2 g_i(\hat{x}) s\rangle + o(\tau^2). \end{aligned} \tag{3.68}$$

We have used here the fact that $g_i(\hat{x}) = 0$ for $i \in I^0(\hat{x})$. Two cases may occur. If $i \notin I^{00}(\hat{x}, s)$ then $\langle \nabla g_i(\hat{x}), s\rangle < 0$. It follows that for every $w \in \mathbb{R}^n$ we have $g_i(x(\tau)) < 0$ for all sufficiently small $\tau > 0$. In the second case, when $i \in I^{00}(\hat{x}, s)$, the first order term vanishes. But if w is an element of the set defined in (3.67), then the second order term is nonpositive. Similar arguments apply to equality constraints. We conclude that if w satisfies the conditions on the right hand side of (3.67), then

$$\begin{aligned} g_i(x(\tau)) &\le o(\tau^2), \quad i \in I^0(\hat{x}), \\ \left|h_i(x(\tau))\right| &\le o(\tau^2), \quad i = 1, \ldots, p, \\ \operatorname{dist}(x(\tau), X_0) &= o(\tau^2). \end{aligned} \tag{3.69}$$

The last relation follows from the fact that $s \in T_{X_0}(\hat{x})$ and $w \in T^2_{X_0}(\hat{x}, s)$.

By Theorem A.10, it follows from Robinson's condition that the system of constraints is metrically regular at $\hat{x}$. Using Definition 3.14, we deduce that there exist $C \ge 0$ and $\tau_0 > 0$ such that for all $\tau \in [0, \tau_0]$ we can find a point $\varphi(\tau) \in X$ satisfying the inequality:

$$\|\varphi(\tau) - x(\tau)\| \le C\Bigg[\operatorname{dist}(x(\tau), X_0) + \sum_{i \in I^0(\hat{x})} \max\big(0, g_i(x(\tau))\big) + \sum_{i=1}^{p} \left|h_i(x(\tau))\right|\Bigg].$$

By virtue of (3.69),

$$\|\varphi(\tau) - x(\tau)\| \le o(\tau^2).$$

Therefore $\operatorname{dist}(x(\tau), X) \le o(\tau^2)$, and indeed $w \in T^2_X(\hat{x}, s)$.

To prove the converse, assume that $w \in T_X^2(\hat{x}, s)$. Then we have (3.69) for some sequence $\tau_k \downarrow 0$. Inspecting the expansions (3.68) we conclude that w must be an element of the right hand side of (3.67). □

It follows from the last two lemmas that for a polyhedral X_0 and under Robinson's condition, the second order tangent set to the set defined by (3.65) is a closed convex cone.

3.7.2 Optimality Conditions

We are now ready to develop the second order counterpart of the fundamental optimality condition of Theorem 3.24. We consider the general problem formulation

$$\operatorname*{minimize}_{x \in X} f(x). \tag{3.70}$$

We assume that X is a set in $\mathbb{R}^n$, and $f : \mathbb{R}^n \to \mathbb{R}$ is twice continuously differentiable.

THEOREM 3.45. *Assume that $\hat{x}$ is an optimal solution of problem* (3.70). *Then for every $s \in T_X(\hat{x})$ such that $\langle \nabla f(\hat{x}), s \rangle = 0$ we have*

$$\langle \nabla f(\hat{x}), w \rangle + \langle s, \nabla^2 f(\hat{x}) s \rangle \geq 0 \quad \textit{for all} \quad w \in T_X^2(\hat{x}, s). \tag{3.71}$$

Proof. Consider the function $x(\tau)$ defined in (3.63) with $x_0 = \hat{x}$. If $s \in T_X(\hat{x})$ and $w \in T_X^2(\hat{x}, s)$, there exist sequences of points $x^k \in X$, $x^k \to \hat{x}$, and scalars $\tau_k \downarrow 0$, such that

$$\left\| x^k - \hat{x} - \tau_k s - \frac{\tau_k^2}{2} w \right\| = o(\tau_k^2).$$

Since $\hat{x}$ is a local minimum, $f(x^k) \geq f(\hat{x})$ for all sufficiently large k. Therefore

$$f(x(\tau_k)) - f(\hat{x}) \geq o(\tau_k^2).$$

The second order expansion of $f(x(\tau))$ at $\tau = 0$ renders

$$f(x(\tau_k)) - f(\hat{x}) = \tau_k \langle \nabla f(\hat{x}), s \rangle + \frac{\tau_k^2}{2} \langle \nabla f(\hat{x}), w \rangle + \frac{\tau_k^2}{2} \langle s, \nabla^2 f(\hat{x}) s \rangle + o(\tau_k^2).$$

Hence

$$\tau_k \langle \nabla f(\hat{x}), s \rangle + \frac{\tau_k^2}{2} \langle \nabla f(\hat{x}), w \rangle + \frac{\tau_k^2}{2} \langle s, \nabla^2 f(\hat{x}) s \rangle \geq o(\tau_k^2).$$

By assumption, the first term on the left hand side equals 0. Dividing both sides by τ_k^2 and letting $\tau_k \downarrow 0$, we obtain inequality (3.71). □

This general result allows for the formulation of second order conditions for a variety of nonlinear optimization problems. Most important is the problem with constraints involving inequalities and equations:

$$\begin{aligned} &\text{minimize} \;\; f(x) \\ &\text{subject to} \;\; g_i(x) \le 0, \quad i = 1, \dots, m, \\ &\qquad\qquad\;\; h_i(x) = 0, \quad i = 1, \dots, p, \\ &\qquad\qquad\;\; x \in X_0. \end{aligned} \tag{3.72}$$

We assume that all functions are twice continuously differentiable at a local minimum $\hat{x}$. The set X_0 is convex and closed. The associated Lagrangian has the form

$$L(x, \lambda, \mu) = f(x) + \sum_{i=1}^{m} \lambda_i g_i(x) + \sum_{i=1}^{p} \mu_i h_i(x).$$

It follows from Lemma 3.44 and Theorem 3.45 that, under Robinson's condition, for every $s \in T_X(\hat{x})$ such that $\langle \nabla f(\hat{x}), s \rangle = 0$, the optimal value of the problem

$$\begin{aligned} &\text{minimize} \;\; \langle \nabla f(\hat{x}), w \rangle + \langle s, \nabla^2 f(\hat{x}) s \rangle \\ &\text{subject to} \;\; w \in T^2_{X_0}(\hat{x}, s), \\ &\qquad\qquad\;\; \langle \nabla g_i(\hat{x}), w \rangle \le -\langle s, \nabla^2 g_i(\hat{x}) s \rangle, \; i \in I^{00}(\hat{x}, s), \\ &\qquad\qquad\;\; \langle \nabla h_i(\hat{x}), w \rangle = -\langle s, \nabla^2 h_i(\hat{x}) s \rangle, \; i = 1, \dots, p, \end{aligned} \tag{3.73}$$

is nonnegative. In order to analyze this problem in more detail, we make an additional assumption that X_0 is a convex polyhedron.

Optimality conditions for problem (3.73) provide the second order conditions for problem (3.72). We use X to denote the feasible set of this problem.

Theorem 3.46. *Assume that X_0 is a convex polyhedron and that a point $\hat{x} \in X$ is a local minimum of problem* (3.72) *satisfying Robinson's condition. Then for every $s \in T_X(\hat{x})$ such that $\langle \nabla f(\hat{x}), s \rangle = 0$ the following condition holds true:*

$$\max_{(\lambda,\mu) \in \hat{\Lambda}(\hat{x})} \left\langle s, \nabla^2_{xx} L(\hat{x}, \lambda, \mu) s \right\rangle \ge 0, \tag{3.74}$$

where $\hat{\Lambda}(\hat{x})$ is the set of optimal values of Lagrange multipliers $(\hat{\lambda}, \hat{\mu})$ in problem (3.72).

Proof. Before starting the proof, we note that under Robinson's condition the set $\hat{\Lambda}(\hat{x})$ is compact (Lemma 3.26), and therefore the maximum in (3.74) exists.

Consider problem (3.73). By virtue of Lemma 3.43,

$$T^2_{X_0}(\hat{x}, s) = T_{T_{X_0}(\hat{x})}(s).$$

Furthermore, Example 2.21 yields that

$$T_{T_{X_0}(\hat{x})}(s) \supset T_{X_0}(\hat{x}).$$

Therefore the following problem

$$\begin{aligned} \text{minimize} \ & \langle \nabla f(\hat{x}), w \rangle + \langle s, \nabla^2 f(\hat{x}) s \rangle \\ \text{subject to} \ & w \in T_{X_0}(\hat{x}), \\ & \langle \nabla g_i(\hat{x}), w \rangle \le -\langle s, \nabla^2 g_i(\hat{x}) s \rangle, \ i \in I^{00}(\hat{x}, s), \\ & \langle \nabla h_i(\hat{x}), w \rangle = -\langle s, \nabla^2 h_i(\hat{x}) s \rangle, \ i = 1, \dots, p, \end{aligned} \tag{3.75}$$

is a restriction of problem (3.73), and its optimal value must be nonnegative as well. It follows from Robinson's condition that the feasible set of problem (3.75) is nonempty. Problem (3.75) has a linear objective function and linear and polyhedral constraints. Therefore it has an optimal solution: $\hat{w}$.

Denote by $\bar{\lambda}_i, i \in I^{00}(\hat{x}, s)$ and $\bar{\mu}_i, i = 1, \dots, p$, the Lagrange multipliers associated with the constraints. The necessary and sufficient conditions of optimality of Theorem 3.25 from page 116 for problem (3.75) read

$$-\nabla f(\hat{x}) - \sum_{i \in I^{00}(\hat{x},s)} \bar{\lambda}_i \nabla g_i(\hat{x}) - \sum_{i=1}^{p} \bar{\mu}_i \nabla h_i(\hat{x}) \in \left[T_{T_{X_0}(\hat{x})}(w) \right]^{\circ},$$
$$\bar{\lambda}_i \left[\langle \nabla g_i(\hat{x}), \hat{w} \rangle + \langle s, \nabla^2 g_i(\hat{x}) s \rangle \right] = 0, \quad i \in I^{00}(\hat{x}, s),$$
$$\bar{\lambda} \ge 0.$$

Therefore the set of optimal values of Lagrange multipliers for problem (3.75) is included in the set $\hat{\Lambda}(\hat{x})$ of optimal values of Lagrange multipliers of problem (3.72) (we formally define $\bar{\lambda}_i = 0$ for $i \notin I^{00}(\hat{x}, s)$). Moreover, the optimal value of problem (3.75) is equal to the value of its Lagrangian:

$$\begin{aligned} L_1(\hat{w}, \bar{\lambda}, \bar{\mu}) = {} & \langle \nabla f(\hat{x}), \hat{w} \rangle + \langle s, \nabla^2 f(\hat{x}) s \rangle \\ & + \sum_{i \in I^{00}(\hat{x},s)} \bar{\lambda}_i \left[\langle \nabla g_i(\hat{x}), \hat{w} \rangle + \langle s, \nabla^2 g_i(\hat{x}) s \rangle \right] \\ & + \sum_{i=1}^{p} \bar{\mu}_i \left[\langle \nabla h_i(\hat{x}), \hat{w} \rangle + \langle s, \nabla^2 h_i(\hat{x}) s \rangle \right]. \end{aligned}$$

Recall from Example 2.21 on page 27 that

$$T_{T_{X_0}(\hat{x})}(w) = T_{X_0}(\hat{x}) + \{ \alpha w : \alpha \in \mathbb{R} \}.$$

If the minimum of the Lagrangian $L_1(\cdot, \bar{\lambda}, \bar{\mu})$ in $T_{X_0}(\hat{x})$ occurs at $\hat{w}$, then

$$\Big\langle \nabla f(\hat{x}) + \sum_{i \in I^{00}(\hat{x},s)} \bar{\lambda}_i \nabla g_i(\hat{x}) + \sum_{i=1}^{p} \bar{\mu}_i \nabla h_i(\hat{x}), \hat{w} \Big\rangle = 0.$$

The first order terms of the Lagrangian disappear and we conclude that

$$L_1(\hat{w}, \bar{\lambda}, \bar{\mu}) = \langle s, \nabla^2 f(\hat{x}) s \rangle + \sum_{i \in I^{00}(\hat{x},s)} \bar{\lambda}_i \langle s, \nabla^2 g_i(\hat{x}) s \rangle$$
$$+ \sum_{i=1}^{p} \bar{\mu}_i \langle s, \nabla^2 h_i(\hat{x}) s \rangle = \big\langle s, \nabla^2_{xx} L(\hat{x}, \bar{\lambda}, \bar{\mu}) s \big\rangle.$$

The optimal value is nonnegative, and therefore inequality (3.74) is valid. □

Another way to prove this theorem is to apply to problem (3.75) the duality theory in conic programming, which we discuss in Section 4.3.

We can mention here that the assumption that X_0 is a polyhedron allowed us to use the explicit form of its second order tangent set. It is also possible to develop more general second order conditions, for a convex X_0, but they are more involved. One has to replace in problem (3.73) the second order tangent set by its convex subsets, and involve general convex duality to analyze problem (3.75).

In the development of the second order *sufficient* conditions for problem (3.72) we do not assume explicitly any constraint qualification condition, but we assume the existence of optimal Lagrange multipliers. The set (3.66) may be larger than the tangent cone, but we still denote it by T_X. We also do not assume any special structure of the set X_0, except for its convexity and closedness.

THEOREM 3.47. *Assume that a point $\hat{x}$ satisfies the first order optimality conditions for problem* (3.72)*, and let $\hat{\Lambda}(\hat{x})$ be the set of Lagrange multipliers $(\hat{\lambda}, \hat{\mu})$ for this problem. Assume that for every nonzero s in the set* (3.66) *such that $\langle \nabla f(\hat{x}), s \rangle = 0$, we have*

$$\sup_{(\hat{\lambda},\hat{\mu}) \in \hat{\Lambda}(\hat{x})} \big\langle s, \nabla^2_{xx} L(\hat{x}, \hat{\lambda}, \hat{\mu}) s \big\rangle > 0. \tag{3.76}$$

Then $\hat{x}$ is a local minimum of (3.72).

Proof. We argue by contradiction. Suppose $\hat{x}$ is not a local minimum. Then there exists a sequence of feasible points y^k such that $y^k \to \hat{x}$ and

$$f(y^k) < f(\hat{x}) \quad \text{for all} \quad k. \tag{3.77}$$

We define the sequence

$$s^k = \frac{y^k - \hat{x}}{\|y^k - \hat{x}\|}$$

and denote its accumulation point by s. It exists, because $\{s^k\}$ is bounded. For $i \in I^0(\hat{x})$ we can expand the function $g_i(\cdot)$ around $\hat{x}$:

$$0 \geq g_i(y^k) = \langle \nabla g_i(\hat{x}), y^k - \hat{x} \rangle + o(y^k - \hat{x}),$$

with $o(z)/\|z\| \to 0$ for $z \to 0$. Dividing both sides by $\|y^k - \hat{x}\|$ and passing to the limit over the subsequence for which $s^k \to s$, we obtain

$$\langle \nabla g_i(\hat{x}), s \rangle \leq 0, \quad i \in I^0(\hat{x}).$$

A similar analysis of equality constraints yields the relations

$$\langle \nabla h_i(\hat{x}), s \rangle = 0, \quad i = 1, \dots, p.$$

Therefore s is an element of the cone (3.66).

It follows from the necessary conditions of optimality that $\langle \nabla f(\hat{x}), s \rangle \geq 0$. In view of (3.77), $\langle \nabla f(\hat{x}), s \rangle = 0$. By assumption (3.76), we can choose $(\hat{\lambda}, \hat{\mu}) \in \hat{\Lambda}(\hat{x})$ such that

$$\left\langle s, \nabla^2_{xx} L(\hat{x}, \hat{\lambda}, \hat{\mu}) s \right\rangle > 0. \tag{3.78}$$

By Taylor's formula

$$f(y^k) = f(\hat{x}) + \langle \nabla f(\hat{x}), y^k - \hat{x} \rangle + \frac{1}{2} \langle y^k - \hat{x}, \nabla^2 f(\hat{x})(y^k - \hat{x}) \rangle + \delta(y^k - \hat{x}), \tag{3.79}$$

with $\frac{\delta(y^k - \hat{x})}{\|y^k - \hat{x}\|^2} \to 0$. Similarly,

$$\begin{aligned} g_i(y^k) =& g_i(\hat{x}) + \langle \nabla g_i(\hat{x}), y^k - \hat{x} \rangle \\ &+ \frac{1}{2} \langle y^k - \hat{x}, \nabla^2 g_i(\hat{x})(y^k - \hat{x}) \rangle + \gamma_i(y^k - \hat{x}), \\ h_i(y^k) =& h_i(\hat{x}) + \langle \nabla h_i(\hat{x}), y^k - \hat{x} \rangle \\ &+ \frac{1}{2} \langle y^k - \hat{x}, \nabla^2 h_i(\hat{x})(y^k - \hat{x}) \rangle + \delta_i(y^k - \hat{x}), \end{aligned} \tag{3.80}$$

where γ_i and δ_i are infinitely small with respect to $\|y^k - \hat{x}\|^2$. Let us add (3.80) multiplied by $\hat{\lambda}_i$ and $\hat{\mu}_i$, respectively, to (3.79). The sum of the first order terms on the right hand side satisfies the first order optimality condition:

$$\left\langle \nabla f(\hat{x}) + \sum_{i=1}^{m} \hat{\lambda}_i \nabla g_i(\hat{x}) + \sum_{i=1}^{m} \hat{\mu}_i \nabla h_i(\hat{x}), y^k - \hat{x} \right\rangle \geq 0,$$

because $y^k \in X_0$. Therefore the sum of (3.79)–(3.80) renders the inequality

$$f(y^k) \geq f(\hat{x}) + \frac{1}{2}\big\langle y^k - \hat{x}, \nabla^2_{xx} L(\hat{x}, \hat{\lambda}, \hat{\mu})(y^k - \hat{x})\big\rangle + \bar{\delta}(y^k - \hat{x}),$$

in which $\bar{\delta}(y^k - \hat{x})$ is infinitely small with respect to $\|y^k - \hat{x}\|^2$. Using (3.77) and dividing both sides by $\|y^k - \hat{x}\|^2$, we obtain for large k the relation

$$\frac{1}{2}\big\langle s^k, \nabla^2_{xx} L(\hat{x}, \hat{\lambda}, \hat{\mu}) s^k\big\rangle + \frac{\bar{\delta}(y^k - \hat{x})}{\|y^k - \hat{x}\|^2} < 0.$$

When $k \to \infty$ (over an appropriate subsequence), then $s^k \to s$. It follows that

$$\big\langle s, \nabla^2_{xx} L(\hat{x}, \hat{\lambda}, \hat{\mu}) s\big\rangle \leq 0,$$

which contradicts (3.78). □

Theorem 3.47 can be used to obtain simpler versions of the second order sufficient conditions. For example, a point $\hat{x}$ is said to satisfy the *semi-strong second order sufficient conditions*, if there exist multipliers $(\hat{\lambda}, \hat{\mu}) \in \hat{\Lambda}(\hat{x})$ such that

$$\big\langle s, \nabla^2_{xx} L(\hat{x}, \hat{\lambda}, \hat{\mu}) s\big\rangle > 0, \tag{3.81}$$

for all nonzero s included in the plane:

$$\hat{D}(\hat{x}) = \big\{s \in R^n : \langle \nabla g_i(\hat{x}), s\rangle = 0 \text{ for all } i \text{ such that } \hat{\lambda}_i > 0, \\ \langle \nabla h_i(\hat{x}), s\rangle = 0,\ i = 1, \dots, p\big\}.$$

We shall show that these conditions imply the assumptions of Theorem 3.47. Suppose s is a nonzero element of $T_X(\hat{x})$ such that $\langle \nabla f(\hat{x}), s\rangle = 0$. The first order necessary conditions imply that

$$\Big\langle \nabla f(\hat{x}) + \sum_{i \in I^0(\hat{x})} \hat{\lambda}_i \nabla g_i(\hat{x}) + \sum_{i=1}^{p} \hat{\mu}_i \nabla h_i(\hat{x}), s\Big\rangle \geq 0.$$

Since $\langle \nabla f(\hat{x}), s\rangle = 0$ and $\langle \nabla h_i(\hat{x}), s\rangle = 0$, we deduce that

$$\sum_{i \in I^0(\hat{x})} \hat{\lambda}_i \langle \nabla g_i(\hat{x}), s\rangle \geq 0.$$

All terms on the left hand side are nonpositive for $s \in T_X(\hat{x})$, and therefore they must be equal to 0. This implies that $\langle \nabla g_i(\hat{x}), s\rangle = 0$ whenever $\hat{\lambda}_i > 0$. Thus $s \in \hat{D}(\hat{x})$. Therefore the assumptions of Theorem 3.47 are satisfied.

Example 3.48. Let Q be a symmetric matrix of dimension n. Consider the problem of minimizing the quadratic form on the unit sphere S:

$$\begin{aligned} &\text{minimize } \langle x, Qx\rangle \\ &\text{subject to } 1 - \|x\|^2 = 0. \end{aligned}$$

We analyzed this problem in Example 3.29, where we found that the first order necessary conditions of optimality,

$$Qx - \mu x = 0,$$

are satisfied by each eigenvector of Q, and the corresponding eigenvalue is the value of the Lagrange multiplier μ. We shall show that the second order conditions allow us to find the optimal solution.

Let $\mu_1 \le \mu_2 \le \cdots \le \mu_n$ be the eigenvalues of Q, and let $z_1, \dots, z_n$ be the corresponding orthogonal eigenvectors of unit length. Suppose $\hat{x} = z_k$ and $\hat{\mu} = \mu_k$. The tangent plane to the sphere S at $\hat{x}$ has the form:

$$\hat{D}(\hat{x}) = \{d \in R^n : \langle \hat{x}, d\rangle = 0\} = \{d \in R^n : \langle z_k, d\rangle = 0\}.$$

Therefore every tangent direction d is a combination of the remaining eigenvectors

$$d = \sum_{i \ne k} \alpha_i z_i.$$

The Lagrangian has the form

$$L(x, \mu) = \langle x, Qx\rangle - \mu\|x\|^2 = \langle x, (Q - \mu I)x\rangle,$$

and its Hessian at z_k and μ_k becomes

$$\nabla^2 L(x, \hat{\mu}) = Q - \mu_k I.$$

The second order necessary condition of optimality of Theorem 3.46 requires that

$$\langle d, (Q - \mu_k I)d\rangle \ge 0$$

for every tangent direction d. In particular, setting $d = z_i$ for $i \ne k$ we obtain

$$\langle z_i, (Q - \mu_k I)z_i\rangle = \mu_i - \mu_k \ge 0.$$

Therefore the local minimum is an eigenvector corresponding to the smallest eigenvalue of Q.

3.8 SENSITIVITY

Let us now consider the problem with parameters $b \in \mathbb{R}^m$ and $c \in \mathbb{R}^p$:

$$\begin{aligned} &\text{minimize } f(x) \\ &\text{subject to } g_i(x) \le b_i, \quad i = 1, \dots, m, \\ &\qquad\qquad\quad h_i(x) = c_i, \quad i = 1, \dots, p. \end{aligned} \tag{3.82}$$

We assume that $f : \mathbb{R}^n \to \mathbb{R}$, $g_i : \mathbb{R}^n \to \mathbb{R}$, $i = 1, \ldots, m$, and $h_i : \mathbb{R}^n \to \mathbb{R}$, $i = 1, \ldots, p$, are twice continuously differentiable functions.

We are interested in the dependence of the solution and of the optimal value of this problem, on the parameters b and c.

Assume that x^0 is a local minimum of (3.82) for $b = b^0$ and $c = c^0$. As before, define the set of active inequality constraints

$$I^0(x^0) = \{1 \le i \le m : g_i(x^0) = b_i^0\}.$$

We assume that x^0 satisfies a constraint qualification condition and we denote the vector of Lagrange multipliers at x^0 by (λ^0, μ^0). We can analyze the sensitivity of problem (3.82) by considering the system of the first order conditions of optimality, under the semi-strong second order sufficient condition (3.81) discussed on page 149.

THEOREM 3.49. *Assume that the gradients of all active constraints are linearly independent at x^0, and that $\lambda_i^0 > 0$ for all $i \in I^0(x^0)$. Furthermore, let the semi-strong second order sufficient condition* (3.81) *be satisfied at x^0. Then there exists a neighborhood $\mathcal{U}$ of (b^0, c^0) in which we can define functions $\hat{x}(b, c)$, $\hat{\lambda}_i(b, c)$, $i \in I^0(x^0)$, and $\hat{\mu}_i(b, c)$, $i = 1, \ldots, p$, such that $\hat{x}(b, c)$ is a local minimum of problem* (3.82)*, and $(\hat{\lambda}(b, c), \hat{\mu}(b, c))$ is the corresponding vector of Lagrange multipliers. Moreover, the functions $\hat{x}(b, c)$, $\hat{\lambda}(b, c)$, and $\hat{\mu}(b, c)$ are differentiable at (b^0, c^0) and*

$$\frac{df(\hat{x}(b^0, c^0))}{db} = -\lambda^0,$$
$$\frac{df(\hat{x}(b^0, c^0))}{dc} = -\mu^0.$$

Proof. For simplicity of notation, we assume that $p = 0$ (no equality constraints) and that $I^0(x^0) = \{1, \ldots, m_0\}$. Define the Jacobian of the active constraint functions:

$$A(x) = \begin{bmatrix} (\nabla g_1(x))^T \\ \vdots \\ (\nabla g_{m_0}(x))^T \end{bmatrix}.$$

If we ignore the constraints that are not active at x^0, the first order necessary condition reads:

$$\nabla f(x) + A(x)^T \lambda = 0. \tag{3.83}$$

We postulate that the inequalities that are active at x^0 be satisfied as equations for b close to b^0:

$$g_i(x) = b_i, \quad i = 1, \ldots, m_0. \tag{3.84}$$

The Jacobian of the system of equations (3.83)–(3.84) at (x^0, λ^0) has the form

$$J^0 = \begin{bmatrix} \nabla^2_{xx} L(x^0, \lambda^0) & [A(x^0)]^T \\ A(x^0) & 0 \end{bmatrix}. \tag{3.85}$$

Owing to the strong second order sufficient condition and the linear independence condition, it is a nonsingular matrix. Indeed, suppose

$$J^0 \begin{bmatrix} s \\ u \end{bmatrix} = 0.$$

Then $A(x^0)s = 0$. Moreover,

$$\begin{aligned} \begin{bmatrix} s^T & u^T \end{bmatrix} J^0 \begin{bmatrix} s \\ u \end{bmatrix} &= s^T \nabla^2_{xx} L(x^0, \lambda^0)s + 2s^T [A(x^0)]^T u \\ &= s^T \nabla^2_{xx} L(x^0, \lambda^0)s. \end{aligned}$$

If $s \neq 0$, then the second order condition implies the positivity of the last expression, a contradiction. Thus $s = 0$. But then

$$J^0 \begin{bmatrix} s \\ u \end{bmatrix} = [A(x^0)]^T u = 0,$$

and the linear independence of the rows of $A(x^0)$ implies that $u = 0$.

Since the Jacobian is nonsingular, system (3.83)–(3.84) defines in a neighborhood of b^0 an implicit function $(\hat{x}(b), \hat{\lambda}(b))$ whose derivative at b^0 satisfies the equation:

$$J^0 \begin{bmatrix} \frac{d\hat{x}(b^0)}{db} \\ \frac{d\hat{\lambda}(b^0)}{db} \end{bmatrix} = \begin{bmatrix} 0 \\ I \end{bmatrix}.$$

Using (3.85) we obtain the relations:

$$\begin{aligned} \nabla^2_{xx} L(x^0, \lambda^0) \frac{d\hat{x}(b^0)}{db} + [A(x^0)]^T \frac{d\hat{\lambda}(b^0)}{db} &= 0, \\ A(x^0) \frac{d\hat{x}(b^0)}{db} &= I. \end{aligned}$$

We can use these equations to calculate the derivatives of $\hat{x}(b)$ and $\hat{\lambda}(b)$ at b^0. We leave these manipulations to the reader (Exercise 3.17) and we concentrate only on the second equation, which simply follows from the fact that $g_i(\hat{x}(b)) = b$, $i = 1, \dots, m_0$. Multiplying the last displayed equation from the left by $[\lambda^0]^T$ we get

$$[\lambda^0]^T A(x^0) \frac{d\hat{x}(b^0)}{db} = [\lambda^0]^T.$$

Using (3.83) at b^0 we obtain

$$\left[\nabla f(x^0)\right]^T \frac{d\hat{x}(b^0)}{db} = -\left[\lambda^0\right]^T.$$

The left hand side of this equation is the transpose of the derivative of the composite function $f(\hat{x}(b))$ at b^0 and therefore

$$\frac{df(\hat{x}(b^0))}{db} = \left[\frac{d\hat{x}(b^0)}{db}\right]^T \nabla f(x^0) = -\lambda^0,$$

as required. As the implicit function $(\hat{x}(b), \hat{\lambda}(b))$ is continuous, we have

$$g_i(\hat{x}(b)) < 0, \quad i \notin I^0(x^0),$$
$$\hat{\lambda}_i(b) > 0, \quad i \in I^0(b^0),$$

for all b sufficiently close to b^0. Therefore the pair $(\hat{x}(b), \hat{\lambda}(b))$ is feasible and satisfies the first order optimality conditions.

It remains to show that $\hat{x}(b)$ is indeed a local minimum. To this end we shall prove that $\hat{x}(b)$ satisfies the strong second order sufficient conditions, for all b sufficiently close to b^0. Suppose the opposite: for every $\varepsilon > 0$ we can find b and $s \neq 0$ such that $\|b - b^0\| < \varepsilon$ and

$$\begin{aligned} &A(\hat{x}(b))s = 0, \\ &\left\langle s, \nabla^2_{xx} L(\hat{x}(b), \hat{\lambda}(b))s \right\rangle \le 0. \end{aligned} \tag{3.86}$$

We can always normalize s to get $\|s\| = 1$. Let us consider a sequence $\varepsilon_k \downarrow 0$ and the corresponding sequences $\{b^k\}$ and $\{s^k\}$ satisfying the above conditions. By construction, we have $b^k \to b^0$. The continuity of the implicit function renders $\hat{x}(b^k) \to x^0$ and $\hat{\lambda}(b^k) \to \lambda^0$. By choosing a subsequence (if needed) we also have the convergence of the sequence $\{s^k\}$ to some limit $\bar{s}$ of length 1. Passing to the limit in (3.86) we obtain a contradiction with the strong second order condition at x^0. □

If the linear independence condition or the semi-strong second order sufficient condition are not satisfied, the value function $f(\hat{x}(b, c))$ may be nondifferentiable. The sensitivity analysis then requires specialized tools of nonsmooth calculus, and exceeds the scope of this book.

In Section 4.6 we return to the sensitivity analysis in the convex case, without the strong second order sufficient condition.

EXERCISES

3.1. Let

$$X = \{x \in \mathbb{R}^n : \langle a_i, x \rangle \le b_i, \ i = 1, \dots, m\},$$

with given $a_i \in \mathbb{R}^n$ and $b_i \in \mathbb{R}$, $i = 1, \dots, m$. The *analytic center* of X is defined as the point at which the function

$$f(x) = \big(b_1 - \langle a_1, x\rangle\big)\big(b_2 - \langle a_2, x\rangle\big) \cdots \big(b_m - \langle a_m, x\rangle\big)$$

attains its maximum. Find an algebraic characterization of the analytic center.

3.2. We have a discrete real random variable Y with realizations y_k, $k = 1, \dots, n$, attained with probabilities $p_k > 0$, $\sum_{k=1}^{n} p_k = 1$.

(a) Find the minimum of the function

$$f(x) = \mathbb{E}\big[(Y - x)^2\big].$$

(b) Find the minimum of the function

$$f(x) = \mathbb{E}\big[|Y - x|\big].$$

(c) For $\alpha \in (0, 1)$ find the minimum of the function

$$f(x) = \mathbb{E}\big[\max\big(\alpha(Y - x), (1 - \alpha)(x - Y)\big)\big].$$

3.3. Consider the problem:

$$\begin{aligned} \text{minimize} \quad & (x_1 - 2)^2 + (x_2 - 1)^2 \\ \text{subject to} \quad & x_1 + x_2 \le b, \\ & x_2 \ge 0. \end{aligned}$$

How do the solution $\hat{x}(b)$ and the optimal value $f(\hat{x}(b))$ depend on the parameter b? Calculate the derivative

$$\frac{df(\hat{x}(b))}{db}$$

and relate it to the value of the Lagrange multiplier.

3.4. Consider the problem:

$$\begin{aligned} \text{minimize} \quad & -x_1 + x_2 \\ \text{subject to} \quad & (x_1 - 2)^3 + x_2 \le b, \\ & x_2 \ge 0. \end{aligned}$$

How do the solution $\hat{x}(b)$ and the optimal value $f(\hat{x}(b))$ depend on the parameter b? Calculate the derivative

$$\frac{df(\hat{x}(b))}{db}$$

and relate it to the value of the Lagrange multiplier.

3.5. Consider the problem

$$\begin{aligned} &\text{maximize } x_1 x_2 \\ &\text{subject to } x_1 + 2x_2 = 4. \end{aligned}$$

Is the objective function convex or concave? Is the feasible set convex? Using necessary conditions of optimality find the optimal solution. Prove its optimality by second order sufficient conditions.

3.6. Describe the orthogonal projection on the set

$$X = \{x \in \mathbb{R}^n : \|x\| \le 1,\ x \ge 0\}.$$

3.7. A ship at sea is observed from three stations located on the coastline 500 meters apart. At each station i, the angle α_i between the line to the ship and the normal to the coastline is measured (see Figure 3.8). The observed angles (which are subject

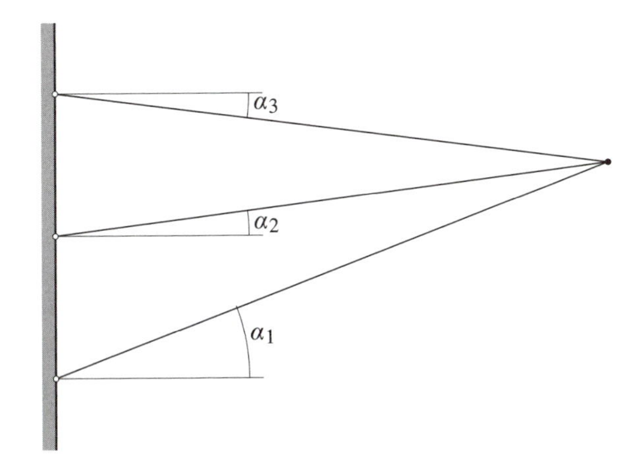

Figure 3.8. Three stations observe the ship (Exercise 3.7).

to measurement errors) are

$$\tilde{\alpha}_1 = 0.083, \quad \tilde{\alpha}_2 = 0.024, \quad \tilde{\alpha}_3 = -0.017$$

(in radians). Correct these results in a consistent way, so that the lines given by the angles cross at one point, and the sum of the squares of the differences of the errors is minimized. Then determine the location of the ship. Use the approximation $\tan\alpha \approx \alpha$ for small angles. *Hint:* Express the condition that the lines cross at one point as a linear constraint on the angles.

3.8. The output variable $y \in \mathbb{R}$ of a system depends on the input variable $x \in \mathbb{R}$. We have collected observations $\tilde{y}_i$, $i = 1, \ldots, N$, for N different settings x_i of the input variable x, with $x_1 < x_2 < \cdots < x_N$.

(a) We know that the dependence of y on x should be nondecreasing. Formulate the problem of finding new values y_i such that $y_1 \le y_2 \le \cdots \le y_N$ and the sum of the squares of adjustments is minimized.

(b) We know that the dependence of y on x should be convex and nondecreasing. Formulate the problem of finding new values y_i such that this condition is satisfied and the sum of the squares of adjustments is minimized.

(c) Solve both problems for $N = 4$, input values x_i equal to 1, 2, 5, and 10, and the corresponding observations $\tilde{y}_i$ equal to 2, 1, 5, and 9.

3.9. We have three assets with uncorrelated returns R_i, $i = 1, 2, 3$. They all have variance 1, and the mean values are 1, 2, and 3, respectively. We consider nonnegative portfolio allocations x_1, x_2, x_3 totaling one, and the portfolio return

$$R(x) = x_1 R_1 + x_2 R_2 + x_2 R_3$$

as a function of x. Let $r(x)$ denote its mean, and $V(x)$ its variance. We considered this example in the Introduction (see (1.4)).

(a) Find the minimum variance portfolio as the solution of the problem

$$\begin{aligned} &\text{minimize} \quad V(x) \\ &\text{subject to} \quad x \in X_0, \end{aligned}$$

where X_0 is the set of possible allocations. Denote by m_0 its expected return.

(b) The mean–variance efficient portfolios are computed by solving the problem

$$\begin{aligned} &\text{minimize} \quad V(x) \\ &\text{subject to} \quad r(x) = m, \\ &\qquad\qquad\quad x \in X_0, \end{aligned}$$

for all $m \ge m_0$ for which this problem has a solution. Find the set of all mean–variance efficient portfolios (so-called *efficient frontier*).

(c) Assume that in addition we have a risk-free asset with a guaranteed return of 0.5. Describe the efficient frontier in this case.

3.10. Let $p > 1$ and let

$$\|x\|_p = \Big(\sum_{j=1}^{n} |x_j|^p \Big)^{\frac{1}{p}}.$$

(a) Solve the problem

$$\begin{aligned} &\text{minimize} \quad c^T x \\ &\text{subject to} \quad \|x\|_p \le 1. \end{aligned}$$

(b) Prove that the point $\hat{x}$ found in (a) solves the problem

$$\begin{aligned} &\text{minimize } \|x\|_p \\ &\text{subject to } c^T x \le c^T \hat{x}. \end{aligned}$$

3.11. Derive the necessary conditions of optimality for the problem

$$\begin{aligned} &\text{minimize } c^T x - \varepsilon \sum_{j=1}^{n} \ln x_j \\ &\text{subject to } Ax = b, \end{aligned}$$

where $\varepsilon > 0$, A is an $m \times n$ matrix of rank m, and c and b are given.

3.12. A dynamical system evolves in time according to the equation

$$x(t+1) = \frac{1}{2}x(t) + u(t), \quad t = 0, 1, \dots, T-1,$$

in which $x(t) \in \mathbb{R}$ denotes the state value at time t and $u(t) \in \mathbb{R}$ is the control value at time t. The initial state is $x(0) = 1$. Find the sequence of controls $u(0), \dots, u(T-1)$ and the resulting sequence of states $x(0), \dots, x(T)$ that minimize the cost function of the system

$$F(x, u) = \frac{1}{2}\sum_{t=0}^{T-1} \big[u(t)\big]^2 + \frac{1}{2}\big[x(T)\big]^2.$$

How will the solution will change if we additionally impose the requirement that $x(T) = 0$?

3.13. The performance of a technical device is described by m quality measures $f_j : \mathbb{R}^n \to \mathbb{R}$, $j = 1, \dots, m$. The smaller their values the better the performance is. The requirements of technical feasibility have been formulated in the form of s inequalities:

$$g_i(x) \le 0, \quad i = 1, \dots, s.$$

The vector $x \in \mathbb{R}^n$ represents design decisions and $g_i : \mathbb{R}^n \to \mathbb{R}$, $i = 1, \dots, s$. All functions $f_j(\cdot)$ and $g_i(\cdot)$ are continuously differentiable.

The builder is considering three different approaches to the design problem.

Approach 1: She selects one of the performance measures, for example $f_1(x)$, to minimize, while keeping all the other measures $f_j(x)$ below some specified target levels b_j, $j = 2, \dots, m$.

Approach 2: The builder creates an aggregate objective function

$$f(x) = \sum_{j=1}^{m} w_j f_j(x),$$

where $w_j \ge 0$ are some selected weights, at least one of which is positive.

Approach 3: She uses target values b_j and weights w_j to create an aggregate objective function

$$F(x) = \max_{1 \le j \le m} w_j \big(f_j(x) - b_j\big)$$

representing the worst weighted excess over the targets.

(a) For each of these three approaches formulate the resulting nonlinear optimization problem.

(b) Assuming that constraint qualification is satisfied, formulate the first order necessary conditions of optimality for the corresponding problem.

(c) Prove that for every solution obtained by Approach 2 we can find target levels b_j such that the same solution is optimal in Approach 1 and in Approach 3.

(d) Prove that if all functions are convex and a constraint qualification condition is satisfied, for every solution obtained by Approach 1 we can find weights w_j such that the same solution is optimal in Approach 2 and Approach 3.

(e) Prove that for every solution obtained by Approach 3 we can find a performance measure to minimize so that the same solution is optimal in Approach 1. Prove that if the functions are convex then this optimal solution is also optimal in Approach 2.

3.14. A large retailer wants to establish n distribution centers to serve m stores located at points $s^1, \dots, s^m$. The average monthly demand of the jth store is D_j, $j = 1, \dots, m$, and the cost of supplying a store from a center is proportional to the distance between them, and to the amount supplied. For simplicity we do not distinguish different products. Formulate this problem as a nonlinear optimization problem with a smooth objective function and convex constraints. Write and analyze the necessary conditions of optimality.

3.15. Consider problem (1.7) of optimal processing the signal from n receivers. Formulate for it the necessary and sufficient conditions of optimality. Assume that the matrix Θ is positive definite.

3.16. Consider the nonlinear optimization problem

$$\begin{aligned} &\text{minimize} \;\; f(x, z) \\ &\text{subject to} \;\; g_i(x, z) \le 0, \quad i = 1, \dots, m, \end{aligned}$$

with the decision vector $x \in \mathbb{R}^n$ and some model parameters $z \in \mathbb{R}^l$. We know that $z \in Z$ for some compact set $Z \subset \mathbb{R}^l$, but the exact value of z is not known.

We assume that the functions $f : \mathbb{R}^n \times \mathbb{R}^l \to \mathbb{R}$ and $g_i : \mathbb{R}^n \times \mathbb{R}^l \to \mathbb{R}$, $i = 1, \dots, m$, are convex with respect to the first argument, for all values of the second argument $z \in Z$.

The *principle of guaranteed result* requires determination of the values of decision variables that minimize the worst possible value of $f(\cdot)$, subject to the requirement that the constraints are satisfied for all possible values of z in Z. Formulate this problem as a convex optimization problem, and formulate necessary conditions of optimality for it.

3.17. Consider problem (3.82) under the conditions of Theorem 3.49. Calculate the derivatives of the optimal solution and the Lagrange multipliers with respect to the parameters b and c.

3.18. Consider the convex programming problem

$$\text{minimize } f(x), \quad \text{subject to } x \in \bigcap_{i=1}^{m} X_i,$$

where $f : \mathbb{R}^n \to \mathbb{R}$ is a convex function, and the sets X_i, $i = 1, \dots, m$ are closed convex subsets of $\mathbb{R}^n$. Suppose it has an optimal solution $\hat{x}$. Prove that there exist at most n sets $X_{i_1}, \dots, X_{i_k}$, $k \le n$, such that $\hat{x}$ is also an optimal solution of the problem

$$\text{minimize } f(x), \quad \text{subject to } x \in \bigcap_{j=1}^{k} X_{i_j}.$$

Hint: Suppose the statement is not true, and apply Helly's theorem (Theorem 3.10 on page 97) to obtain a contradiction.

Chapter Four

Lagrangian Duality

4.1 THE DUAL PROBLEM

We start from the general nonlinear optimization problem

$$\begin{aligned} \text{minimize} \quad & f(x) \\ \text{subject to} \quad & g_i(x) \le 0, \quad i = 1, \dots, m, \\ & h_i(x) = 0, \quad i = 1, \dots, p, \\ & x \in X_0, \end{aligned} \tag{4.1}$$

with some functions $f : \mathbb{R}^n \to \mathbb{R}$, $g_i : \mathbb{R}^n \to \mathbb{R}$, $i = 1, \dots, m$, $h_i : \mathbb{R}^n \to \mathbb{R}$, $i = 1, \dots, p$, and with a set $X_0 \subset \mathbb{R}^n$. At this point we do not make any convexity assumptions about the functions of this problem and about the set X_0.

The Lagrangian has the form:

$$L(x, \lambda, \mu) = f(x) + \sum_{i=1}^{m} \lambda_i g_i(x) + \sum_{i=1}^{p} \mu_i h_i(x). \tag{4.2}$$

We view it as a function of *both* $x \in X_0$ and $(\lambda, \mu) \in \Lambda_0$, where

$$\Lambda_0 = \mathbb{R}^m_+ \times \mathbb{R}^p.$$

We define the *primal function* associated with problem (4.1) as

$$L_P(x) \triangleq \sup_{(\lambda,\mu)\in\Lambda_0} L(x, \lambda, \mu), \tag{4.3}$$

and the *dual function* as

$$L_D(\lambda, \mu) \triangleq \inf_{x\in X_0} L(x, \lambda, \mu). \tag{4.4}$$

If the supremum in (4.3) is $+\infty$ we set $L_P(x) = +\infty$. Similarly, if the infimum in (4.4) is $-\infty$ we set $L_D(\lambda, \mu) = -\infty$. Thus, we consider the primal function $L_P(\cdot)$ and the dual function $L_D(\cdot)$ as functions attaining values in the extended real line $\overline{\mathbb{R}} = \mathbb{R} \cup \{+\infty\} \cup \{-\infty\}$.

The *primal problem* is to find

$$\min_{x \in X_0} L_P(x), \tag{4.5}$$

and the *dual problem* is to find

$$\max_{(\lambda,\mu)\in\Lambda_0} L_D(\lambda, \mu). \tag{4.6}$$

The duality theory investigates relations between the primal problem and the dual problem.

We can easily calculate the primal function at each $x \in X_0$. If x satisfies all constraints of problem (4.1) then the terms in (4.2) that depend on λ are nonpositive and the μ-terms are identically 0. Consequently, $L_P(x) = f(x)$. Suppose at least one constraint is violated, e.g., $g_j(x) > 0$. Then increasing λ_j in (4.2) we can obtain arbitrarily large values of $\lambda_j g_j(x)$. Thus $L_P(x) = +\infty$ in this case. A similar argument can be made for a violated equality constraint, and thus

$$L_P(x) = \begin{cases} f(x) & \text{if } x \text{ is feasible for (4.1)}, \\ +\infty & \text{otherwise}. \end{cases} \tag{4.7}$$

We conclude that the primal problem (4.5) is equivalent to the original problem (4.1).

The dual function, however, is more difficult to calculate in an explicit form. In some cases, like linear or quadratic programming, a closed form of the dual problem can be derived. Before proceeding to the analysis of the properties of the dual problem, we present several examples.

Example 4.1. Consider the projection problem,

$$\begin{aligned} &\text{minimize} \ \ \frac{1}{2}\|x - z\|^2 \\ &\text{subject to} \ \ Ax = 0, \end{aligned}$$

where $z \in \mathbb{R}^n$ is fixed, and A is a matrix of dimension $m \times n$. We considered this problem in Section 2.1.2.

Introducing multipliers $\mu \in \mathbb{R}^m$ we can write the Lagrangian:

$$L(x, \mu) = \frac{1}{2}\|x - z\|^2 + \langle \mu, Ax \rangle.$$

Differentiation with respect to x yields the minimizer in the problem at the right hand side of (4.4):

$$\hat{x}(\mu) = z - A^T \mu.$$

This can be substituted to the Lagrangian to obtain the dual function. The dual problem takes on the form:

$$\text{maximize} \; -\frac{1}{2}\|A^T\mu\|^2 + \langle \mu, Az \rangle.$$

It is an unconstrained quadratic optimization problem. If we add and subtract the constant $\frac{1}{2}\|z\|^2$ to complement to the full square, we can write the dual problem as follows:

$$\text{maximize} \; \frac{1}{2}\big[-\|A^T\mu - z\|^2 + \|z\|^2 \big].$$

Its solution, $A^T\hat{\mu}$, is the point of the range of A^T that is closest to z: the projection of z onto the orthogonal subspace to the null space of A.

We can also note that the optimal values of the original and the dual problems are equal, because of the Pythagorean Theorem:

$$\|z\|^2 = \|\hat{x} - z\|^2 + \|A^T\hat{\mu} - z\|^2.$$

The equality of the optimal values of both problems is not coincidental. We explain it in the next section.

Example 4.2. Consider the linear programming problem

$$\begin{aligned} \text{minimize} \;& \langle c, x \rangle \\ \text{subject to} \;& Ax \geq b, \\ & x \geq 0, \end{aligned} \tag{4.8}$$

with a matrix A of dimension $m \times n$, $c \in \mathbb{R}^n$ and $b \in \mathbb{R}^m$. In (4.1) we have $f(x) = \langle c, x \rangle$, $g(x) = b - Ax$ and $X_0 = \mathbb{R}^n_+$. We formulate the Lagrangian:

$$L(x, \lambda) = \langle c, x \rangle + \langle \lambda, b - Ax \rangle = \langle c - A^T\lambda, x \rangle + \langle b, \lambda \rangle.$$

The dual function has the form

$$L_D(\lambda) = \langle b, \lambda \rangle + \inf_{x \geq 0} \langle c - A^T\lambda, x \rangle.$$

By Lemma 2.26 on page 145, the infimum above is finite if and only if $c - A^T\lambda \geq 0$. In this case the infimum value is just zero. We conclude that

$$L_D(\lambda) = \begin{cases} \langle b, \lambda \rangle & \text{if } A^T\lambda \leq c, \\ -\infty & \text{otherwise.} \end{cases}$$

The dual problem can therefore be formulated as the linear programming problem

$$\begin{aligned} \text{maximize} \;& \langle b, \lambda \rangle \\ \text{subject to} \;& A^T\lambda \leq c, \\ & \lambda \geq 0. \end{aligned} \tag{4.9}$$

A similar analysis can be carried out for a formulation of a linear programming problem with equality constraints.

It is instructive to consider the dual problem to the dual problem (4.9). We change the sign of the objective function to obtain a minimization problem and we use $u \in \mathbb{R}^n_+$ to denote the Lagrange multipliers associated with the constraints. We define the Lagrangian:

$$l(\lambda, u) = -\langle b, \lambda \rangle + \langle u, A^T\lambda - c \rangle = \langle Au - b, \lambda \rangle - \langle c, u \rangle.$$

The dual function can be calculated similarly to the previous case:

$$l_D(u) = \inf_{\lambda \geq 0} l(\lambda, u) = \begin{cases} -\langle c, u \rangle & \text{if } Au \geq b, \\ -\infty & \text{otherwise.} \end{cases}$$

We can thus write the dual problem as

$$\begin{aligned} \text{maximize} \quad & -\langle c, u \rangle \\ \text{subject to} \quad & Au \geq b, \\ & u \geq 0. \end{aligned}$$

Changing the sign of the objective function again, we arrive at the primal problem (4.8).

It should be stressed that the fact that the dual problem to the dual problem coincides with the original problem is due to the bilinear form of the Lagrangian. This is true for linear programming and, more generally, for problems with linear objective, linear constraints and additional cone constraint, as in the next example.

Example 4.3. Given symmetric matrices $A_1, \ldots, A_m$ and C of dimension n, the *semidefinite programming problem* is formulated as follows:

$$\begin{aligned} \text{minimize} \quad & \operatorname{tr}(CX) \\ \text{subject to} \quad & \operatorname{tr}(A_i X) = b_i, \quad i = 1, \ldots, m, \\ & X \in \mathbb{S}^n_+. \end{aligned} \tag{4.10}$$

The last condition means that the decision variable X in this problem is a symmetric positive semidefinite matrix of dimension n. We have considered this problem in Example 3.36 on page 129. We show now that its dual problem can be written explicitly.

Our problem has form (4.1) with

$$\begin{aligned} f(X) &= \langle C, X \rangle_{\mathbb{S}} \\ h_i(X) &= b_i - \langle A_i, X \rangle_{\mathbb{S}}, \quad i = 1, \ldots, m, \\ X_0 &= \mathbb{S}^n_+, \end{aligned}$$

with the symbol $\langle \cdot, \cdot \rangle_{\mathbb{S}}$ denoting the Frobenius inner product (see page 31). The functions $f(\cdot)$ and $h_i(\cdot)$ are affine. Assume Slater's constraint qualification condition (see page 109). It means that there exists a positive definite matrix X_S satisfying the equality constraints.

The Lagrangian has the form

$$L(X,\mu) = \langle C, X\rangle_{\mathbb{S}} + \sum_{i=1}^{m} \mu_i \big(b_i - \langle A_i, X\rangle_{\mathbb{S}}\big)$$
$$= \Big\langle C - \sum_{i=1}^{m} \mu_i A_i, X\Big\rangle_{\mathbb{S}} + \langle b, \mu\rangle.$$

Let us calculate the dual function

$$L_D(\mu) = \inf_{X \in \mathbb{S}^n_+} \left[\Big\langle C - \sum_{i=1}^{m} \mu_i A_i, X\Big\rangle_{\mathbb{S}} + \langle b, \mu\rangle \right].$$

Consider for a fixed μ the problem

$$\underset{X \in \mathbb{S}^n_+}{\text{minimize}} \ \Big\langle C - \sum_{i=1}^{m} \mu_i A_i, X\Big\rangle_{\mathbb{S}}. \tag{4.11}$$

For the infimum to be finite, it is necessary and sufficient that the Frobenius scalar product of $C - \sum_{i=1}^{m} \mu_i A_i$ and X be bounded below for all X belonging to the cone $\mathbb{S}^n_+$. By Lemma 2.26 on page 29, this is equivalent to

$$C - \sum_{i=1}^{m} \mu_i A_i \in -\big[\mathbb{S}^n_+\big]^\circ.$$

The polar cone is calculated in Example 2.31: $\big[\mathbb{S}^n_+\big]^\circ = -\mathbb{S}^n_+$. Therefore $L_D(\mu) > -\infty$ if and only if

$$C - \sum_{i=1}^{m} \mu_i A_i \in \mathbb{S}^n_+. \tag{4.12}$$

Under this condition the lower bound of the scalar product in (4.11) is zero, and it is attained at $\hat{X}(\mu) = 0$. Then $L_D(\mu) = \langle b, \mu\rangle$.

We have arrived at the following formulation of the dual problem:

$$\begin{aligned} &\text{maximize} \ \langle b, \mu\rangle \\ &\text{subject to} \ C - \sum_{i=1}^{m} \mu_i A_i \in \mathbb{S}^n_+. \end{aligned} \tag{4.13}$$

It is a semidefinite programming problem again.

Let us return to the general nonlinear optimization problem (4.1) and the associated dual function (4.4). The dual function has many remarkable properties facilitating the solution of the dual problem.

LEMMA 4.4. *The dual function $L_D(\lambda, \mu)$ is concave.*

Proof. Since the Lagrangian $L(x, \lambda, \mu)$ is affine in (λ, μ) for every $x \in X_0$, the dual function is an infimum of a family of affine functions. Thus $-L_D(\cdot)$ is a supremum of a family of affine functions and is convex by Lemma 2.58 on page 46. □

Our second observation uses the affine function at which the minimum is achieved.

LEMMA 4.5. *Assume that for (λ^0, μ^0) we can find $x^0 \in X_0$ such that*

$$L_D(\lambda^0, \mu^0) = L(x^0, \lambda^0, \mu^0).$$

Then for all (λ, μ) we have

$$L_D(\lambda, \mu) \le L_D(\lambda^0, \mu^0) + \langle g(x^0), \lambda - \lambda^0 \rangle + \langle h(x^0), \mu - \mu^0 \rangle, \qquad (4.14)$$

where $g(x)$ and $h(x)$ are the vectors with coordinates $g_i(x)$, $i = 1, \dots, m$, and $h_i(x)$, $i = 1, \dots, p$, respectively.

Proof. By the definition of the dual function

$$L_D(\lambda, \mu) \le L(x^0, \lambda, \mu) = L(x^0, \lambda^0, \mu^0) + \langle g(x^0), \lambda - \lambda^0 \rangle + \langle h(x^0), \mu - \mu^0 \rangle,$$

which was what we set out to prove. □

We can define the subdifferential of the dual function at (λ^0, μ^0) as the set of vectors $(s_\lambda, s_\mu) \in \mathbb{R}^{m+p}$ such that

$$L_D(\lambda, \mu) \le L_D(\lambda^0, \mu^0) + \langle s_\lambda, \lambda - \lambda^0 \rangle + \langle s_\mu, \mu - \mu^0 \rangle \text{ for all } (\lambda, \mu) \in \mathbb{R}^{m+p}.$$

Note that $L_D(\cdot)$ is concave, and therefore we appropriately modify Definition 2.72 of the subdifferential of a convex function. Our definition corresponds to the set $-\partial\big[-L_D(\lambda^0, \mu^0)\big]$, and thus all properties of the subdifferential of a convex function can be easily translated to properties of a subdifferential of a concave function. We still use the notation $\partial L_D(\lambda^0, \mu^0)$ for the subdifferential; it never leads to any confusion.

If the set X_0 is compact, Theorem 2.87 provides the representation of the subdifferential:

$$\partial L_D(\lambda^0, \mu^0) = \operatorname{conv}\left(\bigcup_{x^0 \in \hat{X}(\lambda^0, \mu^0)} \begin{bmatrix} g(x^0) \\ h(x^0) \end{bmatrix} \right), \qquad (4.15)$$

where $\hat{X}(\lambda^0, \mu^0)$ is the set of the minimizers of the Lagrangian:

$$\hat{X}(\lambda^0, \mu^0) = \{x \in X_0 : L(x^0, \lambda^0, \mu^0) = L_D(\lambda^0, \mu^0)\}.$$

If the set X_0 is not compact, we can still describe the entire subdifferential of the dual function by using the fact that it is defined as a minimum of *affine* functions, rather than just concave functions. We present these calculations at the end of Section 4.6.

Formula (4.14) provides us with means to apply nonsmooth optimization methods to the solution of the dual problem. We discuss these methods in Chapter 7. But the main question is, of course: why should we solve the dual problem at all? In the next section we show that the solutions of the dual problem and the solutions of the primal problem are very closely related.

4.2 DUALITY RELATIONS

Let us return to the Lagrangian of problem (4.1):

$$L(x, \lambda, \mu) = f(x) + \sum_{i=1}^{m} \lambda_i g_i(x) + \sum_{i=1}^{p} \mu_i h_i(x).$$

Recall that we consider it as a function of both $x \in X_0$ and $(\lambda, \mu) \in \Lambda_0$, where $\Lambda_0 = \mathbb{R}^m_+ \times \mathbb{R}^p$. The following concept of a saddle point is central to the duality theory.

DEFINITION 4.6. A point $(\tilde{x}, (\tilde{\lambda}, \tilde{\mu})) \in X_0 \times \Lambda_0$ is called a *saddle point* of the Lagrangian, if for all $x \in X_0$ and all $(\lambda, \mu) \in \Lambda_0$ the following inequalities are satisfied:

$$L(\tilde{x}, \lambda, \mu) \le L(\tilde{x}, \tilde{\lambda}, \tilde{\mu}) \le L(x, \tilde{\lambda}, \tilde{\mu}). \tag{4.16}$$

In other words, a saddle point is such a point at which the maximum of the Lagrangian with respect to $(\lambda, \mu) \in \Lambda_0$ and the minimum with respect to $x \in X_0$ are attained:

$$\max_{(\lambda,\mu)\in\Lambda_0} L(\tilde{x}, \lambda, \mu) = L(\tilde{x}, \tilde{\lambda}, \tilde{\mu}) = \min_{x\in X_0} L(x, \tilde{\lambda}, \tilde{\mu}).$$

For convex optimization problems the optimal solution and its Lagrange multipliers (if they exist) constitute such a saddle point. The theorem below uses the first order optimality conditions in the subdifferential form (Theorem 3.34). An identical result holds for problem (4.1) with smooth convex functions, analyzed in Theorem 3.25.

THEOREM 4.7. *Assume that the functions $f(\cdot)$ and $g_i(\cdot)$, $i = 1, \dots, m$, in problem* (4.1) *are convex, the functions $h_i(\cdot)$, $i = 1, \dots, p$, are affine, and the set X_0 is convex. Then a point $\hat{x}$ satisfies the first order optimality conditions of Theorem* 3.34 *with Lagrange multipliers $(\hat{\lambda}, \hat{\mu})$ if and only if $(\hat{x}, (\hat{\lambda}, \hat{\mu}))$ is a saddle point of the Lagrangian.*

Proof. We provide the proof for the convex nonsmooth case. Readers who prefer to deal with the smooth case of Theorem 3.25 can substitute the gradients $\nabla_x L$, ∇f, and ∇g_i for the subdifferentials $\partial_x L$, ∂f, and ∂g_i.

Assume that $\hat{x}$ is an optimal solution satisfying the optimality conditions with multipliers $(\hat{\lambda}, \hat{\mu})$. For fixed values of $(\hat{\lambda}, \hat{\mu}) \in \Lambda_0$, the Lagrangian is a convex function of x. By Theorem 2.85, its subdifferential at $\hat{x}$ can be calculated as follows:

$$\partial_x L(\hat{x}, \hat{\lambda}, \hat{\mu}) = \partial f(\hat{x}) + \sum_{i=1}^{m} \hat{\lambda}_i \partial g_i(\hat{x}) + \sum_{i=1}^{p} \hat{\mu}_i \nabla h_i(\hat{x}).$$

The first order optimality condition of Theorem 3.33 has the form

$$0 \in \partial_x L(\hat{x}, \hat{\lambda}, \hat{\mu}) + N_{X_0}(\hat{x}). \tag{4.17}$$

It follows from Theorem 3.46 (or Theorem 3.24 in the smooth case) that

$$L(\hat{x}, \hat{\lambda}, \hat{\mu}) = \min_{x \in X_0} L(x, \hat{\lambda}, \hat{\mu}).$$

Thus the right inequality in (4.16) holds true for all $x \in X_0$. We shall prove the left inequality. By the complementarity condition (3.50), we have

$$L(\hat{x}, \hat{\lambda}, \hat{\mu}) = f(\hat{x}).$$

Since $g_i(\hat{x}) \leq 0$, $i = 1, \ldots, m$, and $h_i(\hat{x}) = 0$, $i = 1, \ldots, p$, for every $(\lambda, \mu) \in \Lambda_0$ we have the inequality

$$L(\hat{x}, \lambda, \mu) = f(\hat{x}) + \sum_{i=1}^{m} \lambda_i g_i(\hat{x}) + \sum_{i=1}^{p} \mu_i h_i(\hat{x}) \leq f(\hat{x}).$$

Combining the last two displayed relations we obtain the left inequality in (4.16) for all $(\lambda, \mu) \in \Lambda_0$. Therefore the point $(\hat{x}, (\hat{\lambda}, \hat{\mu}))$ is a saddle point of the Lagrangian.

Let us now prove the converse. Suppose $(\hat{x}, (\hat{\lambda}, \hat{\mu}))$ is a saddle point of the Lagrangian. By virtue of Theorem 3.46, the right inequality in (4.16) is equivalent to (4.17). This is identical with (3.49).

The left of the saddle point conditions (4.16) implies that $\sum_{i=1}^{m} \lambda_i g_i(\hat{x})$ is bounded from above for all $\lambda \geq 0$. Hence $g_i(\hat{x}) \leq 0$, $i = 1, \ldots, m$. Similarly, $h_i(\hat{x}) = 0$, $i = 1, \ldots, p$. Consequently, the point $\hat{x}$ is feasible. As the maximum of $\sum_{i=1}^{m} \lambda_i g_i(\hat{x})$ is attained at $\hat{\lambda}$, we have $\hat{\lambda}_i g_i(\hat{x}) = 0$ for all $i = 1, \ldots, m$. Thus, (3.50) holds true as well. □

The relation between saddle points and the primal and dual problems is straightforward.

Theorem 4.8. *If the Lagrangian has a saddle point $(\tilde{x}, \tilde{\lambda}, \tilde{\mu})$, then $\tilde{x}$ is a solution of the primal problem, $(\tilde{\lambda}, \tilde{\mu})$ is a solution of the dual problem, and the* duality relation *holds true:*

$$\min_{x \in X_0} L_P(x) = \max_{(\lambda,\mu)\in\Lambda_0} L_D(\lambda, \mu). \tag{4.18}$$

Proof. By the saddle point relations, for every $x \in X_0$ and every $(\lambda, \mu) \in \Lambda_0$ we have

$$L(\tilde{x}, \lambda, \mu) \leq L(\tilde{x}, \tilde{\lambda}, \tilde{\mu}) \leq L(x, \tilde{\lambda}, \tilde{\mu}). \tag{4.19}$$

This means that $L_P(\tilde{x}) = L_D(\tilde{\lambda}, \tilde{\mu})$. The minimization of the left hand side of (4.19) in x and the maximization of the right hand side in (λ, μ) may only strengthen the inequalities, so

$$L_D(\lambda, \mu) \leq L(\tilde{x}, \lambda, \mu) \leq L_D(\tilde{\lambda}, \tilde{\mu}) = L_P(\tilde{x}) \leq L(x, \tilde{\lambda}, \tilde{\mu}) \leq L_P(x),$$

which completes the proof of the duality relation. □

Note that no explicit convexity assumptions are made in this theorem. The duality relation is illustrated in Figure 4.1.

We can also state the converse implication to Theorem 4.8.

Theorem 4.9. *Assume that the duality relation* (4.18) *is satisfied with finite values of the primal and dual functions. Then for every solution $\hat{x}$ of the primal problem and for every solution $(\hat{\lambda}, \hat{\mu})$ of the dual problem the point $(\hat{x}, \hat{\lambda}, \hat{\mu})$ is a saddle point of the Lagrangian.*

Proof. The definition of the primal function yields

$$L_P(\hat{x}) = \max_{(\lambda,\mu)\in\Lambda_0} L(\hat{x}, \lambda, \mu) \geq L(\hat{x}, \hat{\lambda}, \hat{\mu}).$$

Similarly, from the definition of the dual function we obtain

$$L_D(\hat{\lambda}, \hat{\mu}) = \min_{x \in X} L(x, \hat{\lambda}, \hat{\mu}) \leq L(\hat{x}, \hat{\lambda}, \hat{\mu}).$$

Since by (4.18)

$$L_P(\hat{x}) = L_D(\hat{\lambda}, \hat{\mu}),$$

we must have

$$L(\hat{x}, \lambda, \mu) \leq L_P(\hat{x}) = L(\hat{x}, \hat{\lambda}, \hat{\mu}) = L_D(\hat{\lambda}, \hat{\mu}) \leq L(x, \hat{\lambda}, \hat{\mu}) \tag{4.20}$$

for all $x \in X_0$ and all $(\lambda, \mu) \in \Lambda_0$. Thus $(\hat{x}, \hat{\lambda}, \hat{\mu})$ is a saddle point. □

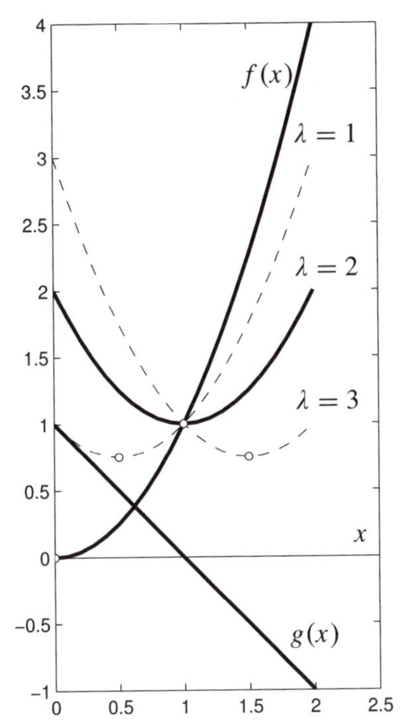

Figure 4.1. Illustration of the duality relation for the problem of minimizing $f(x) = x^2$ subject to $g(x) = 1 - x \le 0$. The graphs represent the Lagrangian, as a function of x, for several fixed values of λ. The minimum of the Lagrangian, i.e., the dual function, is maximized when it is equal to the minimum of the constrained problem.

Based on these results we can look for a solution to the primal problem by first solving the dual problem to get $(\hat{\lambda}, \hat{\mu})$ and then determining the primal solution $\hat{x}$ from the saddle point conditions.

THEOREM 4.10. *Assume that the duality relation* (4.18) *holds true. If* $(\bar{\lambda}, \bar{\mu}) \in \Lambda_0$ *is a feasible point of the dual problem, then every point* $\hat{x} \in X_0$ *such that*

(i) $L(\hat{x}, \bar{\lambda}, \bar{\mu}) = \min_{x \in X_0} L(x, \bar{\lambda}, \bar{\mu})$;

(ii) *all constraints of* (4.1) *are satisfied at* $\hat{x}$;

(iii) $\bar{\lambda}_i g_i(\hat{x}) = 0, \ i = 1, \dots, m,$

is a solution of problem (4.1).

Proof. Conditions (i), (ii) and (iii) imply that $(\hat{x}, \bar{\lambda}, \bar{\mu})$ is a saddle point of the Lagrangian, and the result follows from Theorem 4.8. In fact, the point $(\bar{\lambda}, \bar{\mu})$ is then the optimal solution of the dual problem. □

The following example from game theory motivated the development of the duality theory.

Example 4.11. A two-player game is defined as follows. We have a matrix A of dimension $m \times n$. Player C chooses a column j of the matrix, that is, a number between 1 and n. Player R chooses a row of the matrix, a number i between 1 and m. None of them knows the opponent's decision. Then the decisions are revealed and the matrix entry a_{ij} in row i and column j is the amount that C pays R (if it is negative, R pays C).

It is difficult to decide on the course of action in such a game. The key concept that radically clarifies the problem of choosing the best play is that of a *mixed strategy*. A mixed strategy of Player C is a probability distribution x on the set of columns $\{1, \dots, n\}$. In other words, we assume that Player C chooses her column at random, according to some probabilities x_j, $j = 1, \dots, n$. The set of all possible mixed strategies is equal to

$$X_0 = \{x \in \mathbb{R}^n : \sum_{j=1}^{n} x_j = 1,\ x_j \ge 0,\ j = 1, \dots, n\}.$$

If Player R chooses row i then the expected amount that C pays R equals

$$\sum_{j=1}^{n} a_{ij} x_j = \langle a_i, x \rangle,$$

with a_i denoting the ith row of A. To minimize the worst possible outcome, Player C solves the following problem:

$$\operatorname*{minimize}_{x \in X_0} \max_{1 \le i \le m} \langle a_i, x \rangle. \tag{4.21}$$

As usual in min-max problems, we can reformulate it as a constrained problem:

$$\begin{aligned} &\operatorname*{minimize}_{v,x}\ v \\ &\text{subject to}\ \langle a_i, x \rangle - v \le 0, \quad i = 1, \dots, m, \\ &\qquad\qquad\ x \in X_0. \end{aligned}$$

The set X_0 is a compact polyhedron and the constraints are affine. Therefore, problem (4.21) has an optimal solution and the first order necessary conditions of optimality are satisfied.

Denote by y the vector of Lagrange multipliers associated with the constraints. We use the notation y instead of λ because of the specific interpretation of the multipliers in our context. The Lagrangian has the form

$$L(v, x, y) = v + \sum_{i=1}^{m} y_i\big(\langle a_i, x\rangle - v\big) = v\Big(1 - \sum_{i=1}^{m} y_i\Big) + \langle y, Ax\rangle.$$

Let us calculate the dual function:

$$L_D(y) = \inf_{\substack{v\in\mathbb{R}\\ x\in X_0}} L(v, x, y) = \inf_{v\in\mathbb{R}} v\Big(1 - \sum_{i=1}^{m} y_i\Big) + \min_{x\in X_0}\langle y, Ax\rangle.$$

We see that the infimum with respect to v is finite if and only if $\sum_{i=1}^{n} y_i = 1$. Thus y is an element of the set

$$Y_0 = \{y \in \mathbb{R}^m : \sum_{i=1}^{n} y_i = 1,\ y_i \geq 0,\ i = 1, \ldots, m\}.$$

The dual function becomes

$$L_D(y) = \min_{x\in X_0}\langle y, Ax\rangle = \min_{x\in X_0}\langle A^T y, x\rangle.$$

If we use a^j to denote the jth column of A, we can rewrite the dual function as follows:

$$L_D(y) = \min_{x\in X_0} \sum_{j=1}^{n} x_j \langle a^j, y\rangle = \min_{1\leq j\leq n}\langle a^j, y\rangle.$$

The dual problem takes on the form

$$\underset{y\in Y_0}{\text{maximize}} \min_{1\leq j\leq n}\langle a^j, y\rangle. \tag{4.22}$$

A comparison with (4.21) reveals that the dual problem is the problem of finding the best mixed strategy of Player R. By using the fact that

$$\max_{1\leq i\leq m}\langle a_i, x\rangle = \max_{y\in Y_0}\langle y, Ax\rangle,$$

we can rewrite both problems as follows. The *primal problem* is

$$\underset{x\in X_0}{\text{minimize}} \max_{y\in Y_0}\langle y, Ax\rangle,$$

and the *dual problem* is

$$\underset{y\in Y_0}{\text{maximize}} \min_{x\in X_0}\langle y, Ax\rangle.$$

The Lagrangian for every $x \in X_0$ and $y \in Y_0$ has the form

$$L(x, y) = \langle y, Ax\rangle.$$

From the duality theory it follows that the optimal values of both problems are identical and that their solutions, $\hat{x}$ and $\hat{y}$, form a saddle point of the Lagrangian

$$\langle y, A\hat{x}\rangle \le \langle \hat{y}, A\hat{x}\rangle \le \langle \hat{y}, Ax\rangle \quad \text{for all} \quad x \in X_0, \ y \in Y_0.$$

It is the equilibrium of the game: if both players follow the mixed strategies $\hat{x}$ and $\hat{y}$, it is not profitable for any of them to deviate from their solutions.

Our second example is from the area of finance.

Example 4.12. We have n securities with present prices $c_1, \dots, c_n$. The securities can be either bought or sold *short*, that is, borrowed and sold for cash. Any amounts can be traded.

At some future time one of m states may occur. The price of security j in state i will be equal to a_{ij}, $i = 1, \dots, m$, $j = 1, \dots, n$. At this time we will liquidate our holdings: the securities held will be sold, and the short positions will be covered by purchasing and returning the amounts borrowed.

An *arbitrage* is the existence of $\bar{x} \in \mathbb{R}^n$ such that $\langle c, \bar{x}\rangle < 0$ and $A\bar{x} \ge 0$. Such a portfolio $\bar{x}$ can be "purchased" with an immediate profit and then liquidated at a future point in time with no additional loss, whatever the state. We analyzed this problem in Example 2.30 on page 31. Now we can interpret the results in the context of the duality theory.

Consider the linear programming problem

$$\begin{aligned} &\text{minimize} \ \langle c, x\rangle \\ &\text{subject to} \ Ax \ge 0. \end{aligned}$$

If this problem has a feasible solution $\bar{x}$ with $\langle c, \bar{x}\rangle < 0$, then no optimal solution exists, because the value of the objective goes to $-\infty$ at feasible points $M\bar{x}$, with $M \to \infty$ (this is the essence of the concept of arbitrage). Thus, an optimal solution exists if and only if the optimal value is 0.

Using $p \in \mathbb{R}^n$, $p \ge 0$, to denote the Lagrange multipliers, we can write the Lagrangian in the form

$$L(x, p) = \langle c, x\rangle - \langle p, Ax\rangle = \langle c - A^T p, x\rangle.$$

The infimum of this function with respect to $x \in \mathbb{R}^n$ is $-\infty$, unless $c - A^T p = 0$, in which case the infimum is simply zero. It follows that

$$L_D(p) = \inf_{x \in \mathbb{R}^n} L(x, p) = \begin{cases} 0 & \text{if } A^T p = c, \\ -\infty & \text{otherwise.} \end{cases}$$

The dual problem, therefore, has the simple form of a system of equations

$$A^T p = c.$$

This is the non-arbitrage condition of Example 2.30, and the Lagrange multipliers p are the state prices found there.

Finally, we have an illustration of relations of duality theory to expected utility theory in economics.

Example 4.13. Our decisions $x \in X \subset \mathbb{R}^n$ affect a certain random outcome $F(x)$. We assume that the number of possible elementary events is finite and we denote it by m. For every elementary event $i \in \{1, \dots, m\}$, we denote its probability by p_i. Obviously, $p_i > 0$ and $\sum_{i=1}^m p_i = 1$.

The realization of the random outcome $F(x)$ in event i is denoted by $f_i(x)$. To focus attention, we assume that we prefer larger values of $f_i(x)$ over smaller values (for example, $f_i(x)$ may represent profit in event i). We also assume that the functions $f_i : \mathbb{R}^n \to \mathbb{R}$ are concave, and the set X is convex.

One way to formulate a meaningful optimization problem, in the presence of uncertainty and with risk aversion, is to consider a *benchmark* random outcome Y, with realizations y_i, $i = 1, \dots, m$. We require that our random outcome $F(x)$ has an expected shortfall below each target level $t \in \mathbb{R}$ at most as large as that of Y:

$$\mathbb{E}\big[\max\big(0, t - F(x)\big)\big] \le \mathbb{E}\big[\max\big(0, t - Y\big)\big] \quad \text{for all} \quad t \in \mathbb{R}. \tag{4.23}$$

The symbol $\mathbb{E}[\cdot]$ denotes the expected value. In the theory of stochastic optimization models, constraint (4.23) is called the *second order stochastic dominance constraint*.

Observe that the functions $t \mapsto \mathbb{E}\big[\max\big(0, t - F(x)\big)\big]$ and $t \mapsto \mathbb{E}\big[\max\big(0, t - Y\big)\big]$ are convex. The second one is also piecewise linear, with break points at $y_1, \dots, y_m$. Thus it is sufficient to enforce constraint (4.23) at the break points $t = y_j$:

$$\mathbb{E}\big[\max\big(0, y_j - F(x)\big)\big] \le \mathbb{E}\big[\max\big(0, y_j - Y\big)\big] \quad \text{for all} \quad j = 1, \dots, m.$$

We formulate the optimization problem

$$\begin{aligned}
&\text{maximize} \quad \sum_{i=1}^m p_i f_i(x) \\
&\text{subject to} \quad \sum_{i=1}^m p_i \max\big(0, y_j - f_i(x)\big) \le \sum_{i=1}^m p_i \max\big(0, y_j - y_i\big), \ j = 1, \dots, m, \\
&\qquad\qquad\quad x \in X.
\end{aligned} \tag{4.24}$$

The objective function represents our desire to have the average value of the outcome $F(x)$ large, while the constraint represents risk aversion: we want to perform better with respect to each target level than the benchmark Y.

We can develop the dual problem to problem (4.24). Denoting by λ_j, $j = 1, \dots, m$, the nonnegative Lagrange multipliers associated with the shortfall constraints, we write the Lagrangian:

$$L(x, \lambda) = \sum_{i=1}^m p_i f_i(x) - \sum_{j=1}^m \lambda_j \sum_{i=1}^m p_i \big[\max\big(0, y_j - f_i(x)\big) - \max\big(0, y_j - y_i\big)\big]. \tag{4.25}$$

Define the function $u : \mathbb{R} \to \mathbb{R}$ as

$$u(t) = -\sum_{j=1}^{m} \lambda_j \max\big(0, y_j - t\big). \tag{4.26}$$

By changing the order of summation in (4.25) and using (4.26) we obtain the identity

$$-\sum_{j=1}^{m} \lambda_j \sum_{i=1}^{m} p_i \max\big(0, y_j - f_i(x)\big) = -\sum_{i=1}^{m} p_i \sum_{j=1}^{m} \lambda_j \max\big(0, y_j - f_i(x)\big)$$
$$= \sum_{i=1}^{m} p_i u\big(f_i(x)\big) = \mathbb{E}\big[u(F(x))\big].$$

Similarly,

$$\sum_{j=1}^{m} \lambda_j \sum_{i=1}^{m} p_i \max\big(0, y_j - y_i\big) = -\sum_{i=1}^{m} p_i u\big(y_i\big) = -\mathbb{E}\big[u(Y)\big].$$

The function $u(\cdot)$ is concave and piecewise linear with break points at $y_1, \dots, y_m$. It vanishes to the right of the largest realization of Y. Every nonnegative vector λ of Lagrange multipliers can be associated with a function $u(\cdot)$ defined in (4.26), and every function of such a form defines a vector of Lagrange multipliers $\lambda \geq 0$.

Thus we can rewrite the Lagrangian (4.25) as a function of x and $u(\cdot)$:

$$L(x, u) = \mathbb{E}\big[F(x)\big] + \mathbb{E}\big[u(F(x))\big] - \mathbb{E}\big[u(Y)\big]. \tag{4.27}$$

We keep the symbol L for the Lagrangian, although it is a different function now. Its arguments are $x \in \mathbb{R}^n$ and a concave function $u(\cdot)$. All duality results for the Lagrangian (4.25) can now be formulated in terms of the Lagrangian (4.27). The dual function has the form:

$$D(u) = \sup_{x \in X} \Big\{\mathbb{E}\big[F(x)\big] + \mathbb{E}\big[u(F(x))\big] - \mathbb{E}\big[u(Y)\big]\Big\}. \tag{4.28}$$

Observe that in this problem we do not maximize the expected value of the outcome, but rather the expected value of a certain *utility function* $\psi(t) = t + u(t)$,

$$\mathbb{E}\big[\psi(F(x))\big] = \mathbb{E}\big[F(x) + u(F(x))\big].$$

The dual problem has the form

$$\underset{u(\cdot)}{\text{minimize}}\ D(u),$$

where the function $u(\cdot)$ has to be chosen from the class of piecewise linear concave functions with break points at $y_1, \dots, y_m$, and equal to zero for all sufficiently large values of the argument. Its solution, $\hat{u}(\cdot)$, is such that a solution $\hat{x}$ of (4.28) satisfies the stochastic dominance constraint, and

$$\mathbb{E}\big[\hat{u}(F(\hat{x}))\big] = \mathbb{E}\big[\hat{u}(Y)\big].$$

4.3 CONIC PROGRAMMING

We now focus on the following problem structure:

$$\begin{aligned} &\text{minimize} \ \langle c, x\rangle \\ &\text{subject to} \ Ax = b, \\ &\qquad\qquad x \in K, \end{aligned} \tag{4.29}$$

where $x \in \mathbb{R}^n$, $c \in \mathbb{R}^n$, $b \in \mathbb{R}^m$, and A is an $m \times n$ matrix. The set K is a closed convex cone in $\mathbb{R}^n$.

The linear programming problem of Example 4.2 fits into this model with $K = \mathbb{R}^n_+$. Another special case is the semidefinite programming problem of Example 4.3, with x denoting a symmetric matrix of dimension d (so $n = d^2$), and the scalar product $\langle \cdot, \cdot \rangle$ understood in the Frobenius sense: $\langle c, x\rangle = \operatorname{tr}(cx)$. In this case A is a linear operator from the space of square matrices to $\mathbb{R}^m$ and therefore has the form

$$Ax = \begin{bmatrix} \operatorname{tr}(A_1 x) \\ \operatorname{tr}(A_2 x) \\ \dots \\ \operatorname{tr}(A_m x) \end{bmatrix},$$

with A_i, $i = 1, \dots, m$, being square matrices of dimension d. The cone K is the semidefinite cone $\mathbb{S}^n_+$.

The development of the dual problem follows the argument of Examples 4.2 and 4.3. We introduce the Lagrangian

$$L(x, \mu) = \langle c, x\rangle + \langle \mu, b - Ax\rangle$$

and the dual function

$$L_D(\mu) = \inf_{x\in K} L(x, \mu) = \inf_{x\in K} \langle c - A^*\mu, x\rangle + \langle \mu, b\rangle. \tag{4.30}$$

Here A^* denotes the adjoint operator to A and is defined by the equation $\langle \mu, Ax\rangle = \langle A^*\mu, x\rangle$. The specific algebraic form of A^* depends on the scalar product employed. For example, in linear programming we have $A^* = A^T$, and in semidefinite programming $A^*\mu = \sum_{i=1}^m \mu_i A_i$.

By Lemma 2.26, the infimum in (4.30) is finite if and only if $c - A^*\mu \in -K^\circ$, and then $L_D(\mu) = \langle \mu, b\rangle$. This leads to the following formulation of the dual problem:

$$\begin{aligned} &\text{maximize} \ \langle b, \mu\rangle \\ &\text{subject to} \ A^*\mu - c \in K^\circ. \end{aligned} \tag{4.31}$$

We are now ready to state the duality theorem for conic optimization problems.

THEOREM 4.14. *Assume that there exists a point* $x_s \in \operatorname{int} K$ *such that* $Ax_s = b$. *If the primal problem* (4.29) *has an optimal solution then the dual problem* (4.31) *has an optimal solution and the optimal values of both problems coincide.*

Proof. Recall from page 109 that the assumed existence of the point x_s is nothing but Slater's constraint qualification condition for problem (4.29). Under this condition, Theorem 3.34 implies that the optimal solution satisfies the first order conditions of optimality. Then it follows from Theorem 4.7 that the Lagrangian has a saddle point, and from Theorem 4.8 that the dual problem (4.31) has an optimal solution, and that the optimal values of both problems are equal. □

We can also conclude that under Slater's condition for every solution $\hat{x}$ of the primal problem and every solution $\hat{\mu}$ of the dual problem, the complementarity condition

$$\langle \hat{x}, A^*\hat{\mu} - c \rangle = 0 \tag{4.32}$$

holds true. Indeed, the constraint of the dual problem yields $\langle A^*\mu - c, \hat{x} \rangle \le 0$. Hence

$$\begin{aligned} \langle b, \hat{\mu} \rangle &\le \langle b, \hat{\mu} \rangle - \langle A^*\hat{\mu} - c, \hat{x} \rangle \\ &= \langle c, \hat{x} \rangle + \langle b - A\hat{x}, \hat{\mu} \rangle = \langle c, \hat{x} \rangle, \end{aligned}$$

with the equality occurring if and only if (4.32) is satisfied.

In linear programming the cone K is polyhedral, and we do not need Slater's condition for Theorem 4.14 to hold true.

If we develop the dual problem to the dual problem (4.31) we obtain the primal problem again (see Exercise 4.8). In linear programming, the existence of an optimal solution to the dual problem implies the duality relation and the existence of an optimal solution of the primal problem, by virtue of Theorem 4.14 (without Slater's condition). In semidefinite programming and conic programming, in general, application of Theorem 4.14 to the dual problem requires the dual Slater's condition: *there exists* $\mu_s \in \mathbb{R}^m$ *such that* $A^*\mu_s - c \in \operatorname{int}(K^\circ)$. We thus obtain the stronger version of Theorem 4.14.

THEOREM 4.15. *Assume that both primal and dual Slater's conditions are satisfied. Then the primal problem* (4.29) *has an optimal solution if and only if the dual problem* (4.31) *has an optimal solution, in which case the optimal values of both problems coincide.*

Example 4.16. Let A be an $m \times n$ matrix and $b \in \mathbb{R}^m$. Consider the approximation problem:

$$\text{minimize } \|b - Ax\|_\diamond, \tag{4.33}$$

where $\|\cdot\|_\diamond$ is a norm in $\mathbb{R}^m$. If $\|\cdot\|_\diamond = \|\cdot\|$, the above problem is simply the projection problem of Section 2.1.2, while for $\|\cdot\|_\diamond = \|\cdot\|_\infty$ we obtain the Chebyshev approximation problem of Example 3.32.

Introducing new variables $y \in \mathbb{R}^m$ and $v \in \mathbb{R}$, as in Example 3.39, we can equivalently formulate the problem as follows:

$$\begin{aligned} &\text{minimize } v \\ &\text{subject to } Ax + y = b, \\ &\qquad\qquad \|y\|_\diamond \le v. \end{aligned} \tag{4.34}$$

Since $K \triangleq \text{epi}(\|\cdot\|_\diamond)$ is a convex cone, we can compactly write the last constraint as

$$(y, v) \in K.$$

It follows that problem (4.34) is a conic programming problem. Specializing the derivation of (4.31), we arrive at the dual problem

$$\begin{aligned} &\text{maximize } \langle b, \mu \rangle \\ &\text{subject to } A^T\mu = 0, \\ &\qquad\qquad (\mu, -1) \in K^\circ. \end{aligned}$$

The polar cone to K has the form

$$K^\circ = \{(\lambda, \alpha) \in \mathbb{R}^m \times \mathbb{R} : \alpha \le -\|\lambda\|_*\},$$

where $\|\cdot\|_*$ is the dual norm (see Exercise 2.13). Therefore the dual problem can be equivalently written as follows:

$$\begin{aligned} &\text{maximize } \langle b, \mu \rangle \\ &\text{subject to } A^T\mu = 0, \\ &\qquad\qquad \|\mu\|_* \le 1. \end{aligned} \tag{4.35}$$

Problem (4.34) satisfies the constraint qualification condition, because v can be arbitrarily large. Therefore Theorem 4.14 applies and we conclude that the dual problem (4.35) has a solution, and that the optimal values of both problems coincide.

It might be instructive to notice that the optimality conditions for any of the problems (4.33) or (4.34) are special cases of the general conditions derived in Example 3.39 for smooth nonlinear mappings A.

Example 4.17. Consider the semidefinite programming problem (4.10) discussed in Example 4.3. Suppose the constraints $\langle A_i, X \rangle_{\mathbb{S}} = b_i$ imply that $\text{tr}(X) = k$. This is the case, for example, when (4.10) is a semidefinite relaxation of a 0–1 quadratic optimization problem, as in Exercise 4.11.

Problem (4.10) augmented with the redundant constraint has the form:

$$\begin{aligned} &\text{minimize } \langle C, X\rangle_{\mathbb{S}} \\ &\text{subject to } \langle A_i, X\rangle_{\mathbb{S}} = b_i, \quad i = 1, \dots, m, \\ &\qquad \langle I, X\rangle_{\mathbb{S}} = k, \\ &\qquad X \in \mathbb{S}^n_+. \end{aligned} \tag{4.36}$$

Associating with its constraints multipliers $\mu \in \mathbb{R}^m$ and $\mu_0 \in \mathbb{R}$, we obtain the following form of the dual problem (4.13):

$$\begin{aligned} &\text{maximize } \langle b, \mu\rangle + k\mu_0 \\ &\text{subject to } C - \sum_{i=1}^{m} \mu_i A_i - \mu_0 I \in \mathbb{S}^n_+. \end{aligned} \tag{4.37}$$

If problem (4.36) satisfies Slater's constraint qualification, we can use Theorem 4.14.

The constraint $\text{tr}(X) = k$ implies that every optimal solution $\hat{X}$ of problem (4.36) is nonzero. In view of the complementarity condition (4.32), every optimal solution of the dual problem must satisfy the equation

$$\left\langle C - \sum_{i=1}^{m} \mu_i A_i - \mu_0 I, \hat{X} \right\rangle_{\mathbb{S}} = 0.$$

Since $\hat{X} \neq 0$, the matrix $C - \sum_{i=1}^{m} \mu_i A_i - \mu_0 I$ is singular. As it is positive semidefinite, we must have

$$\lambda_{\min}\left(C - \sum_{i=1}^{m} \mu_i A_i - \mu_0 I\right) = 0.$$

We use the notation $\lambda_{\min}(A)$ and $\lambda_{\max}(A)$ to denote the smallest and largest eigenvalues of a symmetric matrix A. The last relation is equivalent to

$$\lambda_{\min}\left(C - \sum_{i=1}^{m} \mu_i A_i\right) = \mu_0. \tag{4.38}$$

It follows that the dual problem (4.37) is equivalent to

$$\underset{\mu \in \mathbb{R}^n}{\text{maximize}}\ \langle b, \mu\rangle + k\lambda_{\min}\left(C - \sum_{i=1}^{m} \mu_i A_i\right).$$

Since $\lambda_{\min}(A) = -\lambda_{\max}(-A)$, we can write this problem equivalently as an unconstrained convex optimization problem:

$$\underset{\mu \in \mathbb{R}^n}{\text{minimize}}\ k\lambda_{\max}\left(\sum_{i=1}^{m} \mu_i A_i - C\right) - \langle b, \mu\rangle. \tag{4.39}$$

We know from Example 2.89 on page 74 how to calculate subgradients of the maximum eigenvalue function, and thus we can apply methods of convex nonsmooth optimization to this problem.

By Lemma 2.83 on page 67, each subgradient of function (4.39) has the form

$$g = \begin{bmatrix} kA_1W - b_1 \\ kA_2W - b_2 \\ \vdots \\ kA_mW - b_m \end{bmatrix},$$

where $W \in \partial\big[\lambda_{\max}(Q)\big]$ at $Q = \sum_{i=1}^{m} \hat{\mu}_i A_i - C$. The last relation can be made more explicit by employing the results of Example 2.89:

$$\begin{gathered} \Big\langle \sum_{i=1}^{m} \hat{\mu}_i A_i - C, W \Big\rangle_{\mathbb{S}} = \lambda_{\max}\Big(\sum_{i=1}^{m} \hat{\mu}_i A_i - C \Big), \\ \operatorname{tr}(W) = 1, \\ W \in \mathbb{S}^n_+. \end{gathered} \tag{4.40}$$

At the optimal solution we have $g = 0$ and thus there exists $\widehat{W}$ satisfying the above three conditions, such that

$$kA_i\widehat{W} = b_i, \quad i = 1, \dots, m.$$

We shall show that $\hat{X} = k\widehat{W}$ is the optimal solution of the original semidefinite programming problem (4.36). As we have just noticed, it is feasible. Define

$$\hat{\mu}_0 = \Big\langle C - \sum_{i=1}^{m} \hat{\mu}_i A_i, \widehat{W} \Big\rangle_{\mathbb{S}}.$$

Together with the second relation in (4.40) this implies that

$$\Big\langle C - \sum_{i=1}^{m} \hat{\mu}_i A_i - \hat{\mu}_0 I, W \Big\rangle_{\mathbb{S}} = 0.$$

Recalling that $\hat{X} = k\widehat{W}$, we can replace $\widehat{W}$ by $\hat{X}$ in the last equation to obtain the complementarity condition for problem (4.36).

We conclude that optimality conditions (3.51) from page 130 are satisfied for problem (4.36), with Lagrange multipliers $(\hat{\mu}, \hat{\mu}_0)$. Consequently, $\hat{X}$ is the optimal solution.

4.4 DECOMPOSITION

Consider the nonlinear optimization problem

$$\begin{aligned} \text{minimize} \quad & f(x) \\ \text{subject to} \quad & g_i(x) \le b_i, \quad i = 1, \dots, m, \\ & h_i(x) = c_i, \quad i = 1, \dots, p, \\ & x \in X_0, \end{aligned}$$

with $b \in \mathbb{R}^m$ and $c \in \mathbb{R}^p$. Assume that we can partition the vector x as

$$x = (x^1, \dots, x^K), \quad x^k \in \mathbb{R}^{n_k}, \quad \sum_{k=1}^{K} n_k = n,$$

in such a way that the objective and constraint functions can be represented for all x as sums:

$$f(x) = \sum_{k=1}^{K} f^k(x^k),$$

$$g_i(x) = \sum_{k=1}^{K} g_i^k(x^k), \quad i = 1, \dots, m,$$

$$h_i(x) = \sum_{k=1}^{K} h_i^k(x^k), \quad i = 1, \dots, p.$$

Moreover, we assume that

$$X_0 = X_0^1 \times \dots X_0^K, \quad X_0^k \subset \mathbb{R}^{n_k}, \quad k = 1, \dots, K.$$

The Lagrangian has the form

$$\begin{aligned} L(x, \lambda, \mu) &= f(x) + \sum_{i=1}^{m} \lambda_i (g_i(x) - b_i) + \sum_{i=1}^{p} \mu_i (h_i(x) - c_i) \\ &= \sum_{k=1}^{K} \Big(f^k(x^k) + \sum_{i=1}^{m} \lambda_i g_i^k(x^k) + \sum_{i=1}^{p} \mu_i h_i^k(x^k) \Big) - \langle \lambda, b \rangle - \langle \mu, c \rangle. \end{aligned}$$

Therefore the dual function can be calculated as follows:

$$\begin{aligned} L_D(\lambda, \mu) &= \min_{x \in X_0} L(x, \lambda, \mu) \\ &= \sum_{k=1}^{K} \min_{x^k \in X_0^k} \Big[f^k(x^k) + \sum_{i=1}^{m} \lambda_i g_i^k(x^k) + \sum_{i=1}^{p} \mu_i h_i^k(x^k) \Big] - \langle \lambda, b \rangle - \langle \mu, c \rangle. \end{aligned}$$

It follows that the calculation of the dual function decomposes into K smaller problems, each for the subvector x^k, $k = 1, \dots, K$:

$$L_D^k(\lambda, \mu) = \min_{x^k \in X_0^k} \Big[f^k(x^k) + \sum_{i=1}^m \lambda_i g_i^k(x^k) + \sum_{i=1}^p \mu_i h_i^k(x^k) \Big].$$

In some cases these problems can be solved in a closed form, as in the examples below. In other cases, efficient numerical methods can be employed for their solution. In any case, solving the dual problem

$$\underset{\substack{\lambda \geq 0 \\ \mu \in \mathbb{R}^p}}{\text{maximize}} \sum_{k=1}^K L_D^k(\lambda, \mu) - \langle \lambda, b \rangle - \langle \mu, c \rangle$$

may be much easier than solving the primal problem. If the duality relation is satisfied, the above decomposition approach, together with Theorem 4.10, provides the optimal solution of the primal problem. If duality does not hold true, a lower bound for the optimal value of the primal problem can be obtained (see the next section).

Example 4.18. Suppose n power plants have to satisfy jointly some demand $b > 0$ for power. We assume that each plant j can generate power x_j between 0 and some upper bound u_j at the cost (per time unit) equal to

$$f_j(x_j) = c_j x_j + \frac{q_j}{2}(x_j)^2, \quad j = 1, \dots, n.$$

The coefficients c_j and q_j are assumed to be positive for all j, and thus the cost functions are strictly convex. Moreover, we assume that $\sum_{j=1} u_j \geq b$, in order to be able to satisfy the demand. To satisfy it at a minimal cost, we formulate the optimization problem

$$\begin{aligned} &\text{minimize} \ \sum_{j=1}^n f_j(x_j) \\ &\text{subject to} \ \sum_{j=1}^n x_j \geq b, \\ &\qquad 0 \leq x_j \leq u_j, \quad j = 1, \dots, n. \end{aligned}$$

The feasible set is compact and nonempty, because $\sum_{j=1} u_j \geq b$. Therefore an optimal solution must exist. All constraint functions are affine and thus the optimal solution has to satisfy the necessary conditions of optimality. Since the problem is convex, the duality relation holds true.

Denoting by λ the Lagrange multiplier associated with the demand constraint, we can write the Lagrangian as follows:

$$L(x, \lambda) = \sum_{j=1}^n f_j(x_j) + \lambda\Big(b - \sum_{j=1}^n x_j\Big).$$

Hence the dual function has the form

$$L_D(\lambda) = b\lambda + \min_{0\le x\le u} \sum_{j=1}^{n} \big(f_j(x_j) - \lambda x_j\big)$$
$$= b\lambda + \sum_{j=1}^{n} \min_{0\le x_j\le u_j} \big(f_j(x_j) - \lambda x_j\big).$$

Consider the subproblem:

$$L_D^j(\lambda) = \min_{0\le x_j\le u_j} \Big(c_j x_j + \frac{q_j}{2}(x_j)^2 - \lambda x_j\Big). \tag{4.41}$$

If we interpret the Lagrange multiplier λ as the unit price for energy paid to the plants, this subproblem has a clear meaning: choose the production level x_j to maximize the profit of plant j.

We can minimize the quadratic function of one variable in (4.41) by simple calculus. The optimal solution is:

$$\hat{x}_j(\lambda) = \begin{cases} 0 & \text{if } 0 \le \lambda \le c_j, \\ (\lambda - c_j)/q_j & \text{if } c_j \le \lambda \le c_j + q_j u_j, \\ u_j & \text{if } \lambda \ge c_j + q_j u_j. \end{cases}$$

We see that the power output $\hat{x}_j(\lambda)$ of plant j is a nondecreasing function of the multiplier (price) λ. It follows from the duality theory that there exists an optimal value $\hat{\lambda}$ of the price, for which the vector $\hat{x}$ with coordinates $\hat{x}_j(\hat{\lambda})$ is an optimal solution of the problem. We must, therefore, have

$$\sum_{j=1}^{n} \hat{x}_j(\hat{\lambda}) = b.$$

The equality here is necessary because of the complementarity condition (iii) of Theorem 4.10. The optimal price $\hat{\lambda}$ can thus be calculated as the price for which the power plants, driven by their profits, jointly satisfy the demand.

Each dual function can be calculated explicitly,

$$L_D^j(\lambda) = \begin{cases} 0 & \text{if } 0 \le \lambda \le c_j, \\ -(\lambda - c_j)^2/(2q_j) & \text{if } c_j \le \lambda \le c_j + q_j u_j, \\ (c_j - \lambda)u_j + q_j(u_j)^2/2 & \text{if } \lambda \ge c_j + q_j u_j. \end{cases}$$

It is nonpositive for all $\lambda \ge 0$, which means that no plant has losses (we cannot force them to have losses), and the plants that produce power have positive profits. The dual problem

$$\underset{\lambda\ge 0}{\text{maximize}}\ b\lambda + \sum_{j=1}^{n} L_D^j(\lambda)$$

can be interpreted as follows: find the price that maximizes the value of power sold, $b\lambda$, minus the profits of the plants (recall that $-L_D^j$ is the profit of plant j).

If the cost functions of the plants are not convex, or if their feasible sets are disconnected, for example, $x_j = 0$ or $l_j \le x_j \le u_j$, the optimal solution of the dual problem provides a lower bound for the optimal value of the primal problem.

Example 4.19. There are n horses in a race. For every horse k we know the probability p_k that it wins and the amount s_k that the rest of the public is betting on it. The track keeps a certain proportion $C \in (0, 1)$ of the total amount bet and distributes the rest among the public in proportion to the amounts bet on the winning horse. We want to place bets totaling b dollars to maximize the expected net return.

Let us denote by x_k the amount bet on horse k. If horse k wins the race, we gain

$$F_k(x) = \frac{Ax_k}{x_k + s_k},$$

where $A \triangleq (1 - C)(b + \sum_{i=1}^{n} s_i)$ is the total amount to be split among the winners. We can now write the corresponding optimization problem as follows:

$$\text{minimize} \quad -A \sum_{k=1}^{n} \frac{p_k x_k}{x_k + s_k} \tag{4.42}$$

$$\text{subject to} \quad \sum_{k=1}^{n} x_k = b, \tag{4.43}$$

$$x \ge 0. \tag{4.44}$$

Note that each function

$$-\frac{x_k}{x_k + s_k} = \frac{s_k}{x_k + s_k} - 1$$

is convex in x_k. Therefore, (4.42)–(4.44) is a convex problem. Clearly its feasible set is nonempty and bounded, and hence it has an optimal solution. Since, in fact, the objective function is strictly convex, problem (4.42)–(4.44) possesses a unique optimal solution.

The constraints are affine and thus the dual problem of (4.42)–(4.44) has a nonempty set of optimal solutions. No duality gap exists between these problems. We will be able to write their optimal solutions explicitly.

Denoting by μ the multiplier associated with the budget constraint (4.43), we can write the Lagrangian:

$$\begin{aligned} L(x, \mu) &= -A \sum_{k=1}^{n} \frac{p_k x_k}{x_k + s_k} + \mu\Big(\sum_{k=1}^{n} x_k - b\Big) \\ &= \sum_{k=1}^{n} \Big(\mu x_k - A\frac{p_k x_k}{x_k + s_k}\Big) - b\mu. \end{aligned}$$

To calculate the dual function we observe that

$$L_D(\mu) = \min_{x \geq 0} \sum_{k=1}^{n} \left(\mu x_k - A \frac{p_k x_k}{x_k + s_k} \right) - b\mu$$
$$= \sum_{k=1}^{n} \min_{x_k \geq 0} \left(\mu x_k - A \frac{p_k x_k}{x_k + s_k} \right) - b\mu.$$

The minimization can be carried out for each k separately. Two cases are possible: if the unconstrained minimum of the expression

$$L_k(x_k, \mu) = \mu x_k - A \frac{p_k x_k}{x_k + s_k}$$

is attained at a positive x_k, then x_k is the constrained minimum as well. Otherwise the constrained minimum is zero. Simple calculations yield the minimum as a function of μ:

$$\hat{x}_k(\mu) = \max \left(0, \sqrt{\frac{A p_k s_k}{\mu}} - s_k \right), \quad k = 1, \ldots, n.$$

This means that $\hat{x}_k(\mu) > 0$ if and only if

$$\sqrt{\frac{A p_k s_k}{\mu}} > s_k,$$

which can be rewritten as follows:

$$\frac{p_k}{s_k} > \frac{\mu}{A}. \tag{4.45}$$

Ordering the horses (and scenarios) in such a way that

$$\frac{p_1}{s_1} \geq \frac{p_2}{s_2} \geq \cdots \geq \frac{p_n}{s_n}$$

we see that there must exist l (which depends on μ) such that

$$\hat{x}_k(\mu) = \begin{cases} \sqrt{\dfrac{A p_k s_k}{\mu}} - s_k, & k = 1, \ldots, l, \\ 0, & \text{otherwise.} \end{cases} \tag{4.46}$$

At the optimal value $\hat{\mu}$ of the Lagrange multiplier, the corresponding solution $\hat{x}(\hat{\mu})$ is feasible for (4.43). For every l this yields the candidate value of the multiplier:

$$\hat{\mu} = \frac{A \left(\sum_{k=1}^{l} \sqrt{p_k s_k} \right)^2}{\left(b + \sum_{k=1}^{l} s_k \right)^2}. \tag{4.47}$$

This value should guarantee that (4.45) is true only for $k = 1, \dots, l$. To ensure that, we find l as the smallest integer for which

$$\sqrt{\frac{p_l}{s_l}} > \frac{\sum_{k=1}^{l} \sqrt{p_k s_k}}{b + \sum_{k=1}^{l} s_k} \geq \sqrt{\frac{p_{l+1}}{s_{l+1}}}.$$

Note that the left inequality holds for $l = 1$. If such an integer does not exist, we set $l = n$. Once l has been determined, we can calculate the Lagrange multiplier by (4.47). The substitution into (4.46) renders the optimal bets.

Example 4.20. The idea of decomposition can also be applied, in a slightly more involved fashion, to dynamic problems. Consider the optimal control problem with discrete time

$$\text{minimize} \sum_{t=1}^{T-1} f_t(u_t) \tag{4.48}$$

$$\text{subject to } x_{t+1} = A_t x_t + B_t u_t, \quad t = 1, \dots, T-1, \tag{4.49}$$

$$x_T = d, \tag{4.50}$$

$$u_t \in U_t, \quad t = 1, \dots, T-1. \tag{4.51}$$

Here $x_t \in \mathbb{R}^n$ denotes the state of the system at time t, and $u_t \in \mathbb{R}^m$ denotes the control at time t. The initial state x_1 is given. The matrices A_t and B_t have dimensions n and $n \times m$, respectively.

The functions $f_t : \mathbb{R}^m \to \mathbb{R}$, $t = 1, \dots, T-1$, are assumed to be convex and the sets $U_t \subset \mathbb{R}^m$ are also assumed to be convex. Therefore, problem (4.48)–(4.51) is a convex optimization problem.

Assume that Slater's constraint qualification condition holds true: there exists a sequence of controls $\bar{u}_t$, $t = 1, \dots, T-1$, such that $\bar{u}_t \in \operatorname{int} U_t$, $t = 1, \dots, T-1$, and the corresponding system's trajectory $\bar{x}_t$, $t = 1, \dots, T$, obtained from (4.49), satisfies (4.50). Under this condition, if an optimal control exists, it satisfies the first order optimality conditions. Moreover, the duality relation holds true.

Let us associate Lagrange multipliers $\mu_t \in \mathbb{R}^n$, $t = 1, \dots, T-1$, with the state equation (4.49), and a Lagrange multiplier $\psi \in \mathbb{R}^n$ with the terminal state condition (4.50). Consider the Lagrangian:

$$L(x, u, \mu, \psi) = \sum_{t=1}^{T-1} f_t(u_t) + \sum_{t=1}^{T-1} \langle \mu_t, x_{t+1} - A_t x_t - B_t u_t \rangle + \langle \psi, d - x_T \rangle.$$

The unconstrained minimum of the Lagrangian with respect to all x_t, $t = 2, \dots, T$, is finite if and only if

$$\mu_{t-1} = A_t^* \mu_t, \quad t = 2, \dots, T-1, \quad \text{and} \quad \mu_{T-1} = \psi. \tag{4.52}$$

We use A^* to denote the transpose of A. These equations are called *adjoint equations*. The control function u_t is a solution for each t of the problem

$$\underset{u_t \in U_t}{\text{minimize}}\; f_t(u_t) - \langle \mu_t, B_t u_t \rangle, \quad t = 1, \dots, T-1. \tag{4.53}$$

Suppose this problem has a unique solution $\hat{u}_t(\mu_t)$ for each μ_t. Then for every $\psi \in \mathbb{R}^n$ we can solve the adjoint equations (4.52) and obtain the dual trajectory μ_t, $t = T-1, \dots, 1$. Then (4.53) defines the controls $u_t = \hat{u}_t(\mu_t)$, $t = 1, \dots, T-1$. The substitution of the control function to the state equation (4.49) determines the state trajectory $\hat{x}_t(\mu)$, $t = 1, \dots, T$. It follows from Theorem 4.10 that if $x_T = d$, then the solution is optimal. Otherwise, we have to change ψ to maximize the dual function,

$$L_D(\mu, \psi) = \sum_{t=1}^{T-1} \left[f_t(\hat{u}_t(\mu_t)) - \langle \mu_t, B_t \hat{u}_t(\mu_t) \rangle \right] + \langle d, \psi \rangle,$$

subject to equations (4.52). We can eliminate the variables μ_t from this problem by solving the adjoint equations and getting

$$\mu_t = \left(\prod_{s=t+1}^{T-1} A_s^* \right) \psi, \quad t = 1, \dots, T-1.$$

After the substitution into the dual function, we obtain a function of the terminal state multiplier, ψ. The dual problem is equivalent to the problem of unconstrained maximization of this function with respect to ψ.

4.5 CONVEX RELAXATION OF NONCONVEX PROBLEMS

If a saddle point of the Lagrangian does not exist, which is typical for nonconvex problems, the duality relation (4.18) is not valid. Nevertheless, we can still use the dual problem to bound the optimal value of the primal problem from below.

LEMMA 4.21. *For every* $(\lambda, \mu) \in \Lambda_0$ *and for every* $x \in X_0$

$$L_D(\lambda, \mu) \le L_P(x).$$

Proof. The result follows from the chain of inequalities:

$$L_D(\lambda, \mu) \le L(x, \lambda, \mu) \le L_P(x).$$

□

The difference

$$\delta = \min_{x \in X_0} L_P(x) - \max_{(\lambda, \mu) \in \Lambda_0} L_D(\lambda, \mu) \ge 0$$

is called the *duality gap*.

In many applications the lower bound of Lemma 4.21 is very useful. We discuss here its relations to conjugate duality theory of sections 2.6.1 and 2.6.2.

Consider an optimization problem with linear constraints:

$$\begin{aligned} \text{minimize} \quad & f(x) \\ \text{subject to} \quad & Ax \geq b, \\ & x \in X_0. \end{aligned} \tag{4.54}$$

Here $f(\cdot)$ is some function from $\mathbb{R}^n$ to $\overline{\mathbb{R}}$, A is a matrix of dimension $m \times n$ and $b \in \mathbb{R}^m$. The set X_0 is an arbitrary subset of $\mathbb{R}^n$. No convexity assumptions are made about $f(\cdot)$ or X_0. We write the inequality constraints in the "$\geq$" form for the convenience of notation only.

Define the function

$$f_{X_0}(x) = \begin{cases} f(x) & \text{if } x \in X_0, \\ +\infty & \text{otherwise.} \end{cases}$$

Then problem (4.54) can be rewritten as

$$\begin{aligned} \text{minimize} \quad & f_{X_0}(x) \\ \text{subject to} \quad & Ax \geq b. \end{aligned} \tag{4.55}$$

Together with it we consider the *convexified problem*

$$\begin{aligned} \text{minimize} \quad & f_{X_0}^{**}(x) \\ \text{subject to} \quad & Ax \geq b. \end{aligned} \tag{4.56}$$

Here $f_{X_0}^{**}$ is the biconjugate function to f_{X_0}, as defined in Section 2.6.2. We know from Theorem 2.95 that if f_{X_0} has at least one affine minorant then $f_{X_0}^{**}$ is the largest convex minorant of f_{X_0}. Thus problem (4.56) is indeed a convex relaxation of (4.55).

Let us return to the original nonconvex problem (4.55). Its Lagrangian has the form

$$L(x, \lambda) = f_{X_0}(x) + \langle \lambda, b - Ax \rangle.$$

We can calculate the dual function as follows:

$$\begin{aligned} L_D(\lambda) &= \inf_x \left\{ f_{X_0}(x) + \langle \lambda, b - Ax \rangle \right\} \\ &= -\sup_x \left\{ \langle A^T\lambda, x \rangle - f_{X_0}(x) \right\} + \langle \lambda, b \rangle \\ &= -f_{X_0}^{*}(A^T\lambda) + \langle b, \lambda \rangle. \end{aligned}$$

Here $f_{X_0}^*$ denotes the conjugate function to f_{X_0}. The dual problem therefore can be written as follows:

$$\sup_{\lambda \geq 0} \left\{ \langle b, \lambda \rangle - f_{X_0}^*(A^T \lambda) \right\}. \tag{4.57}$$

This transformation allows us to relate the dual problem (4.57) and the convexified problem (4.56).

THEOREM 4.22. *Assume that the duality relation* (4.18) *holds true for the convexified problem* (4.56). *Then the dual problem* (4.57) *has an optimal solution and its optimal value is equal to the optimal value of the convexified problem* (4.56).

Proof. As the convexified primal problem has a solution, the function $f_{X_0}^{**}$ is proper. This means that $f_{X_0}^*$ is proper. It is always convex and lower semicontinuous. Theorem 2.95 implies

$$f_{X_0}^* = \left[f_{X_0}^* \right]^{**} = \left[f_{X_0}^{**} \right]^*.$$

It follows that problem (4.57) can be rewritten as

$$\sup_{\lambda \geq 0} \left\{ \langle b, \lambda \rangle - \left[f_{X_0}^{**} \right]^* (A^T \lambda) \right\},$$

which is the dual problem of the convexified problem (4.56). By assumption, the duality relation for the convexified problem holds true. This implies the assertion of the theorem. □

The above result can be used to generate lower bounds for nonconvex optimization problems by selecting some constraints that enter the Lagrangian and specifying the set X_0 by all remaining constraints. Such an approach is widely used in integer programming.

Example 4.23. A retailer wants to establish b distribution centers to serve m stores. Possible locations of the centers are fixed at some given n points. The cost of supplying store i from the center located at j equals c_{ij}. We have already considered a similar situation in Exercise 3.14, but with free locations of the centers. When the locations have to be chosen from a finite set, the problem has purely combinatorial character.

Denote by x_i the binary variable representing the decision to locate a center at a possible site i, where $i = 1, \ldots, n$:

$$x_i = \begin{cases} 1 & \text{if a center is located at site } i, \\ 0 & \text{otherwise.} \end{cases}$$

In the absence of capacity restrictions on the centers, it is best to cover the demand of each store from exactly one center assigned to it. Introduce the variables:

$$y_{ij} = \begin{cases} 1 & \text{if store } j \text{ is supplied from the center located at } i, \\ 0 & \text{otherwise.} \end{cases}$$

We obtain the following optimization problem:

$$\begin{aligned}
\text{minimize} \quad & \sum_{i=1}^{n} \sum_{j=1}^{m} c_{ij} y_{ij} \\
\text{subject to} \quad & \sum_{i=1}^{n} x_i = b, && (4.58) \\
& \sum_{i=1}^{n} y_{ij} = 1, \quad j = 1, \dots, m, && (4.59) \\
& y_{ij} \le x_i, \quad i = 1, \dots, n, \quad j = 1, \dots, m, && (4.60) \\
& x_i \in \{0, 1\}, \; y_{ij} \in \{0, 1\}, \quad i = 1, \dots, n, \quad j = 1, \dots, m.
\end{aligned}$$

Constraint (4.59) reflects the requirement that each store has to be supplied, and constraint (4.60) prevents assignments to centers which do not exist.

Since the feasible set is disconnected, the above problem is not a convex optimization problem. Still, the duality theory provides useful Lagrangian relaxations of this problem. Define

$$X_0^1 = \{0, 1\}^n \times \{0, 1\}^{mn}.$$

Assigning Lagrange multipliers $\psi \in \mathbb{R}$, $\mu \in \mathbb{R}^m$ and $\lambda \in \mathbb{R}_+^{nm}$ to the constraints (4.58), (4.59), and (4.60), respectively, we formulate the Lagrangian

$$\begin{aligned}
& L^1(x, y, \psi, \mu, \lambda) = \\
& \sum_{i=1}^{n} \sum_{j=1}^{m} c_{ij} y_{ij} + \psi\Big(\sum_{i=1}^{n} x_i - b\Big) + \sum_{j=1}^{m} \mu_j \Big(1 - \sum_{i=1}^{n} y_{ij}\Big) + \sum_{i=1}^{n} \sum_{j=1}^{m} \lambda_{ij} (y_{ij} - x_i) \\
& \quad = \sum_{i=1}^{n} \sum_{j=1}^{m} (c_{ij} - \mu_j + \lambda_{ij}) y_{ij} + \sum_{i=1}^{n} \Big(\psi - \sum_{j=1}^{m} \lambda_{ij}\Big) x_i - b\psi + \sum_{j=1}^{m} \mu_j.
\end{aligned}$$

The minimum of L^1 with respect to $(x, y) \in X_0^1$ can be calculated by inspection

$$\begin{aligned}
L_D^1(\psi, \mu, \lambda) = {} & \sum_{i=1}^{n} \sum_{j=1}^{m} \min(0, c_{ij} - \mu_j + \lambda_{ij}) \\
& + \sum_{i=1}^{n} \min\Big(0, \psi - \sum_{j=1}^{m} \lambda_{ij}\Big) - b\psi + \sum_{j=1}^{m} \mu_j.
\end{aligned}$$

It is a concave and piecewise linear function. By Theorem 4.22 it is equal to the dual function of the convexified problem. This problem is simply the linear programming relaxation of the original problem. This can be verified directly by noticing that the minimum of the Lagrangian with respect to the variables restricted to the set $\{0, 1\}$ is the same as the minimum with respect to the variables allowed to take any values in the interval $[0, 1]$. In fact, this property holds true for all integer linear programming problems. Therefore the optimal value of the dual problem

$$\text{maximize } L_D^1(\psi, \mu, \lambda) \quad \text{subject to} \quad \lambda \geq 0, \tag{4.61}$$

is the same as the optimal value of the linear programming relaxation.

A more interesting situation occurs when the definition of the set X_0 also involves some inequality or equality constraints. Usually, we choose constraints that are easy to handle directly, once the other constraints are moved to the objective function via Lagrangian terms. In our case it is convenient to define

$$X_0^2 = \Big\{(x, y) \in \{0, 1\}^n \times \{0, 1\}^{mn} : y_{ij} \leq x_i,\ i = 1, \dots, n,\ j = 1, \dots, m\Big\}.$$

The corresponding Lagrangian has the form

$$L^2(x, y, \psi, \mu) = \sum_{i=1}^{n} \sum_{j=1}^{m} (c_{ij} - \mu_j) y_{ij} + \psi \sum_{i=1}^{n} x_i - b\psi + \sum_{j=1}^{m} \mu_j.$$

The minimum of the Lagrangian with respect to $(x, y) \in X_0^2$ is more difficult to calculate than before. Recalling the idea of decomposition discussed in the preceding section, we notice that the problem of minimizing L^2 splits into independent subproblems for each potential location $i = 1, \dots, n$. Define the sets

$$Z_i = \Big\{(x_i, y_i) \in \{0, 1\} \times \{0, 1\}^m : y_{ij} \leq x_i,\ j = 1, \dots, m\Big\}.$$

We have $X_0^2 = Z_1 \times \cdots \times Z_n$ and

$$\min_{(x,y) \in X_0^2} L^2(x, y, \psi, \mu) = \sum_{i=1}^{n} \min_{(x_i, y_i) \in Z_i} \Big(\sum_{j=1}^{m} (c_{ij} - \mu_j) y_{ij} + \psi x_i\Big) - b\psi + \sum_{j=1}^{m} \mu_j.$$

Each of the functions

$$L_i(\psi, \mu) \triangleq \min_{(x_i, y_i) \in Z_i} \Big(\sum_{j=1}^{m} (c_{ij} - \mu_j) y_{ij} + \psi x_i\Big),$$

can be calculated by examining two cases. If $x_i = 0$ then all $y_{ij} = 0$, $j = 1, \dots, m$, and the value is 0. If $x_i = 1$ then we choose $y_{ij} = 1$ for all j such that $c_{ij} < \mu_j$. Hence

$$L_i(\psi, \mu) = \min\Big(0, \sum_{j=1}^{m} \min(0, c_{ij} - \mu_j) + \psi\Big), \quad i = 1, \dots, n. \tag{4.62}$$

The optimal value of the dual problem,

$$\text{maximize} \sum_{i=1}^{n} L_i(\psi, \mu) - b\psi + \sum_{j=1}^{m} \mu_j, \tag{4.63}$$

provides a new lower bound for the optimal value of the original facility location problem.

To analyze the relation between the two bounds (4.61) and (4.63) we notice that

$$X_0^2 = X_0^1 \cap Y,$$

with

$$Y = \Big\{(x, y) \in \mathbb{R}^n \times \mathbb{R}^{mn} : y_{ij} \le x_i,\ i = 1, \dots, n,\ j = 1, \dots, m\Big\}.$$

For the linear objective function

$$f(x, y) = \sum_{i=1}^{n} \sum_{j=1}^{m} c_{ij} y_{ij}$$

we have

$$f^{**}_{X_0^1}(x, y) = \begin{cases} f(x, y) & \text{if } (x, y) \in \overline{\text{conv}}(X_0^1), \\ +\infty & \text{otherwise.} \end{cases}$$

Similarly,

$$f^{**}_{X_0^2}(x, y) = \begin{cases} f(x, y) & \text{if } (x, y) \in \overline{\text{conv}}(X_0^2), \\ +\infty & \text{otherwise.} \end{cases}$$

Since

$$\overline{\text{conv}}(X_0^2) = \overline{\text{conv}}\left(X_0^1 \cap Y\right) \subset \overline{\text{conv}}(X_0^1) \cap Y,$$

we always have the inequality

$$f^{**}_{X_0^2}(x, y) \ge f^{**}_{X_0^1}(x, y).$$

Therefore the lower bound (4.63) is at least as good as (4.61).

4.6 THE OPTIMAL VALUE FUNCTION

Let us now consider the problem with parameters $b \in \mathbb{R}^m$ and $c \in \mathbb{R}^p$:

$$\begin{aligned} \text{minimize}\ \ & f(x) \\ \text{subject to}\ \ & g_i(x) \le b_i, \quad i = 1, \dots, m, \\ & h_i(x) = c_i, \quad i = 1, \dots, p, \\ & x \in X_0. \end{aligned} \tag{4.64}$$

We assume that $f : \mathbb{R}^n \to \mathbb{R}$ and $g_i : \mathbb{R}^n \to \mathbb{R}$, $i = 1, \dots, m$, are convex functions, $h_i : \mathbb{R}^n \to \mathbb{R}$, $i = 1, \dots, p$, are affine, and the set $X_0 \subset \mathbb{R}^n$ is convex.

We are interested in the dependence of the optimal value of this problem, which we denote by $v(b, c)$, on the parameters b and c. To define the optimal value function formally, let

$$X(b, c) = \{x \in X_0 : g(x) \le b,\ h(x) = c\}$$

be the feasible set of (4.64). If $X(b, c) = \emptyset$ we define $v(b, c) = +\infty$. Otherwise,

$$v(b, c) = \inf_{x \in X(b,c)} f(x).$$

We have considered the sensitivity of the optimal value function in Section 3.8, under rather restrictive linear independence condition and semi-strong second order sufficient condition. In the convex case we can avoid considering the dependence of the optimal solution on the parameters and we can concentrate directly on the optimal value function.

LEMMA 4.24. *The optimal value function $v(\cdot)$ is convex.*

Proof. Let

$$b = \alpha b^1 + (1 - \alpha) b^2, \quad c = \alpha c^1 + (1 - \alpha) c^2,$$

with $\alpha \in (0, 1)$. Then it follows from the convexity of $g_i(\cdot)$ that for every

$$x^1 \in X(b^1, c^1), \quad x^2 \in X(b^2, c^2)$$

we have

$$\alpha x^1 + (1 - \alpha) x^2 \in X(b, c).$$

The convexity of $f(\cdot)$ implies that we also have

$$f(\alpha x^1 + (1 - \alpha) x^2) \le \alpha f(x^1) + (1 - \alpha) f(x^2).$$

Therefore

$$\begin{aligned} v(b, c) = \inf_{x \in X(b,c)} f(x) &\le \inf_{\substack{x^1 \in X(b^1, c^1) \\ x^2 \in X(b^2, c^2)}} \left[\alpha f(x^1) + (1 - \alpha) f(x^2)\right] \\ &= \alpha \inf_{x^1 \in X(b^1, c^1)} f(x^1) + (1 - \alpha) \inf_{x^2 \in X(b^2, c^2)} f(x^2) \\ &= \alpha v(b^1, c^1) + (1 - \alpha) v(b^2, c^2), \end{aligned}$$

as required. □

To understand the differential properties of the optimal value function we employ basic results of the theory of conjugate duality described in Section 2.6.

Assume for simplicity that there are no equality constraints and consider the Lagrangian

$$L(x,\lambda) = f(x) + \langle \lambda, g(x) - b^0 \rangle.$$

It is convenient here to extend the dual function by assigning the value $-\infty$ to the argument values outside of the dual feasible set, that is

$$\overline{L}_D(\lambda) = \begin{cases} \inf_{x \in X_0} \big[f(x) + \langle \lambda, g(x) - b^0 \rangle \big], & \text{if } \lambda \geq 0, \\ -\infty, & \text{otherwise}. \end{cases}$$

If $\lambda \geq 0$, then

$$\langle \lambda, g(x) - b^0 \rangle = \inf_{b \geq g(x)} \langle \lambda, b - b^0 \rangle.$$

Thus we can transform the dual function as follows:

$$\begin{aligned} \overline{L}_D(\lambda) &= \inf_{\substack{x \in X_0 \\ b \geq g(x)}} \big[f(x) + \langle \lambda, b - b^0 \rangle \big] \\ &= \inf_{b \in \mathbb{R}^m} \inf_{x \in X(b)} \big[f(x) + \langle \lambda, b - b^0 \rangle \big] \\ &= \inf_{b \in \mathbb{R}^m} \big[v(b) + \langle \lambda, b - b^0 \rangle \big]. \end{aligned}$$

Using $v^*(\cdot)$ to denote the conjugate function to the optimal value function $v(\cdot)$, as in Definition 2.90 on page 76, we obtain:

$$\begin{aligned} \overline{L}_D(\lambda) &= -\sup_{b \in \mathbb{R}^m} \big[-v(b) - \langle \lambda, b \rangle \big] - \langle \lambda, b^0 \rangle \\ &= -v^*(-\lambda) - \langle \lambda, b^0 \rangle. \end{aligned} \tag{4.65}$$

If $\lambda \not\geq 0$, then

$$-v^*(-\lambda) = \inf_{b \in \mathbb{R}^m} \big[v(b) + \langle \lambda, b \rangle \big] = -\infty,$$

because $v(\cdot)$ is nonincreasing and we can arbitrarily increase any component b_j corresponding to $\lambda_j < 0$. Therefore, formula (4.65) holds true for all $\lambda \in \mathbb{R}^m$.

LEMMA 4.25. *A vector (λ^0, μ^0) is a solution of the dual problem to* (4.64) *at $b = b^0$ and $c = c^0$ if and only if*

$$v(b^0, c^0) + v^*(-\lambda^0, -\mu^0) + \langle \lambda^0, b^0 \rangle + \langle \mu^0, c^0 \rangle = 0.$$

Proof. To simplify notation we assume that $p = 0$. A vector $\lambda^0 \geq 0$ is a solution of the dual problem if and only if

$$\overline{L}_D(\lambda^0) = v(b^0).$$

Combining the last equation with (4.65) we obtain

$$-v^*(-\lambda^0) - \langle \lambda^0, b^0 \rangle = v(b^0),$$

which is the same as the postulated formula. If $p > 0$ the calculations are almost identical. □

This useful transformation allows us to completely describe the subdifferential of $v(\cdot)$ at (b^0, c^0) as the negative of the set of optimal solutions of the dual problem.

THEOREM 4.26.

$$\partial v(b^0, c^0) = -\big\{(\lambda^0, \mu^0) \in \Lambda_0 : \overline{L}_D(\lambda^0, \mu^0) = v(b^0, c^0)\big\}.$$

Proof. In view of Lemma 4.24, the optimal value function $v(\cdot)$ is convex. By Theorem 2.98, (s, u) is a subgradient of $v(\cdot)$ at (b^0, c^0) if and only if

$$v(b^0, c^0) + v^*(s, u) = \langle s, b^0 \rangle + \langle u, c^0 \rangle.$$

By Lemma 4.25, this is equivalent to the fact that $(-s, -u)$ is a solution to the dual problem. □

It follows from Lemma 2.77 that the Lagrange multiplier vector of minimum norm is the direction of steepest descent of the optimal value function.

Example 4.27. Consider Example 4.18 and suppose $\hat{\lambda}$ is the optimal value of the Lagrange multiplier associated with the demand constraint:

$$-\sum_{j=1}^{n} x_j \leq -b.$$

If $\sum_{j=1}^{n} u_j > -b$, that is, if there is capacity reserve in the system, we can easily verify that the Lagrange multiplier $\hat{\lambda}$ is unique. Indeed, at least one of the variables $\hat{x}_j(\lambda)$ is strictly increasing for λ about $\hat{\lambda}$ and thus the equation

$$\sum_{j=1}^{n} \hat{x}_j(\lambda) = b$$

has a unique solution, $\hat{\lambda}$. Theorem 4.26 implies that the minimum cost of satisfying demand b, denoted by $v(b)$, is differentiable at b. Its derivative equals $\hat{\lambda}$ (note the sign change due to the fact that we have a "greater than or equal to" constraint). In other words, the cost of increasing the power output within the next time unit by a small amount δ is equal to $\hat{\lambda}\delta$. This explains our interpretation of the multiplier $\hat{\lambda}$ as the price of energy.

Formula (4.65) also allows us to calculate the entire subdifferential of the dual function. If the optimal value function is lower semicontinuous, Theorem 2.98 implies that

$$\partial v^*(s) = \{b \in \mathbb{R}^m : v(b) + v^*(s) = \langle b, s \rangle\}.$$

Setting $s = -\lambda$ and using (4.65) we obtain

$$\begin{aligned}\partial v^*(-\lambda) &= \{b \in \mathbb{R}^m : v(b) + v^*(-\lambda) = -\langle b, \lambda \rangle\} \\ &= \{b \in \mathbb{R}^m : v(b) + \langle \lambda, b - b^0 \rangle = \overline{L}_D(\lambda)\}.\end{aligned}$$

This can be re-stated in a more explicit way: $b \in v^*(-\lambda)$ if and only if

$$\inf_{x \in X(b)} \big[f(x) + \langle \lambda, b - b^0 \rangle\big] = \inf_{x \in X_0} \big[f(x) + \langle \lambda, g(x) - b^0 \rangle\big]. \tag{4.66}$$

For $\lambda \geq 0$ we always have

$$\begin{aligned}\inf_{\substack{x \in X_0 \\ g(x) \leq b}} \big[f(x) + \langle \lambda, b - b^0 \rangle\big] &\geq \inf_{\substack{x \in X_0 \\ g(x) \leq b}} \big[f(x) + \langle \lambda, g(x) - b^0 \rangle\big] \\ &\geq \inf_{x \in X_0} \big[f(x) + \langle \lambda, g(x) - b^0 \rangle\big].\end{aligned} \tag{4.67}$$

Suppose $b \in \partial v^*(-\lambda)$. Equation (4.66) means that all inequalities in (4.67) become equations as well. Consider a minimizer $\hat{x}$ of the left member of (4.67). Since the left member is equal to the middle one, substituting $\hat{x}$ into the middle member cannot result in a smaller value. That is,

$$f(\hat{x}) + \langle \lambda, b - b^0 \rangle \leq f(\hat{x}) + \langle \lambda, g(\hat{x}) - b^0 \rangle.$$

This simplifies into $\langle \lambda, b - g(\hat{x}) \rangle \leq 0$. Since both vectors in this product are nonnegative, we conclude that

$$\langle \lambda, b - g(\hat{x}) \rangle = 0 \quad \text{and} \quad b \geq g(\hat{x}). \tag{4.68}$$

Substituting $\hat{x}$ into the right member of (4.67) we obtain the same value of the minimized function. Thus $\hat{x}$ is also the minimizer of the Lagrangian in X_0.

We claim that

$$\partial v^*(-\lambda) = \bigcup_{\hat{x} \in \hat{X}(\lambda)} \{b \in \mathbb{R}^m : g(\hat{x}) \leq b \text{ and } \langle \lambda, b - g(\hat{x}) \rangle = 0\}.$$

We have already proved that every subgradient is a member of this set. Consider now a minimizer $\hat{x}$ of the Lagrangian. If b satisfies conditions (4.68) then $\hat{x}$ is feasible for all three problems in (4.67) and all three optimal values are equal. Relation (4.66) implies that $b \in \partial v^*(-\lambda)$.

Using (4.65) and the chain rule of subdifferentiation we get $\partial \overline{L}_D(\lambda) = \partial v^*(-\lambda) - b^0$, which renders the final formula for the subdifferential of the dual function:

$$\partial \overline{L}_D(\lambda) = \bigcup_{\hat{x} \in \hat{X}(\lambda)} \big\{ d \in \mathbb{R}^m : g(\hat{x}) - b^0 \le d \text{ and } \langle \lambda, d + b^0 - g(\hat{x}) \rangle = 0 \big\}. \tag{4.69}$$

We leave to the reader the verification that this set is in fact convex and closed.

Formula (4.15) (for a compact set X_0) yields:

$$\partial L_D(\lambda) = \bigcup_{\hat{x} \in \hat{X}(\lambda)} \{ d \in \mathbb{R}^m : g(\hat{x}) - b^0 = d \}.$$

The slight difference results from the fact that in (4.69) we consider the *extended* dual function $\overline{L}_D(\cdot)$, and the elements of the normal cone to its domain $\Lambda = \mathbb{R}^m_+$ enter the subdifferential when λ is a boundary point of the domain.

4.7 THE AUGMENTED LAGRANGIAN

Consider the nonlinear optimization problem with equality constraints

$$\begin{aligned} &\text{minimize } f(x) \\ &\text{subject to } h_i(x) = 0, \quad i = 1, \dots, p, \\ &\qquad\qquad x \in X_0. \end{aligned} \tag{4.70}$$

We assume that the functions $f : \mathbb{R}^n \to \mathbb{R}$ and $h_i : \mathbb{R}^n \to \mathbb{R}$ are twice continuously differentiable, and the set X_0 is convex and closed. In general, when the problem does not satisfy the assumptions of Theorem 4.7, we cannot expect that its Lagrangian has a saddle point. However, in some cases a local saddle point of an *augmented Lagrangian function* exists. We define this function as follows:

$$L_\varrho(x, \mu) = f(x) + \sum_{i=1}^p \mu_i h_i(x) + \frac{\varrho}{2} \sum_{i=1}^p \big[h_i(x)\big]^2. \tag{4.71}$$

Here $\varrho > 0$ is a fixed parameter of the function.

Theorem 3.25 from page 116 provides the first order optimality conditions for problem (4.70). If Robinson's condition is satisfied, then there exist multipliers $\hat{\mu} \in \mathbb{R}^p$ such that

$$0 \in \nabla f(\hat{x}) + \sum_{i=1}^{p} \hat{\mu}_i \nabla h_i(\hat{x}) + N_{X_0}(\hat{x}).$$

These are also the necessary conditions of the local optimality for the problem

$$\underset{x \in X_0}{\text{minimize}}\, L_\varrho(x, \hat{\mu}),$$

evaluated at the point $\hat{x}$. To verify whether sufficient conditions of a local minimum are satisfied at the point $(\hat{x}, \hat{\mu})$, we calculate the Hessian of the augmented Lagrangian with respect to x:

$$\nabla^2_{xx} L_\varrho(\hat{x}, \hat{\mu}) = \nabla^2 f(\hat{x}) + \varrho \sum_{i=1}^{p} \hat{\mu}_i \nabla^2 h_i(\hat{x}) + \varrho \sum_{i=1}^{p} \hat{\mu}_i \nabla h_i(\hat{x}) \big[\nabla h_i(\hat{x})\big]^T.$$

Our intention is to show that for all sufficiently large ϱ the Hessian is a positive definite matrix, and that the point $\hat{x}$ is therefore a local minimum of the augmented Lagrangian. To this end we first show a simple algebraic property.

Lemma 4.28. *Assume that a symmetric matrix Q of dimension n and a matrix B of dimension $m \times n$ are such that*

$$\langle x, Qx \rangle > 0 \;\; \textit{for all} \;\; x \neq 0 \;\; \textit{such that} \;\; Bx = 0.$$

Then there exists ϱ_0 such that for all $\varrho > \varrho_0$ the matrix $Q + \varrho B^T B$ is positive definite.

Proof. Suppose the assertion is false. Then for every ϱ_k we can find $\bar{\varrho}_k > \varrho_k$ and some x_k such that

$$\big\langle x_k, (Q + \bar{\varrho}_k B^T B) x_k \big\rangle < 0. \tag{4.72}$$

We can always normalize x_k in such a way that $\|x_k\| = 1$. Consider a sequence $\{\varrho_k\}$ diverging to $+\infty$ and the corresponding sequence $\{x_k\}$. By choosing a subsequence, if necessary, we can assume that the sequence $\{x_k\}$ is convergent. Let z be its limit.

Dividing both sides of (4.72) by $\bar{\varrho}_k$ we obtain

$$\frac{1}{\bar{\varrho}_k} \langle x_k, Qx_k \rangle + \|Bx_k\|^2 \leq 0.$$

Passing to the limit, as $k \to \infty$, we conclude that $\|Bz\| \le 0$, that is, $Bz = 0$. Skipping the term $\|Bx_k\|^2$ entirely, we also have

$$\langle x_k, Qx_k \rangle \le 0.$$

Passing to the limit we get

$$\langle z, Qz \rangle \le 0,$$

which contradicts the assumption. $\square$

We use this result to establish local convexity of the augmented Lagrangian.

THEOREM 4.29. *Assume that a point $\hat{x}$ satisfies the second order sufficient conditions of optimality of Theorem 3.47 with the Lagrange multipliers $\hat{\mu}$. Then there exists $\varrho_0 \ge 0$ such that for all $\varrho > \varrho_0$ the point $\hat{x}$ is a local minimum over X_0 of the augmented Lagrangian with $\mu = \hat{\mu}$.*

Proof. The result follows directly from Lemma 4.28 with

$$Q = \nabla^2 f(\hat{x}), \qquad B = h'(\hat{x}) = \begin{bmatrix} [\nabla h_1(\hat{x})]^T \\ \vdots \\ [\nabla h_p(\hat{x})]^T \end{bmatrix}.$$

$\square$

It follows that the point $(\hat{x}, \hat{\mu})$ is a *local saddle point* of the augmented Lagrangian: there exists $\varrho_0 \ge 0$ and a neighborhood U of $\hat{x}$ such that for all $\varrho > \varrho_0$

$$\max_{\mu \in \mathbb{R}^n} L_\varrho(\hat{x}, \mu) = L_\varrho(\hat{x}, \hat{\mu}) = \min_{x \in X_0 \cap U} L_\varrho(x, \hat{\mu}).$$

The converse is also true.

THEOREM 4.30. *Assume that a point $(\bar{x}, \bar{\mu})$ is a local saddle point of the augmented Lagrangian* (4.71) *for some $\varrho \ge 0$. Then $\bar{x}$ is a local minimum of problem* (4.70) *and $\bar{\mu}$ is the vector of Lagrange multipliers associated with the equality constraints.*

Proof. Since $\bar{\mu}$ is a maximum of the function $L_\varrho(\bar{x}, \cdot)$,

$$h_i(\bar{x}) = 0, \quad i = 1, \dots, p.$$

The point $\bar{x}$ is a local minimum over X_0 of the function

$$L_\varrho(x, \bar{\mu}) = f(x) + \sum_{i=1}^{p} \bar{\mu}_i h_i(x) + \frac{\varrho}{2} \sum_{i=1}^{p} [h_i(x)]^2,$$

which means that there exists a neighborhood U of $\bar{x}$ such that

$$f(\bar{x}) = L_\varrho(\bar{x}, \bar{\mu}) \le L_\varrho(x, \bar{\mu}) \quad \text{for all} \quad x \in X_0 \cap U.$$

In particular, if x is feasible for problem (4.70) we obtain

$$f(\bar{x}) \le f(x),$$

as required. □

Example 4.31. The problem

$$\begin{aligned} &\text{minimize} \ \frac{1}{2}(x_1)^2 - x_1 x_2 \\ &\text{subject to} \ x_1 + x_2 - 1 = 0 \end{aligned}$$

has a nonconvex objective function. It has solution $\hat{x}_1 = \frac{1}{3}$ and $\hat{x}_2 = \frac{2}{3}$, as can be verified directly, by eliminating one variable and the constraint. The Lagrange multiplier corresponding to the equality constraint equals $\hat{\mu} = \frac{1}{3}$.

The augmented Lagrangian has the form

$$L_\varrho(x_1, x_2, \mu) = \frac{1}{2}(x_1)^2 - x_1 x_2 + \mu(x_1 + x_2 - 1) + \frac{\varrho}{2}(x_1 + x_2 - 1)^2.$$

Its Hessian can be derived easily:

$$\nabla^2 L_\varrho(x_1, x_2, \mu) = \begin{bmatrix} 1 & -1 \\ -1 & 0 \end{bmatrix} + \varrho \begin{bmatrix} 1 & 1 \\ 1 & 1 \end{bmatrix}.$$

It is positive semidefinite for $\varrho \ge \frac{1}{3}$ and positive definite for $\varrho > \frac{1}{3}$. By setting $\mu = \hat{\mu} = \frac{1}{3}$ we can easily verify that the point $\hat{x}$ is indeed the minimum of the augmented Lagrangian.

Now consider the problem with inequality and equality constraints:

$$\begin{aligned} \text{minimize} \ & f(x) \\ \text{subject to} \ & g_i(x) \le 0, \quad i = 1, \dots, m, \\ & h_i(x) = 0, \quad i = 1, \dots, p, \\ & x \in X_0. \end{aligned} \tag{4.73}$$

We assume that all functions involved in this problem are twice continuously differentiable, and that the set X_0 is convex and closed. The Lagrangian associated with (4.73) has the form

$$L(x, \lambda, \mu) = f(x) + \sum_{i=1}^{m} \lambda_i g_i(x) + \sum_{i=1}^{p} \mu_i h_i(x).$$

Suppose $\hat{x}$ is a local minimum of problem (4.73) and that the set $\hat{\Lambda}(\hat{x})$ of the Lagrange multipliers associated with the inequality and equality constraints, respectively, is nonempty.

Consider the set of active constraints,

$$I^0(\hat{x}) = \{1 \le i \le m : g_i(\hat{x}) = 0\}.$$

DEFINITION 4.32. Problem (4.73) satisfies the *strong second order sufficient condition* if there exist multipliers $(\hat{\lambda}, \hat{\mu}) \in \hat{\Lambda}(\hat{x})$ such that

(i) $\hat{\lambda}_i > 0, \quad i \in I^0(\hat{x})$;

(ii) $\langle s, \nabla^2_{xx} L(\hat{x}, \hat{\lambda}, \hat{\mu}) s\rangle > 0$, for all nonzero s satisfying the equations:

$$\langle \nabla g_i(\hat{x}), s\rangle = 0, \quad i \in I^0(\hat{x}),$$
$$\langle \nabla h_i(\hat{x}), s\rangle = 0, \quad i = 1, \dots, p.$$

The above condition implies the semi-strong sufficient conditions of optimality (3.81) of page 149.

In order to define the augmented Lagrangian for the inequality constrained problem, we transform (4.73) to the equality constrained problem:

$$\begin{aligned} &\text{minimize} && f(x) \\ &\text{subject to} && g_i(x) + (z_i)^2 = 0, \qquad i = 1, \dots, m, \\ &&& h_i(x) = 0, \qquad i = 1, \dots, p, \\ &&& x \in X_0. \end{aligned} \tag{4.74}$$

It is apparent that $\hat{x}$ and $\hat{z}_i^2 = -g_i(\hat{x})$, $i = 1, \dots, m$, constitute a local minimum of this problem. The set of Lagrange multipliers $\hat{\Lambda}(\hat{x})$ remains unchanged.

The Lagrangian of problem (4.74) has the form

$$\bar{L}(x, z, \lambda, \mu) = f(x) + \sum_{i=1}^{m} \lambda_i \big[g_i(x) + (z_i)^2\big] + \sum_{i=1}^{p} \mu_i h_i(x). \tag{4.75}$$

The augmented Lagrangian (4.71) for problem (4.74) can be written as follows:

$$\bar{L}_\varrho(x, z, \lambda, \mu) = \bar{L}(x, z, \lambda, \mu) + \frac{\varrho}{2} \sum_{i=1}^{m} \big[g_i(x) + (z_i)^2\big]^2 + \frac{\varrho}{2} \sum_{i=1}^{p} \big[h_i(x)\big]^2. \tag{4.76}$$

Here $\varrho \ge 0$ is a fixed parameter of the function.

THEOREM 4.33. *Assume that problem (4.73) satisfies the strong second order sufficient condition. Then there exists $\varrho_0 \geq 0$ such that for all $\varrho > \varrho_0$ the point $(\hat{x}, \hat{z})$ is a local minimum over X_0 of the augmented Lagrangian $\bar{L}_\varrho(x, z, \hat{\lambda}, \hat{\mu})$.*

Proof. The main change from Theorem 4.71 is the presence of inequality constraints. Because of that, and for simplicity of notation, we assume that $p = 0$, that is, that there are no equality constraints in problem (4.73). Let us calculate the Hessian of the Lagrangian (4.75), with respect to (x, z), at the point $(\hat{x}, \hat{z})$ and $\hat{\lambda}$:

$$\nabla^2 \bar{L}(\hat{x}, \hat{z}, \hat{\lambda}) = \begin{bmatrix} \nabla^2 f(\hat{x}) + \sum_{i=1}^m \hat{\lambda}_i \nabla^2 g_i(\hat{x}) & 0 \\ 0 & 2\,\mathrm{diag}\{\hat{\lambda}_i\} \end{bmatrix}.$$

To verify the sufficient conditions of optimality of Theorem 3.47, consider the set of nonzero directions $(s, w) \in \mathbb{R}^n \times \mathbb{R}^m$ that are orthogonal to the gradients of the the equality constraints in (4.74):

$$\langle \nabla g_i(\hat{x}), s \rangle + 2\hat{z}_i w_i = 0, \quad i = 1, \dots, m. \tag{4.77}$$

Suppose $s = 0$. Since $\hat{z}_i > 0$ for all $i \notin I^0(\hat{x})$, it follows from the last equations that $w_i = 0$ for all $i \notin I^0(\hat{x})$. But $(s, w) \neq 0$, and thus at least one of w_i, $i \in I^0(\hat{x})$, must be nonzero. This implies that

$$\begin{bmatrix} s^T & w^T \end{bmatrix} \nabla^2 \bar{L}(\hat{x}, \hat{z}, \hat{\lambda}) \begin{bmatrix} s \\ w \end{bmatrix} = 2 \sum_{i \in I^0(\hat{x})} \hat{\lambda}_i w_i^2 > 0,$$

owing to part (i) of the strong second order sufficient condition.

If $s \neq 0$, then it follows from (4.77) that

$$\langle \nabla g_i(\hat{x}), s \rangle = 0, \quad i \in I^0(\hat{x}).$$

Therefore part (ii) of the strong second order sufficient condition implies that

$$\begin{bmatrix} s^T & w^T \end{bmatrix} \nabla^2 \bar{L}(\hat{x}, \hat{z}, \hat{\lambda}) \begin{bmatrix} s \\ w \end{bmatrix} \geq s^T \nabla^2 L(\hat{x}, \hat{\lambda}) s > 0.$$

Consequently, problem (4.74) satisfies the assumptions of Theorem 4.29, which implies our assertion. □

It follows that the point $(\hat{x}, \hat{z}, \hat{\lambda}, \hat{\mu})$ is a *local saddle point* of the augmented Lagrangian: there exists $\varrho_0 \geq 0$ and a neighborhood U of $(\hat{x}, \hat{z})$ such that for all $\varrho > \varrho_0$

$$\max_{\substack{\lambda \geq 0 \\ \mu \in \mathbb{R}^n}} \bar{L}_\varrho(\hat{x}, \hat{z}, \lambda, \mu) = \bar{L}_\varrho(\hat{x}, \hat{z}, \hat{\lambda}, \hat{\mu}) = \min_{(x,z) \in (X_0 \times \mathbb{R}^m) \cap U} \bar{L}_\varrho(x, z, \hat{\lambda}, \hat{\mu}).$$

In a way identical to Theorem 4.72 we obtain the following result.

THEOREM 4.34. *Assume that a point* $(\bar{x}, \bar{z}, \bar{\lambda}, \bar{\mu})$ *is a local saddle point of the augmented Lagrangian* (4.76) *for some* $\varrho \geq 0$. *Then* $(\bar{x}, \bar{z})$ *is a local minimum of problem* (4.74), *and* $(\bar{\lambda}, \bar{\mu})$ *is the vector of Lagrange multipliers associated with its constraints.*

In practice, we rarely deal with the augmented Lagrangian in its full form (4.76). The dependence of this function of z is simple, and the minimum with respect to z can be calculated in a closed form. Substituting $v_i = z_i^2$ and minimizing the function

$$p_i(v_i) = \lambda_i\big[g_i(x) + v_i\big] + \frac{\varrho}{2}\big[g_i(x) + v_i\big]^2$$

with respect to $v_i \geq 0$, we obtain

$$\hat{v}_i = \max\Big(0, -g_i(x) - \frac{\lambda_i}{\varrho}\Big).$$

This yields

$$\begin{aligned} p_i(\hat{v}_i) &= \lambda_i\big[g_i(x) + \hat{v}_i\big] + \frac{\varrho}{2}\big[g_i(x) + \hat{v}_i\big]^2 \\ &= \lambda_i \max\Big(g_i(x), -\frac{\lambda_i}{\varrho}\Big) + \frac{\varrho}{2}\Big(\max\Big(g_i(x), -\frac{\lambda_i}{\varrho}\Big)\Big)^2. \end{aligned} \tag{4.78}$$

Thus the minimum of the augmented Lagrangian with respect to z takes on the form:

$$\begin{aligned} L_\varrho(x, \lambda, \mu) = f(x) &+ \sum_{i=1}^{m} \lambda_i \max\Big(g_i(x), -\frac{\lambda_i}{\varrho}\Big) + \sum_{i=1}^{p} \mu_i h_i(x) \\ &+ \frac{\varrho}{2}\sum_{i=1}^{m}\Big(\max\Big(g_i(x), -\frac{\lambda_i}{\varrho}\Big)\Big)^2 + \frac{\varrho}{2}\sum_{i=1}^{p}\big[h_i(x)\big]^2. \end{aligned} \tag{4.79}$$

It is rather involved, but at close inspection we see that the terms corresponding to the inequality constraints are similar to those associated with the equality constraints. The only difference is that each function $g_i(x)$ is replaced by $\max\big(g_i(x), -\frac{\lambda_i}{\varrho}\big)$. Although these terms are nonsmooth, because of the presence of the maximum operation here, the entire function is smooth, as the following manipulation shows. Complementing (4.78) to the full square we obtain:

$$
\begin{aligned}
&p_i(\hat{v}_i)\\
&= \frac{1}{2\varrho}\Big[\lambda_i^2 + 2\lambda_i\varrho \max\big(g_i(x), -\frac{\lambda_i}{\varrho}\big) + \varrho^2\Big(\max\big(g_i(x), -\frac{\lambda_i}{\varrho}\big)\Big)^2\Big] - \frac{\lambda_i^2}{2\varrho}\\
&= \frac{1}{2\varrho}\Big[\lambda_i + \varrho \max\big(g_i(x), -\frac{\lambda_i}{\varrho}\big)\Big]^2 - \frac{\lambda_i^2}{2\varrho}\\
&= \frac{\varrho}{2}\Big[\max\big(0, g_i(x) + \frac{\lambda_i}{\varrho}\big)\Big]^2 - \frac{\lambda_i^2}{2\varrho}.
\end{aligned}
$$

Similar (but simpler) manipulations can be carried out for equality constraints. Therefore the augmented Lagrangian function (4.79) can be equivalently written as follows:

$$
\begin{aligned}
L_\varrho(x,\lambda,\mu) = f(x) &+ \frac{\varrho}{2}\sum_{i=1}^{m}\Big[\max\big(0, g_i(x) + \frac{\lambda_i}{\varrho}\big)\Big]^2\\
&+ \frac{\varrho}{2}\sum_{i=1}^{p}\Big[h_i(x) + \frac{\mu_i}{\varrho}\Big]^2 - \frac{1}{2\varrho}\sum_{i=1}^{m}\lambda_i^2 - \frac{1}{2\varrho}\sum_{i=1}^{p}\mu_i^2.
\end{aligned}
\tag{4.80}
$$

If all functions involved are continuously differentiable, then the augmented Lagrangian L_ϱ is continuously differentiable, because the function $t \mapsto [\max(0,t)]^2$ is smooth. Appropriate versions of Theorems 4.33 and 4.34 remain valid for this formulation.

EXERCISES

4.1. Consider the quadratic programming problem

$$
\begin{aligned}
&\text{minimize } \frac{1}{2}\langle x, Qx\rangle + \langle c, x\rangle\\
&\text{subject to } Ax \ge b.
\end{aligned}
$$

Here Q is a positive definite matrix of dimension n, A is a matrix of dimension $m \times n$, $c \in \mathbb{R}^n$, and $b \in \mathbb{R}^m$. Derive the dual problem.

4.2. The matrix game has the payoff matrix

$$
A = \begin{bmatrix} 2 & -5 & 2 \\ -3 & 4 & -1 \end{bmatrix}.
$$

Find its equilibrium in mixed strategies.

4.3. Solve the problem

$$\begin{aligned} &\text{minimize} \ \sum_{j=1}^{n}(x_j)^2 \\ &\text{subject to} \ \sum_{j=1}^{n} x_j = 1, \\ &\qquad 0 \le x_j \le u_j, \quad j = 1, \dots, n. \end{aligned}$$

4.4. Formulate the dual problem to the support vector machine problem from the Introduction (page 11).

4.5. We are given vectors x and $y^1, \dots, y^m$ in $\mathbb{R}^n$. Our objective is to check whether $x \in \operatorname{conv}\{y^1, \dots, y^m\}$. Formulate an appropriate convex optimization problem and its dual problem.†

4.6. A model represents the output variable, $y \in \mathbb{R}$, as a linear function,

$$y = \sum_{i=1}^{n} x_i u_i,$$

of input variables $u_1, \dots, u_n$. The quantities $x_1, \dots, x_n$ are unknown model coefficients. We have N observations of input and output variables: (u^j, y^j), $j = 1, \dots, N$. One way to determine the values of the coefficients $x_1, \dots, x_n$ is to minimize the maximum absolute error:

$$f(x) = \max_{1 \le j \le N} \left| y^j - \sum_{i=1}^{n} x_i u_i^j \right|.$$

This problem can be written as a constrained optimization problem

$$\begin{aligned} &\text{minimize} \ v \\ &\text{subject to} \ -v \le y^j - \sum_{i=1}^{n} x_i u_i^j \le v, \quad j = 1, \dots, N. \end{aligned}$$

Formulate and analyze the dual problem.

4.7. Consider the convex optimization problem

$$\text{minimize} \ \frac{1}{2}\|x - z\|^2 + f(x)$$

---

†Problems of this form arise in *data envelopment analysis*, which aims at verifying whether the characteristics of a specific company's branch (represented by x) can be reproduced by other branches, represented by $y^1, \dots, y^m$.

where

$$f(x) = \max_{1 \le i \le m} \big(\langle a_i, x \rangle + b_i \big).$$

The vectors $z \in \mathbb{R}^n$, $a_i \in \mathbb{R}^n$ and the scalars b_i are fixed.

The problem can be converted to a quadratic programming problem by introducing a new variable $v \in \mathbb{R}$:

$$\begin{aligned} &\text{minimize } \frac{1}{2}\|x - z\|^2 + v \\ &\text{subject to } v \ge \langle a_i, x \rangle + b_i, \quad i = 1, \dots, m. \end{aligned}$$

Formulate the dual problem.

4.8. Consider the conic programming problem

$$\begin{aligned} &\text{maximize } \langle c, x \rangle \\ &\text{subject to } Ax - b \in K, \end{aligned}$$

where K is a closed convex cone in $\mathbb{R}^n$. The vectors $c \in \mathbb{R}^n$, $b \in \mathbb{R}^m$ and the matrix A of dimension $m \times n$ are fixed. Derive the dual problem and formulate the duality theorem.

4.9. Assume that C is a closed convex cone in $\mathbb{R}^n$. The problem of projecting a point $z \in \mathbb{R}^n$ on C can be formulated as follows:

$$\begin{aligned} &\text{minimize } \frac{1}{2}\|y\|^2 \\ &\text{subject to } x + y = z, \\ &\qquad\qquad x \in C. \end{aligned}$$

Formulate the dual problem and prove that $y = \Pi_{C^\circ}(z)$ and that $y \perp x$.

4.10. Consider the integer programming problem

$$\begin{aligned} &\text{minimize } \langle c, x \rangle \\ &\text{subject to } Ax \ge b, \\ &\qquad\qquad x \ge 0, \quad x - \text{integer}, \end{aligned}$$

in which the matrix A has integer entries. Prove that its optimal value is not smaller than the optimal value of the linear programming problem

$$\begin{aligned} &\text{maximize } \langle \lceil b \rceil, \lambda \rangle \\ &\text{subject to } A^T \lambda \ge c, \\ &\qquad\qquad \lambda \ge 0. \end{aligned}$$

The symbol $\lceil b \rceil$ denotes the *roundup* of the vector b: the smallest integer vector greater than or equal to b.

4.11. Consider the integer optimization problem

$$\begin{aligned} &\text{minimize} \;\; \langle x, Cx \rangle \\ &\text{subject to} \;\; x_j \in \{0, 1\}, \quad j = 1, \ldots, m. \end{aligned}$$

The symmetric matrix C has dimension n and is not assumed to be positive semidefinite (otherwise 0 is an optimal solution).

(a) Prove that the following semidefinite programming problem provides a lower bound for the optimal value of the original problem:

$$\begin{aligned} &\text{minimize} \;\; \operatorname{tr}(CX) \\ &\text{subject to} \;\; X_{ii} = 1, \quad i = 1, \ldots, n, \\ &\qquad\qquad\;\; X \in \mathbb{S}^n_+. \end{aligned}$$

(b) Formulate the dual problem to the problem in (a).

4.12. Consider the convex two-stage stochastic optimization problem

$$\begin{aligned} &\text{minimize} \;\; f_0(x) + \sum_{j=1}^{N} p_j f_j(y_j) \\ &\text{subject to} \;\; g_0(x) + g_j(y_j) \le 0, \quad j = 1, \ldots, N, \\ &\qquad\qquad\;\; x \in X_0, \quad y_j \in Y_j, \quad j = 1, \ldots, N, \end{aligned}$$

in which $X_0 \subset \mathbb{R}^{n_0}$ and $Y_j \subset \mathbb{R}^{n_j}$, $j = 1, \ldots, N$, are convex sets, and the functions $f_0 : \mathbb{R}^{n_0} \to \mathbb{R}$, $f_j : \mathbb{R}^{n_j} \to \mathbb{R}$, $g_0 : \mathbb{R}^{n_0} \to \mathbb{R}^m$, $g_j : \mathbb{R}^{n_j} \to \mathbb{R}^m$, $j = 1, \ldots, N$, are convex. The convexity of the functions $g_j(\cdot)$ is understood component-wise. The coefficients p_j are nonnegative and total 1.

The *splitting approach* to this problem replaces it by a larger problem

$$\begin{aligned} &\text{minimize} \;\; \sum_{j=1}^{N} p_j \big[f_0(x_j) + f_j(y_j) \big] \\ &\text{subject to} \;\; g_0(x_j) + g_j(y_j) \le 0, \quad j = 1, \ldots, N, \\ &\qquad\qquad\;\; x_j \in X_0, \quad y_j \in Y_j, \quad j = 1, \ldots, N, \\ &\qquad\qquad\;\; x_1 = x_2 = \cdots = x_N. \end{aligned}$$

(a) Prove the equivalence of both problems.

(b) Each of the following two systems is equivalent to the splitting constraint:

$$\begin{aligned} x_1 &= x_2, \\ x_2 &= x_3, \\ &\;\vdots \\ x_{N-1} &= x_N, \end{aligned}$$

and

$$x_k = \sum_{j=1}^{N} p_j x_j, \quad k = 1, \dots, N.$$

Formulate and analyze the dual problems resulting from introducing Lagrangian terms associated with the splitting constraints, for both systems above. Show that the calculation of the dual function decomposes into N independent subproblems.

4.13. Consider the optimal control problem

$$\begin{aligned} \text{minimize } & \frac{1}{2} \sum_{t=1}^{T-1} \langle u_t, R_t u_t \rangle \\ \text{subject to } & x_{t+1} = A_t x_t + B_t u_t, \quad t = 1, \dots, T-1, \\ & x_T = d. \end{aligned}$$

The initial state x_0 is fixed, and the final state x_T is required to be equal to d. Here $x_t \in \mathbb{R}^n$, $u_t \in \mathbb{R}^m$, and A_t, B_t, and R_t are matrices of appropriate dimension. The matrices R_t are positive definite. Formulate and analyze the dual problem.

4.14. Consider the linear programming problem with parameters $b \in \mathbb{R}$ and $u \in \mathbb{R}$:

$$\begin{aligned} \text{minimize } & \langle c, x \rangle \\ \text{subject to } & \langle a, x \rangle \geq b, \\ & 0 \leq x_j \leq u, \quad j = 1, \dots, n. \end{aligned}$$

The vectors $c \in \mathbb{R}^n$ and $a \in \mathbb{R}^n$ are fixed.

Describe the optimal value of this problem as a function of (b, u).

4.15. The linear two-stage stochastic programming problem is defined as follows:

$$\begin{aligned} \text{minimize } & \langle c, x \rangle + \sum_{j=1}^{N} p_j \langle q_j, y_j \rangle \\ \text{subject to } & T_j x + W y_j = h_j, \quad j = 1, \dots, N, \\ & x \in X_0, \quad y_j \in Y_j, \quad j = 1, \dots, N. \end{aligned}$$

We assume that the sets X_0 and Y_j, $j = 1, \dots, N$, are convex closed polyhedra. The coefficients p_j are nonnegative and total 1.

One way to approach this problem is to consider the functions

$$Q_j(x) = \inf \left\{ \langle q_j, y \rangle : W y = h_j - T_j x, \ y \in Y_j \right\}, \quad j = 1, \dots, N,$$

and to define the problem

$$\underset{x \in X_0}{\text{minimize}} \ \langle c, x \rangle + \sum_{j=1}^{N} p_j Q_j(x).$$

Assuming that $Q_j(x) > -\infty$, $j = 1, \dots, N$, for all x, and that there exist points x such that all $Q_j(x) < \infty$, describe the subdifferential of the objective function of this problem.

4.16. Consider the nonlinear optimization problem

$$\begin{aligned} \text{minimize} \quad & -x_1 x_2 + x_3 \\ \text{subject to} \quad & x_1 + x_2 + (x_3)^2 \le 2 \\ & x_1 \ge 0, \ x_2 \ge 0. \end{aligned}$$

(a) Find the optimal solution and the multiplier corresponding to the nonlinear constraint.

(b) Does the Lagrangian of this problem have a saddle point?

(c) Form the augmented Lagrangian. For which values of the penalty parameter does the duality relation hold true for the augmented Lagrangian in a neighborhood of the optimal solution?

PART 2

Methods

Chapter Five

Unconstrained Optimization of Differentiable Functions

5.1 INTRODUCTION TO ITERATIVE ALGORITHMS

Consider the unconstrained optimization problem

$$\underset{x\in\mathbb{R}^n}{\text{minimize}}\ f(x), \tag{5.1}$$

where $f : \mathbb{R}^n \to \mathbb{R}$ is a continuously differentiable function. We know from Theorem 3.1 on page 88 that if a point $\hat{x}$ is a local minimum of problem (5.1) then

$$\nabla f(\hat{x}) = 0. \tag{5.2}$$

If the function $f(\cdot)$ has a simple form, we can find all its stationary points and, among them, all local minima, by solving this system of equations. But if the description of the function $f(\cdot)$ is involved, as is usual in applications, the approach by the necessary conditions of optimality may fail. This is always the case when $f(\cdot)$ is defined *algorithmically*, that is, we have some procedure for calculating $f(x)$ at each point x (and perhaps the derivatives of $f(\cdot)$ at x), but no closed-form expression of $f(\cdot)$ is available.

In this situation, an *iterative method* for solving (5.1) is needed. Such a method constructs a sequence of points x^k in $\mathbb{R}^n$, $k = 0, 1, 2, \dots$, so that it converges (in some sense) to a solution of problem (5.1). We consider the points x^k as approximations of a solution to (5.1). In practical applications of an iterative method we stop the calculations at some iteration k^* and we accept x^{k^*} as a sufficiently good approximation of a solution.

To be more specific, suppose we have a certain point $x^k \in \mathbb{R}^n$ and we want to find the next point, x^{k+1}, such that

$$f(x^{k+1}) < f(x^k). \tag{5.3}$$

Although this condition is *neither necessary nor sufficient* for a convergent iterative method for solving (5.1), it is clearly the first idea that needs to be addressed.

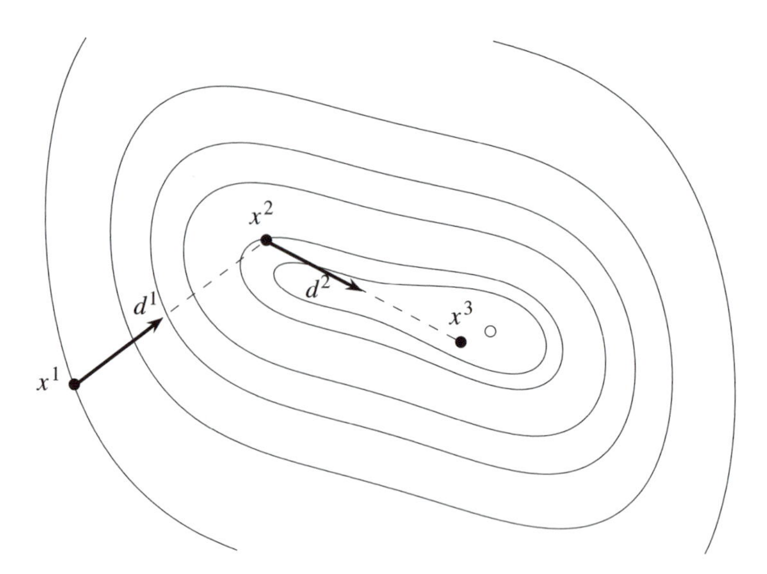

Figure 5.1. Search in directions.

One way to achieve progress, as defined in (5.3), is to choose a direction $d^k \in \mathbb{R}^n$ and then to make a step in this direction,

$$x^{k+1} = x^k + \tau_k d^k, \tag{5.4}$$

with some step size coefficient $\tau_k > 0$. This is illustrated in Figure 5.1. After that, a new direction d^{k+1} can be chosen at the point x^{k+1}, and the method continues.

Several fundamental questions arise in connection with this general idea. How should the direction d^k and step size τ_k be chosen? When is the sequence $\{x^k\}$ convergent to a solution? In which sense is it convergent? Is the solution a local or a global minimum? What is the speed of convergence, do we need many or few iterations to reach a reasonably good approximation of a solution? How can we detect that our current approximation is sufficiently good?

While there have been many attempts to construct a general theory of iterative algorithms of optimization, we believe that it is best to discuss these questions in the context of particular methods. Our next section on line search techniques and the following sections on the method of steepest descent bring more specificity to these topics.

5.2 LINE SEARCH

As discussed in the preceding section, line search algorithms are an essential part of most iterative methods of unconstrained minimization. Formally, given a point $x \in \mathbb{R}^n$ and direction $d \in \mathbb{R}^n$, these methods look for an unconstrained minimum of the function

$$\varphi(\tau) = f(x + \tau d), \quad \tau \in \mathbb{R}.$$

As we shall soon see, finding the exact minimum is not so crucial here, because the real goal is to find the minimum of $f(\cdot)$ rather than of $\varphi(\cdot)$, but a substantial improvement should be achieved.

Golden Section

The main idea of the method is to generate a sequence of intervals $[\alpha_k, \delta_k]$ containing the minimum point $\hat{\tau}$ of $\varphi(\cdot)$:

$$\alpha_0 \le \alpha_1 \le \cdots \le \alpha_k \le \hat{\tau} \le \delta_k \le \cdots \le \delta_1 \le \delta_0.$$

This is achieved as follows. At first, by expanding or contracting the step size we construct four points

$$\alpha_0 < \beta_0 < \gamma_0 < \delta_0$$

such that

$$\varphi(\alpha_0) > \varphi(\beta_0) \quad \text{and} \quad \varphi(\gamma_0) < \varphi(\delta_0).$$

Then, for a continuous $\varphi(\cdot)$, we are sure that a local minimum is contained in the interval $[\alpha_0, \delta_0]$.

Such relations will be maintained for all iterations k. It is convenient to satisfy the proportions

$$\frac{\gamma_k - \alpha_k}{\delta_k - \alpha_k} = \frac{\delta_k - \beta_k}{\delta_k - \alpha_k} = q \tag{5.5}$$

with the *golden ratio*

$$q = \frac{\sqrt{5} - 1}{2} \approx 0.618.$$

Then it is easy to calculate that also

$$\frac{\beta_k - \alpha_k}{\gamma_k - \alpha_k} = \frac{\delta_k - \gamma_k}{\delta_k - \beta_k} = q. \tag{5.6}$$

The next operation is the comparison of the values $\varphi(\beta_k)$ and $\varphi(\gamma_k)$.

If $\varphi(\beta_k) < \varphi(\gamma_k)$ then we conclude that a local minimum is located in the interval $[\alpha_k, \gamma_k]$. We discard the point δ_k and we make the substitutions

$$\begin{aligned}
\alpha_{k+1} &:= \alpha_k, \\
\delta_{k+1} &:= \gamma_k, \\
\gamma_{k+1} &:= \beta_k, \\
\beta_{k+1} &:= q\alpha_k + (1-q)\gamma_k.
\end{aligned}$$

Due to (5.6), relations (5.5) hold true for $k+1$ as well.

If $\varphi(\beta_k) \geq \varphi(\gamma_k)$, a local minimum is located in the interval $[\beta_k, \delta_k]$. We discard the point α_k and we make appropriate substitutions:

$$\begin{aligned}
\delta_{k+1} &:= \delta_k, \\
\alpha_{k+1} &:= \beta_k, \\
\beta_{k+1} &:= \gamma_k, \\
\gamma_{k+1} &:= q\delta_k + (1-q)\beta_k.
\end{aligned}$$

Because of (5.6), relations (5.5) remain true for $k+1$ in this case too. In this way we guarantee that the length of the interval $[\alpha_k, \delta_k]$ containing a local minimum is multiplied by q at each step. The operation of the method is illustrated in Figure 5.2.

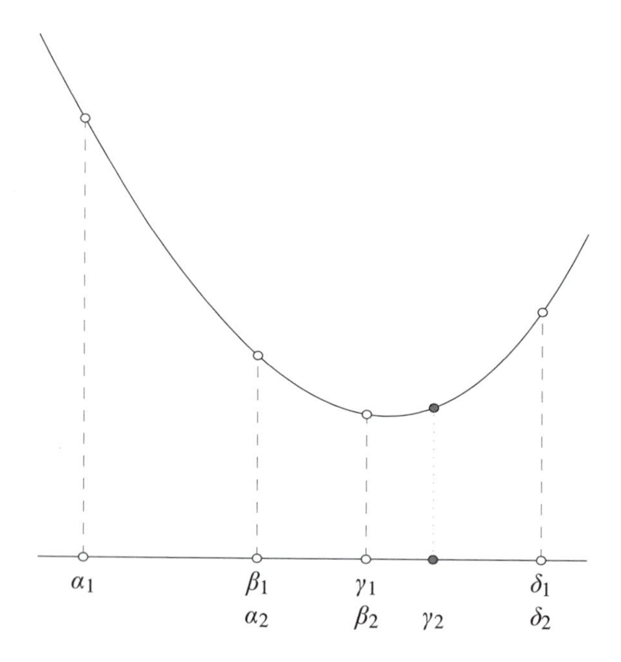

Figure 5.2. Golden section.

After a sufficiently small length of the interval is achieved, the point with the smallest value of the function found so far is considered to be the approximation of a minimum of $\varphi(\cdot)$.

Interpolation

The golden section method, although elegant and applicable to all continuous functions, has rather slow convergence. If the function $\varphi(\cdot)$ is twice continuously differentiable, we can approximate the minimum much faster, by using interpolation techniques.

At iteration k of the method we have three points

$$\alpha_k < \tau_k < \beta_k$$

such that

$$\varphi(\alpha_k) > \varphi(\tau_k) < \varphi(\beta_k).$$

Then the function $\varphi(\cdot)$ has a minimum in the interval $[\alpha_k, \beta_k]$.

The initial three points can be found, for example, by scanning the values $\varphi(0)$ (which is given), $\varphi(\tau_0)$, $\varphi(2\tau_0)$, $\varphi(4\tau_0)$, ..., or $\varphi(-\tau_0)$, $\varphi(-2\tau_0)$, $\varphi(-4\tau_0)$, ..., depending on whether $\varphi(\tau_0) < \varphi(0)$.

Next, we interpolate $\varphi(\cdot)$ with the second order Lagrange polynomial, determined from the values of $\varphi(\cdot)$ at the nodes α_k, τ_k, and β_k. The minimum of this polynomial is denoted by γ_k. Simple algebra yields

$$\gamma_k = \frac{\varphi(\alpha_k)\big[\beta_k^2 - \tau_k^2\big] + \varphi(\tau_k)\big[\alpha_k^2 - \beta_k^2\big] + \varphi(\beta_k)\big[\tau_k^2 - \alpha_k^2\big]}{2\big(\varphi(\alpha_k)[\beta_k - \tau_k] + \varphi(\tau_k)[\alpha_k - \beta_k] + \varphi(\beta_k)[\tau_k - \alpha_k]\big)}.$$

It is evident that γ_k is in the interior of the interval (α_k, β_k).

After that, we evaluate $\varphi(\gamma_k)$ and we remove one of the end points, so that out of the remaining three points the middle one is best. This is illustrated in Figure 5.3. In general, we cannot prove that this method generates a sequence convergent to a local minimum of $\varphi(\cdot)$, unless we make strong and unrealistic assumptions about the second derivative of the function $\varphi(\cdot)$. In practice, the method of interpolation is used with many technical safeguards guaranteeing that the interval $[\alpha_k, \beta_k]$ shrinks sufficiently fast.

A variation of the interpolation method is obtained in the situation when the derivative $\varphi'(0)$ is known. This occurs when the line search method is used in conjunction with *gradient methods* of minimization. In this case $\varphi'(0) < 0$. At first, we search for a point $\beta > 0$ such that $\varphi(\beta) \geq \varphi(0)$. The information about the values of φ at 0 and at β, and about the derivative at 0, is sufficient to construct a quadratic interpolating polynomial. Elementary

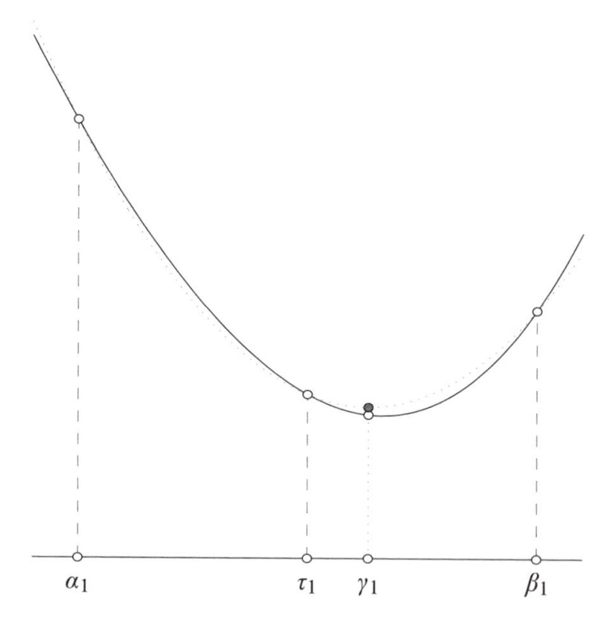

Figure 5.3. Quadratic interpolation.

manipulations yield the formula for its minimum:

$$\tau = \frac{-\varphi'(0)\beta^2}{2[\varphi(\beta) - \varphi(0) - \varphi'(0)\beta]}.$$

Since $\varphi(\beta) \geq \varphi(0)$, we have $0 < \tau \leq \beta/2$. If we still have $\varphi(\tau) > \varphi(0)$, then we replace β with τ and we repeat the above interpolation.

If $\varphi(\tau) < \varphi(0)$, then we have three points with the best one in the middle, and we can apply the interpolation method discussed before.

There are many variations of these techniques, using interpolation with third order polynomials, and many heuristic techniques safeguarding convergence in practical applications. The specificity of one-dimensional minimization problems is fully exploited in these techniques.

Two-Slope Test

When a line search method is employed in conjunction with a gradient method for minimization, we frequently use a rather crude line search satisfying some basic improvement conditions.

In the *two-slope test*,† which is widely used in practical optimization al-

†The two-slope test is also called *Goldstein's test*.

gorithms, we specify two coefficients, α_1 and α_2, satisfying the relations:

$$1 > \alpha_1 > \alpha_2 > 0.$$

A line search procedure terminates its operation if the following conditions are satisfied at the current point τ:

$$\varphi(0) + \alpha_1 \tau \varphi'(0) \le \varphi(\tau) \le \varphi(0) + \alpha_2 \tau \varphi'(0). \tag{5.7}$$

Recall that $\varphi'(0) < 0$. The right inequality in (5.7) ensures that the progress made is proportional to the derivative at 0, while the left inequality prevents τ from becoming too small. In the next section we discuss the application of the two-slope test within the method of steepest descent, where the roles of the two inequalities in (5.7) become apparent. The two-slope test is illustrated in Figure 5.4.

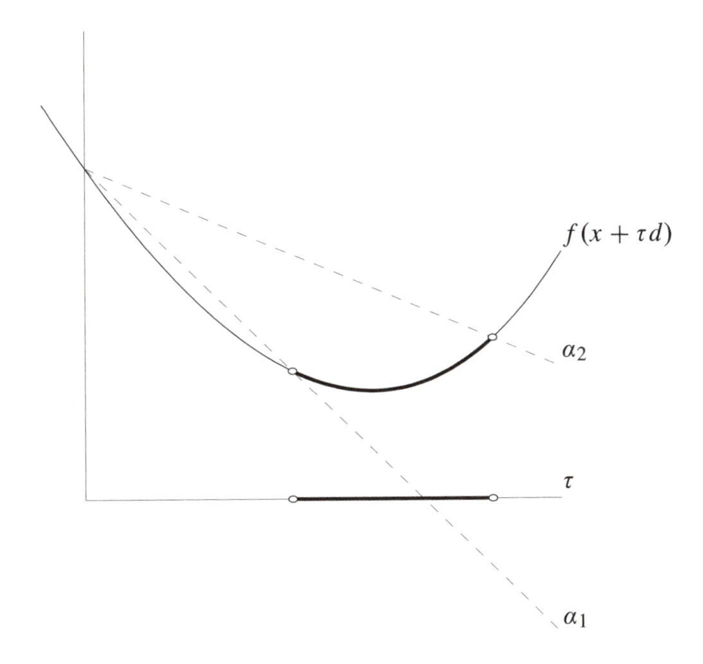

Figure 5.4. Two-slope test.

For a quadratic function $\varphi(\cdot)$ its value at the minimum $\hat{\tau}$ satisfies the relation

$$\varphi(\hat{\tau}) = \varphi(0) + \frac{1}{2}\hat{\tau}\varphi'(0), \tag{5.8}$$

and therefore it is most reasonable to use

$$\alpha_1 > \frac{1}{2} > \alpha_2.$$

5.3 THE METHOD OF STEEPEST DESCENT

5.3.1 The Direction of Steepest Descent

Consider the iterative formula (5.4) and suppose the direction d^k is fixed. We can construct the first order expansion of $f(x^k + \tau d^k)$ about the point x^k:

$$f(x^k + \tau d^k) = f(x^k) + \tau \langle \nabla f(x^k), d^k \rangle + o(\tau),$$

with $o(\tau)/\tau \to 0$ as $\tau \to 0$. Suppose $\nabla f(x^k) \neq 0$. If we want to ensure that inequality (5.3) holds true for all sufficiently small step sizes $\tau_k > 0$, we must make the first order term in the expansion above negative:

$$\langle \nabla f(x^k), d^k \rangle < 0.$$

Recall that this term is the directional derivative $f'(x^k; d^k)$ of $f(\cdot)$ at the point x^k in the direction d^k. It seems reasonable to choose the direction d^k in such a way that the directional derivative is as negative as possible. However, the directional derivative of a differentiable function is linear in the direction, and has no minimum, unless $\nabla f(x^k) = 0$. In order to make the problem of choosing the direction meaningful, we need to restrict the length of the directions under consideration. Any bound does well here, in particular a bound by $\|\nabla f(x^k)\|$, as in the problem below:

$$\begin{aligned} &\text{minimize} \ \ \langle \nabla f(x^k), d \rangle \\ &\text{subject to} \ \ \|d\|^2 \leq \|\nabla f(x^k)\|^2. \end{aligned} \tag{5.9}$$

By using the necessary conditions of optimality of Theorem 3.25 on page 116, we can easily find the optimal solution of this problem:

$$d^k = -\nabla f(x^k). \tag{5.10}$$

It is called the *direction of steepest descent*, and the resulting iterative method

$$x^{k+1} = x^k - \tau_k \nabla f(x^k), \quad k = 0, 1, 2, \ldots, \tag{5.11}$$

is called the *method of steepest descent*. The reader can easily verify that imposing another bound on the length of the direction in problem (5.9) leads to a direction of the form $d = -\alpha \nabla f(x^k)$ with some $\alpha > 0$. Since the direction is multiplied in (5.11) by a step size coefficient anyway, it is sufficient to consider formula (5.10).

In (5.11) the step size coefficients τ_k are positive. Different versions of the method of steepest descent are obtained by detailing the rules for calculating these coefficients.

Before proceeding to the analysis of the method of steepest descent, we should stress that this method is almost never used in its pure form, because it is very slow. However, the understanding of its operation is crucial for the understanding of many iterative methods of optimization. It also allows us to introduce concepts and techniques of analysis of these methods.

5.3.2 Constant Step Size

The simplest version of the method of steepest descent is obtained by fixing all step size coefficients τ_k in (5.11) at some value $\tau > 0$:

$$x^{k+1} = x^k - \tau \nabla f(x^k), \quad k = 0, 1, 2, \ldots. \tag{5.12}$$

THEOREM 5.1. *Assume that the function $f(\cdot)$ is continuously differentiable and its gradient is Lipschitz continuous with some constant M:*

$$\|\nabla f(x) - \nabla f(y)\| \leq M\|x - y\| \quad \textit{for all} \quad x, y \in \mathbb{R}^n.$$

Furthermore, assume that the function $f(\cdot)$ is bounded from below. If the step size τ satisfies the inequalities

$$0 < \tau < \frac{1}{M},$$

then for every x^0 the sequence $\{x^k\}$ generated by algorithm (5.12) *satisfies the condition:*

$$\lim_{k\to\infty} \nabla f(x^k) = 0. \tag{5.13}$$

Proof. Consider iteration k. Applying the mean value theorem we infer that

$$f(x^{k+1}) = f(x^k + \tau d^k) = f(x^k) + \tau \langle \nabla f(\bar{x}), d^k \rangle,$$

where $\bar{x} = x^k + \theta \tau d^k$ and $\theta \in [0, 1]$. Therefore

$$\begin{aligned} f(x^{k+1}) &= f(x^k) + \tau \langle \nabla f(x^k), d^k \rangle + \tau \langle \nabla f(\bar{x}) - \nabla f(x^k), d^k \rangle \\ &\leq f(x^k) + \tau \langle \nabla f(x^k), d^k \rangle + \tau \|\nabla f(x^k) - \nabla f(\bar{x})\| \cdot \|d^k\|. \end{aligned}$$

Setting $d^k = -\nabla f(x^k)$ and using the Lipschitz continuity of the gradient we get

$$f(x^{k+1}) \leq f(x^k) - \tau \|\nabla f(x^k)\|^2 + \tau M \|x^k - \bar{x}\| \cdot \|\nabla f(x^k)\|.$$

But $\|x^k - \bar{x}\|$ can be bounded above by $\|x^k - x^{k+1}\| = \tau \|\nabla f(x^k)\|$ and the last inequality yields

$$\begin{aligned} f(x^{k+1}) &\leq f(x^k) - \tau \|\nabla f(x^k)\|^2 + \tau^2 M \|\nabla f(x^k)\|^2 \\ &= f(x^k) - \tau(1 - \tau M)\|\nabla f(x^k)\|^2. \end{aligned} \tag{5.14}$$

Since $\tau(1 - \tau M) > 0$ by assumption, we conclude that $f(x^{k+1}) \le f(x^k)$ for all k.

The sequence $\{f(x^k)\}$ is nonincreasing and bounded from below by assumption, and thus it has a limit. Hence

$$\lim_{k\to\infty} \big[f(x^k) - f(x^{k+1})\big] = 0. \tag{5.15}$$

Combining the last two relations we see that

$$0 \le \tau(1 - \tau M)\|\nabla f(x^k)\|^2 \le \big[f(x^k) - f(x^{k+1})\big] \to 0,$$

and the assertion of the theorem follows. □

If we can additionally ensure that the sequence $\{x^k\}$ is bounded, we can obtain convergence to a stationary point. Let us observe that all points x^k generated by the method belong to the set

$$X_0 = \{x \in \mathbb{R}^n : f(x) \le f(x^0)\},$$

because $f(x^{k+1}) \le f(x^k)$ for all k.

Corollary 5.2. *In addition to the conditions of Theorem* 5.1, *assume that the set X_0 is bounded. Then the sequence $\{x^k\}$ is bounded and every accumulation point x^* of this sequence satisfies the equation $\nabla f(x^*) = 0$.*

It should be stressed that the above result guarantees only that the accumulation points are *stationary points*. Nothing more can be obtained in this setting, because if $\nabla f(x^0) = 0$ then the method cannot leave the starting point.

5.3.3 Two-Slope Test

The main disadvantage of the method with constant step size is the need to know the Lipschitz constant M of the gradient. A simple way around this difficulty is the use of an adaptive line search procedure. It is easiest, and most instructive, to analyze application of the two-slope test here. Recall that it uses two coefficients:

$$1 > \alpha_1 > \alpha_2 > 0,$$

and accepts every step size τ_k satisfying conditions (5.7). Since the function $\varphi(\tau) = f(x^k - \tau\nabla f(x^k))$ has derivative at 0 equal to

$$\varphi'(0) = -\|\nabla f(x^k)\|^2,$$

the two-slope test is equivalent to the inequalities

$$f(x^k) - \alpha_1 \tau_k \|\nabla f(x^k)\|^2 \le f(x^k - \tau_k \nabla f(x^k)) \le f(x^k) - \alpha_2 \tau_k \|\nabla f(x^k)\|^2. \tag{5.16}$$

The convergence of the method of steepest descent follows immediately from these conditions.

THEOREM 5.3. *Assume that the function $f(\cdot)$ is continuously differentiable and its gradient is Lipschitz continuous with constant M. Furthermore, assume that the set X_0 is bounded. Then the sequence $\{x^k\}$ generated by the method of steepest descent with the two-slope test is bounded and every accumulation point x^* of this sequence satisfies the equation $\nabla f(x^*) = 0$.*

Proof. The right inequality in (5.16) implies that

$$f(x^{k+1}) \le f(x^k), \quad k = 0, 1, 2, \ldots.$$

As the set X_0 is bounded, the sequence $\{f(x^k)\}$ is bounded from below and therefore it is convergent. Moreover, the right inequality of (5.16) yields

$$\tau_k \|\nabla f(x^k)\|^2 \le \frac{f(x^{k+1}) - f(x^k)}{\alpha_2},$$

and thus

$$\lim_{k\to\infty} \tau_k \|\nabla f(x^k)\|^2 = 0. \tag{5.17}$$

We shall now investigate the implications of the left inequality in (5.16). Inequality (5.14) remains valid:

$$f(x^{k+1}) \le f(x^k) - \tau_k(1 - \tau_k M)\|\nabla f(x^k)\|^2.$$

This combined with the left inequality of (5.16) yields

$$f(x^k) - \alpha_1 \tau_k \|\nabla f(x^k)\|^2 \le f(x^k) - \tau_k(1 - \tau_k M)\|\nabla f(x^k)\|^2.$$

Consequently, either $\nabla f(x^k) = 0$, or

$$\tau_k \ge \frac{1 - \alpha_1}{M}.$$

These conditions and (5.17) imply that

$$\lim_{k\to\infty} \|\nabla f(x^k)\| = 0,$$

as required. □

Our proof reveals the reasons for using two slopes in the line search procedure. The right slope ensures that the sequence $\{f(x^k)\}$ is monotonically decreasing, while the left one guarantees that diminishing improvements occur only if $\|\nabla f(x^k)\| \to 0$.

5.3.4 Directional Minimization

Another way is to choose the step size τ_k by minimizing the function

$$\varphi_k(\tau) = f(x^k + \tau d^k)$$

with respect to $\tau > 0$. Such a task can be carried out numerically in a very effective way, because the function $\varphi(\cdot)$ is a function of one variable. We discussed several directional minimization algorithms in Section 5.2.

THEOREM 5.4. *Assume that the function $f(\cdot)$ is continuously differentiable and the set X_0 is bounded. Then the method of steepest descent with directional minimization generates a sequence of points $\{x^k\}$ such that each of its accumulation points x^* satisfies the equation $\nabla f(x^*) = 0$.*

Proof. By the construction of the method, we have $f(x^{k+1}) \le f(x^k)$ for all k. Furthermore, all points x^k belong to the set X_0, which is compact. It follows that the sequence $\{f(x^k)\}$ is bounded. Thus it is convergent.

Moreover, the sequence $\{x^k\}$ has an accumulation point. Consider any accumulation point x^* and let $\mathcal{K}$ be the infinite ordered set of iteration numbers such that

$$\lim_{\substack{k\to\infty \\ k\in\mathcal{K}}} x^k = x^*.$$

Since the sequence $\{f(x^k)\}$ is convergent, we have

$$f(x^*) = \lim_{k\to\infty} f(x^k). \tag{5.18}$$

We prove the theorem by contradiction.

Suppose the point x^* is not stationary, i.e., $\nabla f(x^*) \neq 0$. Consider the direction of steepest descent $-\nabla f(x^*)$ and the points

$$y(\tau) = x^* - \tau \nabla f(x^*), \quad \tau \ge 0.$$

Since $\nabla f(x^*) \neq 0$, the problem

$$\operatorname*{minimize}_{\tau\ge 0} f(y(\tau)) \tag{5.19}$$

has a solution $\tau_* > 0$, and $f(y(\tau_*)) < f(x^*)$ for every solution τ_*. Of course, the method does not carry out the directional minimization at x^*, but we shall show that the outcome is similar, if the directional minimization is carried out at points which are close to x^*.

Consider the points x^k for $k \in \mathcal{K}$, and the immediately following points x^{k+1}. By the directional minimization condition,

$$f(x^{k+1}) \le f(x^k - \tau\nabla f(x^k)) \quad \text{for all} \quad \tau \ge 0. \tag{5.20}$$

When $k \to \infty$, $k \in \mathcal{K}$, then $x^k \to x^*$ and $\nabla f(x^k) \to \nabla f(x^*)$. Define $\varepsilon = \|\nabla f(x^*)\|$. We have

$$x^{k+1} = x^k - \tau_k \nabla f(x^k), \tag{5.21}$$

and $\|\nabla f(x^k)\| \geq \varepsilon/2$ for all sufficiently large $k \in \mathcal{K}$. Therefore, for all sufficiently large $k \in \mathcal{K}$,

$$0 \leq \tau_k = \frac{\|x^{k+1} - x^k\|}{\|\nabla f(x^k)\|} \leq \frac{2}{\varepsilon} \operatorname{diam}(X_0),$$

with $\operatorname{diam}(X_0)$ denoting the largest distance between two points of X_0. Consequently, the step sizes τ_k for all sufficiently large $k \in \mathcal{K}$ are uniformly bounded. We can therefore choose an infinite subset $\mathcal{K}_1$ of $\mathcal{K}$ such that the sequence $\{\tau_k\}$, $k \in \mathcal{K}_1$, has a limit. We denote this limit by $\bar{\tau}$. Passing to the limit in (5.21) yields

$$\lim_{\substack{k\to\infty \\ k\in\mathcal{K}_1}} x^{k+1} = x^* - \bar{\tau} \nabla f(x^*).$$

We can now pass to the limit in inequalities (5.20), when $k \to \infty$, $k \in \mathcal{K}_1$. We obtain:

$$f(x^* - \bar{\tau}\nabla f(x^*)) \leq f(x^* - \tau \nabla f(x^*)) \quad \text{for all} \quad \tau \geq 0.$$

We conclude that the step size $\bar{\tau}$ is an optimal solution of problem (5.19). Hence

$$\lim_{\substack{k\to\infty \\ k\in\mathcal{K}_1}} f(x^{k+1}) = f(y(\tau_*)) < f(x^*).$$

Then also

$$\lim_{k\to\infty} f(x^k) \leq \lim_{\substack{k\to\infty \\ k\in\mathcal{K}_1}} f(x^{k+1}) < f(x^*),$$

which contradicts (5.18). □

5.3.5 Speed of Convergence

An important issue associated with the analysis of any iterative method is its speed of convergence. Usually, the analysis of the speed of convergence can be carried out under much stronger conditions than the analysis of convergence alone. In the case of the method of steepest descent we assume that the function $f(\cdot)$ is twice continuously differentiable and that its Hessian satisfies the condition

$$mI \preceq \nabla^2 f(x) \preceq MI \quad \text{for all} \quad x \in \mathbb{R}^n, \tag{5.22}$$

where $0 < m \leq M$. As in the preceding chapters, the relation $A \preceq B$ means that $B - A$ is positive semidefinite.

Observe that the left hand side of (5.22) implies that the set $X_0 = \{x \in \mathbb{R}^n : f(x) \leq f(x^0)\}$ is bounded. Moreover, the function $f(\cdot)$ is strictly convex, and thus it has a unique minimum point $\hat{x}$. The method of steepest descent, in any of the versions discussed before, generates a sequence convergent to the unique minimum $\hat{x}$ of problem (5.1). But now we are also able to estimate its speed of convergence.

Let us start from the analysis of the version with constant step size, considered in Theorem 5.1.

Theorem 5.5. *Assume that the function $f(\cdot)$ is twice continuously differentiable and satisfies condition (5.22). Then the method of steepest descent with fixed step size $\tau \in (0, 2/M)$ generates a sequence $\{x^k\}$ satisfying the relation*

$$\|x^{k+1} - \hat{x}\| \leq (q)^k \|x^0 - \hat{x}\|,$$

where

$$q = \max\big(|1 - \tau m|, |1 - \tau M|\big) < 1.$$

Proof. We have the obvious relation

$$x^{k+1} - \hat{x} = x^k - \hat{x} - \tau \nabla f(x^k). \tag{5.23}$$

Using the fact that $\nabla f(\hat{x}) = 0$ we can represent the gradient as follows:

$$\nabla f(x^k) = \int_0^1 \big[\nabla^2 f(\hat{x} + \theta(x^k - \hat{x}))\big](x^k - \hat{x})\, d\theta.$$

Define

$$Q_k = \int_0^1 \big[\nabla^2 f(\hat{x} + \theta(x^k - \hat{x}))\big]\, d\theta.$$

With this notation we can write (5.23) as

$$x^{k+1} - \hat{x} = x^k - \hat{x} - \tau Q_k(x^k - \hat{x}) = (I - \tau Q_k)(x^k - \hat{x}).$$

Therefore

$$\|x^{k+1} - \hat{x}\| \leq \|I - \tau Q_k\| \cdot \|x^k - \hat{x}\|.$$

The norm of the matrix $I - \tau Q_k$ is given by the maximum absolute value of its eigenvalue. It follows from condition (5.22) and the definition of Q_k that

$$mI \preceq Q_k \preceq MI.$$

Thus all eigenvalues of Q_k lie between m and M. Therefore the matrix $I - \tau Q_k$ has all eigenvalues contained in the interval

$$[1 - \tau M, 1 - \tau m]$$

and the result follows. □

It follows from the last theorem that the sequence $\{x^k\}$ approaches the solution at the speed of a geometric progression with ratio q. An iterative method that has this property is called *linearly convergent*.

We can choose the value of τ to make the ratio q as small as possible. This occurs when the interval $[1 - \tau M, 1 - \tau m]$ is located symmetrically around 0. Thus the best step size equals

$$\tau = \frac{2}{M + m}$$

and the resulting ratio is

$$q = \frac{M - m}{M + m}.$$

We notice that the key role is played here by the ratio

$$\varkappa = \frac{M}{m}.$$

It is called the *condition index* of the problem. It is always greater than or equal to 1. We have

$$q = \frac{\varkappa - 1}{\varkappa + 1}$$

and if $\varkappa$ is close to 1 then q is close to 0, and convergence is very fast. However, if $\varkappa$ is very large, then q is close to 1, and convergence is slow.

The main disadvantage of the analysis above is the assumption that we know the constants m and M and that we can use a constant step size. The method of steepest descent is almost never used in this form. We focus our attention, therefore, on the method with directional minimization, as a convenient theoretical model of the method with line search procedures.

In general, to derive rate of convergence results, we have to establish some relation between the improvements made by the method and a measure of nonoptimality of the current point. This is the main motivation of the next lemma.

Lemma 5.6. *Assume that the function $f(\cdot)$ is twice continuously differentiable and its Hessian at every point x satisfies (5.22). Then for all x we have the estimate*

$$\|\nabla f(x)\|^2 \geq m\big(1 + \frac{m}{M}\big)\big[f(x) - f(\hat{x})\big],$$

where $\hat{x}$ is the unique minimum point of $f(\cdot)$.

Proof. It follows from the left inequality in (5.22) that the function $f(\cdot)$ is strictly convex and has bounded level sets. Therefore it has a unique minimum point $\hat{x}$. Using Taylor's formula at $\hat{x}$, we get

$$f(x) = f(\hat{x}) + \langle \nabla f(\hat{x}), x - \hat{x}\rangle + \frac{1}{2}\langle x - \hat{x}, \nabla^2 f(\bar{x})(x - \hat{x})\rangle,$$

where $\bar{x} = (1 - \theta)\hat{x} + \theta x$ and $\theta \in [0, 1]$. Since $\nabla f(\hat{x}) = 0$, conditions (5.22) render the estimates

$$\frac{m}{2}\|x - \hat{x}\|^2 \leq f(x) - f(\hat{x}) \leq \frac{M}{2}\|x - \hat{x}\|^2. \tag{5.24}$$

On the other hand, Taylor's formula at x yields

$$f(\hat{x}) = f(x) + \langle \nabla f(x), \hat{x} - x\rangle + \frac{1}{2}\langle \hat{x} - x, \nabla^2 f(\tilde{x})(\hat{x} - x)\rangle,$$

with some intermediate point $\tilde{x} = (1 - \lambda)\hat{x} + \lambda x$ and $\lambda \in [0, 1]$. Applying the left inequality in (5.22) we deduce that

$$\begin{aligned} f(x) - f(\hat{x}) &= \langle \nabla f(x), x - \hat{x}\rangle - \frac{1}{2}\langle \hat{x} - x, \nabla^2 f(\tilde{x})(\hat{x} - x)\rangle \\ &\leq \|\nabla f(x)\| \cdot \|x - \hat{x}\| - \frac{m}{2}\|x - \hat{x}\|^2. \end{aligned} \tag{5.25}$$

This combined with the left inequality in (5.24) renders the relation

$$\frac{m}{2}\|x - \hat{x}\|^2 \leq \|\nabla f(x)\| \cdot \|x - \hat{x}\| - \frac{m}{2}\|x - \hat{x}\|^2,$$

which simplifies into

$$m\|x - \hat{x}\| \leq \|\nabla f(x)\|. \tag{5.26}$$

Substituting this inequality and the right estimate of (5.24) into (5.25) we obtain

$$f(x) - f(\hat{x}) \leq \frac{1}{m}\|\nabla f(x)\|^2 - \frac{m}{M}\big[f(x) - f(\hat{x})\big].$$

This is equivalent to the assertion of the lemma. □

Now we can refine the estimates of Theorem 5.1.

THEOREM 5.7. *Assume that the function $f(\cdot)$ satisfies condition* (5.22). *Then for every $x^0 \in \mathbb{R}^n$ the sequence $\{x^k\}$ generated by the method of steepest descent with directional minimization satisfies the inequalities:*

$$f(x^k) - f(\hat{x}) \le \varrho^k \big[f(x^0) - f(\hat{x})\big], \tag{5.27}$$

$$\|x^k - \hat{x}\| \le \varrho^{k/2}\Big(\frac{M}{m}\Big)^{1/2}\|x^0 - \hat{x}\|, \tag{5.28}$$

where

$$\varrho = 1 - \frac{m}{2M} - \frac{m^2}{2M^2}.$$

Proof. Consider iteration k. Using Taylor's formula at x^k we get

$$f(x^k + \tau d^k) = f(x^k) + \tau\langle\nabla f(x^k), d^k\rangle + \frac{\tau^2}{2}\langle d^k, \nabla^2 f(\bar{x})d^k\rangle,$$

where $\bar{x} = x^k + \theta\tau d^k$ and $\theta \in [0,1]$. Using the right inequality in (5.22) and setting $d^k = -\nabla f(x^k)$ we obtain the estimate:

$$f(x^k - \tau\nabla f(x^k)) \le f(x^k) - \tau\|\nabla f(x^k)\|^2 + \frac{M\tau^2}{2}\|\nabla f(x^k)\|^2.$$

Since the step size τ_k is calculated by minimizing the left hand side of this inequality with respect to τ, the result cannot be worse than just using $\tau = 1/M$, which minimizes the right hand side. Hence

$$f(x^{k+1}) \le f(x^k) - \frac{1}{2M}\|\nabla f(x^k)\|^2.$$

Applying Lemma 5.6 we can rewrite this inequality as follows:

$$\begin{aligned} f(x^{k+1}) - f(\hat{x}) &\le \big[f(x^k) - f(\hat{x})\big] - \frac{m}{2M}\Big(1 + \frac{m}{M}\Big)\big[f(x^k) - f(\hat{x})\big] \\ &= \varrho\big[f(x^k) - f(\hat{x})\big]. \end{aligned}$$

This entails (5.27). Inequalities (5.28) follow now from (5.27) and (5.24). □

We see from the estimates (5.27) that the values of the function $\{f(x^k)\}$ are convergent to the minimum value $f(\hat{x})$ at least linearly, with quotient $\varrho \in (0,1)$. The smaller ϱ the better, and we notice that the key role is again played by the condition index

$$\varkappa = \frac{M}{m}.$$

If $\varkappa$ is very large, then ϱ is close to 1, and convergence is slow.

The estimates of Theorem 5.7 have a global character in the sense that they hold true for all iterations $k = 0, 1, 2, \ldots$. An additional insight may be gained by considering the asymptotic speed of convergence, measured by

$$\limsup_{k\to\infty} \frac{f(x^{k+1}) - f(\hat{x})}{f(x^k) - f(\hat{x})}.$$

The analysis in the proof of Theorem 5.7 provides an upper bound for this limit, but a more precise bound can be found.

At first we analyze the quadratic case

$$f(x) = \frac{1}{2}\langle x, Qx\rangle \tag{5.29}$$

with positive definite Q. It is relevant because every twice differentiable function can be very well approximated by a quadratic function in a neighborhood of its minimum point. The assumptions that the minimum point of (5.29) is 0, and that the minimum value is 0 as well, have been made only for simplicity of notation. They can always be satisfied by shifting the argument and adding a constant to the function.

THEOREM 5.8. *If the method of steepest descent with exact directional minimization is applied to function* (5.29), *then*

$$\frac{f(x^{k+1})}{f(x^k)} \le \left(\frac{\lambda_n - \lambda_1}{\lambda_n + \lambda_1}\right)^2, \quad k = 0, 1, 2, \ldots,$$

where λ_n and λ_1 are the largest and smallest eigenvalues of Q.

Proof. For a given x we denote by

$$g = \nabla f(x) = Qx.$$

Let us calculate the minimum of $f(x - \tau g)$ with respect to τ. We have

$$f(x - \tau g) = f(x) - \tau\|g\|^2 + \frac{\tau^2}{2}\langle g, Qg\rangle. \tag{5.30}$$

Thus the best step size is equal to

$$\hat{\tau} = \frac{\|g\|^2}{\langle g, Qg\rangle}.$$

Substituting it into (5.30) and using the fact that $f(x) = \frac{1}{2}\langle g, Q^{-1}g\rangle$ one obtains the equation

$$f(x - \hat{\tau}g) = f(x) - \frac{\|g\|^2}{2\langle g, Qg\rangle} = f(x)\Big(1 - \frac{\|g\|^2}{\langle g, Qg\rangle\,\langle g, Q^{-1}g\rangle}\Big). \tag{5.31}$$

Application to (5.31) of the *Kantorovich inequality* (see Lemma 5.9 below) yields

$$f(x - \hat{\tau} g) \leq f(x)\Big(1 - \frac{4\lambda_1 \lambda_n}{(\lambda_1 + \lambda_n)^2}\Big),$$

which can be manipulated to the required result. □

It remains to prove the Kantorovich inequality.

LEMMA 5.9. *If Q is positive definite then for every $g \neq 0$ we have*

$$\frac{\|g\|^2}{\langle g, Qg\rangle\langle g, Q^{-1}g\rangle} \geq \frac{4\lambda_1 \lambda_n}{(\lambda_1 + \lambda_n)^2}, \tag{5.32}$$

where λ_1 and λ_n are the minimum and maximum eigenvalues of Q.

Proof. Let $z_1, \dots, z_n$ be orthogonal eigenvectors of Q having length 1, and let $\lambda_1, \dots, \lambda_n$ be the corresponding eigenvalues. We assume that they are ordered from the smallest to the largest. We can express g in the system of coordinates defined by the eigenvectors: $g = \sum_{i=1}^n \xi_i z_i$. Denote by Γ the left hand side of (5.32). We can calculate Γ as a function of the coefficients ξ_i, $i = 1, \dots, n$, as follows:

$$\Gamma = \frac{\|g\|^2}{\langle g, Qg\rangle\, \langle g, Q^{-1}g\rangle} = \frac{(\sum_{i=1}^n \xi_i^2)^2}{(\sum_{i=1}^n \xi_i^2 \lambda_i)(\sum_{i=1}^n \xi_i^2 \frac{1}{\lambda_i})}.$$

Set $\alpha_i = \xi_i^2 / (\sum_{j=1}^n \xi_j^2)$. Dividing each sum in the denominator by $(\sum_{j=1}^n \xi_j^2)$ we obtain

$$\Gamma = \frac{1}{(\sum_{i=1}^n \alpha_i \lambda_i)(\sum_{i=1}^n \alpha_i \frac{1}{\lambda_i})} = \frac{\Phi(\sum_{i=1}^n \alpha_i \lambda_i)}{\sum_{i=1}^n \alpha_i \Phi(\lambda_i)}, \tag{5.33}$$

where $\Phi(\lambda) \triangleq \frac{1}{\lambda}$ for $\lambda > 0$. Since $\Phi(\cdot)$ is convex,

$$\Phi(\lambda_i) \leq \beta_i \Phi(\lambda_1) + (1 - \beta_i)\Phi(\lambda_n),$$

where $\beta_i = \frac{\lambda_n - \lambda_i}{\lambda_n - \lambda_1}$. We can add these inequalities multiplied by α_i to estimate the denominator of (5.33) as follows:

$$\sum_{i=1}^n \alpha_i \Phi(\lambda_i) \leq \alpha \Phi(\lambda_1) + (1 - \alpha)\Phi(\lambda_n),$$

$$\alpha = \sum_{i=1}^n \alpha_i \beta_i.$$

We have used the fact that $\sum_{i=1}^n \alpha_i = 1$,

Let us calculate the numerator of (5.33). By the definition of β_i,

$$(\lambda_n - \lambda_1)\alpha = (\lambda_n - \lambda_1)\sum_{i=1}^n \alpha_i\beta_i = \sum_{i=1}^n \alpha_i(\lambda_n - \lambda_i) = \lambda_n - \sum_{i=1}^n \alpha_i\lambda_i,$$

so

$$\sum_{i=1}^n \alpha_i\lambda_i = \lambda_n - (\lambda_n - \lambda_1)\alpha = \alpha\lambda_1 + (1-\alpha)\lambda_n.$$

Hence

$$\Gamma \geq \frac{\Phi(\alpha\lambda_1 + (1-\alpha)\lambda_n)}{\alpha\Phi(\lambda_1) + (1-\alpha)\Phi(\lambda_n)} = \frac{\lambda_1\lambda_n}{[\alpha\lambda_1 + (1-\alpha)\lambda_n][\alpha\lambda_n + (1-\alpha)\lambda_1]},$$

with some $0 \leq \alpha \leq 1$. We do not know the value of α, but we can calculate the worst case. The minimum of the right hand side with respect to α is attained at $\alpha = \frac{1}{2}$ and this yields the lower bound (5.32). □

Using the estimate of Theorem 5.8 we can easily analyze the general case of a twice continuously differentiable function.

THEOREM 5.10. *Assume that the function $f(\cdot)$ satisfies condition (5.22). Then for every $x^0 \in \mathbb{R}^n$ the sequence $\{x^k\}$ generated by the method of steepest descent with directional minimization satisfies the relation:*

$$\limsup_{k\to\infty} \frac{f(x^{k+1}) - f(\hat{x})}{f(x^k) - f(\hat{x})} \leq \left(\frac{\lambda_n - \lambda_1}{\lambda_n + \lambda_1}\right)^2, \tag{5.34}$$

where λ_n and λ_1 are the largest and smallest eigenvalues of the Hessian $\nabla^2 f(\hat{x})$.

Proof. By virtue of Theorem 5.11, the sequence $\{x^k\}$ is convergent to the unique minimizer $\hat{x}$ of $f(\cdot)$. At every point x^k we have the Taylor expansion:

$$\begin{aligned} f(x^k - \tau\nabla f(x^k)) = f(x^k) - \tau\|\nabla f(x^k)\|^2 \\ + \frac{\tau^2}{2}\langle \nabla f(x^k), [\nabla^2 f(\bar{x}^k)]\nabla f(x^k)\rangle. \end{aligned} \tag{5.35}$$

Denote the Hessian $\nabla^2 f(\hat{x})$ by Q and let $g^k = Q(x^k - \hat{x})$. It follows from the continuous second order differentiability of $f(\cdot)$ that

$$\nabla f(x^k) = \nabla f(\hat{x}) + Q(x^k - \hat{x}) + o(x^k, \hat{x}) = g^k + o(x^k, \hat{x}),$$

where $o(x^k,\hat{x})/\|x^k-\hat{x}\| \to 0$ when $x^k \to \hat{x}$. Noting that $\|g^k\|/\|x^k-\hat{x}\| \le \lambda_n$, we obtain the relation

$$\|\nabla f(x^k)\|^2 \ge \|g^k\|^2 - 2\|g^k\|\cdot\|o(x^k,\hat{x})\| - \|o(x^k,\hat{x})\|^2 \ge \|g^k\|^2 - o_2(x^k,\hat{x}),$$

where $o_2(x^k,\hat{x})/\|x^k-\hat{x}\|^2 \to 0$ when $x^k \to \hat{x}$. Moreover, $\nabla^2 f(\bar{x}^k) \to Q$, when $x^k \to \hat{x}$. Combining this with the last estimates we obtain

$$\big\langle \nabla f(x^k), \big[\nabla^2 f(\bar{x}^k)\big]\nabla f(x^k)\big\rangle \le \langle g^k, Qg^k\rangle + o_3(x^k,\hat{x}),$$

where $o_3(x^k,\hat{x})/\|x^k-\hat{x}\|^2 \to 0$ when $x^k \to \hat{x}$. Furthermore, it is easy to show that the step sizes τ_k are uniformly bounded.

All these estimates allow us to derive from equation (5.35) the inequality

$$f(x^k - \tau\nabla f(x^k)) \le f(x^k) - \tau\|g^k\|^2 + \frac{\tau^2}{2}\langle g^k, Qg^k\rangle + o_4(x^k,\hat{x}),$$

with some second order error satisfying $o_4(x^k,\hat{x})/\|x^k-\hat{x}\|^2 \to 0$, when $x^k \to \hat{x}$. The step size τ_k minimizes the left hand side and thus the minimum cannot be worse than substituting the best step size for the quadratic approximation,

$$\hat{\tau}_k = \frac{\|g^k\|^2}{\langle g^k, Qg^k\rangle}.$$

Therefore, using the Kantorovich inequality, as in the proof of Theorem 5.29, we obtain the estimate

$$f(x^{k+1}) \le f(x^k) - \langle g^k, Q^{-1}g^k\rangle\Big(\frac{4\lambda_n\lambda_1}{\lambda_n+\lambda_1}\Big)^2 + o_4(x^k,\hat{x}).$$

From Taylor's expansion at $\hat{x}$ we get

$$f(x^k) - f(\hat{x}) = \langle x^k-\hat{x}, Q(x^k-\hat{x})\rangle + o_5(x^k,\hat{x}) = \langle g^k, Q^{-1}g^k\rangle + o_5(x^k,\hat{x}),$$

where $o_5(x^k,\hat{x})/\|x^k-\hat{x}\|^2 \to 0$ when $x^k \to \hat{x}$. Combining the last two relations we conclude that

$$\begin{aligned} f(x^{k+1}) - f(\hat{x}) &\le \big[f(x^k) - f(\hat{x})\big] - \big[f(x^k) - f(\hat{x})\big]\Big(\frac{4\lambda_n\lambda_1}{\lambda_n+\lambda_1}\Big)^2 \\ &\quad + o_6(x^k,\hat{x}) \le \big[f(x^k) - f(\hat{x})\big]\Big(\frac{\lambda_n-\lambda_1}{\lambda_n+\lambda_1}\Big)^2 + o_6(x^k,\hat{x}). \end{aligned}$$

Again, $o_6(x^k,\hat{x})/\|x^k-\hat{x}\|^2 \to 0$ when $x^k \to \hat{x}$. From inequality (5.24) we know that $f(x^k) - f(\hat{x}) \ge (m/2)\|x^k-\hat{x}\|^2$. We can, therefore, divide both sides of the last inequality by $\big[f(x^k) - f(\hat{x})\big]$, and pass to the limit when $k \to \infty$. The rest term (associated with o_6) disappears, and we obtain the required result. □

Again, we see the role of the condition index

$$\varkappa = \frac{\lambda_n}{\lambda_1}.$$

The quotient

$$q = \frac{\lambda_n - \lambda_1}{\lambda_n + \lambda_1} = \frac{\varkappa - 1}{\varkappa + 1}$$

becomes smaller, when $\varkappa \downarrow 1$, that is, when the spectrum of the Hessian Q is concentrated within a relatively short interval. If the spectrum is widely spread, though, then $\varkappa \gg 1$ and $q \approx 1$. In this case the speed of convergence is very low. Such problems are called *ill-conditioned*, and more sophisticated methods are needed for their solution.

Example 5.11. Let us consider the function $f : \mathbb{R}^2 \to \mathbb{R}$ defined as follows:

$$f(x_1, x_2) = \frac{1}{2}(x_1)^2 + \frac{a}{2}(x_2)^2,$$

with some $a \geq 1$. Suppose we use for its minimization the method of steepest descent starting from the point $x^0 = \begin{bmatrix} a \\ 1 \end{bmatrix}$. We shall show that the method generates the sequence of points

$$x^k = \left(\frac{a-1}{a+1}\right)^k \begin{bmatrix} a \\ (-1)^k \end{bmatrix}. \tag{5.36}$$

We prove it by induction. For $k = 0$ the formula is obviously true. Suppose it is true for some k. We have

$$\nabla f(x^k) = \left(\frac{a-1}{a+1}\right)^k \begin{bmatrix} a \\ (-1)^k a \end{bmatrix}.$$

The next point, x^{k+1}, has the form

$$\begin{aligned} x^{k+1} &= \left(\frac{a-1}{a+1}\right)^k \begin{bmatrix} a \\ (-1)^k \end{bmatrix} - \tau_k \left(\frac{a-1}{a+1}\right)^k \begin{bmatrix} a \\ (-1)^k a \end{bmatrix} \\ &= \left(\frac{a-1}{a+1}\right)^k \begin{bmatrix} a(1-\tau_k) \\ (-1)^k(1-a\tau_k) \end{bmatrix}. \end{aligned} \tag{5.37}$$

The step size τ_k is obtained by minimizing with respect to τ the function

$$\varphi(\tau) = f(x^k - \tau \nabla f(x^k)) = \frac{1}{2}\left(\frac{a-1}{a+1}\right)^{2k} \left(a^2(1-\tau)^2 + a(1-a\tau)^2\right).$$

Elementary calculus yields

$$\tau_k = \frac{2}{a+1}.$$

This can be substituted into (5.37) to obtain

$$x^{k+1} = \left(\frac{a-1}{a+1}\right)^{k+1} \begin{bmatrix} a \\ (-1)^{k+1} \end{bmatrix}.$$

By induction, formula (5.36) holds true for all k. It follows that in this special case the method with directional minimization uses constant step sizes.

The sequence of points $\{x^k\}$ is illustrated in Figure 5.5 for $a = 4$ (top) and $a = 16$ (bottom).

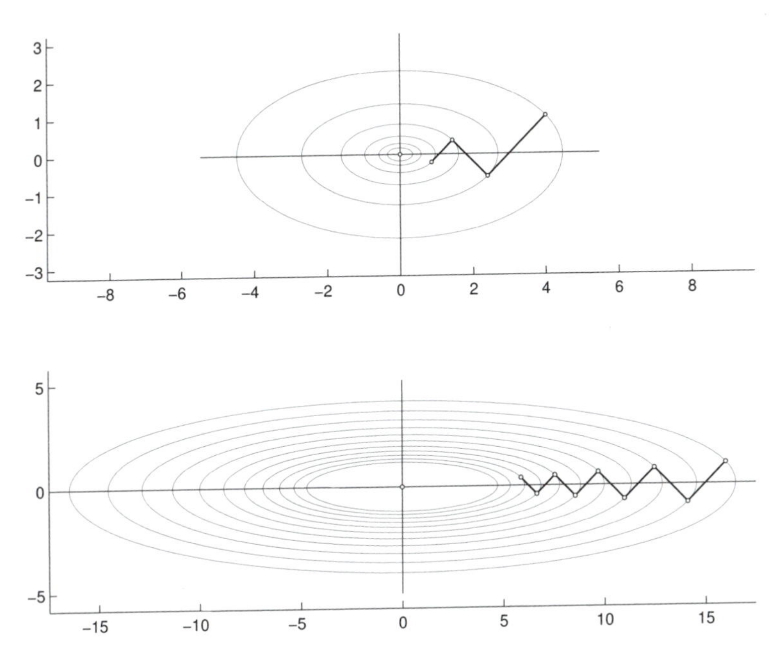

Figure 5.5. Performance of the method of steepest descent.

5.4 NEWTON'S METHOD

5.4.1 Basic Properties

The speed of convergence of the method of steepest descent may be very poor, even for quadratic functions, because the direction of steepest descent minimizes the local rate of descent, rather than the total decrease that can be achieved by moving in a direction. Clearly, for the quadratic function (5.29) the best search direction is equal to

$$d = \hat{x} - x = -Q^{-1}g = -\left[\nabla^2 f(x)\right]^{-1} \nabla f(x).$$

This is the motivation for *Newton's method*, in which we use search directions

$$d^k = -\big[\nabla^2 f(x^k)\big]^{-1} \nabla f(x^k). \tag{5.38}$$

Although for all quadratic functions the method is convergent in one iteration, for more general functions convergence of the method is subject to rather restrictive conditions.

Another useful interpretation, which explains the origins of the name of the method, is to consider the necessary condition of optimality,

$$\nabla f(x) = 0, \tag{5.39}$$

as a system of n equations with n unknowns. Newton's *method of tangents* applied to this system has the form

$$x^{k+1} = x^k - \big[\nabla^2 f(x^k)\big]^{-1} \nabla f(x^k).$$

This is the same as using directions (5.38) with step sizes $\tau_k = 1$. For a quadratic function $f(\cdot)$ the system of necessary conditions of optimality is linear, and the method of tangents is convergent in one iteration, provided that the Hessian is invertible.

However, there is a difference between the system of necessary conditions of optimality (5.39) and a general system of nonlinear equations. In the system of necessary conditions of optimality the vector field $\nabla f(x)$ has a potential: the function $f(x)$. This allows directional minimization of $f(\cdot)$ in directions d^k, which greatly improves the method of tangents.

Before proceeding to the rigorous analysis of convergence, let us observe that for the Newton direction to be a direction of descent at a point x, it is necessary that

$$\big\langle \nabla f(x), \big[\nabla^2 f(x)\big]^{-1} \nabla f(x) \big\rangle > 0.$$

This can be guaranteed if the Hessian $\nabla^2 f(x)$ is positive definite.

Theorem 5.12. *Assume that $f(\cdot)$ is twice continuously differentiable and that its Hessian $\nabla^2 f(x)$ is positive definite at all x in the set $X_0 = \{x : f(x) \le f(x^0)\}$. Moreover, assume that the set X_0 is bounded. Then the sequence $\{x^k\}$ generated by Newton's method with directional minimization is convergent to a minimum x^* of $f(\cdot)$.*

Proof. We argue in a way similar to the proof of Theorem 5.4. Since the Hessian is positive definite, each direction is a direction of descent:

$$\langle \nabla f(x^k), d^k \rangle = \big\langle \nabla f(x^k), \big[\nabla^2 f(x^k)\big]^{-1} \nabla f(x^k) \big\rangle \le 0,$$

with the equality possible only if $\nabla f(x^k) = 0$. Therefore $f(x^{k+1}) \le f(x^k)$ for all k. Furthermore, all points x^k belong to the set X_0, which is compact. It follows that the sequence $\{f(x^k)\}$ is bounded. Consequently, it is convergent.

Consider any accumulation point x^* of $\{x^k\}$ and let $\mathcal{K}$ be the infinite ordered set of iteration numbers such that

$$\lim_{\substack{k\to\infty \\ k\in\mathcal{K}}} x^k = x^*.$$

Since the sequence $\{f(x^k)\}$ is convergent, we have

$$f(x^*) = \lim_{k\to\infty} f(x^k). \tag{5.40}$$

We prove the theorem by contradiction.

Suppose the point x^* is not stationary, i.e., $\nabla f(x^*) \ne 0$. Consider the Newton direction $d^* = -\left[\nabla^2 f(x^*)\right]^{-1}\nabla f(x^*)$ and the points

$$y(\tau) = x^* + \tau d^*, \quad \tau \ge 0.$$

Since $\nabla f(x^*) \ne 0$, the direction d^* is nonzero and the problem

$$\operatorname*{minimize}_{\tau\ge 0} f(y(\tau))$$

has a solution $\tau_* > 0$. Moreover, $f(y(\tau_*)) < f(x^*)$ for every solution τ_*.

Consider the points x^k for $k \in \mathcal{K}$, and the immediately following points x^{k+1}. By the directional minimization condition

$$f(x^{k+1}) \le f(x^k + \tau d^k) \quad \text{for all} \quad \tau \ge 0.$$

When $k \to \infty$, $k \in \mathcal{K}$, then $x^k \to x^*$, $\nabla f(x^k) \to \nabla f(x^*)$ and $\nabla^2 f(x^k) \to \nabla^2 f(x^*)$. Therefore

$$\lim_{\substack{k\to\infty \\ k\in\mathcal{K}}} d^k = d^*.$$

Proceeding in the same way as in the proof of Theorem 5.4, we can show that there exists an infinite subset $\mathcal{K}_1$ of $\mathcal{K}$ such that

$$\lim_{\substack{k\to\infty \\ k\in\mathcal{K}_1}} f(x^{k+1}) = f(y(\tau_*)) < f(x^*).$$

Then

$$\lim_{k\to\infty} f(x^k) \le \lim_{\substack{k\to\infty \\ k\in\mathcal{K}_1}} f(x^{k+1}) < f(x^*),$$

which contradicts (5.40). □

In practice, we usually cannot rely on the general assumption of the positive definiteness of the Hessian. In order to guarantee that d^k is a direction of descent, we use the regularized formula

$$d^k = -\left[\nabla^2 f(x^k) + \Delta_k\right]^{-1} \nabla f(x^k) \tag{5.41}$$

with some diagonal matrix Δ_k of dimension n chosen in such a way that $\nabla^2 f(x^k) + \Delta_k$ is positive definite.†

5.4.2 Speed of Convergence

The main advantage of Newton's method is its rapid convergence in the neighborhood of the minimum. By the construction of the method, the optimal step size for a quadratic function equals 1. Therefore, when using Newton's method for a nonquadratic (but twice continuously differentiable) function, we try $\tau = 1$ at first. If this step size satisfies the two-slope test (5.7) on page 217 with $\alpha_1 > 1/2 > \alpha_2$, then we accept $\tau_k = 1$. Otherwise, line search is used to satisfy the test.

THEOREM 5.13. *Let the assumptions of Theorem* 5.12 *be satisfied and let Newton's method be used together with the two-slope test using* $\alpha_1 > 1/2 > \alpha_2$ *and trying* $\tau_k = 1$ *first. Then Newton's method generates a sequence of points* $\{x^k\}$ *such that:*

$$\limsup_{k\to\infty} \frac{\|x^{k+1} - \hat{x}\|}{\|x^k - \hat{x}\|} = 0, \tag{5.42}$$

where $\hat{x}$ *is the minimum point of* $f(\cdot)$. *If, additionally, the Hessian is Lipschitz continuous in the set* X_0, *then for all* k

$$\|x^{k+1} - \hat{x}\| \le \frac{L}{2m}\|x^k - \hat{x}\|^2, \tag{5.43}$$

where L *is the Lipschitz constant of the Hessian, and* m *is the smallest eigenvalue of the Hessian in* X_0.

Proof. We first prove that $\tau_k = 1$ for all sufficiently large k. Indeed, by Theorem 5.12, the method is convergent, so x^k approaches $\hat{x}$. By Taylor's

---

†Formula (5.41) is never implemented directly, by inverting the Hessian or its modification, and Δ_k is not calculated explicitly. Instead, the Hessian is *factorized*, aiming at the Cholesky representation, $\nabla^2 f(x^k) = L_k D_k L_k^T$, with lower triangular L_k and diagonal D_k. If, during the factorization process, zero or negative diagonal entries of D_k occur, the diagonal elements of the Hessian are increased.

formula, the function

$$\begin{aligned}\phi_k(\tau) &= f(x^k + \tau d^k) \\ &= f(x^k) + \tau \langle \nabla f(x^k), d^k \rangle + \frac{1}{2}\tau^2 \langle d^k, [\nabla^2 f(x^k)] d^k \rangle + o(\tau d^k)\end{aligned}$$

becomes very close to a quadratic function for $\tau \in [0, 2]$, because the rest $o(\tau d^k)$ is infinitely smaller than $\tau^2 \|d^k\|^2$ and $d^k \to 0$. Since for a quadratic function the step size $\tau = 1$ passes the two-slope test with $\alpha_1 = \alpha_2 = 1/2$, and we use $\alpha_1 > 1/2 > \alpha_2$, by continuity we must have $\tau_k = 1$ for all sufficiently large k.

When $\tau_k = 1$ we have

$$x^{k+1} - \hat{x} = x^k - \hat{x} - \left[\nabla^2 f(x^k)\right]^{-1} \nabla f(x^k).$$

Multiplying both sides by $\nabla^2 f(x^k)$ we obtain the equation

$$\left[\nabla^2 f(x^k)\right](x^{k+1} - \hat{x}) = \left[\nabla^2 f(x^k)\right](x^k - \hat{x}) - \nabla f(x^k). \tag{5.44}$$

Consider the function $g(\alpha) = \nabla f(\hat{x} + \alpha(x^k - \hat{x}))$. We have

$$g(0) = 0, \quad g(1) = \nabla f(x^k), \quad g'(\alpha) = \left[\nabla^2 f(\hat{x} + \alpha(x^k - \hat{x}))\right](x^k - \hat{x}).$$

By the Newton–Leibniz fundamental theorem of calculus,

$$g(1) = \nabla f(x^k) = \int_0^1 g'(\alpha)\, d\alpha = \int_0^1 \left[\nabla^2 f(\hat{x} + \alpha(x^k - \hat{x}))\right](x^k - \hat{x})\, d\alpha.$$

Therefore equation (5.44) can be rewritten as follows:

$$\begin{aligned}&\left[\nabla^2 f(x^k)\right](x^{k+1} - \hat{x}) \\ &\qquad = \left(\nabla^2 f(x^k) - \int_0^1 \left[\nabla^2 f(\hat{x} + \alpha(x^k - \hat{x}))\right] d\alpha\right)(x^k - \hat{x}).\end{aligned}$$

Estimating the norm of the right hand side we obtain

$$\begin{aligned}&\left\| \left[\nabla^2 f(x^k)\right](x^{k+1} - \hat{x}) \right\| \\ &\qquad \le \left(\int_0^1 \left\| \nabla^2 f(x^k) - \nabla^2 f(\hat{x} + \alpha(x^k - \hat{x})) \right\| d\alpha\right) \|x^k - \hat{x}\| \\ &\qquad \le O\left(\|x^k - \hat{x}\|\right) \|x^k - \hat{x}\|,\end{aligned} \tag{5.45}$$

with $O(r)$ denoting the upper bound on $\|\nabla^2 f(x) - \nabla^2 f(\hat{x})\|$ for $\|x - \hat{x}\| \leq r$. The Hessian is continuous, and thus $O(r) \to 0$ when $r \to 0$.

The left hand side of (5.45) is bounded from below by $m\|x^{k+1} - \hat{x}\|$ and therefore

$$\|x^{k+1} - \hat{x}\| \leq \frac{O\big(\|x^k - \hat{x}\|\big)}{m} \|x^k - \hat{x}\|.$$

As $x^k \to \hat{x}$, this implies (5.42).

If the Hesssian is Lipschitz continuous with the constant L, estimate (5.45) can be refined:

$$\begin{aligned} m\|x^{k+1} - \hat{x}\| &\leq \big\|\big[\nabla^2 f(x^k)\big](x^{k+1} - \hat{x})\big\| \\ &\leq \bigg(\int_0^1 L\alpha \, d\alpha\bigg)\|x^k - \hat{x}\|^2 = \frac{L}{2}\|x^k - \hat{x}\|^2. \end{aligned}$$

This yields (5.43). □

A convergent iterative method satisfying condition (5.42) is called *superlinearly convergent*, and a method satisfying condition (5.43) is called *quadratically convergent*. Such methods are very efficient in practice; once they reach a neighborhood of the minimum, they dramatically accelerate. This allows for calculating the minimum with very high accuracy in few iterations.

5.4.3 The Gauss–Newton Method

One of the disadvantages of Newton's method is the need to calculate the Hessian of the function $f(\cdot)$ at every iteration. However, for an important class of problems we can retain some benefits of Newton's method, without the need to calculate second derivatives of the functions. These are least-squares problems, having the objective function

$$f(x) = \frac{1}{2}\sum_{i=1}^{m} \big[h_i(x)\big]^2 = \frac{1}{2}\|h(x)\|^2. \tag{5.46}$$

In the formula above, $h_i : \mathbb{R}^n \to \mathbb{R}$, $i = 1, \ldots, m$, are twice continuously differentiable functions. Problems of this type frequently occur in model identification, where x is the vector of unknown parameters in the model

$$y = \varphi(x, u).$$

Here y represents the output variable, and u is the input variable for the model. For a collection of observations (u^i, y^i), $i = 1, \ldots, m$, the functions

$h_i(\cdot)$ represent the differences between the observed values, y^i, and the values predicted by the model:

$$h_i(x) = y^i - \varphi(x, u^i), \quad i = 1, \dots, m.$$

The minimization of function (5.46) amounts to minimizing the sum of squares of model errors.

Simple calculation yields the formulae for the gradient and Hessian of $f(\cdot)$:

$$\nabla f(x) = \sum_{i=1}^{m} h_i(x) \nabla h_i(x), \tag{5.47}$$

$$\nabla^2 f(x) = \sum_{i=1}^{m} [\nabla h_i(x)][\nabla h_i(x)]^T + \sum_{i=1}^{m} h_i(x) \nabla^2 h_i(x). \tag{5.48}$$

The idea of the Gauss–Newton method† is to approximate the Hessian at x by the matrix

$$Q(x) = \sum_{i=1}^{m} [\nabla h_i(x)][\nabla h_i(x)]^T = \big[h'(x)\big]^T h'(x),$$

which is obtained by skipping the second sum in (5.48). As usually, we use $h'(x)$ to denote the Jacobian of the function $h(\cdot)$ at x. This approximation uses the observation that for a reasonable least-squares problem the values of $h_i(x)$ are close to zero near the solution. Moreover, an additional argument for using the approximation arises when the functions h_i are only moderately nonlinear, and thus the Hessians $\nabla^2 h_i(\cdot)$ are small.

The method proceeds as follows. At iteration k we calculate $Q(x^k)$ and $\nabla f(x^k)$. Then we solve the system of equations

$$d^k = -\big[Q(x^k) + \Delta_k\big]^{-1} \nabla f(x^k), \tag{5.49}$$

and we calculate

$$x^{k+1} = x^k + \tau_k d^k,$$

with some step size τ_k. Here Δ_k is a small diagonal matrix guaranteeing that the smallest eigenvalue of $Q(x^k) + \Delta_k$ is sufficiently separated from zero. Again, in practical calculations Δ_k may be determined while factorizing $Q(x^k)$ for the purpose of solving (5.49).

The matrix $Q(x^k)$ is the exact Hessian of the sum of squares of the *linearized* functions $h_i(\cdot)$:

$$Q(x^k) = \nabla^2 f^k(x^k),$$

---

†The name *Levenberg–Marquardt method* is also frequently used.

with

$$f^k(x) = \frac{1}{2}\|h(x^k) + h'(x^k)(x - x^k)\|^2.$$

The gradients of $f(\cdot)$ and $f^k(\cdot)$ coincide at x^k. Thus, the calculation of the direction d^k in (5.49) can be accomplished by solving the linear least-squares problem

$$\operatorname*{minimize}_{d} \; \frac{1}{2}\Big\{\|h(x^k) + h'(x^k)d\|^2 + \langle d, \Delta_k d\rangle\Big\}. \tag{5.50}$$

Many efficient algorithms exist for such problems. We can also set $\Delta_k = 0$ here, and look for a minimum-norm solution, if $Q(x^k)$ is not strictly positive definite.

The convergence of the Gauss–Newton method can be proved by the same argument as the convergence of the method of steepest descent (see, for example, Theorem 5.10). In Section 7.7 we analyze a more general version of this method, where the square norm $\|\cdot\|^2$ is replaced by a general convex function.

5.5 THE CONJUGATE GRADIENT METHOD

5.5.1 Minimization of Quadratic Functions

The minimum of a separable quadratic function,

$$f(x) = \sum_{i=1}^{n} \big[c_i x_i + \frac{1}{2} q_{ii}(x_i)^2\big], \tag{5.51}$$

where $q_{ii} > 0$, can be found by successive directional minimization in the directions

$$d^1 = \begin{bmatrix} 1 \\ 0 \\ \vdots \\ 0 \end{bmatrix}, \quad d^2 = \begin{bmatrix} 0 \\ 1 \\ \vdots \\ 0 \end{bmatrix}, \quad \ldots, \quad d^n = \begin{bmatrix} 0 \\ 0 \\ \vdots \\ 1 \end{bmatrix}.$$

Each direction needs to be used only once. The concept of conjugate directions allows to extend this property to arbitrary quadratic functions.

Definition 5.14. Let Q be a symmetric positive definite matrix of dimension n. Vectors $d^1, d^2, \ldots, d^n$ are called *Q-conjugate* (*Q-orthogonal*) if they are all nonzero and

$$\langle d^i, Qd^j\rangle = 0 \quad \text{for all} \quad i \neq j.$$

Conjugate directions can be used as the basis for a new system of coordinates.

LEMMA 5.15. *Conjugate directions are linearly independent.*

Proof. Suppose

$$\sum_{i=1}^{n} \alpha_i d^i = 0.$$

Take the scalar product of both sides with Qd^j, for an arbitrary $1 \le j \le n$. By the definition of conjugate directions, all products $\langle d^i, Qd^j \rangle$ for $i \neq j$ disappear. Therefore

$$\alpha_j \langle d^j, Qd^j \rangle = 0.$$

Since Q is positive definite and $d^j \neq 0$, we conclude that $\alpha_j = 0$. The index j was arbitrary, and thus all α_j are 0, for $j = 1, \ldots, n$. □

The main reason for introducing the concept of Q-conjugate directions is that they can be used to efficiently minimize a quadratic function having Q as its Hessian,

$$f(x) = \frac{1}{2}\langle x, Qx \rangle + \langle c, x \rangle. \tag{5.52}$$

Two important issues make such simple optimization problems relevant. First, when the dimension n is very large (in thousands or hundreds of thousands), problems of minimizing quadratic functions of the form (5.52) become difficult, because we cannot easily solve the system of equations

$$Qx = -c.$$

Actually, problem (5.52) may be a convenient approach to such a system. Second, any twice continuously differentiable function can be very well approximated by a quadratic model in some neighborhood of its minimum point. A method that does poorly on a quadratic function cannot be expected to perform well on a nonquadratic function. For example, we see from Theorems 5.29 and 5.33 about the method of steepest descent that the performance on a quadratic model was indicative of the speed of convergence in the nonquadratic case. We thus aim at developing a method for solving (5.52) which can be readily extended to nonquadratic problems as well.

THEOREM 5.16. *Assume that $d^1, d^2, \ldots, d^n$ are Q-conjugate and that the sequence $x^1, x^2, \ldots, x^{n+1}$ is obtained by successive minimization of function (5.52) in directions d^k, $k = 1, \ldots, n$:*

$$x^{k+1} = x^k + \tau_k d^k,$$
$$f(x^{k+1}) = \min_{\tau \in \mathbb{R}} f(x^k + \tau d^k).$$

Then for every $k = 1, 2, \ldots, n$ the point x^{k+1} is the minimum of $f(\cdot)$ in the linear manifold

$$L_k = x^1 + \operatorname{lin}\{d^1, d^2, \ldots, d^k\}.$$

Proof. Let us consider the matrix

$$D = \begin{bmatrix} d^1 & d^2 & \ldots & d^n \end{bmatrix}$$

and the linear transformation of variables

$$x = x^1 + Dy = x^1 + d^1 y_1 + d^2 y_2 + \cdots + d^n y_n.$$

As the directions d^i are conjugate,

$$\langle Dy, QDy \rangle = \sum_{i=1}^{n} \sum_{j=1}^{n} \langle d^i, Qd^j \rangle y_i y_j = \sum_{i=1}^{n} \langle d^i, Qd^i \rangle (y_i)^2.$$

Using Taylor's expansion of $f(\cdot)$ at x^1, which is exact, we obtain

$$f(x) = f(x^1 + Dy) = f(x^1) + \frac{1}{2} \langle Dy, QDy \rangle + \langle Qx^1 + c, Dy \rangle$$
$$= f(x^1) + \sum_{i=1}^{n} f_i(y_i),$$

where

$$f_i(y_i) = \frac{1}{2} \langle d^i, Qd^i \rangle (y_i)^2 + \langle Qx^1 + c, d^i \rangle y_i.$$

It follows that in the new system of coordinates the function $f(\cdot)$ is fully separable with respect to the variables y_i and has form (5.51). Thus, the minimum of $f(\cdot)$ with respect to y can be calculated independently for each y_i. The best value of each y_i is equal to the step size τ_i used at iteration i of our method. Consequently, each direction d^i need be used only once, and each x^{k+1} is the minimum of $f(\cdot)$ over L_k. □

COROLLARY 5.17. *The minimum of (5.52) can be found in no more than n steps.*

Proof. The directions d^k, $k = 1, \ldots, n$ are linearly independent, so $L_n = \mathbb{R}^n$, and the result follows from Theorem 5.16. □

Conjugate directions are very good search directions for a quadratic function, but it seems that one needs to know the Hessian Q to construct them. We shall show that this is not necessary at all.

Assume temporarily that we know the Hessian Q. We can construct a sequence of conjugate directions $d^1, d^2, \ldots, d^n$ and a sequence of points $x^1, x^2, \ldots, x^{n+1}$ by the process of successive orthogonalization of the gradients $\nabla f(x^1)$, $\nabla f(x^2)$, ..., $\nabla f(x^n)$.

Q-Orthogonalization Algorithm

Step 0. Set $k = 1$.

Step 1. Calculate $v^k = -\nabla f(x^k)$. If $v^k = 0$ then stop; otherwise continue.

Step 2. Make the vector v^k Q-orthogonal to directions $d^1, \ldots, d^{k-1}$ by the formula[†]

$$d^k := v^k - \sum_{i=1}^{k-1} \frac{\langle v^k, Qd^i \rangle}{\langle d^i, Qd^i \rangle} d^i. \tag{5.53}$$

Step 3. Calculate the next point

$$x^{k+1} = x^k + \tau_k d^k$$

such that

$$f(x^{k+1}) = \min_{\tau \in \mathbb{R}} f(x^k + \tau d^k).$$

Step 4. Increase k by 1 and go to Step 1.

The fact that Step 2 generates Q-orthogonal directions for each sequence $\{v^k\}$ can be proved by induction. Suppose $d^1, \ldots d^{k-1}$ are Q-orthogonal. In the multiplication of d^k and Qd^j, for $j \leq k-1$, all products $\langle d^i, Qd^j \rangle = 0$ for $i \neq j$, and we obtain

$$\langle d^k, Qd^j \rangle = \langle v^k, Qd^j \rangle - \frac{\langle v^k, Qd^j \rangle}{\langle d^j, Qd^j \rangle} \langle d^j, Qd^j \rangle = 0.$$

In our case, owing to the special method of constructing the vectors v^k, the orthogonalization procedure simplifies substantially.

[†]Formula (5.53) is called the Gram–Schmidt orthogonalization procedure.

Suppose the first $k-1$ directions $d^1, \dots, d^{k-1}$ are Q-orthogonal (this is certainly true for $k = 2$). Then, by Theorem 5.16, the point x^k is the minimum of $f(\cdot)$ in the manifold

$$L_{k-1} = x^1 + \mathrm{lin}\{d^1, d^2, \dots, d^{k-1}\}.$$

Consequently, the vector $\upsilon^k = -\nabla f(x^k)$ satisfies the relation

$$\upsilon^k \perp L_{k-1}. \tag{5.54}$$

Since each direction d^i is a linear combination of the previously observed gradients υ^j, for $j \le i$, we can represent the manifold L_{k-1} as follows:

$$L_{k-1} = x^1 + \mathrm{lin}\{\upsilon^1, \upsilon^2, \dots, \upsilon^{k-1}\} = x^1 + \mathrm{lin}\{\upsilon^1, \upsilon^2 - \upsilon^1, \dots, \upsilon^{k-1} - \upsilon^{k-2}\}.$$

Therefore relation (5.54) implies that

$$\upsilon^k \perp (\upsilon^j - \upsilon^{j-1}) \quad \text{for} \quad 2 \le j \le k-1. \tag{5.55}$$

Using the definition of υ^j we obtain

$$\upsilon^j - \upsilon^{j-1} = -Q(x^j - x^{j-1}) = -\tau_{j-1} Q d^{j-1}. \tag{5.56}$$

From (5.53) for $j-1$ it follows that d^{j-1} is a combination of $-\nabla f(x^{j-1})$ and of directions d^i, for $i \le j-2$. Since (5.54) implies that $\nabla f(x^{j-1}) \perp d^i$, for $i \le j-2$, we conclude that

$$\langle \nabla f(x^{j-1}), d^{j-1} \rangle = -\|\nabla f(x^{j-1})\|^2, \tag{5.57}$$

which is negative, if the algorithm did not stop at x^{j-1}. As the direction d^{j-1} is a direction of descent, we conclude that $\tau_{j-1} > 0$. It follows from (5.55) and (5.56) that

$$\upsilon^k \perp Q d^{j-1} \quad \text{for} \quad 2 \le j \le k-1.$$

Therefore, all but the last components of the sum in (5.53) vanish an we get

$$d^k = -\nabla f(x^k) + \alpha_k d^{k-1}, \tag{5.58}$$

with

$$\alpha_k = -\frac{\langle \nabla f(x^k), Q d^{k-1} \rangle}{\langle d^{k-1}, Q d^{k-1} \rangle}.$$

Using the equality $Q d^{k-1} = (\nabla f(x^k) - \nabla f(x^{k-1}))/\tau_{k-1}$ (note that $\tau_{k-1} \ne 0$ by (5.57)) we can transform the last equation as follows:

$$\alpha_k = -\frac{\langle \nabla f(x^k), \nabla f(x^k) - \nabla f(x^{k-1}) \rangle}{\langle d^{k-1}, \nabla f(x^k) - \nabla f(x^{k-1}) \rangle}.$$

Next, by (5.54) (for k and then for $k-1$)

$$\langle d^{k-1}, \nabla f(x^k) - \nabla f(x^{k-1})\rangle = -\langle d^{k-1}, \nabla f(x^{k-1})\rangle = -\|\nabla f(x^{k-1})\|^2,$$

so

$$\alpha_k = \frac{\langle \nabla f(x^k), \nabla f(x^k) - \nabla f(x^{k-1})\rangle}{\|\nabla f(x^{k-1})\|^2}. \tag{5.59}$$

This combined with (5.58) proves that the algorithm can be implemented *without* the knowledge of Q. It is known as the *conjugate gradient method* of Hestenes and Stiefel. Figure 5.6 provides the detailed description of its operation.

Step 0. Set $k := 1$.

Step 1. Calculate $\nabla f(x^k)$. If $\nabla f(x^k) = 0$ then stop; otherwise continue.

Step 2. Calculate

$$d^k = \begin{cases} -\nabla f(x^k) & \text{if } k = 1, \\ -\nabla f(x^k) + \alpha_k d^{k-1} & \text{if } k > 1, \end{cases}$$

with α_k given by (5.59).

Step 3. Calculate the next point

$$x^{k+1} = x^k + \tau_k d^k$$

such that

$$f(x^{k+1}) = \min_{\tau \geq 0} f(x^k + \tau d^k).$$

Step 4. Increase k by 1 and go to Step 1.

Figure 5.6. The conjugate gradient method.

We may observe that for a quadratic function the optimal step size can be calculated explicitly:

$$\tau_k = -\frac{\langle \nabla f(x^k), d^k\rangle}{\langle d^k, Qd^k\rangle} = \frac{\|\nabla f(x^k)\|^2}{\langle d^k, Qd^k\rangle},$$

because of (5.54). However, we prefer to present the method with the operation of directional minimization, to allow direct extension to nonquadratic problems.[†]

The following theorem summarizes our development of the conjugate gradient method.

THEOREM 5.18. *For every positive definite Q, the conjugate gradient method generates a sequence of Q-conjugate descent directions $d^1, d^2, \ldots$ and stops after at most n iterations at the minimum of the function* (5.52).

Proof. The directions are Q-conjugate by the construction of the method. They are descent directions by (5.57). The finite convergence property follows from Theorem 5.16. □

By observing that $\nabla f(x^k) \perp \nabla f(x^{k-1})$, because of the optimality of x^k on the manifold L_{k-1}, we can simplify (5.59) to

$$\alpha_k = \frac{\|\nabla f(x^k)\|^2}{\|\nabla f(x^{k-1})\|^2}. \tag{5.60}$$

Formulas (5.59) and (5.60) are mathematically equivalent, but (5.59) proved to be more robust in the presence of numerical round-off errors in the directional minimization. It also works better when the conjugate gradient method is applied to nonquadratic functions.[‡]

Performance of the conjugate gradient method on the example of Figure 5.5 on page 233 is illustrated in Figure 5.7.

5.5.2 Speed of Convergence. Application to Nonquadratic Functions

While Theorem 5.18 guarantees termination at the minimum point of a quadratic function in at most n iterations, in practical applications to large-scale problems we almost never manage to successfully carry out n steps. Numerical errors in the calculation of the directions and in the directional minimization result in the loss of the property of Q-conjugacy of the directions. At some step, inevitably, the iterations become inefficient, and the directions may even lose the descent property.

Therefore, even for quadratic functions, we periodically *re-initialize* the conjugate gradient method, by setting

$$d^k = -\nabla f(x^k) \tag{5.61}$$

at iterations, at which numerical difficulties arise. Then the kth iteration becomes identical with an iteration of the method of steepest descent. After

---

[†]A specialized version for quadratic problems only is presented on page 281.

[‡]Formulas (5.59) and (5.60) are referred to as the Polak–Ribière and Fletcher–Reeves versions.

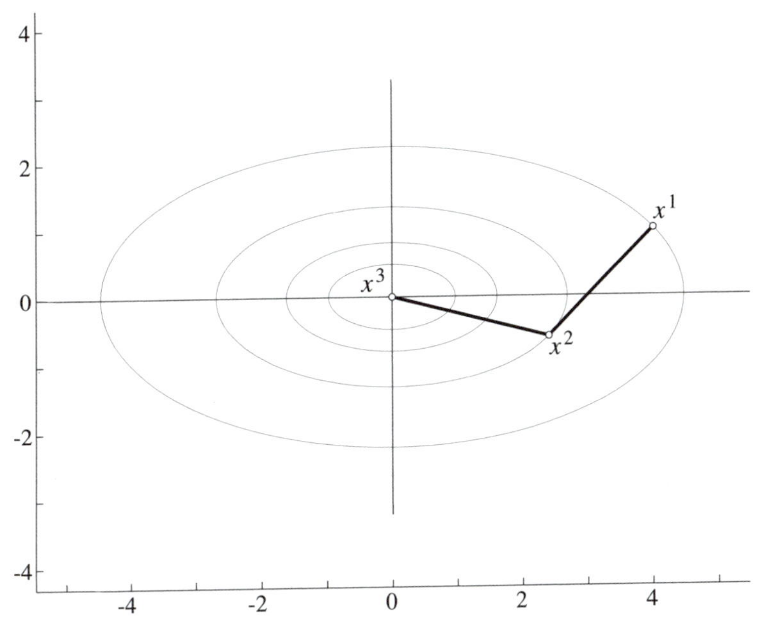

Figure 5.7. Operation of the conjugate gradient method on a quadratic function of two variables.

that, the conjugate directions are built again, until the next re-initialization. The method with periodic re-initialization combines the robustness of the method of steepest descent and the efficacy of the conjugate gradient method. The convergence of the conjugate gradient method with re-initialization is a simple consequence of the convergence of the method of steepest descent, as analyzed in Theorem 5.4.

Much can be said about the speed of convergence of the conjugate gradient method. If we did re-initialization after exactly n steps, we could argue as follows. Assume that the function $f(\cdot)$ is convex and has a positive definite Hessian $\nabla^2 f(\hat{x})$ at the minimum point $\hat{x}$. The sequence $\{x^k\}$ is convergent to the solution $\hat{x}$. After L cycles of n steps, the re-initialization point is x^{Ln+1}. If it is sufficiently close to the solution, the function $f(\cdot)$ in the level set $X_{Ln+1} = \{x : f(x) \le f(x^{Ln+1})\}$ is very close to the quadratic function

$$\tilde{f}(x) = \big\langle x - \hat{x}, \big[\nabla^2 f(\hat{x})\big](x - \hat{x})\big\rangle + f(\hat{x}),$$

with the error of order higher than $\|x^{Ln+1} - \hat{x}\|^2$. Therefore, by straightforward continuity arguments, the sequence of n steps of the conjugate gradient method for $f(\cdot)$ is very close to the sequence generated by the same method for $\tilde{f}(\cdot)$, with some second order error. But the latter sequence terminates at $\hat{x}$, and thus the conjugate gradient method for $f(\cdot)$ must exhibit *n-step quadratic convergence*:

$$\|x^{(L+1)n+1} - \hat{x}\| \le C\|x^{Ln+1} - \hat{x}\|^2.$$

Full mathematical rigor can be put into this argument. However, its relevance is modest. We almost never run full n steps of the conjugate gradient method; on the contrary, re-initialization is employed as frequently as every 10 to 20 iterations, irrespectively of the dimension n of the problem.

Much more relevant is the analysis of the first k steps of the method for small k. It turns out that the location of eigenvalues of the Hessian Q has a decisive impact on the quality of early iterates of the conjugate gradient method. Before proceeding to the general case, it may be instructive to consider an example.

Example 5.19. Suppose the matrix Q has only two distinct eigenvalues, μ_1 and μ_2. Then the vector $x^1 - \hat{x}$, where $\hat{x}$ is the minimum point of $f(\cdot)$, can be represented as

$$x^1 - \hat{x} = \theta_1^1 u^1 + \theta_2^1 u^2, \tag{5.62}$$

where u^1 and u^2 are eigenvectors corresponding to the eigenvalues μ_1 and μ_2, and θ_1^1, θ_2^1 are some real coefficients. Then

$$\nabla f(x^1) = Q(x^1 - \hat{x}) = \mu_1\theta_1^1 u^1 + \mu_2\theta_2^1 u^2.$$

The second point, $x^2 = x^1 - \tau_1 \nabla f(x^1)$, satisfies the relation

$$\begin{aligned} x^2 - \hat{x} &= x^1 - \hat{x} - \tau_1 \nabla f(x^1) \\ &= \left(\theta_1^1 u^1 + \theta_2^1 u^2\right) - \tau_1\left(\mu_1\theta_1^1 u^1 + \mu_2\theta_2^1 u^2\right) \\ &= \theta_1^2 u^1 + \theta_2^2 u^2, \end{aligned}$$

with appropriately defined coefficients θ_1^2 and θ_2^2. The gradient at x^2, therefore, has the form

$$\nabla f(x^2) = Q(x^2 - \hat{x}) = \mu_1\theta_1^2 u^1 + \mu_2\theta_2^2 u^2.$$

By the directional minimization condition, it is orthogonal to $\nabla f(x^1)$, so if it is nonzero, it is linearly independent of $\nabla f(x^1)$. Since both gradients are linear combinations of the eigenvectors u^1 and u^2,

$$\text{lin}\{\nabla f(x^1), \nabla f(x^2)\} = \text{lin}\{u^1, u^2\}.$$

The next point, x^3, is the minimum of $f(\cdot)$ on the two-dimensional manifold

$$L_1 = x^1 + \text{lin}\{\nabla f(x^1), \nabla f(x^2)\} = x^1 + \text{lin}\{u^1, u^2\}.$$

By virtue of (5.62), the global minimum point $\hat{x}$ is an element of this manifold, and thus $x^3 = \hat{x}$. We have a situation similar to that depicted in Figure 5.7 on page 247.

In order to extend the idea of this example, it is convenient to derive another general representation of the linear subspace spanned by the directions generated by the conjugate gradient method,

$$S_k = \operatorname{lin}\{d^1, d^2, \dots, d^k\}.$$

We shall show that

$$S_k = \operatorname{lin}\{\nabla f(x^1), Q\nabla f(x^1), Q^2\nabla f(x^1), \dots, Q^{k-1}\nabla f(x^1)\}. \tag{5.63}$$

Formula (5.63) is trivially satisfied for $k = 1$. Assume that it holds true for some k. We have

$$S_{k+1} = S_k + \operatorname{lin}\{d^{k+1}\}.$$

The direction d^{k+1} is a combination of $\nabla f(x^{k+1})$ and the preceding direction d^k, which is in S_k, and so

$$S_{k+1} = S_k + \operatorname{lin}\{\nabla f(x^{k+1})\}. \tag{5.64}$$

For a quadratic function $f(\cdot)$ we can write

$$\nabla f(x^{k+1}) = \nabla f(x^1) + Q(x^{k+1} - x^1).$$

Since $x^{k+1} - x^1$ is a linear combination of the directions $d^1, \dots, d^k$, it is an element of S_k. It follows from (5.63)–(5.64) that

$$S_{k+1} = S_k + QS_k = \operatorname{lin}\{\nabla f(x^1), Q\nabla f(x^1), Q^2\nabla f(x^1), \dots, Q^k\nabla f(x^1)\}.$$

By induction, formula (5.63) holds true for all k.

We can now easily follow Example 5.19 to show that the conjugate gradient method terminates after at most k steps, if the Hessian Q has exactly k distinct eigenvalues. Representing

$$\nabla f(x^1) = \sum_{j=1}^{k} \theta_j u^j,$$

with eigenvectors u_j, we immediately see from (5.63) that

$$S_m \subset \operatorname{lin}\{u^1, \dots, u^k\}, \quad m = 1, 2, \dots.$$

No more than k distinct nested subspaces can have this property, and therefore $S_{k+1} = S_k$. As $\nabla f(x^{k+1}) \perp S_k$, formula (5.64) implies that $\nabla f(x^{k+1}) = 0$.

This argument can be further extended to cover the case of clustered eigenvalues. We know from Theorem 5.16 that the point x^{k+1}, obtained after k

iterations of the conjugate gradient method, is the minimum of $f(\cdot)$ in the manifold

$$L_k = x^1 + S_k.$$

This combined with formula (5.63) implies that the point x^{k+1} has the lowest value of $f(\cdot)$ among all points of the form

$$\begin{aligned} x(\gamma) &= x^1 + \gamma_1 \nabla f(x^1) + \gamma_2 Q \nabla f(x^1) + \cdots + \gamma_k Q^{k-1} \nabla f(x^1) \\ &= x^1 + \Big(\sum_{j=1}^{k} \gamma_j Q^{j-1} \Big) \nabla f(x^1). \end{aligned}$$

Denoting by $\hat{x}$ the global minimum point of $f(\cdot)$, we can rewrite the last expression as follows:

$$\begin{aligned} x(\gamma) - \hat{x} &= \big[x^1 - \hat{x}\big] + \Big(\sum_{j=1}^{k} \gamma_j Q^{j-1} \Big) Q \big[x^1 - \hat{x}\big] \\ &= \Big(I + \sum_{j=1}^{k} \gamma_j Q^{j} \Big) \big[x^1 - \hat{x}\big]. \end{aligned} \tag{5.65}$$

Useful estimates of the quality of the point x^{k+1} can be obtained by substituting specific values for the coefficients γ_j in formula (5.65). This is usually done by imposing a specific form of the kth degree polynomial

$$P_k(Q) = I + \sum_{j=1}^{k} \gamma_j Q^j. \tag{5.66}$$

We can now show that if the eigenvalues of Q are clustered around k centers, the first k steps of the conjugate gradient method yield a substantial improvement.

THEOREM 5.20. *Assume that the matrix Q of dimension n has all its eigenvalues contained in k nonintersecting intervals*

$$[(1 - \delta_m)\mu_m, (1 + \delta_m)\mu_m], \quad m = 1, \ldots, k,$$

where $0 < \mu_1 < \cdots < \mu_k$ and $\delta_m \geq 0$, $m = 1, \ldots, k$. Then the conjugate gradient method applied to function (5.52) finds after k steps a point x^{k+1} such that[†]

$$\frac{f(x^{k+1}) - f(\hat{x})}{f(x^1) - f(\hat{x})} \leq \max_{1 \leq m \leq k} \delta_m^2 (1 + \delta_m)^{2(m-1)} \prod_{1 \leq i < m} \left(\frac{\mu_m}{\mu_i} \right)^2.$$

[†]If $m = 1$ the product over $1 \leq i < m$ is by definition equal to 1.

Proof. Let $\lambda_1, \dots, \lambda_n$ be the eigenvalues of Q (some of them may be equal), and let u^j, $j = 1, \dots, n$, be orthogonal eigenvectors of length 1 corresponding to these eigenvalues. Define the sets

$$J_m = \{j : |\lambda_j - \mu_m| \le \delta_m \mu_m\}, \quad m = 1, \dots, k.$$

The vector $x^1 - \hat{x}$ can be represented as follows:

$$x^1 - \hat{x} = \sum_{j=1}^{n} \theta_j u^j.$$

Then

$$f(x^1) - f(\hat{x}) = \frac{1}{2}\langle x^1 - \hat{x}, Q(x^1 - \hat{x})\rangle = \frac{1}{2}\sum_{j=1}^{n} \lambda_j \theta_j^2. \tag{5.67}$$

Consider the polynomial

$$P_k(Q) = \prod_{i=1}^{k} \Big(I - \frac{1}{\mu_i} Q\Big). \tag{5.68}$$

It can obviously be written in form (5.66). The corresponding point $x(\gamma) \in L_k$, as defined in (5.65), satisfies the relation

$$x(\gamma) - \hat{x} = \prod_{i=1}^{k} \Big(I - \frac{1}{\mu_i} Q\Big) \sum_{j=1}^{n} \theta_j u^j = \sum_{j=1}^{n} \theta_j \prod_{i=1}^{k} \Big(I - \frac{1}{\mu_i} Q\Big) u_j.$$

Since each u_j is an eigenvector, the term involving u_j in the last formula has the form

$$\theta_j \prod_{i=1}^{k} \Big(I - \frac{1}{\mu_i} Q\Big) u_j = \theta_j \prod_{i=1}^{k} \Big(1 - \frac{\lambda_j}{\mu_i}\Big) u_j.$$

Hence, similarly to (5.67),

$$f(x(\gamma)) - f(\hat{x}) = \frac{1}{2}\sum_{j=1}^{n} \lambda_j \theta_j^2 \prod_{i=1}^{k} \Big(1 - \frac{\lambda_j}{\mu_i}\Big)^2.$$

For $j \in J_m$ we can estimate the product in the sum above as follows:

$$\begin{aligned}\prod_{i=1}^{k} (1 - \frac{\lambda_j}{\mu_i})^2 &\le \delta_m^2 \prod_{i \ne m} \Big(1 - \frac{\lambda_j}{\mu_i}\Big)^2 \le \delta_m^2 \prod_{i < m} \Big(1 - \frac{\lambda_j}{\mu_i}\Big)^2 \\ &\le \delta_m^2 \prod_{i < m} \Big(\frac{\lambda_j}{\mu_i}\Big)^2 \le \delta_m^2 (1 + \delta_m)^{2(m-1)} \prod_{i < m} \Big(\frac{\mu_m}{\mu_i}\Big)^2.\end{aligned}$$

Therefore,

$$f(x(\gamma)) - f(\hat{x}) \le \left[f(x^1) - f(\hat{x})\right] \max_{1\le m\le k} \delta_m^2 (1+\delta_m)^{2(m-1)} \prod_{i<m} \left(\frac{\mu_m}{\mu_i}\right)^2.$$

Since x^{k+1} is a minimizer in L_k, it is at least as good as $x(\gamma)$, and the required estimate follows. □

As a direct application of this theorem, we can develop an estimate for the quality of the $(k+1)$st iterate, when the matrix Q has $n-k+1$ eigenvalues in the interval $[a, b]$ and the remaining $k-1$ eigenvalues above b. We can choose

$$\mu_1 = \frac{1}{2}(a+b), \quad \mu_2 = \lambda_{n-k+1}, \quad \dots \quad \mu_k = \lambda_n$$

and

$$\delta_1 = \frac{b-a}{b+a}, \quad \delta_2 = \dots = \delta_k = 0.$$

Then from Theorem 5.20 we infer that

$$\frac{f(x^{k+1}) - f(\hat{x})}{f(x^1) - f(\hat{x})} \le \left(\frac{b-a}{b+a}\right)^2,$$

which is equivalent to the progress made by one iteration of the method of steepest descent applied to a problem with *all* eigenvalues contained in $[a, b]$ (see Theorem 5.8).

In the even more special case, when there are only k distinct eigenvalues, we have $a = b$ in the above estimate. Thus, the conjugate gradient method finds the minimum point of the function in no more than k steps. This again explains the phenomenon observed in Example 5.19 and in the discussion that follows it.

If the function $f(\cdot)$ is not quadratic, but twice continuously differentiable, with positive definite Hessian $Q = \nabla^2 f(\hat{x})$ at the minimum point $\hat{x}$, much of the discussion from this section can be applied to the case of clustered eigenvalues of Q. Consider the quadratic approximation

$$\tilde{f}(x) = \langle x - \hat{x}, Q(x - \hat{x})\rangle + f(\hat{x}).$$

The conjugate gradient method applied to $\tilde{f}(\cdot)$ generates a sequence of points for which the assertion of Theorem 5.20 holds true. If x^1 is sufficiently close to $\hat{x}$, the first k points generated by the conjugate gradient method for $f(\cdot)$ will be close to the sequence for $\tilde{f}(\cdot)$, with an error of second order. Thus, even for nonquadratic functions, rapid convergence of the conjugate gradient method at the iterations following each re-initialization can be observed.

5.5.3 Pre-Conditioning

An important way of improving performance of the conjugate gradient method, as well as other optimization methods, is a change of the system of independent variables, aimed at decreasing the condition index of the Hessian. Recall from Section 5.3 that the speed of convergence of the method of steepest descent is determined by the condition index

$$\varkappa = \frac{\lambda_{\max}}{\lambda_{\min}},$$

where $\lambda_{\max}$ and $\lambda_{\min}$ are the largest and smallest eigenvalues of the Hessian Q. The conjugate gradient method is much faster, but it is also negatively affected by the condition index.

Suppose we know a symmetric positive definite matrix V, which is a good approximation of the inverse of the Hessian,

$$V \approx Q^{-1}.$$

Let $V^{1/2}$ be its square root: a symmetric matrix such that $V = V^{1/2}V^{1/2}$. Changing the variables

$$x = V^{1/2}y, \tag{5.69}$$

we can express function (5.52) in the new variables as follows:

$$h(y) = f(V^{1/2}y) = \langle y, V^{1/2}QV^{1/2}y\rangle + \langle c, V^{1/2}y\rangle. \tag{5.70}$$

The Hessian of the function in the new coordinates equals

$$\nabla^2 h(y) = V^{1/2}QV^{1/2}.$$

If V were exactly equal to Q^{-1}, we would obtain $\nabla^2 h(y) = I$, and the condition index of (5.70) would be a perfect 1. But knowing Q^{-1} means knowing the minimum $\hat{x} = -Q^{-1}c$, and there is no need for optimization at all. We cannot assume that we know the inverse of the Hessian. Instead, we assume that we can find a matrix V which is close to Q^{-1} in the sense that the condition index of $V^{1/2}QV^{1/2}$ is much closer to 1 than that of Q. The change of variables (5.69) with such a matrix V is called *pre-conditioning*.

Application of pre-conditioning within the conjugate gradient method does not require calculation of the square root $V^{1/2}$ and it does not even require any explicit form of V. Let us denote by r^k the direction used by the conjugate gradient method for (5.70) at iteration k. Application of the basic conjugate gradient method to the function (5.70) yields the relations:

$$r^k = \begin{cases} -\nabla h(y^k) & \text{if } k = 1, \\ -\nabla h(y^k) + \alpha_k r^{k-1} & \text{if } k > 1, \end{cases} \tag{5.71}$$

$$\alpha_k = \frac{\langle \nabla h(y^k), \nabla h(y^k) - \nabla h(y^{k-1}) \rangle}{\| \nabla h(y^{k-1}) \|^2}, \tag{5.72}$$

$$y^{k+1} = y^k + \tau_k r^k,$$

$$h(y^{k+1}) = \min_{\tau \geq 0} h(y^k + \tau r^k). \tag{5.73}$$

Define $x^k = V^{1/2} y^k$. Using the chain rule of differentiation in (5.70) we obtain

$$\nabla h(y^k) = V^{1/2} \nabla f(x^k). \tag{5.74}$$

According to (5.69), the direction of change in the variables x equals

$$d^k = V^{1/2} r^k.$$

Multiplying (5.71) by $V^{1/2}$ and substituting (5.74) we obtain the formula for the search direction:

$$d^k = \begin{cases} -V \nabla f(x^k) & \text{if } k = 1, \\ -V \nabla f(x^k) + \alpha_k d^{k-1} & \text{if } k > 1. \end{cases} \tag{5.75}$$

Furthermore, formulas (5.72) and (5.74) render

$$\alpha_k = \frac{\big\langle \nabla f(x^k), V[\nabla f(x^k) - \nabla f(x^{k-1})] \big\rangle}{\big\langle \nabla f(x^{k-1}), V \nabla f(x^{k-1}) \big\rangle}. \tag{5.76}$$

These formulas use only the multiplication by V. The directional minimization condition (5.73) for y becomes the directional minimization condition for x:

$$f(x^{k+1}) = \min_{\tau \geq 0} f(x^k + \tau d^k).$$

In fact, we do not need to have any explicit form of the pre-conditioner V. The only requirement is to be able to multiply vectors by V. We discuss this in Example 5.21 below.

The conjugate gradient method with pre-conditioning can be formally applied to any continuously differentiable function. Of course, the concept of Q-orthogonality makes little sense there. Nevertheless, good convergence properties of the method for quadratic functions are indicative for the performance in the nonquadratic case.

In the nonquadratic case, as well for large-scale quadratic problems, we employ re-initialization (5.61) by returning to the direction of steepest descent after every m iterations (with $m \ll n$). We also re-initialize in the case of numerical difficulties in achieving improvement in the current direction d^k. Again, the convergence of the conjugate gradient method with re-initialization is a consequence of the convergence of the method of steepest descent. The reader can easily verify that in the pre-conditioned version, with some positive definite V, the argument of Theorem 5.4 applies as well.

Example 5.21. Consider the optimal control problem with discrete time

$$\text{minimize} \sum_{t=1}^{T} f_t(u_t) + f_{T+1}(x_{T+1}) \tag{5.77}$$

$$\text{subject to } x_{t+1} = \varphi_t(x_t, u_t), \quad t = 1, \ldots, T. \tag{5.78}$$

Here $x_t \in \mathbb{R}^n$ denotes the state of the system at time t, and $u_t \in \mathbb{R}^m$ denotes the control vector at time t. The initial state x_1 is given.

The functions $f_t : \mathbb{R}^m \to \mathbb{R}, t = 1, \ldots, T$, and $f_{T+1} : \mathbb{R}^n \to \mathbb{R}$, are assumed to be convex and twice continuously differentiable. The functions $\varphi_t : \mathbb{R}^n \times \mathbb{R}^m \to \mathbb{R}^n$ are continuously differentiable.

For every control sequence

$$u = (u_1, \ldots, u_T),$$

we can calculate from (5.78) the corresponding state trajectory

$$x(u) = \big(x_1, x_2(u_1), x_3(u_1, u_2), \ldots, x_{T+1}(u_1, \ldots, u_T)\big).$$

Thus (5.77) can be expressed as an unconstrained problem in the control variables:

$$\underset{u}{\text{minimize}} \left[F(u) \triangleq \sum_{t=1}^{T} f_t(u_t) + f_{T+1}(x_{T+1}(u)) \right]. \tag{5.79}$$

The gradient of the function F can be calculated by employing chain rules of differentiation. Let us denote (for the current control u and the corresponding trajectory x) the Jacobians of $\varphi(\cdot, \cdot)$ by

$$A_t = \frac{\partial}{\partial x_t} \varphi_t(x_t, u_t), \quad B_t = \frac{\partial}{\partial u_t} \varphi_t(x_t, u_t), \quad t = 1, \ldots, T.$$

Then we have

$$\nabla_{u_t} F(u) = \nabla f_t(u_t) + B_t^* A_{t+1}^* \cdots A_T^* \nabla f_{T+1}(x_{T+1}), \quad t = 1, \ldots, T.$$

In the formula above, the symbol $[\cdot]^*$ denotes transposition. Introduce *adjoint state variables* ψ_t, connected by the equations

$$\psi_{t-1} = A_t^* \psi_t, \quad t = T, T-1, \ldots, 2,$$

with $\psi_T = \nabla f_{T+1}(x_{T+1})$. We can efficiently calculate the gradient of $F(\cdot)$ as follows:

$$\nabla_{u_t} F(u) = \nabla f_t(u_t) + B_t^* \psi_t, \quad t = 1, \ldots, T.$$

Optimal control problems are frequently ill-conditioned, and the method of steepest descent is very slow. The conjugate gradient method is more efficient, and it can be easily implemented despite the very large dimension of the vector u. The condition index can be improved by employing pre-conditioning.

One way to obtain a pre-conditioner is to consider a linear-quadratic optimal control problem in variations:

$$\text{minimize } \frac{1}{2}\sum_{t=1}^{T}\langle \sigma_t, Q_t \sigma_t\rangle + \frac{1}{2}\langle \xi_{T+1}, Q_{T+1}\xi_{T+1}\rangle \tag{5.80}$$

$$\text{subject to } \xi_{t+1} = A_t \xi_t + B_t \sigma_t, \quad t = 1, \ldots, T, \tag{5.81}$$

with $\xi_1 = 0$. The matrices Q_t are assumed to be positive definite and they approximate the Hessians of the corresponding functions $f_t(\cdot)$. The control σ identically equal to zero is the optimal solution of this problem.

The Hessian of the objective function of problem (5.80)–(5.81), considered as a function of controls σ, will serve as an approximation of the Hessian of the function $F(\cdot)$ in (5.79). We do not need to calculate this Hessian, we only need to be able to multiply its inverse V by a vector b. Observe that Vb is the solution of the optimization problem

$$\underset{\sigma}{\text{minimize}} \ \frac{1}{2}\langle \sigma, V^{-1}\sigma\rangle - \langle b, \sigma\rangle.$$

The first part of the quadratic function above is represented by problem (5.80)–(5.81). Thus, Vb is the solution of the linear-quadratic optimal control problem:

$$\text{minimize } \sum_{t=1}^{T}\left[\frac{1}{2}\langle \sigma_t, Q_t\sigma_t\rangle - \langle b_t, \sigma_t\rangle\right] + \frac{1}{2}\langle \xi_{T+1}, Q_{T+1}\xi_{T+1}\rangle$$
$$\text{subject to } \xi_{t+1} = A_t\xi_t + B_t\sigma_t, \quad t = 1, \ldots, T.$$

Such problems can be rapidly solved by dedicated optimal control techniques for linear-quadratic systems.

Our pre-conditioning is effective if the functions $f_t(\cdot)$ are strictly convex and the functions $\varphi_t(\cdot, \cdot)$ are only moderately nonlinear.

This example illustrates the general approach to constructing pre-conditioners: build a tractable linear-quadratic model of the problem at hand and solve it with an appropriate linear term (corresponding to $\langle b, \sigma\rangle$) to multiply the pre-conditioner by a given vector. The linear-quadratic model has to be solved many times, once for each multiplication by V in the method.

5.6 QUASI-NEWTON METHODS

5.6.1 Conjugate Directions by Quasi-Newton Updates

There exists more sophisticated way to generate conjugate directions than the conjugate gradient method. It is based on iterative application of the formula

$$d^k = -V_k \nabla f(x^k), \tag{5.82}$$

where $\{V_k\}$ are symmetric matrices of dimension n constructed in the course of calculation. Formal similarity of (5.82) to Newton's method (5.38) and the rules underlying the construction of the matrices V_k lead to the name *quasi-Newton methods*.

Let us introduce the increments in the decision variables:

$$p_k = x^{k+1} - x^k,$$

and in the gradients

$$q_k = \nabla f(x^{k+1}) - \nabla f(x^k).$$

For a quadratic function

$$f(x) = \frac{1}{2}\langle x, Qx\rangle + \langle c, x\rangle, \tag{5.83}$$

we always have

$$Q^{-1} q_k = p_k.$$

Therefore, by analogy with Newton's method, we construct matrices V_k in such a way that

$$V_{k+1} q_i = p_i, \quad \text{for} \quad i = 1, 2, \ldots, k. \tag{5.84}$$

We call (5.84) the *secant condition*.

There exist many methods for constructing V_k. We describe here the first and most important one, due to Davidon (and later analyzed by Fletcher and Powell).

In the Davidon–Fletcher–Powell (DFP) method the successive matrices V_k are generated by the recursive formula

$$V_{k+1} = V_k + \frac{p_k p_k^T}{p_k^T q_k} - \frac{V_k q_k q_k^T V_k}{q_k^T V_k q_k}, \tag{5.85}$$

where V_1 is some symmetric positive definite matrix. Note that p_k and q_k are understood as column vectors and thus $p_k p_k^T$ and $V_k q_k q_k^T V_k$ are square matrices of dimension n.

At first, we analyze the DFP method applied to the quadratic function (5.83).

LEMMA 5.22. *The DFP method is a descent method.*

Proof. It suffices to prove that all matrices V_k are positive definite. We do it by induction.

V_1 is positive definite by assumption. Suppose V_k is positive definite. We shall prove it for V_{k+1}. For a nonzero z, directly from (5.85) we obtain

$$\begin{aligned} z^T V_{k+1} z &= z^T V_k z + \frac{(z^T p_k)^2}{p_k^T q_k} - \frac{(z^T V_k q_k)^2}{q_k^T V_k q_k} \\ &= \frac{(z^T V_k z)(q_k^T V_k q_k) - (z^T V_k q_k)^2}{q_k^T V_k q_k} + \frac{(z^T p_k)^2}{p_k^T q_k}. \end{aligned} \tag{5.86}$$

By the Cauchy–Schwartz inequality for the scalar product $z^T V_k q_k$ we have

$$(z^T V_k z)(q_k^T V_k q_k) - (z^T V_k q_k)^2 \geq 0,$$

with equality occurring only for $z \parallel q_k$. The second term in (5.86) is nonnegative, too, since

$$p_k^T q_k = p_k^T Q p_k > 0$$

($p_k \neq 0$ for a descent direction d^k). The second term can be zero only for $z \perp p_k$. Consequently, the right hand side of (5.86) is nonnegative and may equal zero only when $q_k \perp p_k$, which never happens for $p_k \neq 0$. □

THEOREM 5.23. *The DFP method with exact directional minimization applied to the function* (5.83) *generates a sequence of Q-conjugate descent directions and stops after at most n iterations at an optimal point of $f(\cdot)$.*

Proof. We shall prove by induction that for every k, for which $\nabla f(x^{k+1}) \neq 0$, the secant condition (5.84) holds true, and that the directions generated by the method are Q-conjugate:

$$p_i^T Q p_j = 0 \quad \text{for all} \quad 1 \leq i < j \leq k. \tag{5.87}$$

Let us note that directly from (5.85) we obtain

$$V_{k+1} q_k = p_k,$$

so the secant condition (5.84) is satisfied for $k = 1$. For $k = 1$, the set of $1 \leq i < j \leq k$ in (5.87) is empty.

Assume now that the secant condition (5.84) and the Q-conjugacy relation (5.87) hold true for $k - 1$. We shall prove them for k. By (5.87) for $k - 1$

and by Theorem 5.16, the point x^k is the minimum of $f(\cdot)$ over the linear manifold

$$L_{k-1} = x^1 + \text{lin}\{p_1, p_2, \ldots, p_{k-1}\},$$

so

$$\nabla f(x^k) \perp p_i, \quad i = 1, \ldots, k-1.$$

By the secant condition (5.84) for $k-1$,

$$V_k q_i = p_i, \quad i = 1, \ldots, k-1.$$

Taking the scalar product with $\nabla f(x^k)$ we obtain

$$(\nabla f(x^k))^T V_k q_i = 0, \quad i = 1, \ldots, k-1.$$

This combined with (5.82) yields

$$(d^k)^T q_i = 0, \quad i = 1, \ldots, k-1.$$

Multiplying both sides by τ_k and using the fact that $q_i = Qp_i$, we obtain the conjugacy property for k:

$$p_k^T Q p_i = 0, \quad i = 1, \ldots, k-1.$$

When (5.87) holds true for k, we can multiply both sides of (5.85) from the right by Qp_i, $i < k$, to get the secant condition (5.84).

Having established the conjugacy of directions, we can now conclude that the method must find the minimum of (8.1) after at most n steps, by virtue of Corollary 5.17. □

The fact that the quasi-Newton method must stop after at most n steps can also be seen from the following corollary.

COROLLARY 5.24. *Suppose none of the points $x^1, \ldots, x^n$ generated by the quasi-Newton method are optimal. Then $V_{n+1} = Q^{-1}$.*

Proof. The secant property (5.84) for $k = n$ reads

$$V_{n+1} \begin{bmatrix} q_1 & q_2 & \ldots & q_n \end{bmatrix} = \begin{bmatrix} p_1 & p_2 & \ldots & p_n \end{bmatrix}. \tag{5.88}$$

The directions $d^1, \ldots, d^n$ are Q-conjugate, and by Lemma 5.15 they are linearly independent. None of the points x^k, $k = 1, \ldots, n$, are optimal and thus the vectors $p_k = \tau_k d^k$, $k = 1, \ldots, n$, are linearly independent as well. Consequently, the vectors $q_k = Qp_k$, $k = 1, \ldots, n$, are linearly independent. It follows that equation (5.88) determines V_{n+1} in a unique way. Since Q^{-1} obviously satisfies this equation, our assertion is true. □

It follows that the nth step of the DFP method is the same as the step of Newton's method. It should be stressed, though, that the main purpose of quasi-Newton methods is to generate good search directions at early iterations, rather than the inversion of the Hessian. If, for some reason, we need the inverse of Q, numerical linear algebra provides much better ways to calculate it.

Not only does the quasi-Newton method generate conjugate directions, but it is in fact equivalent to the conjugate gradient method.

THEOREM 5.25. *The sequence of points generated by the DFP method for function* (5.83)*, starting from a positive definite matrix V, is identical with the sequence generated by the preconditioned conjugate gradient method* (5.75)–(5.76) *with pre-conditioner V.*

Proof. We know from Theorem 5.23 that for every $k = 1, \dots, n+1$ the point x^k is the minimum of $f(\cdot)$ on the linear manifold

$$L_k = x^1 + \operatorname{lin}\{p_1, p_2, \dots, p_{k-1}\}.$$

To prove the equivalence with the pre-conditioned conjugate gradient method, it is sufficient to demonstrate that for all k the manifold L_k is the same as the one generated by the conjugate gradient method:

$$\operatorname{lin}\{p_1, p_2, \dots, p_{k-1}\} = V \operatorname{lin}\{\nabla f(x^1), \nabla f(x^2), \dots, \nabla f(x^{k-1})\}. \quad (5.89)$$

This identity is trivially true for $k = 1$. To establish its validity for $k > 1$, it is convenient to prove, in parallel, that

$$V_k = V + \sum_{i=1}^{k} \sum_{j=1}^{k} \theta_{ij}^k V \nabla f(x^i) \big[\nabla f(x^j)\big]^T V, \quad k = 1, \dots, n+1, \quad (5.90)$$

with some coefficients θ_{ij}^k. Again, for $k = 1$ this formula is trivial.

Supposing that both formulas (5.89)–(5.90) hold true for some $k \le n$, we obtain

$$d^k = -V_k \nabla f(x^k) = V \sum_{i=1}^{k} \gamma_i^k \nabla f(x^i),$$

with some coefficients γ_i^k. This implies that (5.89) holds true for $k+1$. Let us consider the DFP formula (5.85). We have just established that p_k is a linear combination of the pre-conditioned gradients $V \nabla f(x^i)$, $i = 1, \dots, k$. From (5.90) it follows that $V_k q_k$ is a linear combination of $V \nabla f(x^i)$, $i = 1, \dots, k+1$. Thus, formula (5.85) implies that V_{k+1} has form (5.90). This completes the induction step and proves the equivalence of the DFP method and the conjugate gradient method. □

Of course, the main reason for introducing the quasi-Newton method is the need to solve nonquadratic problems. We discuss this issue in the next subsection.

5.6.2 Family of Quasi-Newton Methods. Application to Nonquadratic Functions

Many variants of quasi-Newton methods exist. We shall derive a particular version, which proved to be efficient in practice.

Suppose we want to use the DFP method to construct an approximation of the Hessian Q, rather than of its inverse. To this end we just need to reverse the roles of p_k and q_k in formula (5.85):

$$W_{k+1} = W_k + \frac{q_k q_k^T}{q_k^T p_k} - \frac{W_k p_k p_k^T W_k}{p_k^T W_k p_k}. \tag{5.91}$$

Now, setting $V_k = W_k^{-1}$, we can derive an expression for updating V_k:

$$V_{k+1} = V_k + \left(1 + \frac{q_k^T V_k q_k}{p_k^T q_k}\right) \frac{p_k p_k^T}{p_k^T q_k} - \frac{p_k q_k^T V_k + V_k q_k p_k^T}{p_k^T q_k}. \tag{5.92}$$

It is called the Broyden–Fletcher–Goldfarb–Shanno formula (BFGS). We leave to the reader the tedious multiplication that shows the equality $V_{k+1} W_{k+1} = I$, if $V_k W_k = I$ (Exercise 5.7).

More generally, both versions discussed previously are members of a broader family of updating formulas, given by

$$V_{k+1} = V_k + \frac{p_k p_k^T}{p_k^T q_k} - \frac{V_k q_k q_k^T V_k}{q_k^T V_k q_k} + \omega_k \big(q_k^T V_k q_k\big) v_k v_k^T, \tag{5.93}$$

with

$$v_k = \frac{p_k}{p_k^T q_k} - \frac{V_k q_k}{q_k^T V_k q_k}, \tag{5.94}$$

and with the coefficient $\omega_k \in [0, 1]$. The DFP formula corresponds to $\omega_k = 0$, while the BFGS formula corresponds to $\omega_k = 1$. Again, we leave to the reader the manipulations that support these facts. It follows that for every $0 \le \omega_k \le 1$ the matrix V_{k+1} is a convex combination of the DFP and BFGS matrices, and is, therefore, positive definite, provided that V_k is positive definite.

We show now that all these formulae lead to mathematically equivalent algorithms for *any* continuously differentiable function, if the line search is perfect.

We assume that the function $f(\cdot)$ is continuously differentiable, and that the line search algorithm is *perfect* in the sense that

$$\nabla f(x^{k+1}) \perp d^k,$$

and that the point x^{k+1} is the closest to x^k point on the ray $x^k + \tau d^k$, $\tau \geq 0$, satisfying this condition.

LEMMA 5.26. *Assume that the point x^{k+1} has been obtained by perfect line search from point x^k in the direction $d^k = -V_k \nabla f(x^k)$. Then for every $\omega_k \in [0, 1]$ the direction*

$$d^{k+1} = -V_{k+1} \nabla f(x^{k+1}),$$

where V_{k+1} is calculated by (5.93), is parallel to v_k.

Proof. Let us calculate the direction d^{k+1}. For simplicity, we denote the gradient $\nabla f(x^k)$ by g_k. We have

$$\begin{aligned}
-V_{k+1} g_{k+1} = -V_k g_{k+1} - \frac{p_k^T g_{k+1}}{p_k^T q_k} p_k + \frac{q_k^T V_k g_{k+1}}{q_k^T V_k q_k} V_k q_k \\
+ \omega_k \big(q_k^T V_k q_k\big)\big(v_k^T g_{k+1}\big) v_k.
\end{aligned}$$

The term with ω_k is parallel to v_k, and it remains to consider the DFP update ($\omega_k = 0$). By the perfect line search condition, $g_{k+1}^T V_k g_k = 0$, and the second term in the last displayed expression vanishes. We obtain

$$-V_{k+1} g_{k+1} = -V_k g_{k+1} + \frac{q_k^T V_k g_{k+1}}{q_k^T V_k q_k} V_k q_k.$$

We have the following identities:

$$\begin{aligned}
-V_k g_{k+1} &= -V_k g_k - V_k q_k = \tau_k p_k - V_k q_k, \\
\frac{q_k^T V_k g_{k+1}}{q_k^T V_k q_k} &= \frac{g_{k+1}^T V_k g_{k+1}}{g_{k+1}^T V_k g_{k+1} + g_k^T V_k g_k}.
\end{aligned}$$

Combining the last three relations we get

$$\begin{aligned}
-V_{k+1} g_{k+1} &= \frac{p_k}{\tau_k} + \Big(\frac{g_{k+1}^T V_k g_{k+1}}{g_{k+1}^T V_k g_{k+1} + g_k^T V_k g_k} - 1\Big) V_k q_k \\
&= \frac{p_k}{\tau_k} - \frac{g_k^T V_k g_k}{q_k^T V_k q_k} V_k q_k \qquad (5.95) \\
&= \big(g_k^T V_k g_k\big) v_k + \Big(\frac{1}{\tau_k} - \frac{g_k^T V_k g_k}{p_k^T q_k}\Big) p_k.
\end{aligned}$$

By the perfect line search condition,

$$p_k^T q_k = -p_k^T g_k = \tau_k g_k^T V_k g_k.$$

Thus the last term in (5.95) vanishes and we obtain the required result. □

It follows from the calculations in the proof of the last lemma that, for every $\omega_k \in [0, 1]$, the next direction has the form

$$\begin{aligned} -V_{k+1} g_{k+1} &= \big(g_k^T V_k g_k\big) v_k + \omega_k \big(q_k^T V_k q_k\big)\big(v_k^T g_{k+1}\big) v_k \\ &= \Big(g_k^T V_k g_k - \omega_k g_{k+1}^T V_k g_{k+1}\Big) v_k. \end{aligned}$$

We know that for each $\omega_k \in [0, 1]$ the next matrix, V_{k+1}, is positive definite, and thus the expression above does not vanish, if $g_{k+1} \neq 0$.

We can thus interpret the quasi-Newton formula (5.93) as follows:

$$V_{k+1} = V_{k+1}^D + \gamma_k p_{k+1}^D (p_{k+1}^D)^T, \tag{5.96}$$

with V_{k+1}^D denoting the matrix obtained by the DFP formula, and p_{k+1}^D the step that would be made if the DFP formula was used. The coefficient γ_k is such that the resulting matrix is positive definite.

Lemma 5.27. *Suppose that at iteration k of the DFP method with perfect line search the matrix V_k^D is perturbed to*

$$V_k = V_k^D + \lambda_1 p_k^D (p_k^D)^T, \tag{5.97}$$

while remaining positive definite. Then, after the next iteration of the DFP method, we have

$$V_{k+1} = V_{k+1}^D + \lambda_2 p_{k+1}^D (p_{k+1}^D)^T. \tag{5.98}$$

In both formulas, the vectors p_k^D and p_{k+1}^D denote the steps that would be generated if the perturbation had not occurred.

Proof. The modified matrix is updated using the DFP algorithm, so

$$V_{k+1} = V_k^D + \lambda_1 p_k^D (p_k^D)^T + \frac{p_k p_k^T}{p_k^T q_k} - \frac{V_k q_k q_k^T V_k}{q_k^T V_k q_k}.$$

It follows from Lemma 5.26 that $p_k = p_k^D$ and $q_k = q_k^D$. If we now use the DFP formula for V_{k+1}^D, we can manipulate the last equation to the following form:

$$V_{k+1} = V_{k+1}^D + \lambda_1 p_k^D (p_k^D)^T - \frac{V_k q_k^D (q_k^D)^T V_k}{(q_k^D)^T V_k q_k^D} + \frac{V_k^D q_k^D (q_k^D)^T V_k^D}{(q_k^D)^T V_k^D q_k^D}.$$

Now, substituting for V_k from equation (5.97) and simplifying, we obtain

$$V_{k+1} = V_{k+1}^D + \lambda_1 \frac{\left((p_k^D)^T q_k^D\right)^2 \left((q_k^D)^T V_k^D q_k^D\right)}{\left((q_k^D)^T V_k^D q_k^D + \lambda_1 \left((p_k^D)^T q_k^D\right)^2\right)} \upsilon_k \upsilon_k^T,$$

with υ_k given by (5.94). Since the DFP matrix does not become singular, Lemma 5.26 allows us to rewrite the last equation as (5.98), with some λ_2. □

We can now formulate our main result.

THEOREM 5.28. *Assume that the function $f : \mathbb{R}^n \to \mathbb{R}$ is continuously differentiable and that it is minimized by the quasi-Newton method* (5.93) *with the perfect line search, starting from a positive definite matrix V_1. Then the sequence $\{x^k\}$ generated by the method is identical to the sequence generated by the DFP method starting from V_1.*

Proof. On the first iteration

$$d^1 = -V_1 g_1,$$

and the point x^2 is the same for all versions of the method.

On the second iteration, the conditions of Lemma 5.26 apply, and each formula will generate the same direction (up to a certain scaling coefficient) and the same value of x^3. Formula (5.93) has the form

$$V_2 = V_2^D + \lambda_1 p_1^D (p_1^D)^T \tag{5.99}$$

with some λ_1, for which the matrix remains positive definite. Lemma 5.27 implies

$$V_3 = V_3^D + \lambda_2 p_2^D (p_2^D)^T + \gamma_2 p_2^D (p_2^D)^T,$$

where the term with λ_2 follows from Lemma 5.27, which concerns the DFP method, and the term with γ_2 corresponds to the free choice of $\omega \in [0, 1]$ in formula (5.93), as shown in (5.96). We thus have an equation analogous to (5.99) for the next iteration.

By induction, every matrix V_k generated by the method has the form

$$V_k = V_k^D + \lambda_k p_k^D (p_k^D)^T,$$

with λ_k such that V_k remains positive definite. Moreover, the sequence of points $\{x^k\}$ does not depend on the coefficients λ_k. □

In practice, quasi-Newton methods are much less sensitive to inaccurate line- searches than the conjugate gradient method, which makes them particularly attractive for highly nonlinear problems. The BFGS method has been found to be more efficient and more robust with respect to these inaccuracies, but periodic re-initialization is necessary to guarantee theoretical convergence properties.

5.6.3 Limited Memory Versions

In large-scale optimization, storing and processing $n \times n$ matrices V_k may be difficult. However, we can calculate the direction (5.82) for the BFGS update (5.92) *without* any explicit form of V_k. Recall that V_k is fully determined by V_1 and two sequences of vectors

$$\begin{aligned} p_j &= x^{j+1} - x^j, \\ q_j &= \nabla f(x^{j+1}) - \nabla f(x^j), \quad j = 1, \ldots, k-1. \end{aligned}$$

By using these vectors we can construct the quasi-Newton direction resulting from applying $k - 1$ times the BFGS formula to V_1. It is a question of numerical linear algebra, but for the convenience of the reader we provide here the resulting procedure.

We set $r_k = \nabla f(x^k)$ and for $j = k-1, \ldots, 1$ we calculate

$$\alpha_j = \frac{\langle r_{j+1}, p_j \rangle}{\langle q_j, p_j \rangle} \quad \text{and} \quad r_j = r_{j+1} - \alpha_j q_j. \tag{5.100}$$

After that, we set $s_1 = V_1 r_1$ and for $j = 1, \ldots, k-1$ we calculate

$$\beta_j = \frac{\langle q_j, s_j \rangle}{\langle q_j, p_j \rangle} \quad \text{and} \quad s_{j+1} = s_j + (\alpha_j - \beta_j) p_j. \tag{5.101}$$

Then the quasi-Newton direction (5.82) is simply $d^k = -s_k$ (see Exercise 5.11). This construction allows us to implement the quasi-Newton method in a memory-efficient form. It also provides a convenient framework for the following heuristic modification of the re-initialization step.

Suppose we want to re-initialize the method at iteration k. Instead of returning to some fixed matrix V_1, like the identity matrix, we can use V_1 and the last m increments

$$\begin{aligned} p_j &= x^{j+1} - x^j, \\ q_j &= \nabla f(x^{j+1}) - \nabla f(x^j), \quad j = k-m, \ldots, k-1, \end{aligned}$$

to calculate the mth BFGS update of V_1. Here $m < k$ is some fixed parameter. The matrix obtained in this way is applied at iteration k, and the BFGS algorithm continues. We can even apply this idea at *every* step, always working with the last m vectors p_j and q_j. This version of the BFGS method performs very well in practice, but it does not have theoretical convergence guarantees.

5.7 TRUST REGION METHODS

5.7.1 Main Ideas

Consider the unconstrained optimization problem

$$\operatorname*{minimize}_{x\in\mathbb{R}^n}\ f(x).$$

The Newton method, the Gauss–Newton method, and variable metric methods may also be interpreted as follows. At iteration k we have a quadratic model of the function $f(\cdot)$:

$$f^k(x) = f(x^k) + \langle \nabla f(x^k), x - x^k\rangle + \frac{1}{2}\langle x - x^k, W_k(x - x^k)\rangle, \quad (5.102)$$

with W_k being the Hessian of $f(\cdot)$ at x^k (Newton's method), or an approximation of the Hessian, generated by the variable metric method.

Assume for a moment that the matrix W_k is positive definite. The unconstrained minimum of (5.102) is achieved at the point $x^k - W_k^{-1}\nabla f(x^k)$, which would be obtained by making a step of length 1 in the direction $d^k = -W_k^{-1}\nabla f(x^k)$. In the methods considered so far, line search in the direction d^k is carried out, in order to determine the next point, $x^{k+1} = x^k + \tau_k d^k$.

Another way to use model (5.102) is to restrict the search to a vicinity of the point x^k, in which we consider the model to be sufficiently accurate. Such a vicinity may be defined in many ways, but it is most natural to consider a ball

$$B_k = \{x \in \mathbb{R}^n : \|x - x^k\| \le \Delta_k\},$$

with some radius Δ_k. The set B_k is called the *trust region*. Using the trust region we construct the auxiliary problem:

$$\operatorname*{minimize}_{x\in B_k}\ f^k(x). \quad (5.103)$$

We call problem (5.103) the *trust region subproblem*. We use the Euclidean norm in the definition of the trust region, because it does not prefer any direction.[†] Recall that we have already employed a similar approach in the derivation of the method of steepest descent. Problem (5.9) on page 218 was actually a trust region subproblem. There, after determining the solution of this subproblem, we carried out search in the direction obtained.

We now explore a different idea. We solve subproblem (5.103) and we obtain a point y^k. Then we calculate the value of the objective function at y^k

---

[†]In Section 7.5 we discuss trust region methods for nonsmooth problems, where the use of polyhedral norms $\|\cdot\|_1$ or $\|\cdot\|_\infty$ makes more sense.

and we compare it to the value predicted by the model, $f^k(y^k)$. For a fixed value of $\gamma \in (0, 1)$, we verify the inequality:

$$f(y^k) \leq f(x^k) + \gamma\big[f^k(y^k) - f(x^k)\big]. \tag{5.104}$$

Observe that the difference in brackets on the right hand side is nonpositive, and thus inequality (5.104) requires that the improvement in true function values be comparable to the improvement in model function values. If it is satisfied, we conclude that sufficient progress has been made and we change the current point by setting $x^{k+1} := y^k$. This operation is called the *descent step*. The radius of the trust region either remains unchanged, or is increased.

On the other hand, when inequality (5.104) is not satisfied, we conclude that the model is not sufficiently accurate in the trust region. In this case we decrease the radius of the ball B_k by setting $\Delta_{k+1} := \beta_2 \Delta_k$ with some $\beta_2 \in (0, 1)$. The approximation to the solution remains unchanged: $x^{k+1} := x^k$. This operation is called the *null step*.

The detailed algorithm is presented in Figure 5.8. It starts from a point x^1 with some matrix W_1 in the model (5.102).

Step 0. Set $k := 1$, $\Delta_1 > 0$, $0 \leq \gamma \leq \bar{\gamma} \leq 1$, $\beta_1 \geq 1$, $\beta_2 \in (0, 1/\beta_1)$.

Step 1. Find a solution y^k of subproblem (5.103). If $f^k(y^k) = f(x^k)$ then stop. Otherwise, calculate $f(y^k)$ and continue.

Step 2. If

$$f(y^k) \leq f(x^k) + \gamma\big[f^k(y^k) - f(x^k)\big]$$

then go to Step 3 (descent step). Otherwise go to Step 4 (null step).

Step 3. Set $x^{k+1} := y^k$, calculate $\nabla f(x^{k+1})$ and W_{k+1}. If

$$f(y^k) \leq f(x^k) + \bar{\gamma}\big[f^k(y^k) - f(x^k)\big],$$

then set $\Delta_{k+1} := \beta_1 \Delta_k$. Otherwise, set $\Delta_{k+1} := \Delta_k$. Go to Step 5.

Step 4. Set $x^{k+1} := x^k$, $\Delta_{k+1} := \beta_2 \Delta_k$. Calculate W_{k+1} and continue.

Step 5. Increase k by one and go to Step 1.

Figure 5.8. The trust region algorithm.

It is useful to increase the size of the trust region after some descent steps. The increase makes sense if the model is "too good" in the trust region, which may indicate that longer steps would be successful as well. Such a possibility occurs when inequality (5.104) is satisfied even with some $\bar{\gamma}$ close to 1. In this case the radius of the trust region is multiplied by some $\beta_1 > 1$.

Our classification of steps to descent and null steps is relevant for a broader class of methods, most notably nondifferentiable optimization methods, which we discuss in Chapter 7. The main criterion for this distinction is that at a descent step the current approximation x^k and the model $f^k(\cdot)$ are updated, while at a null step only the model is updated. From this perspective the radius of the trust region is considered a parameter of the model, as are the gradient $\nabla f(x^k)$ and matrix W_k.

5.7.2 The Subproblem

We start from the analysis of the trust region subproblem (5.103). Its solution (for a positive definite matrix W_k) is illustrated in Figure 5.9. We also mark there the trajectory of solutions $y^k(\Delta)$, dependent on the size of the trust region, Δ_k. As we can see, for small Δ_k, the direction $y^k - x^k$ is close to the direction of steepest descent, while for large Δ_k, the direction $y^k - x^k$ is the Newton direction.

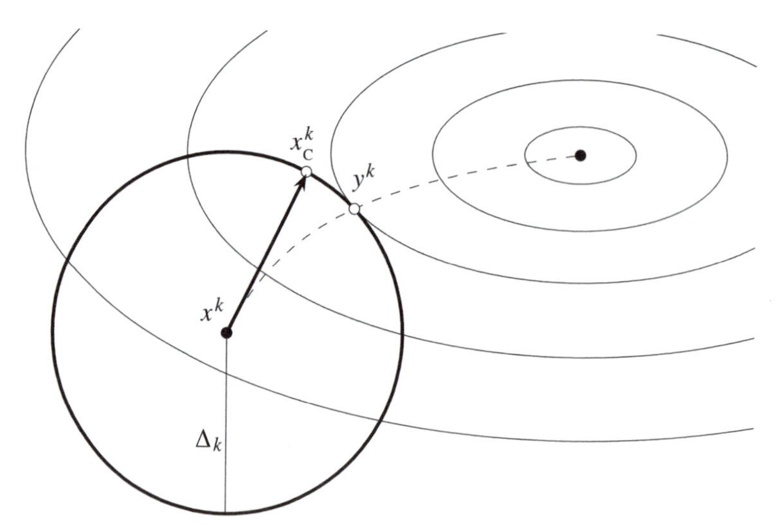

Figure 5.9. The trust region subproblem. The arrow represents the direction of steepest descent and x_C^k is the Cauchy point. The dotted curve represents the solutions of the subproblem for various values of Δ_k.

Let us stress again that in the convergence analysis of the trust region method we do not assume that the matrix W_k is positive semidefinite.

LEMMA 5.29. *A point $y^k \in B_k$ is a global solution to problem (5.103) if and only if there exists $\lambda_k \geq 0$ such that*

$$\begin{aligned} (W_k + \lambda_k I)(y^k - x^k) &= -\nabla f(x^k), \\ \lambda_k(\|y^k - x^k\| - \Delta_k) &= 0, \end{aligned} \tag{5.105}$$

and the matrix $W_k + \lambda_k I$ is positive semidefinite.

Proof. We first establish that these conditions are necessary conditions of optimality. Setting $d = x - x^k$, we can write problem (5.103) as follows:

$$\begin{aligned} &\text{minimize} \;\; f(x^k) + \langle \nabla f(x^k), d\rangle + \frac{1}{2}\langle d, W_k d\rangle \\ &\text{subject to} \;\; \frac{1}{2}\big(\|d\|^2 - \Delta_k^2\big) \leq 0. \end{aligned} \tag{5.106}$$

The Lagrangian has the form:

$$L(d, \lambda) = f(x^k) + \langle \nabla f(x^k), d\rangle + \frac{1}{2}\langle d, W_k d\rangle + \frac{\lambda}{2}\big(\|d\|^2 - \Delta_k^2\big). \tag{5.107}$$

The problem satisfies Robinson's condition at every point, in particular at its solution $d^k = y^k - x^k$. Application of the necessary conditions of optimality of Theorem 3.25 yields equations (5.105). The multiplier λ_k is unique, because the gradient of the constraint function is nonzero, if the constraint is active.

At the optimal solution d^k, with the optimal multiplier λ_k, the Hessian of the Lagrangian has the form:

$$\nabla^2 L(d^k, \lambda_k) = W_k + \lambda_k I.$$

Suppose the constraint is not active at the optimal solution d^k. Then $\lambda_k = 0$, and the point d^k is an unconstrained minimum of the quadratic function $\langle \nabla f(x^k), d\rangle + \frac{1}{2}\langle d, W_k d\rangle$. It follows from the second order necessary condition of optimality of Theorem 3.46 (without any constraints) that the Hessian of the objective function is positive semidefinite, that is, $W_k \succeq 0$.

Suppose $\|d^k\| = \Delta_k$. It follows that the point d^k is also the minimum of the Lagrangian (5.107) on the set $S = \{d \in \mathbb{R}^n : \|d\|^2 = \Delta^2\}$. Thus, for every $d \in S$ we have $L(d, \lambda_k) \geq L(d^k, \lambda_k)$. The function $L(\cdot, \lambda_k)$ is quadratic, and its gradient vanishes at d^k, due to the first order condition (5.105). Therefore

$$L(d, \lambda_k) - L(d^k, \lambda_k) = \frac{1}{2}\langle d - d^k, \nabla^2 L(d^k, \lambda_k)(d - d^k)\rangle.$$

It follows that

$$\langle d - d^k, (W_k + \lambda_k I)(d - d^k)\rangle \geq 0 \quad \text{for all} \quad d \in S.$$

Define the set $H = \operatorname{cone}(S - d^k)$. It is a union of an open halfspace and the point 0. The last inequality can be rewritten as

$$\langle h, (W_k + \lambda_k I)h \rangle \geq 0 \quad \text{for all} \quad h \in H.$$

The same is true for $h \in -H$. Since the closure of $H \cup (-H)$ is the whole $\mathbb{R}^n$, we conclude that the last inequality holds true for all $h \in \mathbb{R}^n$. Hence $W_k + \lambda_k I$ is positive semidefinite.

To prove that the conditions are sufficient for the optimality of y^k, we notice that the fact that the matrix $W_k + \lambda_k I$ is positive semidefinite and the equation $\nabla_d L(d^k, \lambda_k) = 0$ imply that the Lagrangian $L(\cdot, \lambda_k)$ achieves its unconstrained minimum at d^k. From the sufficient condition of Theorem 4.8 we conclude that y^k is the optimal solution of problem (5.103). □

The decrease of the model function, $f^k(x^k) - f^k(y^k)$, can be estimated as follows. The point y^k is at least as good as the optimal solution of the problem

$$\underset{0 \leq \tau \leq \Delta_k}{\text{minimize}} \; f^k\Big(x^k - \tau \frac{\nabla f(x^k)}{\|\nabla f(x^k)\|}\Big), \tag{5.108}$$

because a step of length at most Δ_k in the (normalized) direction of steepest descent does not take us out of the trust region. The point

$$x_{\mathrm{C}}^k = x^k - \tau_{\mathrm{C}} \frac{\nabla f(x^k)}{\|\nabla f(x^k)\|},$$

where τ_{C} is a solution of problem (5.108), is called the *Cauchy point* of the trust region. It is marked in Figure 5.9.

The decrease of the model achieved by the Cauchy point can be estimated easily.

LEMMA 5.30. *Suppose* $\|W_k\| \leq M$. *Then*

$$f^k(x_{\mathrm{C}}^k) - f^k(x^k) \leq -\frac{1}{2}\|\nabla f(x^k)\| \min\Big(\Delta_k, \frac{\|\nabla f(x^k)\|}{M}\Big). \tag{5.109}$$

Proof. The function

$$\varphi(\tau) = f^k\Big(x^k - \tau \frac{\nabla f(x^k)}{\|\nabla f(x^k)\|}\Big),$$

which is minimized in (5.108), is quadratic with the derivatives

$$\varphi'(0) = -\|\nabla f(x^k)\|, \qquad \varphi''(\tau) = \Big\langle \frac{\nabla f(x^k)}{\|\nabla f(x^k)\|}, W_k \frac{\nabla f(x^k)}{\|\nabla f(x^k)\|} \Big\rangle.$$

Suppose $\varphi''(0) \le 0$. Then the solution of (5.108) is $\tau_C = \Delta_k$. We obtain

$$f^k(x_C^k) = \varphi(\tau_C) \le \varphi(0) + \varphi'(0)\Delta_k = f^k(x^k) - \|\nabla f(x^k)\|\Delta_k,$$

and inequality (5.109) is obviously satisfied.

Suppose $\varphi''(0) > 0$. The unconstrained minimum of $\varphi(\cdot)$ is attained at the point

$$\hat{\tau} = -\frac{\varphi'(0)}{\varphi''(0)} = \frac{\|\nabla f(x^k)\|}{\left\langle \frac{\nabla f(x^k)}{\|\nabla f(x^k)\|}, W_k \frac{\nabla f(x^k)}{\|\nabla f(x^k)\|} \right\rangle} \ge \frac{\|\nabla f(x^k)\|}{M}.$$

The best step can thus be estimated as follows:

$$\tau_C = \min\left(\Delta_k, \hat{\tau}\right) \ge \min\left(\Delta_k, \frac{\|\nabla f(x^k)\|}{M}\right).$$

Recall from (5.8) that the slope of the line from $(0, \varphi(0))$ to $(\hat{\tau}, \varphi(\hat{\tau}))$ is $\varphi'(0)/2$. Thus the slope of the line from $(0, \varphi(0))$ to $(\tau_C, \varphi(\tau_C))$ is at most $\varphi'(0)/2$. We conclude that

$$f^k(x_C^k) = \varphi(\tau_C) \le \varphi(0) + \frac{1}{2}\varphi'(0)\tau_C.$$

Combining the last two inequalities we obtain (5.109). □

In the convergence analysis of the trust region methods we use only the fact that the point y^k is at least as good as the Cauchy point x_C^k. This allows for the use of approximate solutions of the trust region subproblem (5.103), as long as they satisfy the inequality

$$f^k(y^k) - f(x^k) \le -\frac{1}{2}\|\nabla f(x^k)\| \min\left(\Delta_k, \frac{\|\nabla f(x^k)\|}{M}\right). \tag{5.110}$$

This should not be understood, though, as some sort of equivalence of different versions of the trust region method. The main purpose of using the quadratic model (5.102) is to perform better than the direction of steepest descent.

The trust region subproblem is an easy problem, its solution does not require expensive function and gradient evaluations, and many efficient algorithms exist for finding good approximations of the solution. In the next chapter we discuss in detail constrained optimization methods. In our case, however, a specialized method, based on the optimality conditions of Lemma 5.29 and on duality theory, is easy to design. Define

$$\lambda_0 = \min\{\lambda \ge 0 : W_k + \lambda I \succeq 0\}$$

and consider the optimization problem

$$\begin{aligned} &\text{minimize} \;\; f(x^k) + \langle \nabla f(x^k), d\rangle + \frac{1}{2}\langle d, (W_k + \lambda_0 I)d\rangle \\ &\text{subject to} \;\; \|d\|^2 \le \Delta^2. \end{aligned} \tag{5.111}$$

It is a convex problem, by construction. By employing the necessary and sufficient conditions of optimality of Lemma 5.29, we immediately notice that problems (5.106) and (5.111) are equivalent: they have the same sets of optimal solutions, and the Lagrange multiplier $\hat{\mu}$ corresponding to the constraint in (5.111) is equal to $\lambda_k - \lambda_0$ (Exercise 5.12). Assuming that we know λ_0, we can apply to problem (5.111) the duality theory.

Consider the Lagrangian (5.107) for $\lambda \ge \lambda_0$. It follows from Theorem 4.3 that if the minimizer $d(\lambda)$ of the Lagrangian satisfies

$$\|d(\lambda)\|^2 = \Delta^2, \tag{5.112}$$

it is an optimal solution of problem (5.106). Equation (5.112) is a simple one-dimensional problem. It can be solved by bi-section, by a method of secants, or by Newton's method. We leave it to the reader to work out the details of these procedures in this simple case (Exercise 5.13).

Other methods of constrained optimization can be specialized to our case as well. Exercise 6.9 discusses application of the sequential quadratic programming method to problem (5.106).

5.7.3 Convergence

We now prove the convergence of the trust region method to stationary points. As mentioned earlier, we only use inequality (5.110) satisfied by the subproblems solution, and thus our analysis covers many practical versions of the method.

Theorem 5.31. *Assume that the function $f(\cdot)$ is continuously differentiable, the set $X_1 = \{x \in \mathbb{R}^n : f(x) \le f(x^1)\}$ is bounded, and that there exists a number M such that $\|W_k\| \le M$, $k = 1, 2, \ldots$. Then every accumulation point x^* of the sequence $\{x^k\}$ generated by the trust region method satisfies the necessary condition of optimality $\nabla f(x^*) = 0$.*

Proof. Our argument is based on an idea similar to the proof of convergence of the method of steepest descent. We prove that the method leaves a neighborhood of a nonstationary point, and that it cannot get back close to this point. The only complication is that we need to consider chains of several iterations, rather than just one step.

As the set X_1 is compact, the sequence $\{x^k\}$ has accumulation points. Suppose a subsequence $\{x^k\}_{k\in\mathcal{K}}$ is convergent to a point x^* with $\nabla f(x^*) \ne 0$.

Our further considerations are carried out under this condition, and we show that such a situation is impossible.

It is convenient to divide the proof into several stages.

The Exit Point

Since the gradient is continuous, we can find $\varepsilon > 0$ and $\delta > 0$ such that $\|\nabla f(x)\| \geq \delta$, whenever x remains in the ball B_ε about x^* having radius ε.

As the subsequence $\{x^k\}_{k\in\mathcal{K}}$ is convergent to x^*, for all sufficiently large $k \in \mathcal{K}$ we have $x^k \in B_\varepsilon$. For every such k consider the iterations $i = k, k+1, \dots$ and define the index

$$l(k,\varepsilon) = \sup\{j \geq k : \|x^i - x^*\| \leq \varepsilon, \text{ for all } i = k, \dots, j\}.$$

In other words, $l(k,\varepsilon)+1$, if it is finite, is the first point of the sequence x^k, $x^{k+1}, \dots,$ outside of the ball B_ε.

Existence of the Exit Point

Our first observation is that $l(k,\varepsilon)$ is finite for each $k \in \mathcal{K}$ such that $x^k \in B_\varepsilon$. Suppose it is not true, and all points x^i, $i \geq k$, remain in the ball B_ε. We shall show that this implies that $\Delta_i \to 0$ and leads to a contradiction.

If the number of descent steps is finite, the number of null steps is infinite, and the convergence of Δ_i to 0 follows from the construction of the algorithm. Consider the case of infinitely many descent steps. If there is a descent step at iteration i, we obtain from condition (5.104) that

$$f(x^{i+1}) - f(x^i) \leq \gamma\left[f^i(x^{i+1}) - f(x^i)\right].$$

This combined with inequality (5.109) yields

$$\begin{aligned} f(x^{i+1}) - f(x^i) &\leq -\frac{\gamma}{2}\|\nabla f(x^i)\| \min\left(\Delta_i, \frac{\|\nabla f(x^i)\|}{M}\right) \\ &\leq -\frac{\gamma\delta}{2}\min\left(\Delta_i, \frac{\delta}{M}\right). \end{aligned} \tag{5.113}$$

Hence, for every $j \geq k$

$$f(x^{j+1}) - f(x^k) \leq -\frac{\gamma\delta}{2}\sum_{\substack{k\leq i\leq j\\ i\text{ - descent}}}\min\left(\Delta_i, \frac{\delta}{M}\right).$$

Passing with j to ∞ we notice that the right hand side in the last estimate can be finite only if $\Delta_i \to 0$ for descent iterations i. But then $\Delta_i \to 0$ for all i. As Δ_i is decreased at null steps only, the number of null steps must be infinite. At every null step i we have the inequality:

$$f(y^i) - f(x^i) > \gamma\left[f^i(y^i) - f(x^i)\right].$$

Observe that $\|y^i - x^i\| \le \Delta_i$. Since all points x^i and y^i belong to a compact set and the gradient $\nabla f(\cdot)$ is continuous, there exists an infinitely small $o(\cdot)$ such that

$$f(y^i) \le f(x^i) + \langle \nabla f(x^i), y^i - x^i \rangle + o(\Delta_i), \quad i = k, k+1, \ldots.$$

As $f^i(\cdot)$ is quadratic, has the same gradient at x^i as $f(\cdot)$, and $\|W_k\| \le M$, it follows from the last inequality that there exists an infinitely small $o_1(\cdot)$ such that

$$f(y^i) \le f^i(y^i) + o_1(\Delta_i), \quad i = k, k+1, \ldots$$

We conclude that

$$(1-\gamma)\big[f^i(y^i) - f(x^i)\big] \ge -o_1(\Delta_i),$$

where $o_1(t)/t \to 0$ as $t \downarrow 0$. This combined with inequality (5.110) yields at all null steps the estimate

$$\frac{1}{2}\|\nabla f(x^i)\| \min\left(\Delta_i, \frac{\|\nabla f(x^i)\|}{M}\right) \le o_1(\Delta_i).$$

But $\|\nabla f(x^i)\| \ge \delta$ for all $i \ge k$, and thus also

$$\frac{\delta}{2} \min\left(\Delta_i, \frac{\delta}{M}\right) \le o_1(\Delta_i).$$

When $i \to \infty$ over null iterations, we obtain

$$\Delta_i \to 0 \quad \text{and} \quad \frac{\delta \Delta_i}{2} \le o_1(\Delta_i),$$

an absurd. We conclude that the index $l(k, \varepsilon)$ is finite, for every $k \in \mathcal{K}$.

Improvement at the Exit Point

We now estimate the value of the function upon exiting the ball B_ε. We have from inequality (5.113) that

$$\begin{aligned} f(x^{l(k,\varepsilon)+1}) - f(x^k) &\le \sum_{\substack{k \le i \le j(k,\varepsilon) \\ i\text{ - descent}}} [f(x^{i+1}) - f(x^i)] \\ &\le -\frac{\gamma\delta}{2} \sum_{\substack{k \le i \le j(k,\varepsilon) \\ i\text{ - descent}}} \min\left(\Delta_i, \frac{\delta}{M}\right). \end{aligned} \tag{5.114}$$

If $\Delta_i > \delta/M$ at some descent step i, then the last inequality yields

$$f(x^{l(k,\varepsilon)+1}) - f(x^k) \le -\frac{\gamma\delta^2}{2M}.$$

If $\Delta_i \le \delta/M$ for all descent steps i between k and $l(k,\varepsilon)$, then

$$f(x^{l(k,\varepsilon)+1}) - f(x^k) \le -\frac{\gamma\delta}{2} \sum_{\substack{k \le i \le j(k,\varepsilon) \\ i\text{ - descent}}} \Delta_i.$$

If $k \in \mathcal{K}$ is sufficiently large, then $\|x^k - x^*\| \le \varepsilon/2$. Consequently, the sum in the expression above is at least $\varepsilon/2$. We obtain

$$f(x^{l(k,\varepsilon)+1}) - f(x^k) \le -\frac{\gamma\delta\varepsilon}{4}.$$

Combining both cases we conclude that

$$f(x^{l(k,\varepsilon)+1}) - f(x^k) \le -\frac{\gamma\delta}{2} \min\left(\frac{\delta}{M}, \frac{\varepsilon}{2}\right).$$

By the construction of the method, the sequence of the function values $\{f(x^k)\}$ is nonincreasing. As it is bounded, it is convergent. When $k \to \infty$, $k \in \mathcal{K}$, we have $f(x^k) \to f(x^*)$. But then also $f(x^{l(k,\varepsilon)+1}) \to f(x^*)$, which contradicts the last displayed inequality. □

5.8 NONGRADIENT METHODS

5.8.1 Numerical Differentiation

If the function $f(\cdot)$ is continuously differentiable, but a closed-form expression for its gradient is not readily available, or cannot be derived by symbolic differentiation software, two courses of action are possible.

First, we may use finite difference approximations of the gradient. Denoting by e^j the jth unit vector in $\mathbb{R}^n$, we may use the approximate formula

$$\frac{\partial f(x)}{\partial x_j} \approx \frac{1}{2\delta}\big[f(x+\delta e^j) - f(x-\delta e^j)\big], \quad j = 1,\dots,n, \tag{5.115}$$

with some small step $\delta > 0$. Other, more economical schemes for gradient estimation are possible as well, and the finite difference step $\delta > 0$ may be updated in the course of calculations.

As an additional benefit of this approach, we obtain an estimate of the second derivative

$$\frac{\partial^2 f(x)}{\partial x_j^2} \approx \frac{1}{\delta^2}\big[f(x+\delta e^j) + f(x-\delta e^j) - 2f(x)\big], \quad j = 1,\dots,n. \tag{5.116}$$

The diagonal matrix having these entries (calculated at a point of re-initialization) can be used as a pre-conditioner in conjugate gradient methods, or as the initial matrix V_1 in quasi-Newton methods, provided that all these estimates are positive.

It should be stressed that finite difference approximations of the derivatives, and even more so of the second derivatives, are very sensitive to numerical errors in calculating the values of the function. Indeed, in (5.115) an error in calculating the value of the function is divided by the small number δ, yielding a much larger error in the estimation of the derivative. In (5.116) we divide by δ^2, and the estimation error of the second derivative is even higher. It is therefore necessary that the errors in the calculation of the values of $f(\cdot)$ are many orders of magnitude smaller than the finite difference step δ.

5.8.2 Coordinate Descent

The second course of action is to use methods that do not rely on the gradient in the calculation of the search directions. These methods have the general form

$$x^{k+1} = x^k + \tau_k d^k, \quad k = 1, 2, \dots, \tag{5.117}$$

with step sizes τ_k calculated via directional minimization

$$f(x^{k+1}) = \min_{\tau \in \mathbb{R}} f(x^k + \tau d^k). \tag{5.118}$$

Note that we cannot be sure that d^k is a direction of descent, and therefore the line search procedure has to allow both positive and negative step sizes.

The simplest method of this form is the *coordinate descent method* in which the unit vectors $e^1, \dots, e^n$ are used as the first n search directions. After that, we use the unit vectors again, in the same order. In the Lth cycle of the method, we thus have

$$d^{Ln+j} = e^j, \quad j = 1, \dots, n, \quad L = 0, 1, 2, \dots.$$

The convergence of the method of coordinate descent[†] can be established similarly to the convergence of the method of steepest descent.

THEOREM 5.32. *Assume that the function $f(\cdot)$ is continuously differentiable and that the set $X_1 = \{x \in \mathbb{R}^n : f(x) \le f(x^1)\}$ is bounded. Moreover, assume that for every $x \in X_1$ and for every direction $e^1, \dots, e^n$ the directional minimization problem*

$$\operatorname*{minimize}_{\tau \in \mathbb{R}} f(x + \tau e^j), \quad j = 1, \dots, n, \tag{5.119}$$

[†]The method of coordinate descent is also called the Gauss–Seidel method.

has a unique solution. Then every accumulation point x^* of the sequence $\{x^k\}$ generated by the method of coordinate descent satisfies the equation $\nabla f(x^*) = 0$.

Proof. Let us consider the sequence of points x^{Ln+1}, $L = 0, 1, 2, \dots$, initiating each cycle of n coordinate steps. It is included in the set X_1 and thus it has accumulation points. Suppose x^* is its accumulation point. We shall show that $\nabla f(x^*) = 0$. Suppose $\nabla f(x^*) \neq 0$. If we initiated the method of coordinate descent from x^*, it would find in at most n steps a point $y^* = x^* + \sum_{j=1}^{n} \tau_j^* e^j$, which has a smaller value of the objective function: $f(y^*) < f(x^*)$. Indeed, if $f(y^*) = f(x^*)$ then the assumption of uniqueness of the solution to the directional minimization problem (5.115) implies that

$$f(x^*) = \min_{\tau \in \mathbb{R}} f(x^* + \tau e^j), \quad j = 1, \dots, n.$$

Therefore

$$\frac{\partial f(x^*)}{\partial x_j} = 0, \quad j = 1, \dots, n,$$

which contradicts our assumption that $\nabla f(x^*) \neq 0$. So, a better point y^* can be found. Suppose x^{Ln+1} is very close to x^*. The results of minimization in the directions e^i,

$$x^{Ln+j+1} = x^{Ln+j} + \tau_{Ln+j} e^i, \quad j = 1, \dots, n,$$

are, by assumption, unique. Using an argument similar to the proof of Theorem 5.4 we can show that the points x^{Ln+j+1} converge to $x^* + \sum_{i=1}^{j} \tau_i^* e^i$, if $x^{Ln+1} \to x^*$. Consequently, $x^{Ln+n+1} \to y^*$, when $L \to \infty$ over a subsequence for which $x^{Ln+1} \to x^*$. But then $f(x^{Ln+n+1}) \to f(y^*)$. This is a contradiction with the monotonicity of $\{f(x^k)\}$, because $f(x^{Ln+1}) \to f(x^*) > f(y^*)$.

The fact that we have focused on points x^{Ln+1}, $L = 0, 1, 2, \dots$, is irrelevant. For a convergent subsequence $\{x^k\}$, $k \in \mathcal{K}$, we can find an infinite sub-subsequence of points of the form $x^{L_j n+m}$ for at least one m. Now we renumber the coordinates in a cyclical fashion: e^{m+1} becomes e^1, ..., e^n becomes e^{n-m}, e^1 becomes e^{n-m+1}, ...e^m becomes e^n. This brings us to the case discussed before. □

It should be stressed that the continuous differentiability of $f(\cdot)$ is essential for the convergence of the coordinate descent method, as well as other "nongradient" methods discussed in this section. If the function is not differentiable, other techniques, which we present in the next chapter, are applicable.

5.8.3 Conjugate Directions without Derivatives

The fact that we use the unit directions $e^1, \dots, e^n$ in the coordinate descent method is not an essential restriction. We may use any system of directions $d^1, \dots, d^n$, provided that they are linearly independent. In fact, the coordinate descent method is mainly used with the accompanying operation of changing the directions.

In order to see the motivation for such a change, let us consider the quadratic function

$$f(x) = \frac{1}{2}\langle x, Qx\rangle + \langle c, x\rangle. \tag{5.120}$$

with a positive definite matrix Q of dimension n.

Lemma 5.33. *Assume that S is a subspace of $\mathbb{R}^n$ and that $x' \in \mathbb{R}^n$ and $x'' \in \mathbb{R}^n$ are such that $x'' - x' \notin S$. If y' is the minimum of $f(\cdot)$ in the linear manifold*

$$L' = x' + S$$

and y'' is the minimum of $f(\cdot)$ in the manifold

$$L'' = x'' + S,$$

then the vector $y'' - y'$ is Q-conjugate to every vector $d \in S$.

Proof. If y' is the minimum in L' then $-\nabla f(y') \perp S$. Similarly, $\nabla f(y'') \perp S$. This implies that for every direction $d \in S$ one has

$$\langle \nabla f(y'') - \nabla f(y'), d\rangle = 0.$$

Since $\nabla f(y'') - \nabla f(y') = Q(y'' - y')$, the last relation is the required Q-conjugacy property. □

Armed with this observation, we can suggest the following modification of the coordinate descent method. We start from some point x^0 with a system of linearly independent directions $d^1, \dots, d^n$ (they may be the unit vectors, as the Gauss–Seidel method). We then proceed as described in Figure 5.10.

Consider the first cycle of the method (Steps 1 and 2). The point $x^{1,1}$ is obtained by minimizing in the direction d^n, and the point $x^{1,n+1}$, by minimizing in directions $d^1, \dots, d^{n-1}$ and d^n again. By virtue of Lemma 5.33, the direction $x^{1,n+1} - x^{1,1}$ is Q-conjugate to d^n. Hence, after the change of the basis at Step 3, the last two directions, d^{n-1} and d^n, become Q-conjugate. Observe that the point $x^{1,n+1}$ is already the result of the minimization in the first of them, d^{n-1} (which was d^n before the change). In the next cycle

Step 0. Set $k = 1$ and $x^{1,0} = x^0$.

Step 1. Minimize the function $f(\cdot)$ in direction d^n to obtain the point

$$x^{k,1} = x^{k,0} + \tau_{k,0}d^n, \ f(x^{k,1}) = \min_{\tau \in \mathbb{R}} f(x^{k,0} + \tau d^n).$$

Step 2. Minimize the function $f(\cdot)$ successively in the directions $d^1, \dots, d^n$ obtaining for $j = 1, \dots, n$ the points

$$x^{k,j+1} = x^{k,j} + \tau_{k,j}d^j, \ f(x^{k,j+1}) = \min_{\tau \in \mathbb{R}} f(x^{k,j} + \tau d^j).$$

Step 3. Change the system of directions by substituting

$$d^1 := d^2, \ d^2 := d^3, \ d^{n-1} := d^n, \ d^n := x^{k,n+1} - x^{k,1}.$$

If the new directions are linearly independent, go to Step 4. Otherwise substitute for $d^1, \dots, d^n$ a known system of linearly independent directions.

Step 4. Set $x^{k+1,0} := x^{k,n+1}$, increase k by one and go to Step 1.

Figure 5.10. The conjugate direction method without derivatives.

($k = 2$) the point $x^{2,1}$ obtained at Step 1 is thus the result of the directional minimization in *two* Q-conjugate directions. By Theorem 5.16, the point $x^{2,1}$ is the minimum in the manifold

$$L' = x^{1,n} + \text{lin}\{d^{n-1}, d^n\}.$$

After Step 2, the point $x^{2,n+1}$ is the minimum in the manifold

$$L'' = x^{2,n-1} + \text{lin}\{d^{n-1}, d^n\}.$$

By Lemma 5.33, the direction $x^{2,n+1} - x^{2,1}$ is Q-conjugate to both d^{n-1} and d^n. After the change of the system of directions, we have three Q-conjugate directions, d^{n-2}, d^{n-1}, and d^n, and a point $x^{2,n+1}$ obtained by minimizing in the first two of them (at the end of Step 2). In this way, each cycle adds one new Q-conjugate direction. If the linear independence condition is satisfied, after n cycles the system $d^1, \dots, d^n$ has n conjugate directions. It follows from Theorem 5.16 that Step 1 of the next cycle yields the minimum of $f(\cdot)$.

The finite convergence property of the method depends on the linear independence of the directions generated, which cannot be guaranteed, in general. Furthermore, even more so than in the case of the conjugate gradient method, numerical errors in the directional minimization result in the loss of the property of conjugacy of the directions before the limit of n cycles is reached. Therefore, we periodically reset the method by returning to the initial system of directions. With periodic re-initialization, the convergence of the method for continuously differentiable (not necessarily quadratic) functions follows from the properties of the Gauss–Seidel method. Still, the change of the system of directions substantially improves the practical performance of the method, between the re-initialization steps.

5.8.4 Newton's Method Without Hessians

Suppose we want to apply Newton's method to find an unconstrained minimum of a twice differentiable function $f : \mathbb{R}^n \to \mathbb{R}$, but only first order derivatives of $f(\cdot)$ are available. A straightforward way to address this difficulty is to approximate the Hessian $\nabla^2 f(x)$ by employing finite differences in the basic directions $e^1, \dots, e^n$. The jth column of the Hessian can be approximated by forward differences:

$$\big[\nabla^2 f(x)\big]e^j = \left[\frac{\partial^2 f(x)}{\partial x_i \partial x_j}\right]_{i=1,\dots,n} \approx \frac{1}{\delta}\big[\nabla f(x+\delta e^j) - \nabla f(x)\big], \quad j = 1, \dots, n.$$

We can also use symmetric differences, as in (5.115). However, it is a very tedious and error-prone procedure.

We can do much better by observing that our goal is not the exact calculation of the Hessian, but rather an approximate calculation of the Newton direction:

$$d^k = -\big[\nabla^2 f(x^k)\big]^{-1}\nabla f(x^k).$$

This requires the solution of the linear system of equations

$$\big[\nabla^2 f(x^k)\big]d^k = -\nabla f(x^k). \tag{5.121}$$

Write, for simplicity, $Q = \nabla^2 f(x^k)$ and $b = -\nabla f(x^k)$. System (5.121) can now be written as

$$Qd = b,$$

and this is equivalent to finding the stationary point of the quadratic function

$$\varphi(d) = -\langle b, d\rangle + \frac{1}{2}\langle d, Qd\rangle.$$

Assume for a moment that the matrix Q is positive definite. The solution of (5.121), which is also the unconstrained minimum of $\varphi(\cdot)$, can be found by the (pre-conditioned) conjugate gradient method of Section 5.5. Since the function $\varphi(\cdot)$ is quadratic, the algorithm (5.75)–(5.76) can be written as follows. We use j to number the iterations of this method, and we use explicit formulas for the gradient of $\varphi(\cdot)$, $g_j = Qd_j - b$, for its Hessian, and for the optimal step size τ_j. The direction of change is now denoted by s_j. The method starts from $d_1 = 0$. The pre-conditioner is denoted by V (recall that V is our approximate model of Q^{-1}).

We calculate the first gradient, $g_1 = Qd_1 - b$, and the first direction is $s_1 = -Vg_1$. Then for $j = 1, \dots, n$, if $g_j \neq 0$, we carry out the calculations:

$$\begin{aligned}\tau_j &= \frac{\langle g_j, Vg_j\rangle}{\langle s_j, Qs_j\rangle},\\ d_{j+1} &= d_j + \tau_j s_j,\\ g_{j+1} &= g_j + \tau_j Q s_j,\\ \alpha_{j+1} &= \frac{\langle g_{j+1}, Vg_{j+1}\rangle}{\langle g_j, Vg_j\rangle},\\ s_{j+1} &= -Vg_{j+1} + \alpha_{j+1}s_j.\end{aligned} \tag{5.122}$$

We know that in at most n iterations this method finds the solution to the system $Qd = b$. In order to implement the method, we do not need to know the Hessian Q; we only need to be able to multiply Q by the directions s_j. Here the techniques of numerical differentiation can be employed. For each s_j we approximate Qs_j as follows:

$$Qs_j = \left[\nabla^2 f(x^k)\right]s_j \approx \frac{1}{\delta}\left[\nabla f(x^k + \delta s_j) - \nabla f(x^k)\right], \quad j = 1, \dots, n,$$

with some $\delta > 0$. The method sketched above is called the *conjugate gradient Newton method*.

It appears that n finite difference approximations are needed, as in the straightforward approach mentioned at the beginning of this subsection. However, we know from Section 5.5.2 that the conjugate gradient method can find a very good approximation to a minimum of a quadratic function in much fewer than n iterations. Therefore, we test at every iteration of algorithm (5.122) whether

$$\|g_j\| \leq \eta \|g_1\|$$

for some small $\eta \in (0, 1)$, and we terminate the iteration, if this condition is satisfied. By Theorem 5.16, the point d_j is the minimum of $\varphi(\cdot)$ in the subspace

$$L_{j-1} = \mathrm{lin}\{Vg_1, Vg_2, \ldots, Vg_{j-1}\}. \tag{5.123}$$

This implies that the direction $d_j - 0$ is the direction of descent for $\varphi(\cdot)$ at 0, that is,

$$\langle \nabla\varphi(0), d_j - 0\rangle = \langle \nabla f(x^k), d_j\rangle < 0.$$

We can thus make a positive step in the direction d_j at x^k and improve the value of the function $f(\cdot)$. Because of the possibility of terminating the conjugate gradient loop, the method is also referred to as the *truncated Newton method.*

The second reason for terminating the conjugate gradient iteration (5.122) is the possible indefiniteness of the Hessian Q. Theoretically, the conjugate gradient method is a method for solving linear systems with positive definite matrices, but we can formally apply algorithm (5.122) to every Q, provided that appropriate safeguarding is employed.

The key issue in ensuring that algorithm (5.122) generates useful search directions is the test of nonpositive curvature of Q in the current direction s_j:

$$\langle s_j, Qs_j\rangle \le 0.$$

If this occurs at the first iteration ($j = 1$), then we abandon algorithm (5.122) and we use the direction of steepest descent at x^k, instead of Newton's direction. If negative curvature is discovered at iteration $j > 1$, then we terminate algorithm (5.122) and we use d_j as the search direction for $f(\cdot)$ at x^k. It remains to verify that d_j is a direction of descent. As s_j is the first direction of nonpositive curvature of Q, the subspace L_{j-1} defined in (5.123) contains only directions of positive curvature. The quadratic form $\langle d, Qd\rangle$ restricted to this subspace is positive definite, and the function $\varphi(\cdot)$ is strictly convex on L_{j-1}. It follows that d_j is the minimum of $\varphi(\cdot)$ in L_{j-1}, and d_j is indeed a direction of descent for $f(\cdot)$ at x^k.

EXERCISES

5.1. The function $f : \mathbb{R}^2 \to \mathbb{R}$ is defined as follows:

$$f(x_1, x_2) = \begin{cases} 5\sqrt{9(x_1)^2 + 16(x_2)^2} & \text{if } x_1 > |x_2|, \\ 9x_1 + 16x_2 & \text{if } x_1 \le |x_2|. \end{cases}$$

Assume that the initial point x^0 satisfies the conditions

$$x_1^0 \ge |x_2^0| \ge \frac{9}{16}x_1^0.$$

Prove that the method of steepest descent with exact directional minimization starting from the point x^0 generates a sequence of points $\{x^k\}$ convergent to the point $(0, 0)$. Is it a minimum point of the function? Explain the failure of the method. *Hint*: Prove that at every iteration k we have the relations

$$x_1^k \geq |x_2^k| \geq \frac{9}{16} x_1^k.$$

5.2. Show that the Newton method is invariant under a nonsingular linear transformation $x = Uy$. Specifically, if we set $g(y) = f(Uy)$ for a twice continuously differentiable function $f : \mathbb{R}^n \to \mathbb{R}$, then the sequence $\{x^k\}$ of points generated by the Newton method for $f(\cdot)$, starting from $x^0 = Uy^0$, can be transformed to the sequence of Newton iterates $\{y^k\}$ for $g(\cdot)$.

5.3. Consider Newton's method with constant step size $\tau_k = 1$ applied to the function $f(x) = \|x\|^3$. Show that it is convergent linearly to the minimum $\hat{x} = 0$. Why does quadratic convergence not occur?

5.4. Let Q be a positive definite matrix of dimension n and let $d^1, \ldots, d^n$ be Q-conjugate. Consider the matrix $D = \left[d^1 \; d^2 \; \ldots \; d^n\right]$ and the diagonal matrix Λ with entries $\lambda_i = \langle d^i, Qd^i \rangle$. Prove that

$$Q^{-1} = D\Lambda^{-1}D^T.$$

5.5. Suppose a positive definite matrix Q of dimension n is expressed as $Q = D\Lambda D^T$, with some $n \times n$ matrix D and a diagonal matrix Λ.

(a) Show that the columns of $M = (D^T)^{-1}$ are Q-conjugate and that $M^T QM = \Lambda$.

(b) Show that the columns of D are Q^{-1}-conjugate.

(c) Discuss the implications of these equations for the Cholesky factorization $Q = L\Lambda L^T$ with a lower triangular L.

5.6. Starting from the point $x^0 = (2, 2)$ minimize the function

$$f(x_1, x_2) = (x_1)^2 + 2(x_2)^2 - 2x_1x_2 - 2x_2 + 2x_1$$

by the following methods:

(a) conjugate gradient method,

(b) DFP method,

(c) BFGS method.

Verify the conjugacy of the directions.

5.7. The quadratic matrix Q of dimension n has the form

$$Q = I + aa^T,$$

with some vector $a \in \mathbb{R}^n$. How many iterations of the conjugate gradient method are needed to find the minimum of the function (5.52)? Generalize this observation to the case when

$$Q = I + \sum_{j=1}^{s} a_j a_j^T$$

with $s < n$ and with arbitrary vectors $a_1, \dots, a_s$ in $\mathbb{R}^n$.

5.8. The quadratic matrix Q of dimension n has the form

$$Q = D + \sum_{j=1}^{s} a_j a_j^T$$

with some positive definite matrix D and with arbitrary vectors $a_1, \dots, a_s$ in $\mathbb{R}^n$, where $s < n$. How many iterations of the pre-conditioned conjugate gradient method with the preconditioner $V = D^{-1}$ are needed to find the minimum of the function (5.52)?

5.9. Suppose the matrix Q of dimension n has all its eigenvalues concentrated in two intervals: $[a, b]$ and $[a + h, b + h]$, with an arbitrary $h > 0$. Prove that after two iterations of the conjugate gradient method for the function (5.52) we have

$$f(x^3) - f(\hat{x}) \le \frac{b-a}{b+a}\big[f(x^1) - f(\hat{x})\big].$$

5.10. Using the identity:

$$[A + ab^T]^{-1} = A^{-1} - \frac{A^{-1}ab^T A^{-1}}{1 + b^T A^{-1} a}$$

derive the BFGS formula (5.92) as the inverse of the dual DFP formula (5.91), under the condition that $V_k W_k = I$. Check the equality $V_{k+1} W_{k+1} = I$ by direct multiplication.

5.11. Prove that algorithm (5.100)–(5.101) calculates the BFGS direction. Use induction on k.

5.12. Prove that the trust region subproblems (5.106) and (5.111) are equivalent.

5.13. Assume that problem (5.112) has a solution λ_k such that the matrix $W_k + \lambda_k I$ is positive definite. Develop specialized versions of the method of bi-section, the method of secants, and Newton's method for this problem.

5.14. Starting from the point $x^0 = (2, 2)$ minimize the function

$$f(x_1, x_2) = (x_1)^2 + 2(x_2)^2 - 2x_1x_2 - 2x_2 + 2x_1$$

by the following methods:

(a) coordinate descent method (first four steps),

(b) conjugate direction method without derivatives.

Verify the conjugacy of directions in (b).

5.15. Consider the problem of minimizing the function $f : \mathbb{R}^2 \to \mathbb{R}$ defined as follows:

$$f(x_1, x_2) = x_1 + x_2 + \max\big(0, (x_1)^2 + (x_2)^2 - 4\big).$$

Suppose we use the coordinate descent (Gauss–Seidel) method. Analyze the operation of the method for several starting points: $x^0 = (0, 0)$, $x^0 = (0, 1)$, $x^0 = (0, 2)$. Explain the reasons for the unreliability of the method.

Chapter Six

Constrained Optimization of Differentiable Functions

6.1 FEASIBLE POINT METHODS

6.1.1 The Projection Method

We focus at first on the set-constrained problem,

$$\underset{x \in X}{\text{minimize}}\ f(x), \tag{6.1}$$

with a continuously differentiable function $f : \mathbb{R}^n \to \mathbb{R}$, and with a convex closed set $X \subset \mathbb{R}^n$. The simplest idea of a descent method for solving (6.1) is to make steps in the direction of steepest descent, and to return to the feasible set, when the result is outside of X. We thus consider the following iterative process:

$$x^{k+1} = \Pi_X(x^k - \tau \nabla f(x^k)), \quad k = 1, 2, \ldots. \tag{6.2}$$

Here $\Pi_X(\cdot)$ is the operation of the orthogonal projection on the set X analyzed in Section 2.1.2, and τ is a positive step size. The method is a direct extension of the method of steepest descent with constant step sizes, and we have a direct analog of Theorem 5.1.

THEOREM 6.1. *Assume that the function $f(\cdot)$ is continuously differentiable and its gradient is Lipschitz continuous with some constant M:*

$$\|\nabla f(x) - \nabla f(y)\| \le M\|x - y\| \quad \textit{for all} \quad x, y \in \mathbb{R}^n.$$

Furthermore, assume that the set $\{x \in X : f(x) \le f(x^1)\}$ is bounded. If the step size τ satisfies the inequalities

$$0 < \tau < \frac{1}{M}, \tag{6.3}$$

then every accumulation point x^* of the sequence $\{x^k\}$ generated by algorithm (6.2) satisfies the necessary condition of optimality:

$$-\nabla f(x^*) \in N_X(x^*). \tag{6.4}$$

Proof. From the differentiability of $f(\cdot)$ we obtain

$$
\begin{aligned}
f(x^{k+1}) &= f(x^k) + \int_0^1 \left\langle \nabla f(x^k + \theta(x^{k+1} - x^k)), x^{k+1} - x^k \right\rangle d\theta \\
&= f(x^k) + \left\langle \nabla f(x^k), x^{k+1} - x^k \right\rangle \\
&\quad + \int_0^1 \left\langle \nabla f(x^k + \theta(x^{k+1} - x^k)) - \nabla f(x^k), x^{k+1} - x^k \right\rangle d\theta \\
&\le f(x^k) + \left\langle \nabla f(x^k), x^{k+1} - x^k \right\rangle + M \int_0^1 \theta \|x^{k+1} - x^k\|^2 \, d\theta \\
&\le f(x^k) + \left\langle \nabla f(x^k), x^{k+1} - x^k \right\rangle + \frac{M}{2} \|x^{k+1} - x^k\|^2. \qquad (6.5)
\end{aligned}
$$

In view of (6.2), Lemma 2.11 yields

$$\langle x^k - x^{k+1}, x^k - \tau \nabla f(x^k) - x^{k+1} \rangle \le 0.$$

Hence

$$\langle \nabla f(x^k), x^{k+1} - x^k \rangle \le -\frac{1}{\tau} \|x^{k+1} - x^k\|^2.$$

Substituting this estimate into (6.5) we obtain

$$f(x^{k+1}) \le f(x^k) - \left(\frac{1}{\tau} - \frac{M}{2}\right) \|x^{k+1} - x^k\|^2, \quad k = 1, 2, \ldots. \qquad (6.6)$$

Assumption (6.3) implies that $f(x^{k+1}) \le f(x^k)$ for all k. Since the sequence $\{x^k\}$ is bounded, the sequence $\{f(x^k)\}$ is convergent. Then it follows from (6.6) that $\|x^{k+1} - x^k\|^2 \to 0$, as $k \to \infty$. This can be rewritten as

$$\lim_{k \to \infty} \left\| \Pi_X(x^k - \tau \nabla f(x^k)) - x^k \right\|^2 = 0. \qquad (6.7)$$

Suppose x^* is an accumulation point of the sequence $\{x^k\}$. The projection operator $\Pi_X(\cdot)$ is continuous, because it is nonexpansive (Theorem 2.13). Passing to the limit in (6.7) over a subsequence for which $x^k \to x^*$, we conclude that

$$\left\| \Pi_X(x^* - \tau \nabla f(x^*)) - x^* \right\|^2 = 0.$$

Thus $x^* = \Pi_X(x^* - \tau \nabla f(x^*))$. By virtue of Lemma 2.11,

$$\langle -\nabla f(x^*), d \rangle \le 0 \quad \text{for every} \quad d \in K_X(x^*).$$

Therefore $-\nabla f(x^*) \in \left[K_X(x^*)\right]^\circ$, which is the required optimality condition. □

Three difficulties are associated with the projection method (6.2). First, the projection operation requires solving an auxiliary optimization problem

$$\underset{x \in X}{\text{minimize}} \ \|x - x^k + \tau \nabla f(x^k)\|^2,$$

which may be hard, unless the set X has some special structure, like a box or a ball. Second, no easy line search is possible in this method, and thus very small step size τ has to be used to ensure convergence. Finally, similarly to the method of steepest descent, the method is very slow for ill-conditioned problems.

For these reasons, the projection method is very rarely used in practice in its pure form. However, it appears as a part of other methods, especially multiplier methods, which we discuss in Section 6.4.

6.1.2 The Reduced Gradient Method

Another way to ensure feasibility of the iterates generated by the method is to consider only *feasible directions* as candidates for search directions. This idea is best explained on optimization problems with linear constraints:

$$\begin{aligned} \text{minimize} \ & f(x) \\ \text{subject to} \ & Ax = b, \\ & x \geq 0. \end{aligned} \tag{6.8}$$

Here $f : \mathbb{R}^n \to \mathbb{R}$ is a continuously differentiable function. The matrix A of dimension $m \times n$, and the vector $b \in \mathbb{R}^m$ are given. We also assume that the rank of A is m.

We choose the above *standard form* of linear constraints only for ease of presentation. Our considerations can be easily extended to the case when inequality constraints and bounds on the vector x are present.

The reduced gradient method extends to the case of the nonlinear problem (6.8) the techniques of the simplex method in linear programming, and readers familiar with linear programming will easily recognize many concepts in our presentation.

Our idea is to generate a sequence of feasible points $\{x^k\}$ by moving within the facets of the feasible polyhedron of (6.8). At the current point x^k we split the decision vector x into three subvectors:

- *nonbasic* variables x_N, which are assumed to be fixed at 0;

- *superbasic* variables x_S, which are nonnegative, and are considered as independent variables; and
- *basic* variables x_B, whose values are calculated to satisfy the equation $Ax = b$.

The partition of x into these three subvectors corresponds to the division of A into three submatrices: N, S, and B. The choice of the subvector of basic variables should be such that the matrix B is square and has rank m.

Let us introduce three index sets: I_B, I_S, and I_N, containing the indices of the basic, superbasic, and nonbasic variables, correspondingly.

After rearranging the components of x and columns of A in such a way that the components (columns) corresponding to I_B come first, those corresponding to I_S after them, and the components and columns associated with I_N last, we may write:

$$x = \begin{bmatrix} x_B \\ x_S \\ x_N \end{bmatrix}, \qquad A = \begin{bmatrix} B & S & N \end{bmatrix}.$$

The equality constraints take on the form:

$$Bx_B + Sx_S + Nx_N = b.$$

By fixing the nonbasic variables at zero, we choose a facet of the feasible set of problem (6.8) defined by the relations:

$$x_B = B^{-1}(b - Sx_S), \tag{6.9}$$

$$x_B \geq 0, \tag{6.10}$$

$$x_S \geq 0, \tag{6.11}$$

$$x_N = 0. \tag{6.12}$$

We shall try to minimize $f(\cdot)$ within this facet, by treating x_S as independent variables, and x_B as dependent variables determined via (6.9). Using the convention that

$$f(x) = f(x_B, x_S, x_N)$$

we can represent the objective function within this facet as

$$\varphi(x_S) = f\big(B^{-1}(b - Sx_S), x_S, 0\big). \tag{6.13}$$

Let us denote by $\nabla_{x_B} f(x)$, $\nabla_{x_S} f(x)$, and $\nabla_{x_N} f(x)$ the three subvectors of the gradient of $f(\cdot)$ corresponding to the partition of the vector x. The gradient of $\varphi(\cdot)$ can now be calculated by chain rules of multivariate calculus:

$$\nabla\varphi(x_S) = \nabla_{x_S} f(x) - S^T\big[B^{-1}\big]^T \nabla_{x_B} f(x). \tag{6.14}$$

We do not need to evaluate the inverse of the matrix B to calculate the gradient. We just need to solve the system of equations

$$B^T\pi = \nabla_{x_B} f(x), \tag{6.15}$$

and we set

$$\nabla\varphi(x_S) = \nabla_{x_S} f(x) - S^T\pi. \tag{6.16}$$

The vector $\nabla\varphi(x_S)$ is called the *reduced gradient* of the objective function.

Since we have an easy way of calculating the gradient of $\varphi(\cdot)$, we can apply an efficient unconstrained optimization method to the problem

$$\text{minimize } \varphi(x_S). \tag{6.17}$$

To be specific, let us assume that it is the conjugate gradient method. Of course, the minimization of $f(\cdot)$ over the facet defined by (6.9)–(6.12) is not equivalent to problem (6.17), because we have ignored inequality constraints (6.10)–(6.11). Therefore, we introduce to the conjugate gradient method a modification within the step size selection rule.

Given a direction of descent d_S^k for $\varphi(\cdot)$ in the space of superbasic variables, we can calculate the direction of change of the basic variables:

$$d_B^k = -B^{-1}Sd_S^k. \tag{6.18}$$

The nonbasic variables remain fixed at $x_N^k = 0$, and thus the resulting direction of change of the vector x is

$$d^k = \begin{bmatrix} d_B^k \\ d_S^k \\ 0 \end{bmatrix}.$$

The step size value τ_k is the solution to the problem

$$\begin{aligned} \text{minimize } & f(x_B^k + \tau d_B^k, x_S^k + \tau d_S^k, x_N^k) \\ \text{subject to } & x_B^k + \tau d_B^k \geq 0, \\ & x_S^k + \tau d_S^k \geq 0, \\ & \tau \geq 0. \end{aligned} \tag{6.19}$$

Consider the point $x^{k+1} = x^k + \tau_k d^k$ obtained. If

$$\langle \nabla f(x^{k+1}), d^k \rangle = 0,$$

the point x^{k+1} is the same as in an unconstrained version of (6.19). In this case, we can continue the operation of the conjugate gradient method in the subspace of the superbasic variables x_S.

If some of the bounds in (6.19) are active, that is

$$\langle \nabla f(x^{k+1}), d^k \rangle < 0,$$

we cannot continue with the unconstrained optimization algorithm in the same subspace. We then select a basic or superbasic variable x_r that hit its bound. It can be identified by the conditions

$$x_r^{k+1} = 0 \quad \text{and} \quad d_r^k < 0. \tag{6.20}$$

The variable x_r is reclassified to the set of nonbasic variables. If $r \in I_S$ then we simply set

$$I_S^{\text{new}} := I_S \setminus \{r\}, \qquad I_N^{\text{new}} := I_N \cup \{r\},$$

and we continue. If x_r is a basic variable, we move the index r from the set I_B to the set I_N of nonbasic variables. In this case, however, we also need to augment the set of basic variables, in order to keep the basis matrix B square and nonsingular. To this end we choose among the superbasic variables a variable x_e to be reclassified to the set of basic variables, and we set

$$I_B^{\text{new}} := I_B \setminus \{r\} \cup \{e\}, \qquad I_S^{\text{new}} := I_S \setminus \{e\}, \qquad I_N^{\text{new}} := I_N \cup \{r\}.$$

The new basis matrix will have columns $\{a^i : i \in I_B^{\text{new}}\}$. We must ensure that it is nonsingular.

LEMMA 6.2. *Suppose* (6.20) *holds true for a basic variable x_r. Then we can find a superbasic variable x_e such that the columns a^i, $i \in I_B^{\text{new}}$, are linearly independent.*

Proof. Let us assume that r is the first index in I_B. The basis matrix has the form

$$B = \begin{bmatrix} b^1 & b^2 & \dots & b^m \end{bmatrix}.$$

Suppose it is impossible to find a column a of S to replace b^1 and to make the new matrix,

$$B^{\text{new}} = \begin{bmatrix} a & b^2 & \dots & b^m \end{bmatrix},$$

nonsingular. This means that the matrix

$$\begin{bmatrix} b^2 & \dots & b^m \mid S \end{bmatrix}$$

is singular. Hence there exists a nonzero vector $z \in \mathbb{R}^m$ such that

$$z^T \begin{bmatrix} b^2 & \dots & b^m \mid S \end{bmatrix} = 0. \tag{6.21}$$

By (6.18),

$$Bd_B + Sd_S = 0.$$

Multiplying from the left by z^T and expanding into coordinates of d_B we obtain

$$(z^T b^1)d_{B1} + \sum_{i=2}^{m}(z^T b^i)d_{Bi} + (z^T S)d_S = 0.$$

By (6.21), all components of the sum above, but the first one, are zero. Then also

$$z^T b^1 d_{B1} = 0.$$

It follows from (6.20) that d_{B1} is nonzero, and thus $z^T b^1 = 0$. Hence $z^T B = 0$, which contradicts the nonsingularity of the matrix B. Consequently, (6.21) cannot hold true for a nonzero z. □

It follows that it is possible to introduce into the basis matrix a column of S and keep the basis matrix nonsingular. The choice of a particular column, and testing whether its introduction yields a nonsingular basis matrix, can be done by specialized techniques of numerical linear algebra, which are well established in linear programming. Usually, the current matrix B is maintained in a *factorized* form,

$$B = LU,$$

with some lower triangular matrix L and an upper triangular matrix U. An exchange of a column of B is equivalent to an exchange of the corresponding column of U. Then, by dedicated refactorization techniques, U is brought back to an upper triangular form (which changes L as well). If this process is successful and yields nonsingular factors

$$B^{\text{new}} = L^{\text{new}} U^{\text{new}},$$

the update is completed. Otherwise, another column of S has to be tried. Lemma 6.2 guarantees that a successful update will eventually occur. We refer the reader to the linear programming literature for the details on factorization and refactorization techniques.

Returning to the reduced gradient method, we notice that every encounter with a nonnegativity bound for basic or superbasic variables results in an increase of the cardinality of the set of nonbasic variables and a decrease of the cardinality of the set of superbasic variables. After such a change, the conjugate gradient method has to be re-initialized in the new facet of the feasible set.

Since the cardinality of the set of superbasic variables can be decreased only finitely many times, after finitely many iterations it remains unchanged (possibly equal to zero). As the feasible set is compact, every accumulation point of the sequence generated by the conjugate gradient method with re-initialization satisfies the necessary condition of a local minimum of problem (6.17), associated with the last classification into basic, superbasic, and non-basic variables. In practical computations an approximation to such a point will be found, but we are not going to introduce these technical details into our analysis. The mechanism of the method is best explained if we assume that a point $\hat{x}_S$ is found such that

$$\nabla\varphi(\hat{x}_S) = 0. \tag{6.22}$$

The corresponding values of basic variables are

$$\hat{x}_B = B^{-1}(b - S\hat{x}_S),$$

and the nonbasic variables are $\hat{x}_N = 0$. If the last partition has no superbasic variables at all, we set $\hat{x}_B = B^{-1}b$. We shall call the point $\hat{x} = (\hat{x}_B, \hat{x}_S, \hat{x}_N)$ a *semi-stationary point*.

The question to be addressed now is whether a semi-stationary point satisfies optimality conditions of a local minimum of problem (6.8). It can be answered by analyzing the vector

$$\bar{g}_N = \nabla_{x_N} f(\hat{x}) - N^T\hat{\pi},$$

with $\hat{\pi}$ calculated by (6.15) at $\hat{x}$:

$$B^T\hat{\pi} = \nabla_{x_B} f(\hat{x}). \tag{6.23}$$

Problem (6.8) is a special case of problem (3.55) from page 133, with no inequality constraints, $h(x) = b - Ax$, and with $X_0 = \mathbb{R}^n_+$ and $Y_0 = \{0\}$. The necessary conditions of optimality for this problem are provided in Theorem 3.25, and we refer to these conditions in the lemma below.

LEMMA 6.3. *If $\bar{g}_N \geq 0$, then the point $\hat{x} = (\hat{x}_B, \hat{x}_S, \hat{x}_N)$ satisfies the necessary conditions of optimality for problem* (6.8).

Proof. The necessary condition (3.30) for problem (6.8) takes on the form

$$0 \in \nabla f(\hat{x}) - A^T\mu + N_{X_0}(\hat{x}).$$

Due to the the explicit form of X_0, the optimality conditions simplify:

$$\begin{aligned} &\nabla f(\hat{x}) - A^T\mu \geq 0, \\ &\langle \hat{x}, \nabla f(\hat{x}) - A^T\mu \rangle = 0. \end{aligned} \tag{6.24}$$

We can now verify that these conditions hold true with $\mu = \hat{\pi}$. Equation (6.23) implies (6.24) for the basic variables. As $\nabla\varphi(\hat{x}_N) = 0$, formula (6.16) yields (6.24) for the superbasic variables. Finally, the assumption that $\bar{g}_N \geq 0$ is equivalent to (6.24) for the nonbasic variables. The complementarity condition holds true as well, because the nonbasic variables are zero. □

If the condition $\bar{g}_N \geq 0$ is not satisfied, the semi-stationary point does not have to satisfy the necessary conditions of optimality.

Let us first analyze the case when the values of all basic variables are strictly positive. After finding the semi-stationary point $\hat{x}$, we enlarge the facet to allow further progress. We define a nonempty set of "promising" nonbasic variables

$$E_N \subset \{i \in I_N : \bar{g}_{Ni} < 0\}$$

and we set

$$I_S^{\text{new}} := I_S \cup E_N, \qquad I_N^{\text{new}} := I_N \setminus E_N.$$

Consider the direction of steepest descent in the space of superbasic variables at $\hat{x}$:

$$d_S = -\nabla\varphi(\hat{x}_S).$$

Since $\hat{x}$ was a semi-stationary point, the components of d_S corresponding to $i \in I_S$ (all previous superbasic variables) are zero, and the components corresponding to $i \in E_N$ (the newly introduced variables) are positive. As the basic variables are strictly positive, the direction d_S is feasible. Since it is nonzero, the cone of feasible directions of descent is nonempty, and the necessary conditions of optimality are not satisfied. A small step in the direction d_S will decrease the value of the objective function, without violating any constraints.

If some of the basic variables are equal to 0, and the corresponding coordinates of the direction of change of the basic variables,

$$d_B = -B^{-1}Sd_S,$$

are *negative*, then no positive step in the direction $d = (d_B, d_S, 0)$ can be made. An iteration in this case will only result in a reclassification of the variables, without any changes in their values. Such a situation is called *degenerate* and is well understood in the theory of linear programming. Actually, when no changes in the values of the variables occur, the situation is the same as if the problem was linear, with the objective function

$$\tilde{f}(x) = f(\hat{x}) + \langle \nabla f(\hat{x}), x - \hat{x} \rangle.$$

In theory, when degeneracy occurs and many ways to reclassify the variables are possible, careless reclassification may lead to cycling. Fortunately, there exist specialized linear programming techniques guaranteeing that the sequence of degenerate steps at $\hat{x}$ is finite, and either a positive step can be made, or optimality conditions are satisfied. The reader is referred to the literature on linear programming for details of these procedures.

We are now ready to prove convergence of the reduced gradient method. As every minimization within a facet ends with a certain point satisfying condition (6.22), we focus on the sequence $\hat{x}^1, \hat{x}^2, \ldots$ of semi-stationary points. Although in practical computations we cannot find any semi-stationary point exactly, it is more instructive to analyze the method under such an idealized assumption.

THEOREM 6.4. *Assume that the function $f(\cdot)$ is convex and the feasible set of problem* (6.8) *is compact. Then the sequence of semi-stationary points generated by the reduced gradient method is finite, and its last element is an optimal solution of problem* (6.8).

Proof. Every semi-stationary point $\hat{x}^k$ corresponds to a certain classification of the index set $\{1, \ldots, n\}$ into I_B^k, I_S^k, and I_N^k. As the function $f(\cdot)$ is convex, the point $\hat{x}^k$ is a minimum point of $f(\cdot)$ within the facet defined by (6.9)–(6.12). If $\hat{x}^k$ does not satisfy the optimality condition of Lemma 6.3, an iteration of the method of steepest descent with respect to the superbasic variables (after some degenerate steps, possibly), results in a decrease of the value of the objective function. Thus $f(\hat{x}^{m+1}) < f(\hat{x}^m), m = 1, 2, \ldots$. The method cannot return to a facet in which a semi-stationary point has already been found. Since the number of facets is finite, the method must stop at an optimal point. □

We can apply the reduced gradient method to linearly constrained problems where $f(\cdot)$ is smooth but not necessarily convex. If the method stops at a semi-stationary point satisfying the condition of Lemma 6.3, we know that it is a stationary point of problem (6.8). However, if the method generates an infinite sequence of semi-stationary points $\{\hat{x}^m\}$, we cannot guarantee that its accumulation point, x^*, will be stationary. The difficulty is in the specific way in which a direction of descent is generated at a semi-stationary point. It depends on the current classification into basic, superbasic, and nonbasic variables. Although at x^* a good classification guaranteeing a decrease of the value of $f(\cdot)$ may exist, we cannot claim that the same classification is used at points $\hat{x}^m$ close to x^*. If this were true, we would be able to obtain a contradiction, as values lower than $f(x^*)$ would be obtained after a re-start from $\hat{x}^m$ for sufficiently large m. Unfortunately, it is theoretically possible that even very close points $\hat{x}^m$ use a different classification than the one that is good at x^*, and very short steps occur. In practice, such behavior has not been recorded.

Theoretically, we can modify the reduced gradient method to guarantee convergence to stationary points of (6.8). We employ at each semi-stationary point one iteration of a very robust but extremely inefficient method: the *feasible direction method*. It linearizes the objective function at $\hat{x}$ and constructs the linear programming problem:

$$\begin{aligned} &\text{maximize } \sigma \\ &\text{subject to } \langle \nabla f(\hat{x}), d \rangle \leq -\sigma, \\ &\qquad\qquad Ad = 0, \\ &\qquad\qquad \hat{x}_j + d_j \geq \sigma, \quad j = 1, \dots, n. \end{aligned} \tag{6.25}$$

The decision variables in this problem are the direction $d \in \mathbb{R}^n$ and the "buffer size" $\sigma \in \mathbb{R}$. Under the condition that a point $x^0 > 0$ satisfying the constraints exists, one can prove that $\hat{\sigma} = 0$ is an optimal solution of this problem if and only if the point $\hat{x}$ is a stationary point of problem (6.8) (see Exercise 6.3). If the optimal solution has $\hat{\sigma} > 0$, the direction $\hat{d}$ is a feasible direction of descent at $\hat{x}$. Most importantly, problem (6.25) is stable in the following sense. If a sequence $\{\hat{x}^m\}$ is convergent to x^*, then every accumulation point of the sequence of solutions $(\hat{\sigma}_m, \hat{d}^m)$ of problem (6.25) (constructed at $\hat{x}^m$) is a solution of this problem constructed at x^*. This can be easily proved by contradiction, by assuming that a better solution at x^* exists. Then, after a minimal adjustment, it is feasible for the problem at $\hat{x}^m$, and is better than $(\hat{\sigma}_m, \hat{d}^m)$. We leave the details of this argument to the reader. The stability property precludes the case of x^* being nonstationary.

A re-start from one iteration of the feasible direction method will make all variables positive, and thus only basic and superbasic variables will be present. After that, many iterations will be needed to generate the classification for which the next semi-stationary point will be found. All this is very time-consuming and complicated, but it guarantees convergence. There is no real need to employ (6.25) at all semi-stationary points, except in some rare situations of jamming in a nearly degenerate solution.

Feasible direction methods, based on the idea of problem (6.25), can be applied to problems involving nonlinear inequality constraints:

$$g_i(x) \leq 0, \quad i = 1, \dots, m.$$

At *each* iteration of the method we construct a linearized problem similar to (6.25), but also with linearized inequality constraints:

$$g_i(\hat{x}) + \langle \nabla g_i(\hat{x}), d \rangle \leq -\sigma, \quad i = 1, \dots, m.$$

Such methods are very stable, but very slow, and have mainly historical importance.

Finally, we may add that the reduced gradient method can be easily adapted to a slightly more general problem formulation:

$$\begin{aligned} \text{minimize} \quad & f(x) \\ \text{subject to} \quad & Ax = b, \\ & l_j \le x_j \le u_j, \quad j \in J_1, \\ & l_j \le x_j, \quad j \in J_2, \\ & x_j \le u_j, \quad j \in J_3, \end{aligned}$$

with lower and upper bound vectors $l \in \mathbb{R}^n$ and $u \in \mathbb{R}^n$, and disjoint sets of indices J_1, J_2 and J_3. The coordinates x_j for $j \in \{1, \dots, n\} \setminus (J_1 \cup J_2 \cup J_3)$ are unrestricted.

The only differences are that the nonbasic variables may be frozen at their lower *or* upper bounds, and that the directional minimization has to take care of all lower and upper bounds on the variables. The unrestricted variables never become nonbasic.

6.2 PENALTY METHODS

6.2.1 General Ideas

The idea of penalty methods is to approximate a constrained optimization problem by an unconstrained optimization problem or by a problem with simple constraints. Consider the problem

$$\underset{x \in X \cap X_0}{\text{minimize}}\ f(x), \tag{6.26}$$

where $f : \mathbb{R}^n \to \mathbb{R}$, and X and X_0 are closed subsets of $\mathbb{R}^n$. We represent the feasible set as an intersection of two sets to allow treating "easy" constraints $x \in X_0$ directly.

We construct a continuous function $P : \mathbb{R}^n \to \mathbb{R}$ having the following property:

$$\begin{aligned} P(x) = 0, \quad & \text{if} \quad x \in X, \\ P(x) > 0, \quad & \text{if} \quad x \notin X. \end{aligned}$$

A function satisfying these conditions is called a *penalty function*. It can be used to formulate an auxiliary problem with simple constraints:

$$\underset{x \in X_0}{\text{minimize}}\ \left[\Phi_\varrho(x) \triangleq f(x) + \varrho P(x)\right]. \tag{6.27}$$

Here $\varrho > 0$ is called a *penalty parameter*. The idea of problem (6.27) is that the term $\varrho P(x)$, which is added to the objective function, introduces a "penalty" for violating the constraints $x \in X$. We hope that if ϱ is sufficiently large, the solution of (6.27) should be close to a solution of (6.26).

Lemma 6.5. *If problem* (6.27) *for some* $\varrho \geq 0$ *has a solution* x^* *which is an element of the set* X*, then* x^* *is an optimal solution of problem* (6.26).

Proof. As x^* solves (6.27), for every $x \in X_0 \cap X$ we have

$$f(x^*) + \varrho P(x^*) \leq f(x) + \varrho P(x).$$

Since both x^* and x are feasible, $P(x^*) = P(x) = 0$. Hence x^* is optimal. □

In general, we cannot expect to obtain a feasible minimizer of (6.27) for a finite value of the penalty parameter. Usually, we consider a sequence of problems (6.27), where the penalty parameter ϱ is increased to $+\infty$. Still, under fairly general conditions, convergence to a solution of (6.26) occurs.

Theorem 6.6. *Assume that problem* (6.26) *has an optimal solution. Let* $\varrho_k \to \infty$*, as* $k \to \infty$*, and assume that problem* (6.27) *has a solution* x^k *for* $\varrho = \varrho_k$*. Then every accumulation point of the sequence* $\{x^k\}$ *is an optimal solution of problem* (6.26).

Proof. Suppose $\hat{x}$ is an optimal solution of (6.26). As it is feasible for (6.27),

$$f(x^k) + \varrho_k P(x^k) \leq f(\hat{x}) + \varrho P(\hat{x}) = f(\hat{x}), \quad k = 1, 2, \ldots. \tag{6.28}$$

Hence

$$P(x^k) \leq \frac{f(\hat{x}) - f(x^k)}{\varrho_k}.$$

Consider a convergent subsequence $\{x^k\}$, $k \in \mathcal{K}$. Let x^∞ be its limit. Passing to the limit with $k \to \infty$, $k \in \mathcal{K}$, in the last inequality we conclude that $P(x^k) \to 0$, as $k \to \infty$, $k \in \mathcal{K}$. It follows from the continuity of $P(\cdot)$ that $P(x^\infty) = 0$, and thus $x^\infty \in X$.

Furthermore, inequality (6.28) implies that

$$f(x^\infty) = \lim_{\substack{k\to\infty \\ k\in\mathcal{K}}} f(x^k) \leq \lim_{\substack{k\to\infty \\ k\in\mathcal{K}}} \left[f(x^k) + \varrho_k P(x^k) \right] \leq f(\hat{x}).$$

The point x^∞ is an optimal solution of (6.26). □

To apply this theorem, we need additional conditions guaranteeing that the sequence $\{x^k\}$ indeed has accumulation points. The easiest is the condition that the set X_0 is compact. More generally, it is sufficient that a feasible point x^0 exists, such that the set $\{x \in X_0 : f(x) \leq f(x^0)\}$ is bounded.

6.2.2 Quadratic Penalty

We can now apply the general ideas of penalty methods to the nonlinear optimization problem:

$$\begin{aligned} &\text{minimize} \ \ f(x) \\ &\text{subject to} \ \ g_i(x) \le 0, \quad i = 1, \dots, m, \\ &\qquad\qquad\ \ h_i(x) = 0, \quad i = 1, \dots, p, \\ &\qquad\qquad\ \ x \in X_0. \end{aligned} \tag{6.29}$$

All functions $f : \mathbb{R}^n \to \mathbb{R}$, $g_i : \mathbb{R}^n \to \mathbb{R}$, $i = 1, \dots, m$, and $h_i : \mathbb{R}^n \to \mathbb{R}$, $i = 1, \dots, p$, are assumed to be continuously differentiable. The set X_0 is convex and closed.

The following function

$$P_2(x) = \frac{1}{2} \sum_{i=1}^{m} \big[\max\big(0, g_i(x)\big)\big]^2 + \frac{1}{2} \sum_{i=1}^{p} \big[h_i(x)\big]^2$$

is called the *quadratic penalty function* for problem (6.29). For the set X defined by the inequality and equality constraints of problem (6.29), the function $P_2(\cdot)$ satisfies the general conditions of the penalty function. Thus, under the condition that X_0 is compact, Theorem 6.6 can be applied.

The main advantage of the penalty function $P_2(\cdot)$ is that it is continuously differentiable, as a composition of the continuously differentiable functions $g_i(\cdot)$ and $h_i(\cdot)$ with the functions $t \mapsto \big[\max(0, t)\big]^2$ and $t \mapsto t^2$, which are continuously differentiable as well.

Consider the sequence of problems

$$\underset{x \in X_0}{\text{minimize}} \ \big\{\Phi_{\varrho_k}(x) \triangleq f(x) + \varrho_k P_2(x)\big\}, \tag{6.30}$$

with $\varrho_k \to \infty$. The corresponding solutions of problem (6.30) are denoted by x^k. If the sequence $\{x^k\}$ is bounded, then it has accumulation points, and Theorem 6.6 implies that each accumulation point is a solution of problem (6.29). Therefore, to focus attention, we can assume that the entire sequence $\{x^k\}$ is convergent to some solution $\hat{x}$ of problem (6.29).

Let us recall Robinson's constraint qualification condition for problem (6.29). It is convenient to introduce the set $I^0(\hat{x})$ of active inequality constraints:

$$I^0(\hat{x}) = \{1 \le i \le m : g_i(\hat{x}) = 0\}.$$

Now we consider (locally about $\hat{x}$) the mapping $\bar{g}(x)$ with coordinates equal to $g_i(x)$, $i \in I^0(\hat{x})$. Robinson's condition takes on the form:

$$0 \in \operatorname{int}\left\{\begin{bmatrix} \bar{g}'(\hat{x})d + y \\ h'(\hat{x})d \end{bmatrix} : d \in K_{X_0}(\hat{x}),\ y \geq 0\right\}. \tag{6.31}$$

The set $K_{X_0}(\hat{x})$ is the set of feasible directions for X_0 at $\hat{x}$.

We know from Theorem 3.25 that Robinson's condition guarantees the existence of Lagrange multipliers at the point $\hat{x}$: vectors $\hat{\lambda} \in \mathbb{R}^m_+$ and $\hat{\mu} \in \mathbb{R}^m$ such that

$$0 \in \nabla f(\hat{x}) + \sum_{i=1}^{m} \hat{\lambda}_i \nabla g_i(\hat{x}) + \sum_{i=1}^{p} \hat{\mu}_i \nabla h_i(\hat{x}) + N_{X_0}(\hat{x}).$$

Moreover, $\hat{\lambda}_i g_i(\hat{x}) = 0$, $i = 1, \dots, m$. It turns out that the quadratic penalty method can approximate the Lagrange multipliers very well.

Theorem 6.7. *Assume that $\varrho_k \to \infty$, as $k \to \infty$, and that the sequence $\{x^k\}$ of solutions of problems* (6.30) *is convergent to some solution $\hat{x}$ of problem* (6.29). *Furthermore, assume that Robinson's constraint qualification condition is satisfied at $\hat{x}$. Then the sequences*

$$\begin{aligned} \lambda_i^k &= \varrho_k \max\left(0, g_i(x^k)\right), \quad i = 1, \dots, m, \\ \mu_i^k &= \varrho_k h_i(x^k), \quad i = 1, \dots, p, \end{aligned} \tag{6.32}$$

are bounded, and each accumulation point $(\hat{\lambda}, \hat{\mu})$ of the sequence $\{(\lambda^k, \mu^k)\}$ is the vector of Lagrange multipliers satisfying the necessary conditions of optimality at $\hat{x}$.

Proof. The optimality conditions for problem (6.30) at x^k have the form

$$0 = \nabla \Phi_{\varrho_k}(x^k) + v^k$$

with some $v^k \in N_{X_0}(x^k)$. Differentiating the function $\Phi_{\varrho_k}(\cdot)$ and using the definition of λ_i^k and μ_i^k we obtain

$$\begin{aligned} 0 &= \nabla f(x^k) + \varrho_k \sum_{i=1}^{m} \left[\max\left(0, g_i(x^k)\right)\right] \nabla g_i(x^k) && (6.33) \\ &\quad + \varrho_k \sum_{i=1}^{p} h_i(x) \nabla h_i(x^k) + v^k \\ &= \nabla f(x^k) + \sum_{i=1}^{m} \lambda_i^k \nabla g_i(x^k) + \sum_{i=1}^{p} \mu_i^k \nabla h_i(x^k) + v^k. && (6.34) \end{aligned}$$

Suppose the sequence $\{(\lambda^k, \mu^k)\}$ is unbounded. Then we can select an infinite subsequence $\mathcal{K}$ such that $\|(\lambda^k, \mu^k)\| \to \infty$, as $k \to \infty$, $k \in \mathcal{K}$. Consider the sequence

$$(\bar{\lambda}^k, \bar{\mu}^k) = \frac{(\lambda^k, \mu^k)}{\|(\lambda^k, \mu^k)\|}, \quad k \in \mathcal{K}.$$

As it is bounded by construction, it has a convergent sub-subsequence. Denote its limit by $(\bar{\lambda}, \bar{\mu})$. It has norm 1. Let us divide both sides of (6.34) by $\|(\lambda^k, \mu^k)\|$:

$$0 = \frac{\nabla f(x^k)}{\|(\lambda^k, \mu^k)\|} + \sum_{i=1}^{m} \bar{\lambda}_i^k \nabla g_i(x^k) + \sum_{i=1}^{p} \bar{\mu}_i^k \nabla h_i(x^k) + \bar{\upsilon}^k, \tag{6.35}$$

with

$$\bar{\upsilon}^k = \frac{\upsilon^k}{\|(\lambda^k, \mu^k)\|} \in N_{X_0}(x^k).$$

The vectors $\bar{\upsilon}^k$ have some limit $\bar{\upsilon}$, when $k \to \infty$ over the sub-subsequence for which everything else has a limit in (6.35). By Lemma 2.42 on page 38, $\bar{\upsilon} \in N_{X_0}(\hat{x})$. By passing to the limit over the convergent sub-subsequence we obtain from (6.35) the relation

$$0 = \sum_{i=1}^{m} \bar{\lambda}_i \nabla g_i(\hat{x}) + \sum_{i=1}^{p} \bar{\mu}_i \nabla h_i(\hat{x}) + \bar{\upsilon}.$$

Observe that $g_i(x^k) < 0$ for $i \notin I^0(\hat{x})$ and for all sufficiently large k, because $g_i(x^k) \to g_i(\hat{x})$. Hence $\bar{\lambda}_i^k = 0$ for these i and we can rewrite the last relation as

$$0 = \sum_{i \in I^0(\hat{x})} \bar{\lambda}_i \nabla g_i(\hat{x}) + \sum_{i=1}^{p} \bar{\mu}_i \nabla h_i(\hat{x}) + \bar{\upsilon}. \tag{6.36}$$

On the other hand, it follows from Robinson's condition that there exists $d \in K_{X_0}(\hat{x})$ such that

$$\begin{aligned} \langle \nabla g_i(\hat{x}), d \rangle &< 0, \quad i \in I^0(\hat{x}), \\ \langle \nabla h_i(\hat{x}), d \rangle &< 0, \quad \text{if} \quad \bar{\mu}_i > 0, \\ \langle \nabla h_i(\hat{x}), d \rangle &> 0, \quad \text{if} \quad \bar{\mu}_i \le 0. \end{aligned}$$

As $\bar{\upsilon} \in N_{X_0}(\hat{x})$, we also have

$$\langle \bar{\upsilon}, d \rangle \le 0.$$

Since the norm of $(\bar{\lambda}, \bar{\mu})$ is 1, some of the multipliers $\bar{\lambda}_i$ or $\bar{\mu}_i$ are nonzero. Multiplying both sides of (6.36) by d we obtain a negative number on the right hand side and zero on the left hand side, a contradiction. Therefore the sequence $\{(\lambda^k, \mu^k)\}$ is bounded. It follows that it has accumulation points. For each such an accumulation point $(\hat{\lambda}, \hat{\mu})$ we can extract a subsequence such that $(\lambda^k, \mu^k) \to (\hat{\lambda}, \hat{\mu})$ over this subsequence. Passing to the limit in (6.34) over this subsequence (an using Lemma 2.42 again) we conclude that

$$0 \in \nabla f(\hat{x}) + \sum_{i=1}^{m} \hat{\lambda}_i \nabla g_i(\hat{x}) + \sum_{i=1}^{p} \hat{\mu}_i \nabla h_i(\hat{x}) + N_{X_0}(\hat{x}).$$

By the definition of λ_i^k, the complementarity condition $\hat{\lambda}_i g_i(\hat{x}) = 0$, $i = 1, \dots, m$, holds true as well. $\square$

Again, we make the interesting observation that the values of the Lagrange multipliers are obtained as a by-product of a constrained optimization method.

If separate penalty coefficients are used for different constraint functions, formulae (6.32) have to be modified accordingly. Theorem 6.32 remains valid, and the proof is almost the same.

A disadvantage of the quadratic penalty method is the increased difficulty of problem (6.30) for large values of the penalty parameter ϱ. To illustrate this issue, let us start with an example.

Example 6.8. Consider the following problem:

$$\begin{aligned} &\text{minimize } \frac{1}{2}\big[(x_1)^2 + (x_2)^2\big] \\ &\text{subject to } 2 - x_1 - x_2 = 0. \end{aligned}$$

Its solution is $\hat{x}_1 = \hat{x}_2 = 1$ and the Lagrange multiplier associated with the constraint equals $\hat{\mu} = 1$. The penalty problem takes on the form

$$\text{minimize } \Big[\Phi_\varrho(x_1, x_2) \triangleq \frac{1}{2}(x_1)^2 + \frac{1}{2}(x_2)^2 + \frac{\varrho}{2}(2 - x_1 - x_2)^2\Big].$$

Its solution is

$$\hat{x}_1(\varrho) = \hat{x}_2(\varrho) = \frac{2\varrho}{2\varrho + 1}.$$

As the general theory guarantees, $\hat{x}(\varrho) \to \hat{x}$, when $\varrho \to \infty$. Moreover,

$$\varrho h(\hat{x}(\varrho)) = \varrho(2 - \hat{x}_1(\varrho) - \hat{x}_2(\varrho)) = \frac{2\varrho}{2\varrho + 1}.$$

This quantity is convergent to $\hat{\mu}$ as $\varrho \to \infty$.

The Hessian of the objective function of the penalty problem equals

$$\nabla^2 \Phi_\varrho(x_1, x_2) = \begin{bmatrix} 1+\varrho & \varrho \\ \varrho & 1+\varrho \end{bmatrix}.$$

The eigenvalues of this matrix are: $\lambda_1 = 1$ and $\lambda_2 = 1 + 2\varrho$, and thus the condition index is

$$\varkappa = \frac{\lambda_2}{\lambda_1} = 1 + 2\varrho.$$

We see that $\varkappa \to \infty$ as $\varrho \to \infty$; the penalty problem becomes more and more ill-conditioned, when the penalty coefficient increases. We illustrate this phenomenon in Figure 6.1.

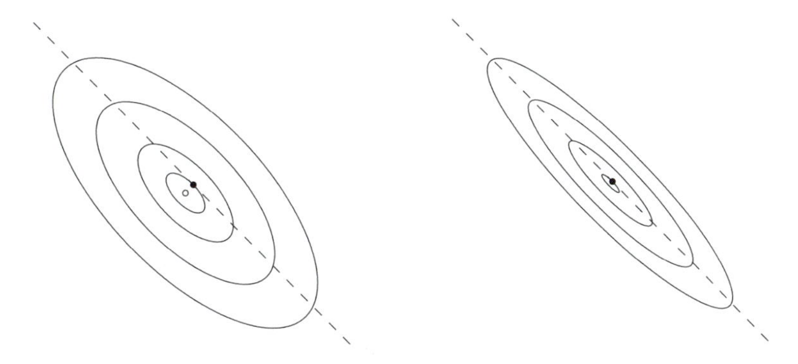

Figure 6.1. Operation of the quadratic penalty function in Example 6.8 for $\varrho = 2$ (left) and $\varrho = 10$ (right). The dashed line represents the feasible set.

More generally, let us analyze a nonlinear optimization problem with equality constraints only:

$$\begin{aligned} &\text{minimize } f(x) \\ &\text{subject to } h_i(x) = 0, \quad i = 1, \ldots, p. \end{aligned}$$

Suppose the functions $f(\cdot)$ and $h_i(\cdot)$ are twice continuously differentiable. Consider an optimal solution $\hat{x}$ and the Hessian of the objective function of problem (6.30):

$$\nabla^2 \Phi_\varrho(\hat{x}) = \nabla^2 f(\hat{x}) + \varrho \sum_{i=1}^{p} \nabla h_i(\hat{x}) \big[\nabla h_i(\hat{x})\big]^T.$$

In order to estimate the condition index of the Hessian, we consider the quadratic form

$$Q(d) = \big\langle d, \big[\nabla^2 \Phi_\varrho(\hat{x})\big] d \big\rangle.$$

Suppose the dimension of the subspace

$$D = \{d \in \mathbb{R}^n : \langle \nabla h_i(\hat{x}), d \rangle = 0,\ i = 1, \dots, p\}$$

is positive. This is a typical situation; otherwise the point $\hat{x}$ is an isolated feasible point. We also assume that the dimension of D is smaller than n, which means that at least one of the gradients of the constraint functions is nonzero.

Consider a direction $d \in D$ of length 1. From the second order necessary conditions of optimality we know that $Q(d) \geq 0$. We also obtain the upper bound

$$Q(d) = \langle d, [\nabla^2 f(\hat{x})]d \rangle \leq \lambda_{\max}(\nabla^2 f(\hat{x})),$$

because the terms with the gradients of the constraints are zero for $d \in D$. Hence

$$\lambda_{\min}(\nabla^2 \Phi_\varrho(\hat{x})) \leq \lambda_{\max}(\nabla^2 f(\hat{x})).$$

Now consider a direction z of length 1 in the complementary subspace $D^\perp$.

$$\begin{aligned} Q(z) &= \langle z, [\nabla^2 f(\hat{x})]z \rangle + \varrho \sum_{i=1}^{p} \langle z, \nabla h_i(\hat{x})[\nabla h_i(\hat{x})]^T z \rangle \\ &= \langle z, [\nabla^2 f(\hat{x})]z \rangle + \varrho \sum_{i=1}^{p} \left([\nabla h_i(\hat{x})]^T z \right)^2 \\ &= \langle z, [\nabla^2 f(\hat{x})]z \rangle + \varrho \| h'(\hat{x}) z \|^2. \end{aligned}$$

As z is a nonzero element of $D^\perp$, it cannot belong to D. Thus $h'(\hat{x})z \neq 0$. The condition index of the Hessian is at least

$$\varkappa \geq \frac{Q(z)}{Q(d)} \geq \frac{\lambda_{\min}(\nabla^2 f(\hat{x})) + \varrho \| h'(\hat{x}) z \|^2}{\lambda_{\max}(\nabla^2 f(\hat{x}))}.$$

The right hand side of this inequality tends to ∞ when $\varrho \to \infty$, and thus problem (6.30) becomes very ill-conditioned for large ϱ.

Similar phenomena occur in problems with inequality constraints. For these reasons the quadratic penalty method is rarely used as a stand-alone method for constrained optimization, but rather is an auxiliary module in more efficient methods.

6.2.3 Exact Penalty Function

Let us return to the nonlinear optimization problem (6.29). Since the quadratic penalty function $P_2(\cdot)$ has zero gradient at the optimal point, it cannot provide an exact solution to the constrained optimization problem, when $-\nabla f(\hat{x}) \notin N_{X_0}(\hat{x})$. In fact, every differentiable penalty function must have zero gradient at the solution of (6.29).

One possibility to overcome this disadvantage is to use the penalty function

$$P_1(x) = \sum_{i=1}^{m} \max\big(0, g_i(x)\big) + \sum_{i=1}^{p} |h_i(x)|. \tag{6.37}$$

As before, we replace (6.29) with the problem

$$\underset{x \in X_0}{\text{minimize}} \ \big[\Phi_\varrho(x) \triangleq f(x) + \varrho P_1(x)\big], \tag{6.38}$$

where $\varrho > 0$. We call the function $P_1(\cdot)$ an *exact penalty function*, because in many cases problem (6.38) can provide an optimal solution of the constrained problem for a finite value of the penalty parameter ϱ. But before we examine this issue, we have to stress that the function $P_1(\cdot)$ is usually nonsmooth, even if the functions $g_i(\cdot)$ and $h_i(\cdot)$ are continuously differentiable. This is due to the fact that the functions $t \mapsto \max(0, t)$ and $t \mapsto |t|$ are nondifferentiable at 0. Thus, the nondifferentiability of $P_1(\cdot)$ at a solution of (6.29) is a rule, except for the degenerate case of zero gradients of the constraint functions.

To analyze the property of delivering exact solutions for a finite value of the penalty coefficient, we use second order sufficient conditions formulated in Theorem 3.47 on page 147.

THEOREM 6.9. *Assume that problem* (6.29) *has a local minimum* $\hat{x}$ *satisfying second order sufficient conditions of optimality with multipliers* $(\hat{\lambda}, \hat{\mu})$. *Then for every* $\varrho > \|(\hat{\lambda}, \hat{\mu})\|_\infty$ *the point* $\hat{x}$ *is a local minimum of problem* (6.38).

Proof. Although the function $P_1(\cdot)$ is nonsmooth, we can show that it has directional derivatives in every direction d. Take any $i \in I^0(\hat{x})$, where $I^0(\hat{x}) = \{i : g_i(\hat{x}) = 0\}$. For $\tau > 0$ we have

$$\max\big(0, g_i(\hat{x} + \tau d)\big) = \max\big(0, \tau\langle \nabla g_i(\hat{x}), d\rangle + o(\tau)\big),$$

with $o(\tau)/\tau \to 0$, as $\tau \downarrow 0$. Dividing by τ and passing to the limit we get

$$\lim_{\tau \downarrow 0} \frac{1}{\tau} \max\big(0, g_i(\hat{x} + \tau d)\big) = \max\big(0, \langle \nabla g_i(\hat{x}), d\rangle\big).$$

We treat the equality constraints in a similar way and we conclude that the penalty function has directional derivatives:

$$P_1'(\hat{x}; d) = \sum_{i \in I^0(\hat{x})} \max\big(0, g_i'(\hat{x}; d)\big) + \sum_{i=1}^{p} |h_i'(\hat{x}; d)|.$$

Therefore,

$$\Phi_\varrho'(\hat{x}; d) = f'(\hat{x}; d) + \varrho \sum_{i \in I^0(\hat{x})} \max\big(0, g_i'(\hat{x}; d)\big) + \varrho \sum_{i=1}^{p} |h_i'(\hat{x}; d)|. \quad (6.39)$$

If $d \in T_{X_0}(\hat{x})$, it follows from the necessary conditions of optimality (3.27) that

$$f'(\hat{x}; d) + \sum_{i \in I^0(\hat{x})} \hat{\lambda}_i g_i'(\hat{x}; d) + \sum_{i=1}^{p} \hat{\mu}_i h_i'(\hat{x}; d) \geq 0.$$

Then also

$$f'(\hat{x}; d) + \sum_{i \in I^0(\hat{x})} \hat{\lambda}_i \max\big(0, g_i'(\hat{x}; d)\big) + \sum_{i=1}^{p} |\hat{\mu}_i| \cdot |h_i'(\hat{x}; d)| \geq 0.$$

Combining the last inequality with (6.39) we get

$$\Phi_\varrho'(\hat{x}; d) \geq \sum_{i \in I^0(\hat{x})} (\varrho - \hat{\lambda}_i) \max\big(0, g_i'(\hat{x}; d)\big) + \sum_{i=1}^{p} (\varrho - |\hat{\mu}_i|) |h_i'(\hat{x}; d)|. \quad (6.40)$$

We shall use this estimate to prove the theorem by contradiction.

Suppose $\hat{x}$ is a local minimum of problem (6.29), but it is not a local minimum of problem (6.38). Then there exists a sequence of feasible points $y^k \in X_0$ such that $y^k \to \hat{x}$ and

$$f(y^k) + \varrho P_1(y^k) < f(\hat{x}) \quad \text{for all} \quad k. \quad (6.41)$$

We shall show that it leads to a contradiction. Define

$$s^k = \frac{y^k - \hat{x}}{\|y^k - \hat{x}\|}.$$

Since s^k have length 1, there exists a convergent subsequence of $\{s^k\}$. Let s be its limit. By construction, $s \in T_{X_0}(\hat{x})$. Defining $\tau_k = \|y^k - \hat{x}\|$, we can write (for the selected subsequence)

$$y^k = \hat{x} + \tau_k s + o(\tau_k).$$

As $\Phi_\varrho(\cdot)$ is Lipschitz continuous, we have (with some Lipschitz constant C) the inequality $\Phi_\varrho(\hat{x} + \tau_k s) \leq \Phi_\varrho(y^k) + Co(\tau_k)$. It then follows from (6.41) that

$$\Phi_\varrho(\hat{x} + \tau_k s) - \Phi_\varrho(\hat{x}) \leq \Phi_\varrho(y^k) - \Phi_\varrho(\hat{x}) + Co(\tau_k) \leq Co(\tau_k).$$

Dividing by τ_k and passing to the limit we conclude that

$$\Phi_\varrho{}'(\hat{x}; s) \le 0.$$

By assumption, $\varrho > \lambda_i$ for all $i = 1, \dots, m$, and $\varrho > |\mu_i|$ for all $i = 1, \dots, p$. Then (6.40) yields

$$\begin{aligned} \langle \nabla g_i(\hat{x}), s \rangle &\le 0, \quad i \in I^0(\hat{x}), \\ \langle \nabla h_i(\hat{x}), s \rangle &= 0, \quad i = 1, \dots, p. \end{aligned} \tag{6.42}$$

We also observe that (6.41) implies that $f(y^k) < f(\hat{x})$. Now we can repeat the proof of Theorem 3.47 from page 147 to obtain a contradiction with the second order sufficient condition. Thus, the sequence $\{y^k\}$ does not exist and $\hat{x}$ is a local minimum of the function $\Phi_\varrho(\cdot)$. □

The existence of Lagrange multipliers and the ability of the penalty function $P_1(\cdot)$ to deliver exact solutions are closely related.

Theorem 6.10. *Assume that X_0 is a closed polyhedron and that $\hat{x}$ is a local minimum of problem* (6.38). *If $\hat{x}$ is feasible for problem* (6.29) *then there exist Lagrange multipliers $(\hat{\lambda}, \hat{\mu})$ such that $\hat{x}$ satisfies the first order necessary conditions of optimality for problem* (6.29).

Proof. Since $\hat{x}$ is a local minimum of $\Phi_\varrho(\cdot)$ in X_0, and $\Phi_\varrho(\cdot)$ has directional derivatives, the necessary condition of optimality (3.27) implies that

$$\Phi_\varrho'(\hat{x}; d) \ge 0, \quad \text{for all} \quad d \in T_{X_0}(\hat{x}).$$

Using (6.39), for all $d \in T_{X_0}(\hat{x})$ we obtain the inequality

$$f'(\hat{x}; d) + \varrho \sum_{i \in I^0(\hat{x})} \max\big(0, g_i'(\hat{x}; d)\big) + \varrho \sum_{i=1}^{p} |h_i'(\hat{x}; d)| \ge 0.$$

In particular, if $s \in T_{X_0}(\hat{x})$ and s satisfies (6.42), we must have $f'(x; s) \ge 0$. It follows that the optimal value of the problem

$$\begin{aligned} \text{minimize} \quad & \langle \nabla f(\hat{x}), s \rangle \\ \text{subject to} \quad & \langle \nabla g_i(\hat{x}), s \rangle \le 0, \quad i \in I^0(\hat{x}), \\ & \langle \nabla h_i(\hat{x}), s \rangle = 0, \quad i = 1, \dots, p, \\ & s \in T_{X_0}(\hat{x}), \end{aligned} \tag{6.43}$$

is zero. As X_0 is a polyhedron, we do not need any constraint qualification conditions for problem (6.43) to guarantee the existence of Lagrange multipliers. We conclude that there exist $\hat{\lambda} \geq 0$ and $\hat{\mu}$ such that

$$0 \in \nabla f(\hat{x}) + \sum_{i \in I^0(\hat{x})} \hat{\lambda}_i \nabla g_i(\hat{x}) + \sum_{i=1}^{p} \hat{\mu}_i \nabla h_i(\hat{x}) + N_{X_0}(\hat{x}).$$

These are the first order necessary conditions of optimality for problem (6.38) at $\hat{x}$. □

The assumption that X_0 is polyhedral is essential here, because otherwise we have no guarantee that the solution of the conic programming problem (6.43) satisfies the necessary conditions of optimality. We discuss conic programming in Section 4.3.

6.3 THE BASIC DUAL METHOD

We return to the general formulation of a nonlinear optimization problem:

$$\begin{aligned} &\text{minimize} \ f(x) \\ &\text{subject to} \ g_i(x) \leq 0, \quad i = 1, \dots, m, \\ &\qquad\qquad h_i(x) = 0, \quad i = 1, \dots, p, \\ &\qquad\qquad x \in X_0. \end{aligned} \tag{6.44}$$

All functions $f : \mathbb{R}^n \to \mathbb{R}$, $g_i : \mathbb{R}^n \to \mathbb{R}$, $i = 1, \dots, m$, and $h_i : \mathbb{R}^n \to \mathbb{R}$, $i = 1, \dots, p$, are assumed to be continuously differentiable. The set X_0 is convex and closed.

Together with problem (6.44) we consider the Lagrangian

$$L(x, \lambda, \mu) = f(x) + \langle \lambda, g(x) \rangle + \langle \mu, h(x) \rangle$$

and the dual function

$$L_D(\lambda, \mu) = \inf_{x \in X_0} L(x, \lambda, \mu). \tag{6.45}$$

The dual problem is defined as follows:

$$\underset{(\lambda,\mu) \in \Lambda_0}{\text{maximize}} \ L_D(\lambda, \mu), \tag{6.46}$$

with $\Lambda_0 = \mathbb{R}^m_+ \times \mathbb{R}^p$.

Assume that the functions $f(\cdot)$ and $g_i(\cdot)$, $i = 1, \dots, m$ are convex, the functions $h_i(\cdot)$ are affine, and Slater's condition is satisfied. By virtue of Theorems 4.7 and 4.8, problem (6.44) has an optimal solution if and only if the dual problem (6.46) has an optimal solution. Also, the solutions of the dual problem are the Lagrange multipliers satisfying the necessary and sufficient conditions of optimality for the primal problem, at each of its optimal solutions.

The idea of the dual methods is to solve the dual problem (6.46) by an iterative method for optimization with simple constraints, and then to recover the primal solution from the conditions of Theorem 4.10.

The dual function $L_D(\cdot)$ is concave, but not necessarily differentiable. Therefore, in the general case, the solution of the dual problem requires application of methods of nonsmooth optimization. We discuss these methods in Chapter 7, and we analyze their application to dual problems. We focus here on the special case, when the dual function $L_D(\cdot)$ is continuously differentiable. We make the following assumptions:

- The set X_0 is bounded;
- The function $f(\cdot)$ is strictly convex.

These conditions ensure that for every (λ, μ) a solution $\hat{x}(\lambda, \mu)$ of problem (6.45) exists and is unique. Furthermore, we know from Section 2.5.3 that the dual function is continuously differentiable, and its gradient is given by

$$\nabla L_D(\lambda, \mu) = \begin{bmatrix} g(\hat{x}(\lambda, \mu)) \\ h(\hat{x}(\lambda, \mu)) \end{bmatrix}.$$

Thus, we can apply to problem (6.46) any method for problems with simple constraints.

Let us consider the projection method (6.2). Since

$$\Pi_{\Lambda_0}(\lambda, \mu) = (\max(0, \lambda), \mu),$$

the projection method takes on the form

$$\begin{aligned} \lambda^{k+1} &= \max(0, \lambda^k + \tau g(\hat{x}(\lambda^k, \mu^k)), \\ \mu^{k+1} &= \mu^k + \tau h(\hat{x}(\lambda^k, \mu^k)), \quad k = 1, 2, \ldots. \end{aligned} \tag{6.47}$$

Note that the dual problem is a maximization problem and therefore we make steps in the direction of the gradient, rather than its negative. Unfortunately, it is hard to determine the right value of the step size τ in this method.

We can also apply the reduced gradient method to the dual problem (6.46). It has a very simple form: some of the multipliers λ_i are nonbasic and fixed at 0, and the others are superbasic and are free to move. There are no basic variables at all, and the reduced gradient is just the gradient of the dual function with respect to the superbasic multipliers. Here the difficulty is that the directional maximization requires the evaluation of the dual function at many points, and this may be expensive.

Example 6.11. Let us return to the optimal power generation problem of Example 4.18 on page 181. It has the form

$$\begin{aligned} &\text{minimize} \quad \sum_{j=1}^{n} f_j(x_j) \\ &\text{subject to} \quad \sum_{j=1}^{n} x_j \geq b, \\ &\qquad\qquad 0 \leq x_j \leq u_j, \quad j = 1, \dots, n. \end{aligned}$$

with the cost functions of the plants

$$f_j(x_j) = c_j x_j + \frac{q_j}{2}(x_j)^2, \quad j = 1, \dots, n.$$

Denoting by λ the Lagrange multiplier associated with the demand constraint, we can write the Lagrangian as follows:

$$L(x, \lambda) = \sum_{j=1}^{n} f_j(x_j) + \lambda\Big(b - \sum_{j=1}^{n} x_j\Big).$$

Its minimization with respect to x_j yields

$$\hat{x}_j(\lambda) = \begin{cases} 0 & \text{if } 0 \leq \lambda \leq c_j, \\ (\lambda - c_j)/q_j & \text{if } c_j \leq \lambda \leq c_j + q_j u_j, \\ u_j & \text{if } \lambda \geq c_j + q_j u_j. \end{cases}$$

The derivative of the dual function has the form:

$$\frac{dL_D(\lambda)}{d\lambda} = b - \sum_{j=1}^{n} \hat{x}_j(\lambda).$$

The dual method with constant step size is simply

$$\lambda^{k+1} = \max\Big(0, \lambda^k + \tau\Big[b - \sum_{j=1}^{n} \hat{x}_j(\lambda^k)\Big]\Big).$$

If $\tau > 0$ is sufficiently small, the method is convergent to the optimal dual solution. Its interpretation is obvious: if there is a power shortage,

$$b > \sum_{j=1}^{n} \hat{x}_j(\lambda^k),$$

then increase the price paid to the plants; if there is a surplus, decrease the price.

6.4 THE AUGMENTED LAGRANGIAN METHOD

6.4.1 General Ideas

We return to the nonlinear programming problem

$$\begin{aligned} \text{minimize} \quad & f(x) \\ \text{subject to} \quad & g_i(x) \le 0, \quad i = 1, \dots, m, \\ & h_i(x) = 0, \quad i = 1, \dots, p, \\ & x \in X_0. \end{aligned} \tag{6.48}$$

The functions $f : \mathbb{R}^n \to \mathbb{R}$, $g_i : \mathbb{R}^n \to \mathbb{R}$, $i = 1, \dots, m$, and $h_i : \mathbb{R}^n \to \mathbb{R}$, $i = 1, \dots, p$, are now assumed to be twice continuously differentiable. The set X_0 is convex and closed.

The augmented Lagrangian for problem (6.48) has form (4.80), which we recall for convenience:

$$\begin{aligned} L_\varrho(x, \lambda, \mu) = f(x) &+ \frac{\varrho}{2} \sum_{i=1}^{m} \Big[\max\Big(0, g_i(x) + \frac{\lambda_i}{\varrho}\Big) \Big]^2 \\ &+ \frac{\varrho}{2} \sum_{i=1}^{p} \Big[h_i(x) + \frac{\mu_i}{\varrho} \Big]^2 - \frac{1}{2\varrho} \sum_{i=1}^{m} \lambda_i^2 - \frac{1}{2\varrho} \sum_{i=1}^{p} \mu_i^2. \end{aligned} \tag{6.49}$$

We notice its similarity with the quadratic penalty function (6.30): the penalty is calculated for the perturbed constraints $g_i(x) + \lambda_i/\varrho$ and $h_i(x) + \mu_i/\varrho$. Because of this relation, the augmented Lagrangian is sometimes referred to as the *shifted penalty function*.

Assume that $\hat{x}$ is a local minimum of problem (6.48) satisfying Robinson's condition. We also assume that the strong second order sufficient condition is satisfied (see Definition 4.32 on page 200) with Lagrange multipliers $(\hat{\lambda}, \hat{\mu})$. By Theorem 4.33 there exists ϱ_0 such that for all $\varrho > \varrho_0$ the pair $\hat{x}$ and $(\hat{\lambda}, \hat{\mu})$ is a local saddle point of the augmented Lagrangian.

If we knew the optimal Lagrange multipliers $(\hat{\lambda}, \hat{\mu})$, we could, for a sufficiently large ϱ, minimize the augmented Lagrangian to obtain the local minimum $\hat{x}$. However, the multipliers are not known, and need to be estimated.

Theorem 6.7 provides a method by which to estimate them. Suppose we minimize the augmented Lagrangian with respect to x in a certain neighborhood U of the optimal point, with given values of multipliers (λ^k, μ^k). If (λ^k, μ^k) are close to the optimal values $(\hat{\lambda}, \hat{\mu})$, the minimum of the augmented Lagrangian will be close to $\hat{x}$. The terms corresponding to $i \notin I^0(\hat{x})$ become zero, and in the neighborhood of $\hat{x}$ the augmented Lagrangian simplifies:

$$
\begin{aligned}
L_\varrho(x,\lambda^k,\mu^k) &= f(x) + \frac{\varrho}{2}\sum_{i\in I^0(\hat{x})}\Big[g_i(x)+\frac{\lambda_i^k}{\varrho}\Big]^2 + \frac{\varrho}{2}\sum_{i=1}^{p}\Big[h_i(x)+\frac{\mu_i^k}{\varrho}\Big]^2 \\
&\quad - \frac{1}{2\varrho}\sum_{i=1}^{m}(\lambda_i^k)^2 - \frac{1}{2\varrho}\sum_{i=1}^{p}(\mu_i^k)^2 \\
&= f(x) + \sum_{i\in I^0(\hat{x})}\Big(\lambda_i^k g_i(x) + \frac{\varrho}{2}\big[g_i(x)\big]^2\Big) \\
&\quad + \sum_{i=1}^{p}\Big(\mu_i^k h_i(x) + \frac{\varrho}{2}\big[h_i(x)\big]^2\Big) - \frac{1}{2\varrho}\sum_{i\notin I^0(\hat{x})}(\lambda_i^k)^2.
\end{aligned}
$$

This is equal (after a translation by the last sum) to the quadratic penalty function for the problem

$$
\begin{aligned}
\text{minimize}\quad & f(x) + \sum_{i\in I^0(\hat{x})}\lambda_i^k g_i(x) + \sum_{i=1}^{p}\mu_i^k h_i(x) \\
\text{subject to}\quad & g_i(x) = 0,\quad i\in I^0(\hat{x}), \\
& h_i(x) = 0,\quad i=1,\dots,p, \\
& x\in X_0.
\end{aligned}
\tag{6.50}
$$

We know from Theorem 6.7 that at the minimum x^k of the quadratic penalty function, the quantities $\varrho g_i(x^k)$, $i\in I^0(\hat{x})$, and $\varrho h_i(x^k)$, $i=1,\dots,p$, become very close to the Lagrange multipliers associated with the constraints of problem (6.50). Since the gradient of the objective function of (6.50) has the form

$$
\nabla f(x) + \sum_{i\in I^0(\hat{x})}\lambda_i^k \nabla g_i(x) + \sum_{i=1}^{p}\mu_i^k \nabla h_i(x),
$$

we conclude that $\lambda_i^k + \varrho g_i(x^k)$ and $\mu_i^k + \varrho h_i(x^k)$ are good approximations of the Lagrange multipliers in the original problem (6.48). As $\hat{\lambda}\ge 0$, we may take the positive part of $\lambda_i^k + \varrho g_i(x^k)$.

We construct, therefore, the following iterative process. At iteration k, for given values of Lagrange multipliers $(\lambda^k,\mu^k)\in\Lambda_0$, we solve the problem

$$
\underset{x\in X_0}{\text{minimize}}\ L_\varrho(x,\lambda^k,\mu^k). \tag{6.51}
$$

After obtaining the solution x^k, we update the multipliers by the formula:

$$\begin{aligned} \lambda_i^{k+1} &= \max\left(0, \lambda_i^k + \varrho g_i(x^k)\right), \quad i = 1, \ldots, m, \\ \mu_i^{k+1} &= \mu_i^k + \varrho h_i(x^k), \quad i = 1, \ldots, p, \end{aligned} \tag{6.52}$$

and the iteration continues. Algorithm (6.51)–(6.52) is called the *augmented Lagrangian method* or the *method of multipliers*.

In the next subsection we make our heuristic considerations precise.

The multiplier method is closely related to the duality theory for augmented Lagrangians presented in Section 4.7. By the strong second order condition, if $\varrho > \varrho_0$, the function $L_\varrho(\cdot, \lambda, \mu)$ is locally strictly convex about $\hat{x}$, if (λ, μ) is close to $(\hat{\lambda}, \hat{\mu})$. We can thus consider the local dual function

$$L_{\varrho D}(\lambda, \mu) = \min_{x \in X_0 \cap U} L_\varrho(x, \lambda, \mu),$$

where U is a small neighborhood of $\hat{x}$. The minimizer is unique and it follows form Theorem 2.87 on page 71 that

$$\begin{aligned} \nabla_\lambda L_{\varrho D}(\lambda^k, \mu^k) &= g(x^k), \\ \nabla_\mu L_{\varrho D}(\lambda^k, \mu^k) &= h(x^k). \end{aligned}$$

Thus the method of multipliers (6.52) may be regarded as a version of the method of steepest ascent with projection applied to the dual problem. An interesting observation is that the penalty parameter ϱ is a good value of the step size in this method. We prove this in the next subsection, but before that we analyze a simple example.

Example 6.12. Let us return to Example 4.31:

$$\begin{aligned} &\text{minimize} \ \frac{1}{2}(x_1)^2 - x_1 x_2 \\ &\text{subject to} \ x_1 + x_2 - 1 = 0 \end{aligned}$$

The solution of this problem is $\hat{x}_1 = \frac{1}{3}$ and $\hat{x}_2 = \frac{2}{3}$, and the Lagrange multiplier equals $\hat{\mu} = \frac{1}{3}$. Similarly to the analysis in Example 4.31, we conclude that the augmented Lagrangian

$$L_\varrho(x_1, x_2, \mu) = \frac{1}{2}(x_1)^2 - x_1 x_2 + \mu(x_1 + x_2 - 1) + \frac{\varrho}{2}(x_1 + x_2 - 1)^2$$

is strictly convex for $\varrho > \frac{1}{3}$. Its minimum, as a function of the multiplier μ, has the form:

$$\hat{x}_1(\mu) = \frac{\varrho - \mu}{3\varrho - 1}, \qquad \hat{x}_2(\mu) = \frac{2(\varrho - \mu)}{3\varrho - 1}.$$

After elementary manipulations, the multiplier method (6.52) can be written as follows:

$$\mu^{k+1} = \mu^k + \varrho\big(\hat{x}_1(\mu^k) + \hat{x}_2(\mu^k) - 1\big) = \frac{\varrho - \mu^k}{3\varrho - 1}.$$

Hence

$$\mu^{k+1} - \frac{1}{3} = -\frac{1}{3\varrho - 1}\Big(\mu^k - \frac{1}{3}\Big).$$

If $\varrho > \frac{2}{3}$ the sequence $\{\mu^k\}$ is linearly convergent to $\hat{\mu} = \frac{1}{3}$. The larger ϱ, the faster the convergence.

6.4.2 Local Convergence

To simplify notation, we concentrate on inequality constraints only ($p = 0$). The considerations for equality constraints are simpler. We assume that the Lagrange multipliers $\hat{\lambda}$ are uniquely defined at $\hat{x}$, and that the strong second order sufficient condition of Definition 4.32 is satisfied. Finally, we assume that the set X_0 is a convex closed polyhedron.

As the multipliers $\hat{\lambda}_i$, $i \in I^0(\hat{x})$, are positive and uniquely defined, it is implicit that the gradients of active constraints, $\nabla g_i(\hat{x})$, $i \in I^0(\hat{x})$, are linearly independent.

Our first result proves local stability of subproblems (6.51).

Lemma 6.13. *There exist constants $\varepsilon_1 > 0$, $\varrho_0 > 0$ and $C_1 > 0$ such that if $\|\lambda^k - \hat{\lambda}\| \le \varepsilon_1$ and $\varrho \ge \varrho_0$ then problem (6.51) has a local minimum x^k such that $\|x^k - \hat{x}\| \le C_1\|\lambda^k - \hat{\lambda}\|$.*

Proof. We can choose $\delta > 0$, $\varrho_0 > 0$, and $\kappa > 0$ such that the Hessian of the augmented Lagrangian has all eigenvalues above κ, provided that $\|x - \hat{x}| \le \delta$ and $\|\lambda - \hat{\lambda}\| \le \delta$. This follows from Theorems 4.29 and 4.79, and from the continuity of the Hessian. Notice that due to the strict positivity of the multipliers $\hat{\lambda}_i$ for $i \in I^0(\hat{x})$, the active pieces of the "max" terms in the augmented Lagrangian do not change in a small neighborhood of $(\hat{x}, \hat{\lambda})$. Furthermore, if ϱ_0 is sufficiently large, and $\varepsilon_1 > 0$ sufficiently small, a local minimum of the augmented Lagrangian will occur in the δ-neighborhood of $\hat{x}$, provided that $\|\lambda^k - \hat{\lambda}\| \le \varepsilon_1$. This can be obtained easily from the second order sufficient conditions, the minimum positive curvature of $L_\varrho(\cdot, \lambda^k, \mu^k)$, and the perturbation of the gradient at $\hat{x}$ resulting from the difference $\lambda^k - \hat{\lambda}$. The distance to this minimum can be bounded from above by linear function of this perturbation, as required. □

From now on we assume that $\varrho \ge \varrho_0$, λ^k is sufficiently close to $\hat{\lambda}$, and that x^k is the local minimum referred to in Lemma 6.13.

Our next step is the analysis of metric regularity of the system of optimality conditions at $\hat{x}$:

$$\begin{aligned} -\nabla_x L(x,\lambda) &\in N_{X_0}(\hat{x}), \\ g_i(x) = 0, \quad i &\in I^0(\hat{x}), \\ x &\in X_0. \end{aligned} \tag{6.53}$$

The cone $N_{X_0}(\hat{x})$ and the set $I^0(\hat{x})$ are fixed in this system. Observe that the pair $(\hat{x}, \hat{\lambda})$ is a solution of (6.53), by virtue of Theorem 3.25 applied to problem (6.48).

LEMMA 6.14. *System* (6.53) *is metrically regular at* $(\hat{x}, \hat{\lambda})$.

Proof. We verify Robinson's condition (3.6) from page 102 for system (6.53). For convenience of notation we assume that $I^0(\hat{x}) = \{1, \dots, m_0\}$ and we introduce the matrices:

$$H = \nabla^2_{xx} L(\hat{x}, \hat{\lambda}), \qquad A^T = \begin{bmatrix} \nabla g_1(\hat{x}) & \dots & \nabla g_{m_0}(\hat{x}) \end{bmatrix}.$$

The Jacobian of system (6.53) at $(\hat{x}, \hat{\lambda})$ has the form:

$$J = \begin{bmatrix} -H & -A^T \\ A & 0 \end{bmatrix}. \tag{6.54}$$

Define $\hat{y} = -\nabla_x L(\hat{x}, \hat{\lambda})$. Suppose Robinson's condition is not satisfied. Then

$$\begin{aligned} 0 \notin \operatorname{int} \Bigg\{ \begin{bmatrix} -H d_x - A^T d_\lambda - d_v \\ A d_x \end{bmatrix} &: d_x \in K_{X_0}(\hat{x}), \\ d_\lambda \in \mathbb{R}^{m_0},\ d_v &\in K_{N_{X_0}(\hat{x})}(\hat{y}) \Bigg\}. \end{aligned} \tag{6.55}$$

The set on the right hand side is convex and we can separate it from 0. It follows from Theorem 2.15 that there exist $z \in \mathbb{R}^n$ and $w \in \mathbb{R}^{m_0}$, $(z, w) \neq 0$, such that for all $d_x \in K_{X_0}(\hat{x})$, $d_\lambda \in \mathbb{R}^{m_0}$ and $d_v \in K_{N_{X_0}(\hat{x})}(\hat{y})$ the following inequalities are satisfied:

$$\begin{aligned} \langle z, -H d_x - A^T d_\lambda - d_v \rangle &\le 0, \\ \langle w, A d_x \rangle &\le 0. \end{aligned} \tag{6.56}$$

The second relation means that

$$A^T w \in N_{X_0}(\hat{x}).$$

Recall that we assume Robinson's condition for problem (6.48). It has the form: $AK_{X_0}(\hat{x}) = \mathbb{R}^{m_0}$. In particular, it is possible to find $d_x \in K_{X_0}(\hat{x})$ such that $Ad_x = w$. But then the last displayed relation implies that

$$\|w\|^2 = \langle w, w\rangle = \langle w, Ad_x\rangle = \langle A^T w, d_x\rangle \le 0.$$

Consequently, $w = 0$ and thus $z \neq 0$ in (6.56).

The first relation in (6.56) implies that $\langle z, A^T d_\lambda\rangle \ge 0$ for all $d_\lambda \in \mathbb{R}^{m_0}$. Therefore,

$$Az = 0.$$

It also implies that $\langle z, d_v\rangle \ge 0$ for all $d_v \in K_{N_{X_0}(\hat{x})}(\hat{y})$. By virtue of Example 2.21 on page 27, $N_{X_0}(\hat{x}) \subset K_{N_{X_0}(\hat{x})}(\hat{y})$. Hence $-z \in \big[N_{X_0}(\hat{x})\big]^\circ$. Since X_0 is a convex closed polyhedron, $\big[N_{X_0}(\hat{x})\big]^\circ = K_{X_0}(\hat{x})$, and we obtain

$$-z \in K_{X_0}(\hat{x}).$$

Setting $d_x = -z$, $d_\lambda = 0$ and $d_v = 0$ in (6.56) we conclude that

$$\langle z, Hz\rangle \le 0.$$

But $Az = 0$ and $z \neq 0$, and the last inequality contradicts the strong second order sufficient condition. □

In the following lemma we denote by $g^0(x^k)$ the subvector of the constraints having indices $i \in I^0(\hat{x})$ (active at $\hat{x}$).

Lemma 6.15. *There exist $\varepsilon_2 > 0$ and $C_2 > 0$ such that if $\|\lambda^k - \hat{\lambda}\| \le \varepsilon_2$, then*

$$\|x^k - \hat{x}\| \le C_2 \|g^0(x^k)\|$$

and

$$\|\lambda^{k+1} - \hat{\lambda}\| \le C_2 \|g^0(x^k)\|.$$

Proof. Observe that at the point x^k the necessary conditions of optimality for problem (6.51) read:

$$-\nabla_x L_\varrho(x^k, \lambda^k) \in N_{X_0}(x^k). \tag{6.57}$$

Since the set X_0 is a convex closed polyhedron, there exists a neighborhood U of $\hat{x}$ such that $X_0 \cap U = \big(\hat{x} + K_{X_0}(\hat{x})\big) \cap U$. Thus, if x^k is sufficiently close to $\hat{x}$, we have

$$N_{X_0}(x^k) = N_{K_{X_0}(\hat{x})}(x^k - \hat{x}).$$

By Example 2.21 on page 27, we have $K_{K_{X_0}(\hat{x})}(x^k - \hat{x}) \supset K_{X_0}(\hat{x})$, and thus $N_{X_0}(x^k) \subset N_{X_0}(\hat{x})$, provided that x^k is sufficiently close to $\hat{x}$. By Lemma 6.44, we can choose ε_2 sufficiently small (but still positive) to guarantee this inclusion. Also, in view of the strict complementarity assumption, we may assume that in this small neighborhood of $\hat{x}$ the active pieces of the "max" terms of the augmented Lagrangian are the same as at $\hat{x}$. Thus, the augmented Lagrangian locally has the form

$$L_\varrho(x, \lambda^k) = f(x) + \sum_{i \in I^0(\hat{x})} \left(\lambda_i^k g_i(x) + \frac{\varrho}{2}\left[g_i(x)\right]^2\right) - \frac{1}{2\varrho} \sum_{i \notin I^0(\hat{x})} (\lambda_i^k)^2. \quad (6.58)$$

This yields (locally) the equation

$$\nabla_x L_\varrho(x^k, \lambda^k) = \nabla f(x^k) + \sum_{i \in I^0(\hat{x})} \left(\lambda_i^k + \varrho g_i(x^k)\right) \nabla g_i(x^k) = \nabla_x L(x^k, \widetilde{\lambda}^k),$$

with $\widetilde{\lambda}_i^k = \lambda_i^k + \varrho g_i(x^k)$ for $i \in I^0(\hat{x})$, and $\widetilde{\lambda}_i^k = 0$ for $i \notin I^0(\hat{x})$. We conclude from (6.57) that the pair $(x^k, \widetilde{\lambda}^k)$ satisfies the relations

$$-\nabla_x L(x^k, \widetilde{\lambda}^k) \in N_{X_0}(\hat{x}),$$
$$x^k \in X_0.$$

This is almost the same as (6.53); the second relation of (6.53) may be violated. System (6.53) is metrically regular, by virtue of Lemma 6.14. Therefore there exist $\varepsilon > 0$ and $C_2 > 0$ such that if $\|x^k - \hat{x}\| \le \varepsilon$ and $\|\widetilde{\lambda}^k - \hat{\lambda}\| \le \varepsilon$ then we can find a solution $(x_{\mathrm{R}}, \lambda_{\mathrm{R}})$ of (6.53) having distance at most $C_2\|g^0(x^k)\|$ to $(x^k, \widetilde{\lambda}^k)$. But $(\hat{x}, \hat{\lambda})$ is an isolated solution of system (6.53). It follows that $x_{\mathrm{R}} = \hat{x}$ and $\lambda_{\mathrm{R}} = \hat{\lambda}$. Hence

$$\|x^k - \hat{x}\| \le C_2\|g^0(x^k)\|$$

and

$$\|\widetilde{\lambda}^k - \hat{\lambda}\| \le C_2\|g^0(x^k)\|,$$

provided that $\varepsilon > 0$ is sufficiently small. By Lemma 6.13, we can choose ε_2 small enough so that the above estimate holds true whenever $\|\lambda^k - \hat{\lambda}\| \le \varepsilon_2$.

Finally, we notice that $\lambda^{k+1} = \Pi_\Lambda(\widetilde{\lambda}^k)$. Since the projection operation is nonexpansive (Theorem 2.13 on page 23), we also have

$$\|\lambda^{k+1} - \hat{\lambda}\| \le C_2\|g^0(x^k)\|,$$

as required. □

We are now ready to prove the main local convergence theorem.

THEOREM 6.16. *There exist $\varepsilon > 0$, $\varrho_0 \geq 0$ and $0 < M < \varrho_0$ such that if $\varrho \geq \varrho_0$ and $\|\lambda^1 - \hat{\lambda}\| \leq \varepsilon$ then the iterates of the multiplier method satisfy the relations:*

$$\|\lambda^{k+1} - \hat{\lambda}\| \leq \frac{M}{\varrho}\|\lambda^k - \hat{\lambda}\|, \quad k = 1, 2, \dots. \tag{6.59}$$

Proof. Consider iteration k. Assume that ϱ is sufficiently large and λ^k is sufficiently close to $\hat{\lambda}$, so that the estimates of Lemmas 6.13 and 6.15 hold true.

Since $\hat{x}$ is a feasible solution of problem (6.51),

$$L_\varrho(x^k, \lambda^k) \leq f(\hat{x}).$$

Using the local representation (6.58) of the augmented Lagrangian we obtain

$$\frac{\varrho}{2} \sum_{i \in I^0(\hat{x})} \big[g_i(x^k)\big]^2 \leq f(\hat{x}) - f(x^k) - \sum_{i \in I^0(\hat{x})} \lambda_i^k g_i(x^k).$$

The inactive inequality constraints do not play any role here. To simplify notation, we assume that all constraints are active, that is, $I^0(\hat{x}) = \{1, \dots, m\}$. Then we can rewrite the last inequality simply as

$$\begin{aligned} \frac{\varrho}{2}\|g(x^k)\|^2 &\leq f(\hat{x}) - f(x^k) - \langle \lambda^k, g(x^k) \rangle \\ &= \langle \hat{\lambda} - \lambda^k, g(x^k) \rangle - \big[L(x^k, \hat{\lambda}) - L(\hat{x}, \hat{\lambda})\big]. \end{aligned} \tag{6.60}$$

Since the Lagrangian $L(x, \hat{\lambda})$ is twice continuously differentiable about $\hat{x}$, there exists a constant $\beta > 0$ such that

$$L(x^k, \hat{\lambda}) - L(\hat{x}, \hat{\lambda}) \geq \langle \nabla L(\hat{x}, \hat{\lambda}), x^k - \hat{x} \rangle - \beta\|x^k - \hat{x}\|^2.$$

Using the necessary condition of optimality at $\hat{x}$ we notice that the first term on the right hand side is nonnegative, and thus

$$L(x^k, \hat{\lambda}) - L(\hat{x}, \hat{\lambda}) \geq -\beta\|x^k - \hat{x}\|^2.$$

Substituting this into (6.60) we obtain

$$\frac{\varrho}{2}\|g(x^k)\|^2 \leq \|g(x^k)\|\|\lambda^k - \hat{\lambda}\| + \beta\|x^k - \hat{x}\|^2.$$

Lemma 6.15 allows us to transform this inequality to

$$\frac{\varrho}{2}\|g(x^k)\|^2 \leq \|g(x^k)\|\|\lambda^k - \hat{\lambda}\| + \beta C_2^2\|g(x^k)\|^2.$$

If $\varrho > 4\beta C_2^2$, we conclude that

$$\|g(x^k)\| \le \frac{4}{\varrho}\|\lambda^k - \hat{\lambda}\|.$$

Lemma 6.15 yields

$$\|\lambda^{k+1} - \hat{\lambda}\| \le C_2\|g(x^k)\| \le \frac{4C_2}{\varrho}\|\lambda^k - \hat{\lambda}\|.$$

If $\varrho > M \ge 4C_2$, then (6.59) is satisfied and $\|\lambda^{k+1} - \hat{\lambda}\| < \|\lambda^k - \hat{\lambda}\|$. Therefore, if $\|\lambda^1 - \hat{\lambda}\|$ is sufficiently small and ϱ is sufficiently large, our estimates hold true for all iterations $k = 1, 2, \ldots$. □

It follows from the theorem that for $\varrho > M$ the sequence $\{\lambda^k\}$ generated by the method of multipliers is linearly convergent to $\hat{\lambda}$ with the coefficient

$$q = \frac{M}{\varrho}.$$

Also, in view of the stability property of Lemma 6.44, the sequence $\{x^k\}$ of the minimum points of the augmented Lagrangian is convergent to $\hat{x}$. Upper bounds on the distances $\|x^k - \hat{x}\|$ are provided by a geometric progression with ratio q. The penalty coefficient ϱ affects the speed of convergence: the larger ϱ, the smaller q. However, for very large values of ϱ, subproblems (6.51) become ill-conditioned, and their solution by an iterative method for unconstrained optimization or for linearly constrained optimization becomes difficult.

In applications, we usually adjust the penalty coefficient in the course of computation. For some fixed parameter $\gamma \in (0, 1)$ we verify whether

$$\max\big(0, g_i(x^{k+1})\big) \le \gamma \max\big(0, g_i(x^k)\big), \quad i = 1, \ldots, m.$$

Similar tests are made for the absolute values of the equality constraints. If the condition is satisfied, we keep ϱ unchanged; otherwise we increase ϱ. In fact, it is better to use different values of the penalty coefficient for different constraints, and to adjust them according to the progress in the violations of these constraints. Our analysis of the quadratic penalty method suggests that for sufficiently large values of the penalty coefficients a neighborhood of the local minimum will be reached. Then, in a sufficiently regular case, the convergence estimates of Theorem 6.16 begin to work.

6.4.3 Application to Convex Problems

We now focus on the case of a *convex* problem (6.48). Specifically, we assume that the functions $f(\cdot)$ and $g_i(\cdot)$, $i = 1, \ldots, m$, are convex, the functions $h_i(\cdot)$, $i = 1, \dots, p$ are affine, and X_0 is a closed convex set. In this case the augmented Lagrangian is a convex function of x, for all nonnegative values of the penalty parameter ϱ.

Consider a fixed value $\varrho > 0$ and the *regularized Lagrangian function*

$$\ell(x, \lambda, \mu; \bar{\lambda}, \bar{\mu}) = f(x) + \langle \lambda, g(x) \rangle + \langle \mu, h(x) \rangle - \frac{1}{2\varrho}\|\lambda - \bar{\lambda}\|^2 - \frac{1}{2\varrho}\|\mu - \bar{\mu}\|^2.$$

Here $(\bar{\lambda}, \bar{\mu}) \in \Lambda_0$ is a fixed regularization center.

Our first observation is that by maximizing this function we obtain the augmented Lagrangian.

LEMMA 6.17. $L_\varrho(x, \bar{\lambda}, \bar{\mu}) = \max_{(\lambda,\mu) \in \Lambda_0} \ell(x, \lambda, \mu; \bar{\lambda}, \bar{\mu})$.

Proof. The problem on the right hand side can be solved in a closed form; its solution is

$$\begin{aligned} \widetilde{\lambda} &= \max\big(0, \bar{\lambda} + \varrho g(x)\big), \\ \widetilde{\mu} &= \bar{\mu} + \varrho h(x). \end{aligned}$$

The maximum above is taken component-wise. Consider a particular constraint function $g_i(\cdot)$. The corresponding term in $\ell(\cdot)$ has the form

$$\begin{aligned} \ell_i(x, \bar{\lambda}_i) &= \widetilde{\lambda}_i g_i(x) - \frac{1}{2\varrho}(\widetilde{\lambda}_i - \bar{\lambda}_i)^2 \\ &= \max\big(0, \bar{\lambda}_i + \varrho g_i(x)\big) g_i(x) - \frac{1}{2\varrho}\big[\max\big(-\bar{\lambda}_i, \varrho g_i(x)\big)\big]^2 \\ &= \varrho \max\big(0, \frac{\bar{\lambda}_i}{\varrho} + g_i(x)\big) g_i(x) - \frac{\varrho}{2}\big[\max\big(-\frac{\bar{\lambda}_i}{\varrho}, g_i(x)\big)\big]^2. \end{aligned}$$

Manipulating in a way similar to the transformations on page 202, we obtain equivalent expressions corresponding to the constraints in the augmented Lagrangian:

$$\begin{aligned} \ell_i(x, \bar{\lambda}_i) &= \bar{\lambda}_i \max\big(g_i(x), -\frac{\bar{\lambda}_i}{\varrho}\big) + \frac{\varrho}{2}\Big[\max\big(g_i(x), -\frac{\bar{\lambda}_i}{\varrho}\big)\Big]^2 \\ &= \frac{\varrho}{2}\Big[\max\big(0, g_i(x) + \frac{\bar{\lambda}_i}{\varrho}\big)\Big]^2 - \frac{1}{2\varrho}\bar{\lambda}_i^2. \end{aligned}$$

This can also be verified directly by considering the case $\bar{\lambda}_i + \varrho g_i(x) > 0$ and its complement. The manipulations for equality constraints are easier, because no cases need to be distinguished. □

In view of Lemma 6.17, problem (6.51), which is solved at each iteration of the multiplier method, may be equivalently written as follows:

$$\underset{x \in X_0}{\text{minimize}} \max_{(\lambda,\mu) \in \Lambda_0} \ell(x, \lambda, \mu; \lambda^k, \mu^k). \tag{6.61}$$

Let us define

$$\begin{aligned} \widetilde{\lambda}^k &= \max\big(0, \lambda^k + \varrho g(x^k)\big), \\ \widetilde{\mu}^k &= \mu^k + \varrho h(x^k). \end{aligned}$$

The key to understanding the augmented Lagrangian method in the convex case is the following lemma.

Lemma 6.18. *The pair $(x^k, (\widetilde{\lambda}^k, \widetilde{\mu}^k))$ is a saddle point of the function $\ell(x, \lambda, \mu; \lambda^k, \mu^k)$ on the set $X_0 \times \Lambda_0$.*

Proof. The function $\ell(x, \lambda, \mu; \lambda^k, \mu^k)$ is convex with respect to x and concave with respect to (λ, μ). By construction,

$$\ell(x^k, \widetilde{\lambda}^k, \widetilde{\mu}^k; \lambda^k, \mu^k) = \max_{(\lambda,\mu) \in \Lambda_0} \ell(x^k, \lambda, \mu; \lambda^k, \mu^k).$$

The point x^k minimizes $L_\varrho(x, \lambda^k, \mu^k)$ over $x \in X_0$. Assume, for simplicity of notation, that there are no equality constraints ($p = 0$). We can calculate the gradient of the augmented Lagrangian:

$$\begin{aligned} \nabla_x L_\varrho(x^k, \lambda^k) &= \nabla f(x^k) + \sum_{i=1}^{m} \max\big(0, \varrho g_i(x^k) + \lambda_i^k\big) \nabla g_i(x^k) \\ &= \nabla f(x^k) + \sum_{i=1}^{m} \widetilde{\lambda}_i^k \nabla g_i(x^k). \end{aligned}$$

This is the same as the gradient of $L(\cdot, \widetilde{\lambda}^k)$ at x^k. Therefore the point x^k is also a minimum of $L(x, \lambda^k)$ over $x \in X_0$. The function $\ell(\cdot)$, as a function of x, is just a translation of the Lagrangian, and we obtain

$$\ell(x^k, \widetilde{\lambda}^k, \widetilde{\mu}^k; \lambda^k \mu^k) = \min_{x \in X_0} \ell(x, \widetilde{\lambda}^k, \widetilde{\mu}^k; \lambda^k \mu^k).$$

This completes the verification of the saddle point conditions. □

This lemma allows us to exchange the "min" and "max" operations in problem (6.61) and to write problem (6.51) equivalently as

$$\underset{(\lambda,\mu) \in \Lambda_0}{\text{maximize}} \min_{x \in X_0} \ell(x, \lambda, \mu; \lambda^k \mu^k).$$

The minimization with respect to x yields the usual dual function:

$$\begin{aligned}\min_{x\in X_0} \ell(x,\lambda,\mu;\lambda^k\mu^k) &= \min_{x\in X_0}\Big[L(x,\lambda,\mu) - \frac{1}{2\varrho}\|\lambda-\lambda^k\|^2 - \frac{1}{2\varrho}\|\mu-\mu^k\|^2\Big]\\ &= L_D(\lambda,\mu) - \frac{1}{2\varrho}\|\lambda-\lambda^k\|^2 - \frac{1}{2\varrho}\|\mu-\mu^k\|^2.\end{aligned}$$

Therefore the multiplier method (6.51)–(6.52) can be equivalently represented as follows. At iteration k we solve the problem

$$\underset{(\lambda,\mu)\in\Lambda_0}{\text{maximize}}\ L_D(\lambda,\mu) - \frac{1}{2\varrho}\|\lambda-\lambda^k\|^2 - \frac{1}{2\varrho}\|\mu-\mu^k\|^2. \tag{6.62}$$

Its solution $(\widetilde{\lambda}^k, \widetilde{\mu}^k)$ is used as the regularization center at the next step:

$$\lambda^{k+1} = \widetilde{\lambda}^k, \qquad \mu^{k+1} = \widetilde{\mu}^k, \tag{6.63}$$

k is replaced by $k+1$, and the iteration continues. Method (6.62)–(6.63) is known in optimization theory as the *proximal point method.* It applies to any convex problem, without assumptions about the differentiability of $L_D(\lambda,\mu)$. We provide the analysis of its convergence in Section 7.3 of the next chapter, which is devoted to nonsmooth optimization methods. Here we briefly summarize the results of that analysis.

It follows from Theorem 7.13 that for each starting point $(\lambda^1,\mu^1)\in\Lambda_0$ the sequence $\{(\lambda^k,\mu^k)\}$ generated by the method is convergent to a solution of the dual problem

$$\underset{(\lambda,\mu)\in\Lambda_0}{\text{maximize}}\ L_D(\lambda,\mu),$$

provided that a solution of this problem exists. This implies that every accumulation point of the sequence $\{x^k\}$ is a solution of the primal problem (6.48). Furthermore, it follows from Theorem 7.14 that the convergence is finite, if the function $L_D(\cdot,\cdot)$ is polyhedral. This is always true, when X_0 is a closed convex polyhedron and all functions of the problem are affine. In practice, however, we rarely treat linear constraints by augmented Lagrangians, except for special decomposition methods.

It should be stressed that the proximal point method (6.62)–(6.63) is in our case not used directly, but rather in an implicit form, by implementing the augmented Lagrangian iterations (6.51)–(6.52).

Example 6.19. Consider the following linear programming problem:

$$\begin{aligned}&\text{minimize}\ x_1+x_2\\ &\text{subject to}\ 1-2x_1-x_2\le 0,\\ &\qquad 0\le x_1\le 1,\\ &\qquad 0\le x_2\le 1.\end{aligned}$$

Its solution is $\hat{x} = (\frac{1}{2}, 0)$ and the optimal value of the multiplier equals $\hat{\lambda} = \frac{1}{2}$.

We shall treat the first inequality as a constraint, and the last two as the definition of the set X_0. The augmented Lagrangian has the form

$$L_\varrho(x_1, x_2, \lambda) = x_1 + x_2 + \frac{\varrho}{2}\Big[\max\Big(0, 1 - 2x_1 - x_2 + \frac{\lambda}{\varrho}\Big)\Big]^2 - \frac{1}{2\varrho}\lambda^2.$$

Suppose $\lambda^1 = 0$ and thus the augmented Lagrangian becomes identical with the quadratic penalty function. Let $\varrho \geq \frac{1}{2}$. Then the minimum is attained at the point

$$x_1^1 = \frac{1}{2} - \frac{1}{4\varrho}, \qquad x_2^1 = 0.$$

The next value of the multiplier equals

$$\lambda^2 = \varrho(1 - 2x_1^1 - x_2^1) = \frac{1}{2}.$$

The minimum of the augmented Lagrangian with $\lambda = \lambda^2$ is attained at $x_1^2 = \frac{1}{2}$, $x_2^2 = 0$, which is the optimal point of the problem. This is illustrated in Figure 6.2, with respect to the first variable, x_1.

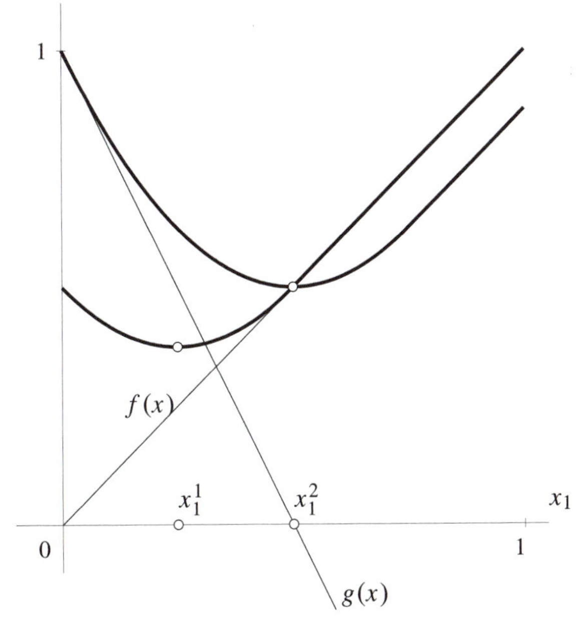

Figure 6.2. Operation of the augmented Lagrangian method in Example 6.19 for $\varrho = 1$.

If $\varrho \in (0, \frac{1}{2})$ then the first minimizer is $x^1 = (0, 0)$. The next multiplier becomes $\lambda^2 = \varrho$. The minimization of the augmented Lagrangian yields

$$x_1^2 = 1 - \frac{1}{4\varrho}, \qquad x_2^2 = 0.$$

The next value of the multiplier becomes $\lambda^3 = \frac{1}{2}$, and the next minimum is $x^3 = (\frac{1}{2}, 0)$.

6.5 NEWTON'S METHOD

6.5.1 Main Ideas

Newton's method, which we analyzed for unconstrained problems in Section 5.4, has its more general form, which is applicable to constrained optimization problems. It can be conveniently introduced for the following form of a nonlinear optimization problem:

$$\begin{aligned} \text{minimize} \ & f(x) \\ \text{subject to} \ & g_i(x) \le 0, \quad i = 1, \dots, m, \\ & h_i(x) = 0, \quad i = 1, \dots, p. \end{aligned} \tag{6.64}$$

We assume that the functions $f : \mathbb{R}^n \to \mathbb{R}$, $g_i : \mathbb{R}^n \to \mathbb{R}$, $i = 1, \dots, m$, and $h_i : \mathbb{R}^n \to \mathbb{R}$, $i = 1, \dots, p$, are twice continuously differentiable. We do not have a general constraint of the form $x \in X_0$ here; rather we use an explicit formulation of such constraints by inequalities and equations.

As usual, we formulate the Lagrangian $L : \mathbb{R}^n \times \mathbb{R}^m \times \mathbb{R}^p \to \mathbb{R}$ in the form

$$L(x, \lambda, \mu) = f(x) + \langle \lambda, g(x) \rangle + \langle \mu, h(x) \rangle.$$

The necessary conditions of optimality for problem (6.64) take on the form of a system of nonlinear equations and inequalities:

$$\begin{aligned} \nabla_x L(x, \lambda, \mu) &= 0, \\ g(x) &\le 0, \\ h(x) &= 0, \\ \lambda &\ge 0, \\ \langle \lambda, g(x) \rangle &= 0. \end{aligned} \tag{6.65}$$

Consider the cone $K = R^m_+ \times \mathbb{R}^p$. System (6.65) can be abstractly expressed as

$$\begin{aligned} &\nabla_x L(x, \lambda, \mu) = 0, \\ &\begin{bmatrix} g(x) \\ h(x) \end{bmatrix} \in N_K(\lambda, \mu). \end{aligned} \tag{6.66}$$

Indeed, the last relation implies that $(\lambda, \mu) \in K$; otherwise, $N_K(\lambda, \mu) = \emptyset$.

Moreover, in view of Example 2.21 on page 27,

$$\begin{aligned} N_K(\lambda,\mu) &= \left[T_K(\lambda,\mu)\right]^\circ = \left[K + \{t(\lambda,0) : t \in \mathbb{R}\}\right]^\circ \\ &= K^\circ \cap \{t(\lambda,0) : t \in \mathbb{R}\}^\perp. \end{aligned}$$

Thus the second relation in (6.66) also implies that $(g(x), h(x)) \in K^\circ = \mathbb{R}^m_- \times \{0\}$ and that $g(x) \perp \lambda$. This proves that (6.66) implies (6.65). The converse implication is apparent, and thus both systems are equivalent.

Motivated by Newton's method of tangents, we construct, at a given point $(\bar{x}, \bar{\lambda})$, an approximation of system (6.66), in which the functions $\nabla_x L(\cdot, \lambda, \mu)$, $g(\cdot)$ and $h(\cdot)$ are linearized:

$$\begin{aligned} &\nabla_x L(\bar{x}, \lambda, \mu) + \nabla^2_{xx} L(\bar{x}, \bar{\lambda}, \bar{\mu})(x - \bar{x}) = 0, \\ &\begin{bmatrix} g(\bar{x}) + g'(\bar{x})(x - \bar{x}) \\ h(\bar{x}) + h'(\bar{x})(x - \bar{x}) \end{bmatrix} \in N_K(\lambda, \mu). \end{aligned} \tag{6.67}$$

Observe that we do not need to linearize the gradient $\nabla_x L(x, \lambda, \mu)$ with respect to λ and μ, because it is already linear with respect to these variables.

We can now show that the linearized system (6.67) is the system of necessary conditions of optimality for the quadratic programming problem:

$$\begin{aligned} \text{minimize}\ \ & \langle \nabla f(\bar{x}), x - \bar{x} \rangle + \frac{1}{2}\big\langle x - \bar{x}, [\nabla^2_{xx} L(\bar{x}, \bar{\lambda}, \bar{\mu})](x - \bar{x}) \big\rangle \\ \text{subject to}\ \ & g(\bar{x}) + g'(\bar{x})(x - \bar{x}) \le 0, \\ & h(\bar{x}) + h'(\bar{x})(x - \bar{x}) = 0. \end{aligned} \tag{6.68}$$

Indeed, the Lagrangian of problem (6.68) has the form

$$\begin{aligned} \bar{L}(x, \lambda, \mu) &= \langle \nabla f(\bar{x}), x - \bar{x} \rangle + \frac{1}{2}\big\langle x - \bar{x}, [\nabla^2_{xx} L(\bar{x}, \bar{\lambda}, \bar{\mu})](x - \bar{x}) \big\rangle \\ &\quad + \langle \lambda, g(\bar{x}) + g'(\bar{x})(x - \bar{x}) \rangle + \langle \mu, h(\bar{x}) + h'(\bar{x})(x - \bar{x}) \rangle. \end{aligned}$$

Its gradient equals

$$\begin{aligned} &\nabla_x \bar{L}(x, \lambda, \mu) \\ &\quad = \nabla f(\bar{x}) + \nabla^2_{xx} L(\bar{x}, \bar{\lambda}, \bar{\mu})(x - \bar{x}) + \left[g'(\bar{x})\right]^T \lambda + \left[h'(\bar{x})\right]^T \mu \\ &\quad = \nabla_x L(\bar{x}, \lambda, \mu) + \nabla^2_{xx} L(\bar{x}, \bar{\lambda}, \bar{\mu})(x - \bar{x}). \end{aligned}$$

It is exactly equal to the expression in the first relation of (6.67). It is now evident that the necessary conditions of optimality for problem (6.68) are exactly conditions (6.67). For this reason we call (6.68) the *tangent quadratic programming problem*.

The Newton method can now be formulated as follows. At iteration k, given the current approximation of the solution x^k and multipliers (λ^k, μ^k), we solve the tangent quadratic programming problem

$$\begin{aligned} \text{minimize} \quad & \langle \nabla f(x^k), d\rangle + \frac{1}{2}\big\langle d, [\nabla^2_{xx} L(x^k, \lambda^k, \mu^k)]d\big\rangle \\ \text{subject to} \quad & g(x^k) + g'(x^k)d \le 0, \\ & h(x^k) + h'(x^k)d = 0. \end{aligned} \tag{6.69}$$

We denote the solution of this problem by d^k and the Lagrange multipliers associated with the constraints by $\hat{\lambda}^k$ and $\hat{\mu}^k$.

Then we update the approximate solution and the multipliers by

$$\begin{aligned} x^{k+1} &= x^k + \tau_k d^k, \\ \lambda^{k+1} &= \hat{\lambda}^k, \\ \mu^{k+1} &= \hat{\mu}^k, \end{aligned} \tag{6.70}$$

and the iteration continues. Here $\tau_k \in (0, 1]$ is a step size coefficient, whose selection is discussed later. Algorithm (6.69)–(6.70) is also called the *sequential quadratic programming method.*

6.5.2 Local Convergence

We start from the analysis of the method with step sizes $\tau_k = 1$ in close proximity to a solution of problem (6.64). To simplify notation, we concentrate on inequality constraints only ($p = 0$). The considerations for equality constraints follow the same line, but are easier.

We assume that problem (6.64) has a local minimum $\hat{x}$, at which the gradients of active constraints, $\nabla g_i(\hat{x})$, $i \in I^0(\hat{x})$, are linearly independent. In this case the Lagrange multipliers $\hat{\lambda}$ exist and are unique. Furthermore, we assume that strong second order sufficient condition (Definition 4.32) is satisfied at the local minimum $\hat{x}$ with the multipliers $\hat{\lambda}$.

Consider the system of nonlinear equations:

$$\begin{aligned} \nabla f(x) + \sum_{i \in I^0(\hat{x})} \lambda_i \nabla g_i(x) = 0, \\ g_i(x) = 0, \quad i \in I^0(\hat{x}). \end{aligned} \tag{6.71}$$

These are necessary conditions of optimality of a modification of problem (6.64), in which the active inequality constraints are treated as equations. By the linear independence condition and by the second order sufficient condition, $(\hat{x}, \hat{\lambda})$ is an isolated solution of this problem. Furthermore, the

Jacobian at this point is given by

$$J = \begin{bmatrix} H & A^T \\ A & 0 \end{bmatrix},$$

with

$$H = \nabla^2_{xx} L(\hat{x}, \hat{\lambda}), \qquad A^T = \begin{bmatrix} \nabla g_1(\hat{x}) & \dots & \nabla g_{m_0}(\hat{x}) \end{bmatrix}.$$

We assume here, for simplicity of notation, that $I^0(\hat{x}) = \{1, \dots, m_0\}$. By virtue of the strong second order sufficient condition, the Jacobian is a nonsingular matrix. Because of that, the classical Newton method for solving systems of nonlinear equations is locally superlinearly convergent. The convergence is quadratic if the Jacobian of the system is Lipschitz continuous about $(\hat{x}, \hat{\lambda})$.

We shall show that the iterates of the sequential quadratic programming method are exactly the iterates of Newton's method for system (6.71). For the current iterate (x^k, λ^k), we introduce the matrices

$$H_k = \nabla^2_{xx} L(x^k, \lambda^k), \qquad A_k^T = \begin{bmatrix} \nabla g_1(x^k) & \dots & \nabla g_{m_0}(x^k) \end{bmatrix}.$$

The Jacobian of system (6.71) at iteration k has the form:

$$J_k = \begin{bmatrix} H_k & A_k^T \\ A_k & 0 \end{bmatrix}. \tag{6.72}$$

It is nonsingular if (x^k, λ^k) is close to $(\hat{x}, \hat{\lambda})$, because the Jacobian at $(\hat{x}, \hat{\lambda})$ is nonsingular (see page 152). Newton's method for the system of nonlinear equations (6.71) takes on the form

$$\begin{bmatrix} x^{k+1} \\ \lambda^{0,k+1} \end{bmatrix} = \begin{bmatrix} x^k \\ \lambda^{0,k} \end{bmatrix} - \begin{bmatrix} H_k & A_k^T \\ A_k & 0 \end{bmatrix}^{-1} \begin{bmatrix} \nabla f(x^k) + \sum_{i \in I^0(\hat{x})} \lambda_i^k \nabla g_i(x^k) \\ g^0(x^k) \end{bmatrix}, \tag{6.73}$$

where $\lambda^{0,k}$ and $g^0(x)$ are vectors with coordinates λ_i^k and $g_i(x^k)$, for $i \in I^0(\hat{x})$. Multiplying from the left by the Jacobian and noticing that

$$\sum_{i \in I^0(\hat{x})} \lambda_i^k \nabla g_i(x^k) = A_k^T \lambda^{0,k}$$

we can simplify this system. We conclude that the direction d^k and the new values of multipliers $\lambda^{0,k+1}$ are the solution of the system

$$\begin{aligned} H_k d^k + A_k^T \lambda^{k+1} &= -\nabla f(x^k), \\ A_k d^k &= 0. \end{aligned}$$

This is the system of necessary conditions of optimality of the tangent quadratic programming problem (with active inequalities treated as equations). Let us observe that for $x^k = \hat{x}$ and $\lambda^k = \hat{\lambda}$ the point $d = 0$ and $\lambda = \hat{\lambda}$ satisfies strong second order sufficient conditions of optimality for the tangent problem (6.69). This follows from the fact that this condition is exactly the same as the assumed strong second order sufficient condition for the original problem (6.64). By the continuity of the Hessian $\nabla^2_{xx} L(x, \lambda)$, we can always choose ε sufficiently small, so that the strong second order sufficient conditions of optimality for problem (6.69) hold true at the solution (d, λ), provided that (x^k, λ^k) are in a ε-neighborhood of $(\hat{x}, \hat{\lambda})$. Therefore, the points generated by the Newton method (6.73) are indeed local minima (and the corresponding multipliers) of the tangent quadratic programming problems (6.69). This proves that tangent quadratic programming problems are just a computationally convenient way of implementing Newton's method.

Our considerations can be extended to the case when the strict complementarity condition is not satisfied, and the problem cannot be locally reduced to an equality-constrained problem. Also, additional set constraints may be included in the problem's formulation. Unfortunately, the analysis in this case draws from the theory of multifunctions, which exceeds the scope of this book. The reader is referred to the literature cited at the end of the book.

6.5.3 Line Search

The sequential quadratic programming method is only a local algorithm, similar to the unconstrained Newton method. In order to enlarge its area of convergence, several numerical techniques can be employed.

It is convenient to consider a modified version of the tangent quadratic programming problem (6.69):

$$\begin{aligned} \text{minimize} \quad & \langle \nabla f(x^k), d \rangle + \frac{1}{2} \langle d, W_k d \rangle \\ \text{subject to} \quad & g(x^k) + g'(x^k) d \le 0, \\ & h(x^k) + h'(x^k) d = 0. \end{aligned} \tag{6.74}$$

In this problem the Hessian of the Lagrangian $\nabla^2_{xx} L(x^k, \lambda^k, \mu^k)$ is replaced by the matrix W_k. This allows us to use a positive definite matrix W_k and to guarantee that problem (6.74) is a convex quadratic programming problem. It can then be reliably solved by specialized methods.

LEMMA 6.20. *Assume that W_k is a positive definite matrix. Then problem* (6.74) *has a solution $(d^k, \hat{\lambda}^k, \hat{\mu}^k)$ with $d^k = 0$ if and only if $(x^k, \hat{\lambda}^k, \hat{\mu}^k)$ satisfy the first order optimality conditions for problem* (6.64).

Proof. The result follows immediately from the fact that the necessary and sufficient conditions of optimality for the quadratic programming problem (6.74) with $d^k = 0$ become identical with the necessary conditions of optimality for problem (6.64). □

Of course, under our modification, the relation of the sequential quadratic programming method to Newton's method becomes remote, and additional modifications are needed to stabilize the method.

Similar to the unconstrained case, we introduce a line search procedure. In the constrained case, however, we cannot simply minimize the objective function $f(\cdot)$ along the ray $x^k + \tau d^k$ over $\tau \geq 0$. We have to mind the constraints.

A way to address this issue is to employ the exact penalty function

$$\Phi_\varrho(x) \triangleq f(x) + \varrho\Big[\sum_{i=1}^{m} \max\big(0, g_i(x)\big) + \sum_{i=1}^{p} |h_i(x)|\Big]. \tag{6.75}$$

We know from Section 6.2.3 that, under mild conditions, a local minimum of problem (6.64) is also a local minimum of the function $\Phi_\varrho(\cdot)$, provided that the penalty coefficient ϱ is larger than the "max" norm of the Lagrange multipliers $(\hat{\lambda}, \hat{\mu})$ at the solution. We can, therefore, use function (6.75) as the *merit function*, to measure the progress toward the local minimum of the problem. Let us stress once again that $\Phi_\varrho(\cdot)$ is a nonsmooth function, and its minimization cannot, in general, be effectively accomplished by numerical methods for minimizing smooth problems. In our case, however, the specific form of the function allows some useful observations about the method.

We know from the proof of Theorem 6.9 that the function $\Phi_\varrho(\cdot)$ has the directional derivative in every direction. The following lemma shows that the quadratic programming problem (6.74) generates a descent direction for the function (6.75).

LEMMA 6.21. *Assume that* W_k *is a positive definite matrix and that problem* (6.74) *has a solution* $(d^k, \hat{\lambda}^k, \hat{\mu}^k)$. $(d^k, \hat{\lambda}^k, \hat{\mu}^k)$. *If* $\varrho > \|(\hat{\lambda}^k, \hat{\mu}^k)\|_\infty$, *then*

$$\Phi'_\varrho(x^k; d^k) \leq -\langle d^k, W_k d^k\rangle.$$

Proof. Consider a function $g_i(x)$. By the convexity of the "max" function, we have

$$\begin{aligned}
&\max\big(0, g_i(x^k) + \tau g_i'(x^k)d^k\big)\\
&\qquad = \max\big(0, (1-\tau)g_i(x^k) + \tau(g_i(x^k) + g_i'(x^k)d^k)\big)\\
&\qquad \leq (1-\tau)\max\big(0, g_i(x^k)\big) + \tau\max\big(0, g_i(x^k) + g_i'(x^k)d^k\big)\\
&\qquad \leq (1-\tau)\max\big(0, g_i(x^k)\big).
\end{aligned}$$

In the last transformation we also used the inequality constraint of problem (6.74). Hence

$$\lim_{\tau\downarrow 0}\frac{1}{\tau}\big[\max\big(0, g_i(x^k+\tau d^k)\big) - \max\big(0, g_i(x^k)\big)\big]$$
$$= \lim_{\tau\downarrow 0}\frac{1}{\tau}\big[\max\big(0, g_i(x^k)+\tau g_i'(x^k)d^k\big) - \max\big(0, g_i(x^k)\big)\big]$$
$$\le -\max\big(0, g_i(x^k)\big).$$

Similarly,

$$|h_i(x^k)+\tau h_i'(x^k)d^k| \le (1-\tau)|h_i(x^k)|,$$

and

$$\lim_{\tau\downarrow 0}\frac{1}{\tau}\big[|h_i(x^k+\tau d^k)| - |h_i(x^k)|\big] \le -|h_i(x^k)|.$$

This allows us to estimate the directional derivative of $\Phi_\varrho(\cdot)$ as follows:

$$\Phi_\varrho'(x^k; d^k) \le \langle\nabla f(x^k), d^k\rangle - \varrho\sum_{i=1}^m \max\big(0, g_i(x^k)\big) - \varrho\sum_{i=1}^p |h_i(x^k)|.$$

Observe that it follows from the optimality conditions of problem (6.74) that

$$\nabla f(x^k) + W_k d^k + \big[g'(x^k)\big]^T\hat{\lambda}^k + \big[h'(x^k)\big]^T\hat{\mu}^k = 0.$$

Thus the estimate of the directional derivative can be rewritten as follows:

$$\Phi_\varrho'(x^k; d^k) \le -\langle d^k, W_k d^k\rangle - \langle g'(x^k)d^k, \hat{\lambda}^k\rangle - \langle h'(x^k)d^k, \hat{\mu}^k\rangle$$
$$-\varrho\sum_{i=1}^m \max\big(0, g_i(x^k)\big) - \varrho\sum_{i=1}^p |h_i(x^k)|.$$

Furthermore, the complementarity condition for the quadratic programming problem (6.74) yields

$$\langle g'(x^k)d^k, \hat{\lambda}^k\rangle = -\langle g(x^k), \hat{\lambda}^k\rangle.$$

Substituting this into the estimate of the directional derivative we obtain

$$\Phi_\varrho'(x^k; d^k) \le -\langle d^k, W_k d^k\rangle + \langle g(x^k), \hat{\lambda}^k\rangle + \langle h(x^k), \hat{\mu}^k\rangle$$
$$-\varrho\sum_{i=1}^m \max\big(0, g_i(x^k)\big) - \varrho\sum_{i=1}^p |h_i(x^k)|.$$

If $\varrho > \hat{\lambda}_i^k$, then $g_i(x^k)\hat{\lambda}_i^k \le \varrho\max\big(0, g_i(x^k)\big)$. Similarly, if $\varrho > |\hat{\mu}_i^k|$, then $h_i(x^k)\hat{\mu}_i^k \le \varrho|h_i(x^k)|$. Consequently, the last displayed inequality implies the assertion of the lemma. □

It follows from the last lemma that for a sufficiently large ϱ the directional minimization of the merit function makes good sense:

$$\underset{\tau \geq 0}{\text{minimize}} \; \Phi_\varrho(x^k + \tau d^k).$$

Nevertheless, this in itself is not sufficient to guarantee convergence of the sequential quadratic programming method to a local minimum of problem (6.64).

Another way of avoiding excessively long steps in bad directions is the use of the trust region idea, as in Section 5.7. We augment problem (6.74) with a constraint $\|d\| \leq \Delta$, where $\Delta > 0$ is our trust region radius. If no good steps can be made, Δ is decreased. Then, however, infeasibility of problem (6.74) may occur, and further modifications are needed.

Existing theoretical results about the properties of the method with line search or with trust regions contain many assumptions, some of which are difficult to verify. The situation becomes more complicated when quasi-Newton updates are used to construct the matrices W_k. Any of the formulas discussed in Section 5.6 may be used here, but the resulting methods have mainly a heuristic character. Their numerical implementation involves many algorithmic tricks aimed at stabilizing the iterates.

6.6 BARRIER METHODS

6.6.1 Main Ideas

We now focus on the problem

$$\begin{aligned} &\text{minimize} \; f(x) \\ &\text{subject to} \; g_i(x) \leq 0, \quad i = 1, \ldots, m. \end{aligned} \tag{6.76}$$

We assume that all functions involved are twice continuously differentiable. For simplicity of presentation we do not include equality constraints. We also do not have additional set constraints of the form $x \in X_0$ here. Although the barrier methods, which we present in this section, can also be defined with such additional set conditions, their numerical implementation becomes more difficult.

Problem (6.76) can be converted to an equivalent problem with linear inequalities and nonlinear equations:

$$\begin{aligned} &\text{minimize} \; f(x) \\ &\text{subject to} \; g_i(x) + z_i = 0, \quad i = 1, \ldots, m, \\ &\qquad\qquad\; z_i \geq 0, \quad i = 1, \ldots, m. \end{aligned} \tag{6.77}$$

Our next step is to eliminate the inequality constraints by formulating an approximate problem

$$\begin{aligned}
&\text{minimize } \ f(x) - \sigma \sum_{i=1}^{m} \ln(z_i) \\
&\text{subject to } \ g_i(x) + z_i = 0, \quad i = 1, \dots, m.
\end{aligned} \tag{6.78}$$

In this problem σ is a positive parameter, and we assume implicitly that we look for a solution with $z_i > 0$. The idea of problem (6.78) is that the functions $-\sigma \ln(z_i)$ create a "barrier" close to the boundary of the cone $\mathbb{R}^m_+$, preventing z_i from becoming too close to this boundary. We hope that when $\sigma \to 0$ the solution of problem (6.78) approaches a solution of problem (6.77). The method of solving a constrained problem by a sequence of approximate problems (6.78) is called the *barrier method.*

For our approach to make any sense at all, we have to be sure that there exists a point x_S such that $g_i(x_S) < 0$ for all $i = 1, \dots, m$. Then problem (6.78) has a feasible solution for all $\sigma > 0$. We assume this condition throughout.

LEMMA 6.22. *Assume that the feasible set of problem* (6.76) *is bounded. Then for every $\sigma > 0$ problem* (6.78) *has a solution* $(x(\sigma), z(\sigma))$.

Proof. Consider the function

$$\Phi_\sigma(x, z) = f(x) - \sigma \sum_{i=1}^{m} \ln(z_i).$$

Denote by X the feasible set of problem (6.76). The feasible set of problem (6.78) has the form:

$$S = \{(x, z) \in X \times \mathbb{R}^m_+ : g(x) = z, \ z_i > 0, \ i = 1, \dots, m\}. \tag{6.79}$$

It is nonempty, because the point $(x_S, g(x_S))$ is its element. Define

$$\beta = \Phi_\sigma(x_S, g(x_S)).$$

The set

$$S^\beta_\sigma = \{(x, z) \in S : \Phi_\sigma(x, z) \le \beta\}$$

is compact, as can be verified directly from the definition. Indeed, take a convergent sequence of points $(x^k, z^k) \in S^\beta_\sigma$, and denote its limit by $(\bar{x}, \bar{z})$. The points x^k are elements of the compact set X and thus $\bar{x} \in X$. The continuity of $g(\cdot)$ and the definition of S imply that $\bar{z} = g(\bar{x})$. From the definition of the set S_σ^β we see that

$$-\sigma \sum_{i=1}^{m} \ln(z_i^k) \leq \beta - f(x^k).$$

As all points x^k belong to the compact set X, there exists a constant C such that $f(x^k) \geq C$ for all k. Hence

$$-\sigma \sum_{i=1}^{m} \ln(z_i^k) \leq \beta - C.$$

The sequence $\{z^k\}$ is bounded as well:

$$z_i^k \leq M_i \triangleq \sup\{g_i(x) : x \in X\}, \quad i = 1, \ldots, m.$$

The last two inequalities imply that

$$-\sigma \ln(z_i^k) \leq \beta - C + \sigma \sum_{j \neq i} \ln(M_j).$$

Therefore we can find $\varepsilon > 0$ such that $z_i^k \geq \varepsilon$ for all k and all $i = 1, \ldots, m$. Then $\bar{z}_i \geq \varepsilon$. The function $\Phi_\sigma(x, z)$ is continuous at $(\bar{x}, \bar{z})$ and thus $(\bar{x}, \bar{z}) \in S_\sigma^\beta$. This proves the compactness of the set S_σ^β. Every optimal solution of the problem

$$\begin{aligned} &\text{minimize} \ \ \Phi_\sigma(x, z) \\ &\text{subject to} \ \ (x, z) \in S_\sigma^\beta \end{aligned} \tag{6.80}$$

is also an optimal solution of problem (6.78). Problem (6.80) has an optimal solution for every $\sigma > 0$, owing to the continuity of its objective function and to the compactness of is feasible set. □

The function $\sigma \mapsto x(\sigma)$ is called the *central path*.

We can now prove a general property of the barrier method, which corresponds to Theorem 6.6 about penalty methods.

The nature of problem (6.78) is that its solution approximates the solution of problem (6.77) "from within," that is, by points (x, z) such that $z_i > 0$ for $i = 1, \ldots, m$. This is the reason barrier methods are called *interior point methods*. In order to prove convergence of the solutions of the approximate problem (6.78) to a solution of problem (6.77) we need to make an assumption that every neighborhood of each feasible point contains an "interior" feasible point, in the sense discussed here. We assume that the closure of the set (6.79) is the entire feasible set of problem (6.77):

$$\overline{S} = \{(x, z) \in X \times \mathbb{R}^m_+ : g(x) = z,\ z \geq 0\}. \tag{6.81}$$

We can now prove a property of the barrier method, which corresponds to Theorem 6.6 about general penalty methods.

THEOREM 6.23. *Assume that the feasible set of problem* (6.76) *is bounded and that condition* (6.81) *is satisfied. If $\sigma_k \to 0$, as $k \to \infty$, then every accumulation point of the sequence $(x(\sigma_k), z(\sigma_k))$ is a solution of problem* (6.77).

Proof. By assumption, the set $\overline{S}$ is the feasible set of problem (6.77). By virtue of Lemma 6.76, the sequence $(x(\sigma_k), z(\sigma_k))$ is well defined. Also, $(x(\sigma_k), z(\sigma_k)) \in S$ for all k. As the sequence $(x(\sigma_k), z(\sigma_k))$ is bounded, with no loss of generality we may assume that it is convergent to some point $(\bar{x}, \bar{z}) \in \overline{S}$. Consider an arbitrary point $(x, z) \in S$. By the optimality of the point $(x(\sigma_k), z(\sigma_k))$,

$$f(x(\sigma_k)) - \sigma_k \sum_{i=1}^{m} \ln(z_i(\sigma_k)) \leq f(x) - \sigma_k \sum_{i=1}^{m} \ln(z_i).$$

If $\bar{z}_i > 0$, then $\lim_{k\to\infty} \sigma_k \ln(z_i(\sigma_k)) = 0$. If $\bar{z}_i = 0$, then $\ln(z_i(\sigma_k)) < 0$ for all sufficiently large k. Thus $\limsup_{k\to\infty} \sigma_k \ln(z_i(\sigma_k)) \leq 0$ in both cases. Passing to the limit when $k \to \infty$ in the last displayed inequality, we conclude that

$$f(\bar{x}) \leq f(x).$$

The point $(x, z) \in S$ was arbitrary and thus the last inequality holds true for all $(x, z) \in S$. By the continuity of $f(\cdot)$, this inequality also holds true for all $(x, z) \in \overline{S}$. By virtue of assumption (6.81), these are all feasible points (x, z) of problem (6.77). □

Theorem 6.23 is valid in a more general version, in which the barrier function in (6.78), $-\sigma \sum_{i=1}^{m} \ln(z_i)$, may be replaced by other functions $B(z)$, having the property that $B(z) \to +\infty$, as z approaches the boundary of the nonnegative orthant $\mathbb{R}^n_+$. However, using logarithmic penalty functions is convenient for the efficient Newton method discussed in the next subsection.

Example 6.24. Let us illustrate the operation of the barrier method on the problem

$$\begin{aligned} &\text{minimize} \ \ f(x) \triangleq \frac{3}{2} - x \\ &\text{subject to} \ \ g(x) \triangleq x - 1 \leq 0. \end{aligned}$$

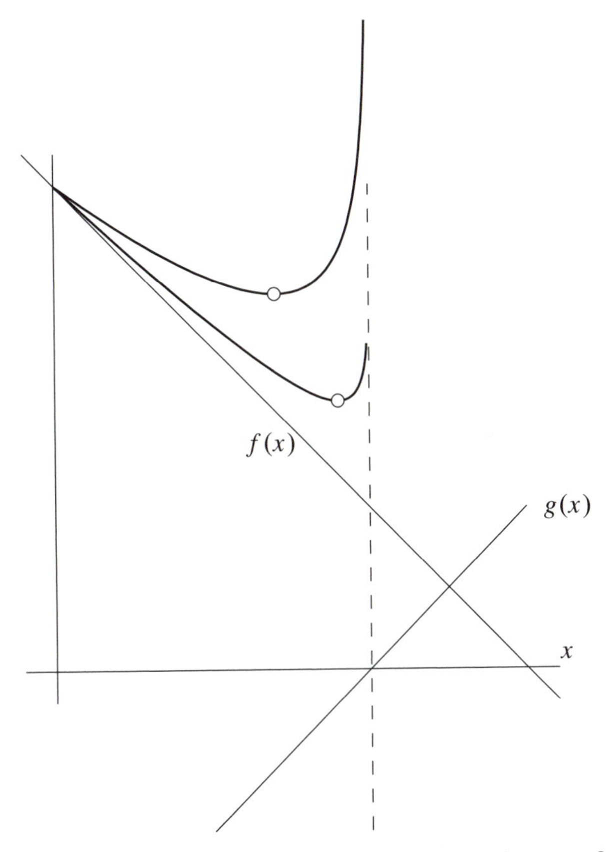

Figure 6.3. Operation of barrier function in Example 6.24 for $\sigma = 0.3$ (top) and $\sigma = 0.1$ (bottom). The vertical dashed line marks the boundary of the feasible set.

Its solution is $\hat{x} = 1$. Problem (6.78) takes on the form

$$\begin{aligned}
&\text{minimize} \ \frac{3}{2} - x - \sigma \ln(z) \\
&\text{subject to} \ x - 1 + z = 0.
\end{aligned}$$

By assigning multiplier λ to the equality constraint we formulate the necessary conditions of optimality

$$\begin{aligned}
-1 + \lambda &= 0, \\
-\frac{\sigma}{z} + \lambda &= 0.
\end{aligned}$$

It follows that $\lambda = 1$ and $z = \sigma$. Then $x = 1 - z = 1 - \sigma$. When $\sigma \downarrow 0$ then $x \to 1$. The operation of the method is illustrated in Figure 6.3, after eliminating z

and converting the problem to the unconstrained problem

$$\text{minimize } f(x) - \sigma \ln(-g(x)) = \frac{3}{2} - x - \sigma \ln(1 - x).$$

Example 6.25. The logarithmic barrier function has an interesting form in the semidefinite programming problem of Example 3.36 on page 129.

The condition that X is a positive semidefinite matrix of dimension n, $X \in \mathbb{S}^n_+$, can be formulated as a linear constraint $X \in \mathbb{S}^n$ together with the system of inequalities

$$\lambda_1(X) \geq 0, \ \lambda_2(X) \geq 0, \ \ldots, \ \lambda_n(X) \geq 0,$$

where $\lambda_1(X), \lambda_2(X), \ldots, \lambda_n(X)$ are the eigenvalues of X. The logarithmic barrier function takes on the form

$$B(X) \triangleq -\sum_{i=1}^{n} \ln\left(\lambda_i(X)\right) = -\ln\left(\prod_{i=1}^{n} \lambda_i(X)\right) = -\ln(\det X).$$

The order of the eigenvalues does not matter here.

We shall prove that the function $B(X)$ is convex and differentiable on the set of positive definite matrices, with

$$\nabla B(X) = -X^{-1}.$$

This can be obtained from the first principles, but we prefer an approach using the properties of the eigenvectors. Consider the optimization problem:

$$\begin{aligned} &\text{maximize} \quad -\sum_{j=1}^{n} \ln\left(\langle u_j, X u_j \rangle\right) \\ &\text{subject to } \langle u_i, u_j \rangle = \begin{cases} 1 & \text{if } i = j, \\ 0 & \text{if } i \neq j, \end{cases} \qquad i, j = 1, \ldots, n. \end{aligned} \tag{6.82}$$

The variables in this problem are the vectors $u_1, \ldots, u_n \in \mathbb{R}^n$. Define the matrix

$$U = \begin{bmatrix} u_1 & u_2 & \cdots & u_n \end{bmatrix}.$$

The objective function of problem (6.82) depends only on the diagonal elements of the matrix $U^T X U$, and the constraint is $U^T U = I$ (U is a unitary matrix). Let

$$Z = \begin{bmatrix} z_1 & z_2 & \cdots & z_n \end{bmatrix}$$

be the matrix of eigenvectors of X of length 1, and let $\lambda_1, \lambda_2, \ldots, \lambda_n$ be the corresponding eigenvalues. We can now change the variables by $U = ZQ$. Since Z is a unitary matrix, U is unitary if and only if Q is unitary. Thus we can equivalently transform problem (6.82) to a problem with a diagonal matrix:

$$\underset{Q^T Q = I}{\text{maximize}} \ -\sum_{j=1}^{n} \ln\left(\langle q_j, \Lambda q_j \rangle\right) \tag{6.83}$$

with q_j denoting the jth column of Q and with $\Lambda = \operatorname{diag}\{\lambda_1, \dots, \lambda_n\}$. As Q is a unitary matrix, $\sum_{i=1}^{n} q_{ij}^2 = 1$. The function $\ln(\cdot)$ is strictly concave, and therefore

$$\ln\big(\langle q_j, \Lambda q_j\rangle\big) = \ln\Big(\sum_{i=1}^{n} q_{ij}^2 \lambda_i\Big) \geq \sum_{i=1}^{n} q_{ij}^2 \ln(\lambda_i),$$

with equality occurring if and only if q_j is a unit vector. Substituting this estimate into the objective function of (6.83) and using the fact that Q is a unitary matrix, we obtain

$$\begin{aligned} -\sum_{j=1}^{n} \ln\big(\langle q_j, \Lambda q_j\rangle\big) &\leq -\sum_{j=1}^{n}\sum_{i=1}^{n} q_{ij}^2 \ln(\lambda_i) \\ &= -\sum_{i=1}^{n} \ln(\lambda_i) \sum_{j=1}^{n} q_{ij}^2 = -\sum_{i=1}^{n} \ln(\lambda_i), \end{aligned}$$

with equality occurring if and only if Q is a permutation matrix. It follows that the set of solutions of problem (6.83) are all permutation matrices and that the optimal value equals

$$-\sum_{i=1}^{n} \ln(\lambda_i) = -\ln\Big(\prod_{i=1}^{n} \lambda_i\Big) = -\ln(\det X).$$

The optimal solutions of (6.82) are all matrices of form $\hat{U} = ZQ$ with a permutation matrix Q, that is, matrices $\hat{U}$ having the eigenvectors $z_1, \dots, z_n$ of X as their columns.

Each function $X \mapsto \langle u_j, Xu_j\rangle$ is linear and the function $-\ln(\cdot)$ is convex. Hence the objective function of (6.82) is a convex function of X. By virtue of Lemma 2.58 on page 46, the optimal value function $-\ln(\det X)$ is a convex function of X.

We have

$$\nabla \ln\big(\langle z_j, Xz_j\rangle\big) = \frac{1}{\lambda_j} z_j z_j^T.$$

It follows from Theorem 2.87 on page 71 that

$$\partial\big[-\ln(\det X)\big] = \operatorname{conv}\Big\{-\nabla\sum_{j=1}^{n} \ln\big(\langle \hat{u}_j, X\hat{u}_j\rangle\big) : \hat{U} = ZQ,\ Q \text{ is a permutation}\Big\}.$$

It does not matter in which order the eigenvectors z_j appear in $\hat{U}$, and we get

$$\partial\big[-\ln(\det X)\big] = \Big\{-\sum_{j=1}^{n} \nabla \ln\big(\langle z_j, Xz_j\rangle\big)\Big\} = \Big\{-\sum_{j=1}^{n} \frac{1}{\lambda_j} z_j z_j^T\Big\} = \{-X^{-1}\}.$$

We conclude that the subdifferential of $-\ln(\det X)$ contains only one element, $-X^{-1}$, which is the gradient of this function (Lemma 2.76 on page 62).

Consequently, $B(X)$ is convex and smooth on the set of positive definite matrices, and its gradient equals $\nabla B(X) = -X^{-1}$.

The main difficulty associated with the barrier method is the fact that the optimization problem (6.78) is ill-conditioned. For these reasons, barrier methods were not considered to be particularly attractive numerical methods for constrained optimization. Recent advances in the numerical implementations of the Newton method significantly changed this opinion.

6.6.2 Primal-Dual Newton's Method

The Lagrangian of problem (6.78) has the form

$$L(x, \lambda) = f(x) - \sigma \sum_{i=1}^{m} \ln(z_i) + \sum_{i=1}^{m} \lambda_i \big(g_i(x) + z_i\big).$$

Since (6.78) has only equality constraints, the necessary conditions of optimality are simply the equations:

$$\begin{aligned} \nabla f(x) + \sum_{i=1}^{m} \lambda_i \nabla g_i(x) &= 0, \\ \lambda_i z_i = \sigma, \quad i &= 1, \dots, m, \\ g(x) + z &= 0. \end{aligned} \tag{6.84}$$

Observe that the second equation is a perturbed complementarity condition for problem (6.76). Our idea is to solve the nonlinear system (6.84) by Newton's method, decrease σ, solve it again, etc. In fact, the decrease of σ may occur before reaching the exact solution of (6.84).

It is convenient to represent the Lagrange multiplier vector and the decision vector z as diagonal matrices

$$Y = \operatorname{diag}\{\lambda_1, \dots, \lambda_m\}, \qquad Z = \operatorname{diag}\{z_1, \dots, z_m\}.$$

We also use $\mathbb{1}$ to denote the vector of ones in $\mathbb{R}^m$: $\mathbb{1}^T = [1\ 1\ \dots\ 1]$, and we denote by $A(x)$ the Jacobian $g'(x)$ of $g(\cdot)$ at x. Then we can write the necessary conditions of optimality (6.84) in a compact form:

$$\begin{aligned} \nabla f(x) + A(x)^T \lambda &= 0, \\ ZYe &= \sigma \mathbb{1}, \\ g(x) + z &= 0. \end{aligned} \tag{6.85}$$

Denote by $H(x, \lambda)$ the Hessian of the ordinary Lagrangian of problem (6.76),

$$H(x, \lambda) = \nabla^2 f(x) + \sum_{i=1}^{m} \lambda_i \nabla^2 g_i(x).$$

The Jacobian of system (6.85) at (x, z, λ) can be calculated as follows:

$$J(x, z, \lambda) = \begin{bmatrix} H(x, \lambda) & 0 & A(x)^T \\ 0 & Y & Z \\ A(x) & I & 0 \end{bmatrix}. \tag{6.86}$$

Owing to the use of the logarithmic barrier function, the Jacobian has a manageable form. We can now apply Newton's method to the nonlinear system (6.85). At the point (x^k, z^k, λ^k) we calculate the directions $(d_x^k, d_z^k, d_\lambda^k)$ by solving the system of linear equations:

$$\begin{bmatrix} H(x^k, \lambda^k) & 0 & A(x^k)^T \\ 0 & Y^k & Z^k \\ A(x^k) & I & 0 \end{bmatrix} \begin{bmatrix} d_x^k \\ d_z^k \\ d_\lambda^k \end{bmatrix} = \begin{bmatrix} -\nabla f(x^k) - A(x^k)^T \lambda^k \\ \sigma \mathbb{1} - Z^k Y^k \mathbb{1} \\ -z^k - g(x^k) \end{bmatrix}.$$

From the second equation we can find the direction in the space of the slack variables:

$$d_z^k = \left(Y^k\right)^{-1}\left(\sigma \mathbb{1} - Z^k Y^k \mathbb{1} - Z^k d_\lambda^k\right). \tag{6.87}$$

We can thus eliminate d_z^k from the Newton system. After simple manipulations we obtain the following reduced system:

$$\begin{bmatrix} H(x^k, \lambda^k) & A(x^k)^T \\ A(x^k) & -\left(Y^k\right)^{-1} Z^k \end{bmatrix} \begin{bmatrix} d_x^k \\ d_\lambda^k \end{bmatrix} = \begin{bmatrix} -\nabla f(x^k) - A(x^k)^T \lambda^k \\ -g(x^k) - \sigma \left(Y^k\right)^{-1} \mathbb{1} \end{bmatrix}. \tag{6.88}$$

The method proceeds as follows. At iteration k, given the current approximation (x^k, z^k, λ^k), we solve system (6.88), which yields directions d_x^k and d_λ^k. The direction d_z^k can then be calculated from (6.87). After that, we update the current solution by

$$\begin{aligned} x^{k+1} &= x^k + \tau_k d_x^k, \\ z^{k+1} &= z^k + \tau_k d_z^k, \\ \lambda^{k+1} &= \lambda^k + \alpha_k d_\lambda^k. \end{aligned}$$

Here τ_k is a positive *primal step size*, such that z^{k+1} remains positive. The quantity α_k is a positive *dual step size*, such that λ^{k+1} is nonnegative. Finally, the barrier parameter σ is decreased, and the iteration continues. Similar derivations can be carried out for a problem involving equality constraints (see Exercise 6.11).

Proper implementation of the interior point method is very difficult. The matrix of system (6.88) is symmetric, but indefinite (neither positive nor negative definite) and ill-conditioned. Its solution requires application of specialized techniques for symmetric indefinite systems. In particular, the regularization techniques, known from the unconstrained Newton method, are important here. The primal regularization replaces the Hessian by the matrix $H(x^k, \lambda^k)+\delta I$, with some small $\delta > 0$, to make the resulting matrix positive definite. Dual regularization replaces the diagonal matrix $-\left(Y^k\right)^{-1} Z^k$ by $-\left(Y^k\right)^{-1} Z^k - \varepsilon I$, with $\varepsilon > 0$.

Selection of regularization parameters, step sizes, and the decrease of the barrier parameter σ have to be coordinated, to ensure practical convergence. These techniques are presently an object of intensive research. In practical computation, the interior methods can be very efficient, if they are implemented skillfully.

EXERCISES

6.1. Describe the operation of projection on the following sets:

(a) $X = \{x \in \mathbb{R}^n : l_j \le x_j \le u_j,\ j = 1, \dots, n\}$.

(b) $X = \{x \in \mathbb{R}^3 : x_1 \le x_2 \le x_3\}$.

(c) $X = \{x \in \mathbb{R}^n : Ax = b\}$, where A is an $m \times n$ matrix of rank m, and $b \in \mathbb{R}^m$.

6.2. Consider the problem

$$\begin{aligned} \text{minimize}\ & x_1 + (x_2)^2 + (x_3)^2 + (x_4)^2 - x_4 \\ \text{subject to}\ & 2x_1 + x_2 + x_3 + 4x_4 = 8 \\ & x_1 + 2x_2 + 2x_3 + x_4 = 6 \\ & x \ge 0. \end{aligned}$$

At the point $x = (1, 1, 1, 1)$ treat (x_1, x_2) as basic variables and (x_3, x_4) as superbasic variables. Carry out one iteration of the reduced gradient method.

6.3. Assume that a point $x^0 > 0$ satisfying its constraints of problem (6.25) exists. Prove that $\hat{\sigma} = 0$ is an optimal solution of this problem if and only if the point $\hat{x}$ is a stationary point of problem (6.8).

6.4. Consider the problem

$$\begin{aligned} \text{minimize}\ & x_1 - x_2 \\ \text{subject to}\ & (x_1)^2 + (x_2)^2 \le 1 \\ & x \ge 0. \end{aligned}$$

At the point $(-1, 0)$ calculate the direction used by the feasible direction method.

6.5. Consider the quadratic programming problem

$$\text{minimize } \langle x, Qx \rangle$$
$$\text{subject to } Ax \le b,$$

in which Q is a symmetric positive definite matrix of dimension n, A is an $m \times n$ matrix of rank m, and $b \in \mathbb{R}^m$.

(a) Prove that the basic dual method (6.47) with fixed step size τ is convergent, provided that τ is sufficiently small.

(b) Consider inequality constraints $Ax \le b$ instead of equations. Extend the analysis of (a) to this case.

6.6. Consider the problem

$$\text{minimize } x_1 x_2$$
$$\text{subject to } x_1 - 2x_2 = 3.$$

(a) For which values of the penalty parameter does the quadratic penalty function have a minimum? Calculate the minimum point as a function of the penalty parameter. What is its limit?

(b) For a sufficiently large value of the penalty parameter, analyze the operation of the augmented Lagrangian method, starting from $\mu = 0$.

(c) Formulate the exact penalty function and find its unconstrained minimum, if it exists.

6.7. For a general penalty method (6.27) prove that if the penalty coefficient increases, then the values of the penalty function at solutions to successive problems decrease.

6.8. Consider the problem

$$\text{minimize } x_1 - x_2$$
$$\text{subject to } (x_1)^2 + (x_2)^2 \le 1$$
$$x \ge 0.$$

(a) Starting from the point $x = (-1, 0)$ and from the initial value of the Lagrange multiplier $\lambda = 1$, carry out two iterations of the sequential quadratic programming method. Explain the reasons for using a positive value of the multiplier at the first iteration.

(b) For the same initial point, carry out two iterations of the interior point method.

6.9. Consider the problem

$$\text{minimize } \langle c, x \rangle + \frac{1}{2}\langle x, Wx \rangle$$
$$\text{subject to } \|x\|^2 \le \Delta.$$

The symmetric matrix W of dimension n, the vector $c \in \mathbb{R}^n$ and scalar $\Delta > 0$ are given. Problems of this form arise as subproblems in trust region methods of Section 5.7. Suppose we know

$$\lambda_0 = \min\{\lambda \geq 0 : W + \lambda I \succeq 0\}.$$

Design a specialized version of the sequential quadratic programming method for solving the problem.

6.10. Consider the quadratic programming problem

$$\begin{aligned} \text{minimize} \quad & \frac{1}{2}\langle x, Qx\rangle + \langle c, x\rangle \\ \text{subject to} \quad & Ax = b, \\ & x \geq 0, \end{aligned}$$

with a positive semidefinite matrix Q. Develop the iterative formula of the primal-dual interior point method, analogous to (6.88). As the inequality constraints are formulated directly for the variables x, you do not need to introduce slacks z here.

6.11. For a nonlinear programming problem with both inequality and equality constraints, develop the iterative formula of the primal-dual interior point method, analogous to (6.88).

6.12. For a semidefinite programming problem of Example 3.36, develop the primal-dual interior point method, analogous to (6.88). Use the barrier function analyzed in Example 6.25.

Chapter Seven

Nondifferentiable Optimization

7.1 THE SUBGRADIENT METHOD

7.1.1 The Basic Version

We start our analysis from the unconstrained minimization problem

$$\underset{x \in \mathbb{R}^n}{\text{minimize}}\ f(x), \tag{7.1}$$

in which the function $f(\cdot)$ is convex, but not necessarily differentiable. The nondifferentiability of the function precludes application of any of the methods discussed in Chapter 5, including methods which do not require the calculation of derivatives. The assumption about the differentiability of the function is essential for their convergence, as illustrated in Exercises 5.1 and 5.15. However, if we abandon the idea of decreasing the value of the function at *every* step of the method, it is possible to construct a convergent method using subgradients of the function $f(\cdot)$.

The simplest such method, called the *subgradient method*, constructs a sequence of points $\{x^k\}$ by the iterative formula:

$$x^{k+1} = x^k - \tau_k \gamma_k g^k, \quad k = 1, 2, \ldots, \tag{7.2}$$

in which g^k is a subgradient of $f(\cdot)$ at the current point x^k:

$$g^k \in \partial f(x^k).$$

The coefficient τ_k is a positive step size, and γ_k is a positive scaling coefficient whose role is to keep the norm of the vector $\gamma_k g^k$ bounded. While the product $\tau_k \gamma_k$ can be considered as one effective step size, it is convenient to separate them in our analysis, to underline the essence of the conditions on these quantities. To be specific, we analyze the convergence of the subgradient method for a particular choice of the scaling coefficients,

$$\gamma_k = \frac{1}{\max\left(c, \|g^k\|\right)}, \quad k = 1, 2, \ldots, \tag{7.3}$$

where $c > 0$ is fixed. It will be obvious how to modify our considerations for other popular versions, such as

$$\gamma_k = \frac{1}{\|g^k\|} \quad \text{or} \quad \gamma_k = \frac{1}{\max\left(c, \|g^k\|^2\right)},$$

or their variations.

The first observation that has to be made is that the direction negative to a subgradient g^k is not necessarily the direction of descent of the function $f(\cdot)$.

Example 7.1. Consider the function

$$f(x_1, x_2) = x_1 + x_2 + \max\left(0, (x_1)^2 + (x_2)^2 - 4\right).$$

At the point $x = (0, -2)$ the subdifferential of $f(x)$ can be calculated by Example 2.80 on page 66:

$$\partial f(x) = \text{conv}\left\{\begin{bmatrix}1\\1\end{bmatrix}, \begin{bmatrix}0\\-4\end{bmatrix}\right\}.$$

In particular, one may take

$$g^1 = \begin{bmatrix}1\\1\end{bmatrix}.$$

The direction $-g$ is not a direction of descent for $f(x)$. Also, the negative of the other extreme point of $\partial f(x)$,

$$g^2 = \begin{bmatrix}1\\-3\end{bmatrix},$$

is not a direction of descent. Figure 7.1 illustrates this situation.

The minimum-norm subgradient is

$$g = \begin{bmatrix}1\\0\end{bmatrix},$$

and its negative is the direction of steepest descent.

The negative of the minimum-norm subgradient is a direction of descent, and actually, it is the direction of steepest descent (see Lemma 2.77). But even then the method is not guaranteed to converge, as it is illustrated in Exercise 5.1.

The key property of the subgradient $g^k \in \partial f(x^k)$ is directly implied by Definition 2.72 on page 57:

$$f(x) - f(x^k) \geq \langle g^k, x - x^k\rangle \quad \text{for all} \quad x \in \mathbb{R}^n.$$

Therefore all points x, which have smaller function values than x^k, lie in the halfspace:

$$H_k = \left\{x \in \mathbb{R}^n : \langle g^k, x - x^k\rangle \leq 0\right\}.$$

It appears reasonable to expect that a small step in the direction negative to g^k will decrease the *distance* to every optimal point. This can be seen in Figure 7.1, and is used in the proofs of the main convergence theorems.

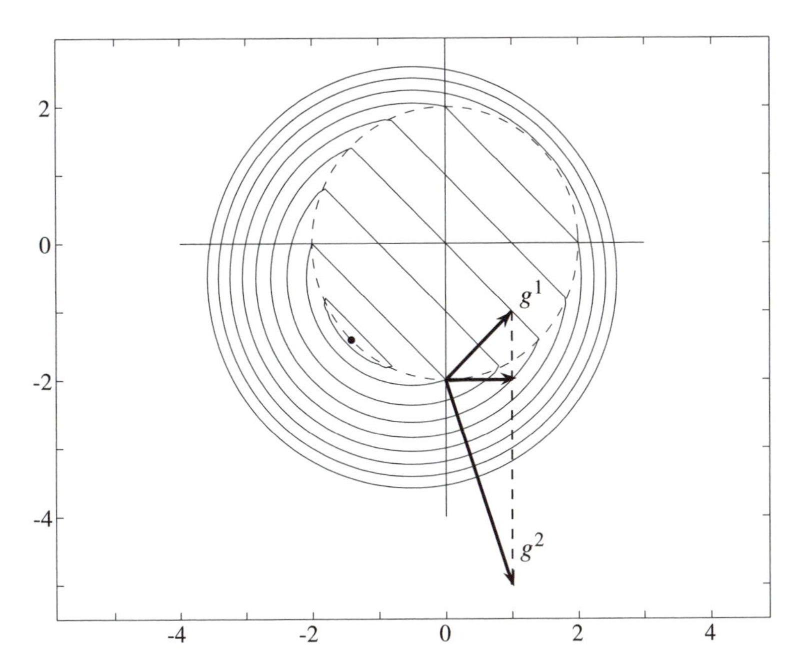

Figure 7.1. Subgradients in Example 7.1.

THEOREM 7.2. *Assume that the subgradient method* (7.2) *uses nonnegative step sizes* $\{\tau_k\}$ *satisfying the conditions*

$$\limsup_{k\to\infty} \tau_k = \bar{\tau} < \infty, \tag{7.4}$$

$$\sum_{k=1}^{\infty} \tau_k = \infty. \tag{7.5}$$

Then the sequence $\{x^k\}$ *generated by the subgradient method is such that for every optimal solution* x^* *of problem* (7.1) *the following relation holds true:*

$$\liminf_{k\to\infty} f(x^k) \le \max_{\|y-x^*\|\le\bar{\tau}/2} f(y). \tag{7.6}$$

Proof. Suppose x^* is an optimal solution. Expansion of the square of the norm yields

$$\begin{aligned} \|x^{k+1} - x^*\|^2 &= \|x^k - x^* - \tau_k\gamma_k g^k\|^2 \\ &= \|x^k - x^*\|^2 - 2\tau_k\gamma_k\langle g^k, x^k - x^*\rangle + \tau_k^2\gamma_k^2\|g^k\|^2. \end{aligned} \tag{7.7}$$

The definition of γ_k implies that $\gamma_k\|g^k\| \le 1$. Consequently, (7.7) becomes

$$\|x^{k+1} - x^*\|^2 \le \|x^k - x^*\|^2 - 2\tau_k\gamma_k\|g^k\|\Big(\big\langle \frac{g^k}{\|g^k\|}, x^k - x^*\big\rangle - \frac{\tau_k}{2}\Big). \tag{7.8}$$

Consider the quantities appearing in the expression above,

$$\delta_k = \Big\langle \frac{g^k}{\|g^k\|}, x^k - x^* \Big\rangle,$$

and the points

$$y^k = x^* + \delta_k \frac{g^k}{\|g^k\|}.$$

Using the definition of a subgradient at x^k and the definitions of y^k and δ_k, we obtain the inequality

$$\begin{aligned} f(y^k) &\geq f(x^k) + \langle g^k, y^k - x^k \rangle \\ &= f(x^k) + \langle g^k, x^* - x^k \rangle + \delta_k \Big\langle g^k, \frac{g^k}{\|g^k\|} \Big\rangle \\ &= f(x^k) - \delta_k \|g^k\| + \delta_k \|g^k\| = f(x^k). \end{aligned} \tag{7.9}$$

Thus x^k is at least as good as the point y^k, which is at distance δ_k to x^*. If

$$\liminf_{k \to \infty} \delta_k \leq \frac{\bar{\tau}}{2},$$

inequality (7.9) immediately implies the assertion of the theorem. It remains to consider the case when

$$\liminf_{k \to \infty} \delta_k > \frac{\bar{\tau}}{2}.$$

Rewriting (7.8) as

$$\|x^{k+1} - x^*\|^2 \leq \|x^k - x^*\|^2 - 2\tau_k \gamma_k \|g^k\| \Big(\delta_k - \frac{\tau_k}{2} \Big),$$

we see that for sufficiently large k the sequence $\big\{ \|x^k - x^*\|^2 \big\}$ is nonincreasing. Let us choose l large enough and a sufficiently small $\varepsilon > 0$, so that $\delta_k - \tau_k/2 \geq \varepsilon$ for all $k \geq l$. Summing the last inequality from $k = l$ to $k = m$ we obtain

$$0 \leq \|x^{m+1} - x^*\|^2 \leq \|x^l - x^*\|^2 - 2\varepsilon \sum_{k=l}^{m} \tau_k \gamma_k \|g^k\|. \tag{7.10}$$

If there exists $\alpha > 0$ such that $\gamma_k \|g^k\| \geq \alpha$ for all $k \geq l$, then assumption (7.5) implies that the right hand side of (7.10) diverges to $-\infty$ as $m \to \infty$, which makes (7.10) an absurd. The only remaining possibility is that

$$\liminf_{k \to \infty} \gamma_k \|g^k\| = 0.$$

Hence there exists an infinite set of iterations $\mathcal{K}$ such that $g^k \to 0$, as $k \to \infty$, $k \in \mathcal{K}$. As the sequence $\{x^k\}$ is bounded, it follows from the subgradient inequality,

$$\langle g^k, x^k - x^* \rangle \geq f(x^k) - f(x^*) \geq 0, \tag{7.11}$$

that $f(x^k) \to f(x^*)$, as $k \to \infty$, $k \in \mathcal{K}$. Consequently, the assertion of the theorem holds true in this case as well. □

In particular, when $\tau_k \to 0$, it follows from the above theorem that

$$\liminf_{k\to\infty} f(x^k) = f(x^*).$$

But we can also use this result to estimate the quality of the best point found by the method using *constant* step sizes, which is of great practical importance. We can simply run method (7.16) with a small constant step size $\bar{\tau}$, and the best point after sufficiently many iterations will have an error (in terms of the value of the objective function) comparable to $\bar{\tau}$, as given in (7.6).

A surprising conclusion is that if the set of optimal solutions contains a ball of diameter $\bar{\tau}/2$, then actually $\liminf f(x^k) = f(x^*)$. Indeed, setting x^* in (7.6) to be the center of this ball, we can make the right hand side equal to $f(x^*)$.

By assuming a little more about the function $f(\cdot)$, we can prove the convergence of the entire sequence of function values to the optimal value.

THEOREM 7.3. *Let the assumptions of Theorem* 7.2 *be satisfied with* $\bar{\tau} = 0$. *If the set of optimal solutions of problem* (7.1) *is bounded, then*

$$\lim_{k\to\infty} f(x^k) = f^*,$$

where f^* *is the optimal value of the problem.*

Proof. Since the optimal set X^* is bounded and $f(\cdot)$ is convex, all level sets of $f(\cdot)$ are bounded (Exercise 2.12). Then for every $\varepsilon > 0$ the number

$$r(\varepsilon) = \sup\left\{ \operatorname{dist}(x, X^*) : f(x) \leq f^* + \varepsilon \right\} \tag{7.12}$$

is finite. In fact, $r(\varepsilon) \downarrow 0$, if $\varepsilon \downarrow 0$. In the remaining part of the proof we just write r for $r(\varepsilon)$.

We know from Theorem 7.2 that there exists an infinite set of iterations $\mathcal{K}_1$ such that $f(x^k) \to f^*$ when $k \to \infty$, $k \in \mathcal{K}_1$. Then $\operatorname{dist}(x^k, X^*) \to 0$ for $k \in \mathcal{K}_1$. Suppose an infinite set of iterations $\mathcal{K}_2$ exists such that $\operatorname{dist}(x^k, X^*) \geq 2r$ for all $k \in \mathcal{K}_2$. Choose $k_1 \in \mathcal{K}_1$ such that $\operatorname{dist}(x^{k_1}, X^*) < r$. Let k_2 be the first element of the set $\mathcal{K}_2$ after k_1:

$$k_2 = \min\{k \in \mathcal{K}_2 : k > k_1\}.$$

All points x^j for $j = k_1, \ldots, k_2 - 1$ have distance to X^* less than $2r$. Thus the subgradients g^j at these points are uniformly bounded: there exists a constant L such that $\|g^j\| \le L$, $j = k_1, \ldots, k_2 - 1$. The constant L depends only on r. We can choose k_1 sufficiently large, so that

$$\tau_j \le \min\left(\frac{r}{2}, \frac{2\varepsilon}{L}\right) \quad \text{for} \quad j \ge k_1. \tag{7.13}$$

Since $\|x^{j+1} - x^j\| \le \tau_j$, the quantity $\operatorname{dist}(x^j, X^*)$ cannot increase by more than $r/2$ for $j \ge k_1$. At least two steps are needed between k_1 and k_2 and the set

$$J = \left\{k_1 \le j < k_2 : \operatorname{dist}(x^j, X^*) \ge r\right\}$$

is nonempty.

Employing the subgradient inequality (7.11) and using the fact that $\gamma_k \|g^k\|$ is bounded by 1, we can manipulate inequality (7.8) to the form

$$\|x^{k+1} - x^*\|^2 \le \|x^k - x^*\|^2 - 2\tau_k \gamma_k \Big(f(x^k) - f^* - \frac{1}{2}\tau_k \|g^k\| \Big).$$

For every $j \in J$ the definition of r implies that $f(x^j) - f^* \ge \varepsilon$. Since $\tau_j \|g^j\| \le 2\varepsilon$, we get

$$\|x^{j+1} - x^*\|^2 \le \|x^j - x^*\|^2, \quad j \in J.$$

Taking the infimum with respect to $x^* \in X^*$ on both sides we obtain

$$\operatorname{dist}(x^{j+1}, X^*) \le \operatorname{dist}(x^j, X^*), \quad j \in J.$$

It follows that the quantity $\operatorname{dist}(x^j, X^*)$ is unable to climb up, once it passes r, and thus it cannot exceed $2r$. Consequently, the set $\mathcal{K}_2$ must be finite and

$$\limsup_{k \to \infty} \big[\operatorname{dist}(x^k, X^*) \big] \le 2r.$$

Since r can be made arbitrarily small by choosing a sufficiently small $\varepsilon > 0$, we conclude that $\operatorname{dist}(x^k, X^*) \to 0$, as $k \to \infty$. This is equivalent to the assertion of the theorem. □

It is possible to carry out a similar analysis for $\bar{\tau} > 0$, but then the bound on $\limsup f(x^k)$ depends on the properties of the function $f(\cdot)$. The interested reader may easily reconstruct it from our considerations here.

Under stronger conditions on the step size coefficients, we can prove the convergence of the entire sequence $\{x^k\}$.

THEOREM 7.4. *Assume that problem* (7.1) *has an optimal solution. Let the subgradient method use nonnegative step sizes* $\{\tau_k\}$ *satisfying condition* (7.5) *and such that*

$$\sum_{k=1}^{\infty} \tau_k^2 < \infty. \tag{7.14}$$

Then the sequence $\{x^k\}$ *generated by the subgradient method is convergent to a solution of problem* (7.1).

Proof. Suppose x^* is an optimal solution of problem (7.1). Equation (7.7) and the bound $\gamma_k \|g^k\| \le 1$ yield

$$\|x^{k+1} - x^*\|^2 \le \|x^k - x^*\|^2 - 2\tau_k \gamma_k \langle g^k, x^k - x^* \rangle + \tau_k^2.$$

It follows from the definition of the subgradient (Definition 2.72 on page 57) that

$$\langle g^k, x^k - x^* \rangle \ge f(x^k) - f(x^*).$$

Combining the last two relations we obtain for $k = 1, 2, \dots$ that

$$\begin{aligned} \|x^{k+1} - x^*\|^2 &\le \|x^k - x^*\|^2 - 2\tau_k \gamma_k \big[f(x^k) - f(x^*) \big] + \tau_k^2 \\ &\le \|x^k - x^*\|^2 + \tau_k^2. \end{aligned} \tag{7.15}$$

We add these inequalities from $k = 1$ to $k = m$, which yields the bound:

$$\begin{aligned} \|x^{m+1} - x^*\|^2 &\le \|x^1 - x^*\|^2 - \sum_{k=1}^{m} \tau_k \gamma_k \big[f(x^k) - f(x^*) \big] + \sum_{k=1}^{m} \tau_k^2 \\ &\le \|x^1 - x^*\|^2 + \sum_{k=1}^{\infty} \tau_k^2. \end{aligned}$$

We have used here the fact that $f(x^k) - f(x^*) \ge 0$, by the optimality of x^*. Furthermore, by virtue of condition (7.14), the last sum on the right hand side is bounded, and thus the sequence $\{x^k\}$ is bounded. By Theorem 7.2 with $\bar{\tau} = 0$, there exists an infinite set of iterations $\mathcal{K}$ such that $f(x^k) \to f(x^*)$ as $k \to \infty$, $k \in \mathcal{K}$. As the sequence $\{x^k\}$ is bounded, we can choose an infinite set $\mathcal{K}_1 \subset \mathcal{K}$ such that the subsequence $\{x^k\}$, $k \in \mathcal{K}_1$, is convergent. Its limit, denoted by $\hat{x}$, must be optimal for problem (7.1). We can thus substitute $\hat{x}$ for x^* in all estimates above.

Choose $l \in \mathcal{K}_1$. Adding inequalities (7.15) from $k = l$ to m we obtain

$$\|x^{m+1} - \hat{x}\|^2 \le \|x^l - \hat{x}\|^2 + \sum_{k=l}^{\infty} \tau_k^2, \quad m = l+1, l+2, \dots.$$

For each $\varepsilon > 0$ we can choose $l \in \mathcal{K}_1$ such that $\|x^l - \hat{x}\|^2 \le \varepsilon$, and $\sum_{k=l}^{\infty} \tau_k^2 \le \varepsilon$. Then $\|x^{m+1} - \hat{x}\|^2 \le 2\varepsilon$ for all $m \ge l$, which proves that the entire sequence $\{x^k\}$ is convergent to $\hat{x}$. □

An example of a sequence of step sizes satisfying conditions (7.5) and (7.14) is

$$\tau_k = \frac{\tau_1}{k}, \quad k = 1, 2, 3, \ldots.$$

It has only theoretical significance, and is rarely used in any practical computations.† Generally, because of the non-monotone character of the sequence $\{f(x^k)\}$ in the subgradient method, it is very difficult to develop reliable and efficient step size rules.

An advantage of the subgradient method is its utmost simplicity. It can be applied to problems of very large dimension.

7.1.2 Projection

The subgradient method can be easily adapted to constrained convex optimization problems of the form

$$\underset{x \in X}{\text{minimize}}\ f(x), \tag{7.16}$$

where $X \subset \mathbb{R}^n$ is a convex closed set, and the function $f : \mathbb{R}^n \to \mathbb{R}$ is convex.

We denote by $\Pi_X(x)$ the orthogonal projection of the point $x \in \mathbb{R}^n$ onto the set X. It is defined as the point $v \in X$ that is closest to x. We know from Section 2.1.2 that for every $x \in \mathbb{R}^n$ such a point exists and is unique. Using the orthogonal projection, we modify the subgradient method (7.2) as follows:

$$x^{k+1} = \Pi_X\big(x^k - \tau_k \gamma_k g^k\big), \quad k = 1, 2, \ldots. \tag{7.17}$$

As before, g^k is a subgradient of $f(\cdot)$ at the current point x^k. When $X = \mathbb{R}^n$, method (7.17) reduces to the basic version (7.2).

Clearly, application of projection is practical, when we can do it easily, for example, when the set X is the nonnegative orthant or a box in $\mathbb{R}^n$. More complicated constraints require application of special methods for constrained problems, discussed in Section 7.6.

The properties of the method with projection are virtually the same as those of the unconstrained version. Indeed, suppose x^* is an optimal solution of problem (7.16). Using the fact that the projection operation $\Pi_X(\cdot)$ is

---

†Condition (7.14) plays an important role in stochastic versions of the subgradient method.

nonexpansive (see Theorem 2.13 on page 23) and that $x^* \in X$, we can write the inequality

$$\|x^{k+1} - x^*\|^2 = \left\|\Pi_X(x^k - \tau_k \gamma_k g^k) - \Pi_X(x^*)\right\|^2 \le \|x^k - \tau_k \gamma_k g^k - x^*\|^2.$$

Hence

$$\|x^{k+1} - x^*\|^2 \le \|x^k - x^*\|^2 - 2\tau_k \gamma_k \langle g^k, x^k - x^* \rangle + \tau_k^2 \gamma_k^2 \|g^k\|^2. \quad (7.18)$$

This can be used in place of (7.7) in the proofs of Theorems 7.2, 7.3, and 7.4. They remain valid for the method with projection, and the proofs can be copied *verbatim*.

If the set X is bounded, there is no need for the scaling coefficients γ_k, because all subgradients g^k are uniformly bounded; we can simply use $\gamma_k = 1$ for all k.

7.1.3 Application to Dual Problems

A particularly important application of the subgradient method, and other nonsmooth optimization methods, is the solution of dual problems associated with constrained optimization problems.

For the general nonlinear programming problem:

$$\begin{aligned} \text{minimize} \ & f(x) \\ \text{subject to} \ & g_i(x) \le 0, \quad i = 1, \dots, m, \\ & h_i(x) = 0, \quad i = 1, \dots, p, \\ & x \in X_0, \end{aligned} \quad (7.19)$$

the Lagrangian has the form

$$L(x, \lambda, \mu) = f(x) + \langle \lambda, g(x) \rangle + \langle \mu, h(x) \rangle.$$

The associated dual function is defined as follows:

$$L_D(\lambda, \mu) = \inf_{x \in X_0} L(x, \lambda, \mu). \quad (7.20)$$

The dual function is always concave, and thus the dual problem,

$$\underset{(\lambda,\mu) \in \Lambda_0}{\text{maximize}} \ L_D(\lambda, \mu),$$

is a convex optimization problem. Recall that its feasible set has the form $\Lambda_0 = \mathbb{R}^m_+ \times \mathbb{R}^p$.

Lemma 4.5 on page 165 provides us with a way to calculate a subgradient of the dual function: for a given pair of multipliers (λ^k, μ^k) solve the optimization problem on the right hand side of (7.20). If it has an optimal solution x^k then the vector

$$d^k = \begin{bmatrix} g(x^k) \\ h(x^k) \end{bmatrix}$$

is a subgradient of $L_D(\cdot)$ at the current point. Therefore, the subgradient method can be written for the dual problem as follows:[†]

$$\begin{aligned} \lambda^{k+1} &= \max\left(0, \lambda^k + \tau_k \gamma_k g(x^k)\right), \\ \mu^{k+1} &= \mu^k + \tau_k \gamma_k h(x^k), \quad k = 1, 2, \ldots. \end{aligned} \tag{7.21}$$

As we discuss in sections 4.4 and 4.5, dual approaches are particularly useful for large-scale problems. Usually, only a subset of constraints is included in the Lagrangian, and the other constraints of the problem are treated directly, as a part of the definition of the set X_0 in (7.20).

The key practical problem associated with the dual method is *primal recovery*, that is, determining the optimal solution of the original problem, after solving the dual problem. Theorem 4.10 on page 169 provides the theoretical answer, but its practical application may encounter difficulties, when the solution of problem (7.20) is not unique (for the optimal values of the multipliers $\hat{\lambda}$ and $\hat{\mu}$). This situation is typical for linear programming problems.

In convex programming it is possible to recover the primal solution in the course of the operation of the subgradient method. Assume that the functions $f(\cdot)$ and $g_i(\cdot)$, $i = 1, \ldots, m$, in problem (7.19) are convex, the functions $h_i(\cdot)$, $i = 1, \ldots, p$ are affine, the set X_0 is convex and compact, and Slater's constraint qualification condition is satisfied.

Owing to the compactness of X_0, the problem has a solution. Due to Slater's condition, the dual problem has a solution as well, and we can apply the dual method (7.21). In view of the compactness of X_0 the subgradients $(g(x^k), h(x^k))$ are uniformly bounded, and we can set $\gamma_k = 1$ for all k in (7.21).

Consider the sequence of weighted averages

$$\bar{x}^\ell = \frac{\sum_{k=1}^{\ell} \tau_k x^k}{\sum_{k=1}^{\ell} \tau_k}, \quad \ell = 1, 2, \ldots. \tag{7.22}$$

THEOREM 7.5. *Assume that the sequence of step sizes $\{\tau_k\}$ satisfies conditions* (7.4)–(7.5), *with $\bar{\tau} = 0$. Then every accumulation point of the sequence of averages $\{\bar{x}^\ell\}$ is an optimal solution of problem* (7.19).

[†]Because the dual problem is a maximization problem, we make steps in the direction of a subgradient, not its negative.

Proof. By construction, $\bar{x}^\ell \in \operatorname{conv}\{x^1, x^2, \dots, x^\ell\}$, and thus $\bar{x}^\ell \in X_0$. For simplicity of notation we assume that only inequality constraints are present ($p = 0$). Since $\lambda^{k+1} \geq \lambda^k + \tau_k g(x^k)$, $k = 1, 2, \dots,$

$$\sum_{k=1}^{\ell} \tau_k g(x^k) \leq \lambda^{\ell+1} - \lambda^1.$$

The inequality here, as usual, is understood component-wise. Dividing both sides by $\sum_{k=1}^{\ell} \tau_k$, using the definition of $\bar{x}^\ell$ and the convexity of $g(\cdot)$, we obtain:

$$g(\bar{x}^\ell) \leq \frac{\sum_{k=1}^{\ell} \tau_k g(x^k)}{\sum_{k=1}^{\ell} \tau_k} \leq \frac{\lambda^{\ell+1} - \lambda^1}{\sum_{k=1}^{\ell} \tau_k}. \tag{7.23}$$

Our first step is to show that the right hand side of this equation converges to 0, as $\ell \to \infty$. Let λ^* be a solution of the dual problem. Similarly to (7.18), we derive from (7.21) with $\gamma_k = 1$ the inequality

$$\|\lambda^{k+1} - \lambda^*\|^2 \leq \|\lambda^k - \lambda^*\|^2 + 2\tau_k \langle \lambda^k - \lambda^*, g(x^k)\rangle + \tau_k^2 \|g(x^k)\|^2. \tag{7.24}$$

Since $g(x^k) \in \partial L_D(\lambda^k)$ (for a concave function $L_D(\cdot)$),

$$\langle \lambda^k - \lambda^*, g(x^k)\rangle \leq L_D(\lambda^k) - L_D(\lambda^*) \leq 0.$$

Let C be such that $\|g(x^k)\| \leq C$ for all k. Then the last two relations imply

$$\|\lambda^{k+1} - \lambda^*\|^2 \leq \|\lambda^k - \lambda^*\|^2 + C^2 \tau_k^2.$$

Adding from $k = 1$ to $k = \ell$ we conclude that

$$\|\lambda^{\ell+1} - \lambda^*\|^2 \leq \|\lambda^1 - \lambda^*\|^2 + C^2 \sum_{k=1}^{\ell} \tau_k^2.$$

Therefore

$$\begin{aligned}
\|\lambda^{\ell+1} - \lambda^1\| &\leq \|\lambda^1 - \lambda^*\| + \|\lambda^{\ell+1} - \lambda^*\| \\
&\leq \|\lambda^1 - \lambda^*\| + \Big(\|\lambda^1 - \lambda^*\|^2 + C^2 \sum_{k=1}^{\ell} \tau_k^2\Big)^{1/2} \\
&\leq 2\|\lambda^1 - \lambda^*\| + C\Big(\sum_{k=1}^{\ell} \tau_k^2\Big)^{1/2}.
\end{aligned}$$

Dividing both sides by $\sum_{k=1}^{\ell} \tau_k$ we get

$$\frac{\|\lambda^{\ell+1} - \lambda^1\|}{\sum_{k=1}^{\ell} \tau_k} \leq 2\frac{\|\lambda^1 - \lambda^*\|}{\sum_{k=1}^{\ell} \tau_k} + C\frac{\big(\sum_{k=1}^{\ell} \tau_k^2\big)^{1/2}}{\sum_{k=1}^{\ell} \tau_k}.$$

Let $\ell \to \infty$. Both fractions on the right hand side converge to 0, due to the conditions $\sum_{k=1}^{\infty} \tau_k = \infty$ and $\tau_k \to 0$. Therefore the left hand side converges to 0. In view of (7.23), we obtain

$$\limsup_{\ell\to\infty} g_i(\bar{x}^\ell) \le 0, \quad i = 1, \dots, m.$$

Consequently, every accumulation point of the sequence $\{\bar{x}^\ell\}$ satisfies the constraints of problem (7.19).

Let us now analyze the quality of these accumulation points. By Lemma 4.21 on page 186, the dual function is a lower bound for the optimal value of the problem, that is,

$$f(x^k) + \langle \lambda^k, g(x^k)\rangle \le f^*, \quad k = 1, 2, \dots .$$

Multiplying by τ_k, summing over $k = 1, \dots, \ell$ and normalizing by $\sum_{k=1}^{\ell} \tau_k$ we obtain

$$\frac{\sum_{k=1}^{\ell} \tau_k f(x^k)}{\sum_{k=1}^{\ell} \tau_k} \le f^* - \frac{\sum_{k=1}^{\ell} \tau_k \langle \lambda^k, g(x^k)\rangle}{\sum_{k=1}^{\ell} \tau_k}. \tag{7.25}$$

Observe that the convexity of the function $f(\cdot)$ and the definition (7.22) of $\bar{x}^\ell$ yield:

$$f(\bar{x}^\ell) \le \frac{\sum_{k=1}^{\ell} \tau_k f(x^k)}{\sum_{k=1}^{\ell} \tau_k}.$$

Thus the right hand side of (7.25) is an upper bound on $f(\bar{x}^\ell)$. We shall show that the fraction on the right hand side of (7.25) has a nonnegative lower limit, as $\ell \to \infty$.

Using the fact that $\lambda = 0$ is feasible for the dual problem, similar to (7.24) we can write the inequality

$$\|\lambda^{k+1}\|^2 \le \|\lambda^k\|^2 + 2\tau_k \langle \lambda^k, g(x^k)\rangle + \tau_k^2 \|g(x^k)\|^2.$$

Hence

$$\tau_k \langle \lambda^k, g(x^k)\rangle \ge \frac{1}{2}\|\lambda^{k+1}\|^2 - \frac{1}{2}\|\lambda^k\|^2 - \frac{C^2}{2}\tau_k^2,$$

where C is a bound on $\|g(x^k)\|$. Summing over $k = 1, \dots, \ell$ and normalizing by the sum of step sizes, we obtain

$$\frac{\sum_{k=1}^{\ell} \tau_k \langle \lambda^k, g(x^k)\rangle}{\sum_{k=1}^{\ell} \tau_k} \ge -\frac{\|\lambda^1\|^2}{2\sum_{k=1}^{\ell} \tau_k} - \frac{C^2 \sum_{k=1}^{\ell} \tau_k^2}{2\sum_{k=1}^{\ell} \tau_k}.$$

When $\ell \to \infty$, both fractions on the right hand side go to 0, by virtue of (7.4) and by the fact that $\tau_k \to 0$. Therefore, the last term on the right hand side of (7.25) has a nonnegative lower limit, as $\ell \to \infty$. We conclude that

$$\limsup_{\ell \to \infty} f(\bar{x}^\ell) \leq f^*.$$

It follows that every accumulation point of the sequence $\{\bar{x}^\ell\}$ is an optimal solution of problem (7.19). □

Finally, we may note that the summation over k in the numerator and denominator of (7.22) may run from some $\ell_0 \geq 1$, rather than from 1, because every point generated by the method may be regarded as the starting point.

7.1.4 Known Optimal Value

We return to the set-constrained problem

$$\underset{x \in X}{\text{minimize}}\ f(x), \tag{7.26}$$

with a convex function $f : \mathbb{R}^n \to \mathbb{R}$ and a convex closed set $X \subset \mathbb{R}^n$. We assume that problem (7.26) has a solution and that the optimal value f^* of its objective function is known.

Such a situation arises when we apply the penalty method to the system of inequalities:

$$\begin{aligned} &f_i(x) \leq 0, \quad i = 1, \dots, m, \\ &x \in X. \end{aligned}$$

We can define there

$$f(x) = \max\big(0, f_1(x), \dots, f_m(x)\big),$$

or

$$f(x) = \sum_{i=1}^{m} \max\big(0, f_i(x)\big),$$

and convert the system of inequalities to the optimization problem (7.26). In both cases $f^* = 0$, provided that the system of inequalities has a solution.

If the optimal value f^* is known, we can use specialized step sizes in the subgradient method with projection

$$x^{k+1} = \Pi_X\big(x^k - \tau_k g^k\big), \quad k = 1, 2, \dots, \tag{7.27}$$

namely,

$$\tau_k = \frac{f(x^k) - f^*}{\|g^k\|^2}. \tag{7.28}$$

THEOREM 7.6. *Assume that problem* (7.26) *has an optimal solution. Then the subgradient method* (7.27)–(7.28) *generates a sequence* $\{x^k\}$ *which is convergent to a solution of* (7.26).

Proof. Suppose x^* is an optimal solution of problem (7.26). Using the fact that the projection operation $\Pi_X(\cdot)$ is nonexpansive (see Theorem 2.13 on page 23) and that $x^* \in X$, we can write the inequality

$$\|x^{k+1} - x^*\|^2 = \left\|\Pi_X(x^k - \tau_k g^k) - \Pi_X(x^*)\right\|^2 \le \|x^k - \tau_k g^k - x^*\|^2.$$

Similarly to (7.7), we obtain the estimate

$$\|x^{k+1} - x^*\|^2 \le \|x^k - x^*\|^2 - 2\tau_k\langle g^k, x^k - x^*\rangle + \tau_k^2\|g^k\|^2.$$

Using the subgradient inequality $\langle g^k, x^k - x^*\rangle \ge f(x^k) - f^*$ and the definition of τ_k in (7.28), we can continue the last estimate as follows:

$$\begin{aligned}\|x^{k+1} - x^*\|^2 &\le \|x^k - x^*\|^2 - 2\tau_k\big[f(x^k) - f^*\big] + \tau_k^2\|g^k\|^2\\ &\le \|x^k - x^*\|^2 - \frac{\big[f(x^k) - f^*\big]^2}{\|g^k\|^2}, \quad k = 1, 2, \dots.\end{aligned} \tag{7.29}$$

This proves that the sequence $\{x^k\}$ is bounded, and thus the subgradients g^k are uniformly bounded. Summing up (7.29) from $k = 1$ to ∞ we obtain

$$\sum_{k=1}^{\infty} \frac{\big[f(x^k) - f^*\big]^2}{\|g^k\|^2} \le \|x^1 - x^*\|^2.$$

Since the subgradients are bounded, $f(x^k) \to f^*$. Therefore, for every convergent subsequence of the sequence $\{x^k\}$, its accumulation point $\hat{x}$ must be optimal: $f(\hat{x}) = f^*$. We can thus substitute $\hat{x}$ for x^* in (7.29) and conclude that the entire sequence is convergent to $\hat{x}$. □

The possibility of using "automatic" step sizes (7.28) is a great advantage of this version of the method. Still, the speed of convergence may be quite slow, because τ_k quickly diminishes when x^k approaches the solution set. For systems of inequalities, discussed at the beginning of this section, the method with *constant* step sizes is a good alternative (see Theorem 7.2 with $\bar{\tau} > 0$). If the solution set of the system of inequalities contains a small ball, then Theorem 7.2 guarantees that the method with step sizes not exceeding the diameter of this ball, and with normalizing coefficients (7.3), is convergent to a solution of this system.

7.2 THE CUTTING PLANE METHOD

7.2.1 The Main Concepts

Let us consider the problem

$$\underset{x \in X}{\text{minimize}}\ f(x), \tag{7.30}$$

in which the function $f : \mathbb{R}^n \to \mathbb{R}$ is convex and the set $X \subset \mathbb{R}^n$ is convex and compact.

The idea of the *cutting plane method* is to use the subgradient inequality,

$$f(y) \geq f(x) + \langle g, y - x \rangle,$$

which holds true for every $y \in \mathbb{R}^n$ and each subgradient $g \in \partial f(x)$, for constructing lower approximations of $f(\cdot)$.

The method starts from a point $x^1 \in X$, calculates $g^1 \in \partial f(x^1)$, and constructs a linear approximation of $f(\cdot)$:

$$f^1(x) = f(x^1) + \langle g^1, x - x^1 \rangle.$$

Clearly, $f(\cdot) \geq f^1(\cdot)$. The function $f^1(\cdot)$ is minimized over X, which is possible due to the compactness of X, and a certain point $x^2 \in X$ is obtained. A subgradient $g^2 \in \partial f(x^2)$ provides a new lower bound for $f(\cdot)$:

$$f(x) \geq f(x^2) + \langle g^2, x - x^2 \rangle.$$

Both lower bounds, obtained at x^1 and at x^2, can be combined into the function

$$f^2(x) = \max\big(f(x^1) + \langle g^1, x - x^1 \rangle, f(x^2) + \langle g^2, x - x^2 \rangle\big).$$

Again, $f^2(\cdot) \leq f(\cdot)$. The minimization of $f^2(x)$ over $x \in X$ yields a new point x^3, and the iteration continues.

In general, at iteration k, having points $x^1, \dots, x^k$, values of the function $f(x^1), \dots, f(x^k)$, and subgradients $g^1, \dots, g^k$, we construct a lower approximation of the function

$$f^k(x) \triangleq \max_{1 \leq j \leq k} \big[f(x^j) + \langle g^j, x - x^j \rangle\big]. \tag{7.31}$$

It is used in the *master problem*:

$$\underset{x \in X}{\text{minimize}}\ f^k(x), \tag{7.32}$$

whose solution, x^{k+1}, is added to the set of points. After evaluating $f(x^{k+1})$ and $g^{k+1} \in \partial f(x^{k+1})$, we increase k by one, and continue the calculations.

If

$$f(x^{k+1}) = f^k(x^{k+1}),$$

we can terminate the calculations, because the point x^{k+1} is optimal, as we shall show in the next subsection.

We call the linear pieces, $f(x^j) + \langle g^j, x - x^j \rangle$, added at the iterations of the method, *cutting planes*, or simply *cuts*. The first three iterations of the cutting plane method are illustrated in Figure 7.2.

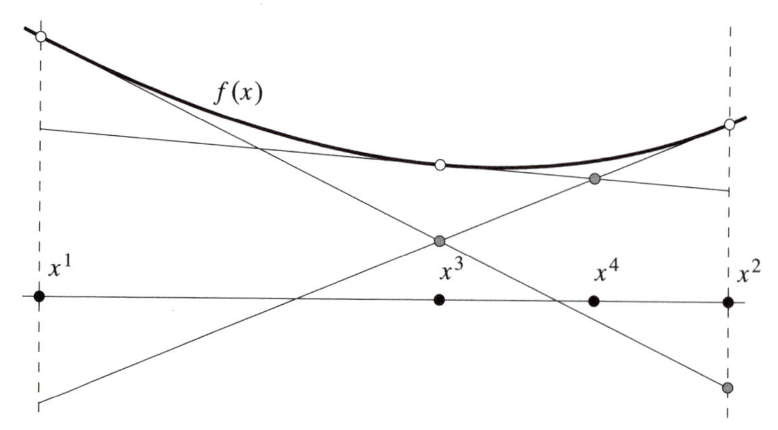

Figure 7.2. Illustration of the cutting plane method in one dimension. The vertical dashed lines bound the feasible set. The gray points are the minima of the cutting plane approximations. The white points are the points of the graph of the function where new cutting planes are derived.

Before proceeding to the convergence analysis, let us note that the master problem (7.32) is equivalent to the following constrained optimization problem:

$$\begin{aligned} &\text{minimize } v \\ &\text{subject to } f(x^j) + \langle g^j, x - x^j \rangle \le v, \quad j = 1, \dots, k, \\ &\qquad\qquad x \in X, \end{aligned} \tag{7.33}$$

whose solution (x^{k+1}, v^{k+1}) is the next approximation to the solution of (7.30) and a lower bound for $f(\cdot)$ on X. Indeed, let us assume that x is fixed in (7.33), so that the optimization is carried out with respect to the variable v. The optimal value of v is then clear from the constraints of this problem:

$$\hat{v}(x) = \max_{1 \le j \le k} \left[f(x^j) + \langle g^j, x - x^j \rangle \right] = f^k(x).$$

Therefore, (7.33) is equivalent to (7.32). The new formulation has the advantage that, after passing to iteration $k + 1$, just one constraint is added to this problem, and re-optimization by a dual method is an attractive option. This is particularly useful when X is a polyhedron, because efficient linear programming techniques can be employed.

7.2.2 Convergence

Let us denote by f^* the optimal value of the original problem (7.30).

The key property of the master problem is that its optimal value provides a lower bound for the optimal value of the original problem:

$$v^{k+1} = \min_{x \in X} f^k(x) \le \min_{x \in X} f(x) = f^*, \quad k = 1, 2, \ldots.$$

Therefore the stopping test is correct; if $f(x^{k+1}) = v^{k+1}$, then $f(x^{k+1}) = f^*$. This may happen if the function is piecewise linear, as we discuss in the next subsection, but generally an infinite number of steps may occur.

THEOREM 7.7. *The cutting plane method generates a sequence of points* $\{x^k\}$ *such that*

$$\lim_{k \to \infty} f(x^k) = f^*.$$

Proof. As we have mentioned earlier, the method can stop only if x^k is optimal. It remains to analyze the case of infinitely many steps. For $\varepsilon > 0$ we define the set

$$\mathcal{K}_\varepsilon = \{k : f^* + \varepsilon < f(x^k) < +\infty\}.$$

Let $k_1, k_2 \in \mathcal{K}_\varepsilon$ with $k_1 < k_2$. Since $f(x^{k_1}) > f^* + \varepsilon$ and $f^* \ge v^{k_1}$, there will be a new cutting plane generated at x^{k_1}. It will stay in the master problem from k_1 on, so it has to be satisfied at (x^{k_2}, v^{k_2}):

$$f(x^{k_1}) + \langle g^{k_1}, x^{k_2} - x^{k_1} \rangle \le v^{k_2} \le f^*.$$

On the other hand, $\varepsilon < f(x^{k_2}) - f^*$, which combined with the last inequality yields

$$\varepsilon < f(x^{k_2}) - f(x^{k_1}) - \langle g^{k_1}, x^{k_2} - x^{k_1} \rangle.$$

The function $f(\cdot)$ is subdifferentiable everywhere (because it is finite) and X is compact, so there exists a constant C such that $f(x_1) - f(x_2) \le C\|x_1 - x_2\|$, for all $x_1, x_2 \in X$. Subgradients on bounded sets are bounded, and thus we can choose C large enough so that $\|g^k\| \le C$ for all k. It follows from the last displayed inequality that

$$\varepsilon < 2C\|x^{k_1} - x^{k_2}\| \quad \text{for all} \quad k_1, k_2 \in \mathcal{K}_\varepsilon,\ k_1 \ne k_2.$$

As the set X is compact, there can exist only finitely many points in X having a distance at least $\varepsilon/(2C)$ from each other. Thus the last inequality implies that the set $\mathcal{K}_\varepsilon$ is finite for each $\varepsilon > 0$. This means that for every $\varepsilon > 0$ all but finitely many points x^k have the objective function values in the interval $[f^*, f^* + \varepsilon]$. Therefore, the sequence $\{f(x^k)\}$ is convergent to f^*. $\square$

Due to the compactness of the set X, the sequence $\{x^k\}$ must have accumulation points. It follows from the above theorem that each such accumulation point is a solution of problem (7.30).

The cutting plane method, when applied to general convex problems, is rather slow. No reliable rule exists for deleting old cuts, even those that are not active at the current solution of problem (7.33). Usually, very many iterations are needed to achieve a satisfactory accuracy of the solution found. However, in the special case when the objective function is piecewise linear, the cutting plane method becomes much more efficient. We discuss this topic in the next section.

7.2.3 Finite Convergence for Piecewise-Linear Functions

Assume that the function $f(\cdot)$ in (7.30) is piecewise linear:

$$f(x) = \max_{i \in I} \{\gamma_i + \langle u_i, x \rangle\} \quad \text{for all} \quad x \in \mathbb{R}^n, \tag{7.34}$$

where I is a finite index set, and $\gamma_i \in \mathbb{R}$ and $u_i \in \mathbb{R}^n$ are fixed. Problems with such objective functions frequently arise in applications, in particular, as dual problems to large scale linear or integer optimization problems.

For purposes of illustration, look at the facility location problem of Example 4.23 on page 188. Each of the dual relaxations discussed there leads to piecewise linear, concave dual function. For instance, each of the functions in (4.62) is concave and piecewise linear, and so is their sum in (4.63). The number of pieces in the equivalent representation (7.34) is huge, and they are not enumerated explicitly. However, whatever the values of the dual variables, it is straightforward to identify an *active* piece i_* of the function: such γ_{i_*} and u_{i_*} that $f(x) = \gamma_{i_*} + \langle u_{i_*}, x \rangle$. We leave to the reader the derivation of an appropriate procedure in this case (Exercise 7.5).

Assume that the cutting plane method is applied to problem (7.30) with the objective function of the form (7.34), and that the subgradients calculated are the gradients of *active* pieces of (7.34). Specifically, we assume that for every x^k

$$g^k = u_{i_k},$$

with

$$i_k \in I(x^k) \triangleq \{i \in I : f(x^k) = \gamma_i + \langle u_i, x \rangle\}.$$

If the method does not stop at iteration k, then $f(x^k) > f^k(x^k)$ and therefore i_k is different than $i_1, i_2, \ldots, i_{k-1}$. Since the number of different active pieces is finite, the method must eventually stop, and we already know that the termination may occur only at an optimal point. Interestingly, the finite convergence property holds true even if arbitrary subgradients of $f(\cdot)$ are calculated.

THEOREM 7.8. *After finitely many iterations, the cutting plane method stops at an optimal solution of* (7.30).

Proof. For each x we can define the set

$$I(x) \triangleq \{i \in I : \gamma_i + \langle u_i, x\rangle = f(x)\}.$$

There exist only finitely many different sets $I(x)$, and we denote them by $I_1, I_2, \ldots, I_L$. Each of the sets I_l defines a cell:

$$C_l \triangleq \{x : \gamma_i + \langle u_i, x\rangle = f(x),\ i \in I_l;\ \gamma_i + \langle u_i, x\rangle < f(x),\ i \in I \setminus I_l\}.$$

Suppose our assertion is false, and $f(x^k) > f^*$ for infinitely many k. Let $f(x^k) > f^*$ for some $x^k \in C_l$. By Example 2.80, the cut generated at x^k has the form

$$f(x^k) + \langle \overline{g}^k, x - x^k\rangle \le v, \tag{7.35}$$

with the subgradient

$$\overline{g}^k = \sum_{i\in I_l} \lambda_i u_i, \quad \lambda_i \ge 0, \quad \sum_{i\in I_l} \lambda_i = 1.$$

For all $x \in C_l$ we have

$$\begin{aligned} f(x) &= \sum_{i\in I_l} \lambda_i\big(\gamma_i + \langle u_i, x\rangle\big) \\ &= \sum_{i\in I_l} \lambda_i\big(\gamma_i + \langle u_i, x^k\rangle + \langle u_i, x - x^k\rangle\big) \\ &= f(x^k) + \langle \overline{g}^k, x - x^k\rangle. \end{aligned}$$

By (7.35), for every $m > k$ such that $x^m \in C_l$ we must have

$$f(x^m) \le v^m,$$

and the method must stop at x^m. Thus, the cell C_l may be visited at most once.

Therefore, after visiting finitely many cells the algorithm must eventually satisfy the stopping test. □

Simplicity is the main advantage of the cutting plane method. However, the number of cuts in the master problem grows and there is no easy way to keep it bounded. A natural idea would be to drop inactive cuts, that is, the cuts that are satisfied as sharp inequalities at the current solution (x^k, v^k) of the master problem.

If we apply the method to piecewise linear problems and gradients of active pieces are used as cuts, we may drop inactive cuts whenever the optimal value of the master program increases.

To prove that we will still have a finitely convergent method, we note at first that when inactive cuts are deleted, no decrease of the optimal value of the master problem may result. Thus the sequence of optimal values of the master problem, $\{v^k\}$, is monotone. To prove that it is convergent to f^*, suppose $v^k < f^*$ for some k and that no increase in the master's objective function occurs for all $m > k$. Then no deletion takes place, and the discussion preceding Theorem 7.8 guarantees finite convergence of $\{v^m\}$ to f^*, a contradiction. Thus, an increase in the master's objective function must occur at some $m \geq k$. The number of different optimal values of the master problem is finite, because a finite number of different sets of active cuts exists. Therefore, an increase in this value can occur only finitely many times, and the method must stop at an optimal solution after finitely many iterations.

In the version with arbitrary subgradient cuts we have no guaranteed finiteness of the sets of cuts, and it is difficult to propose a useful and reliable rule for deleting inactive cuts.

Actually, in both cases, deleting *all* inactive cuts is not a good idea, because experience shows that many of them will have to be reconstructed.

Example 7.9. The *two-stage stochastic programming problem* is formulated, in its simplest form, as follows:

$$
\begin{aligned}
\underset{x, y_1, \dots, y_S}{\text{minimize}} \quad & \langle c, x \rangle + \sum_{s=1}^{S} p_s \langle q_s, y_s \rangle \\
\text{subject to} \quad & Ax = b, \quad x \geq 0, \\
& T_s x + W y_s = h_s, \quad s = 1, \dots, S, \\
& y_s \geq 0, \quad s = 1, \dots, S.
\end{aligned}
\tag{7.36}
$$

Here $x \in \mathbb{R}^{n_1}$ is called the vector of *first stage variables*, and $y_s \in \mathbb{R}^{n_2}$ is the vector of *second stage variables* in scenario $s = 1, \dots, S$. The matrices A, T_s, and W_s, of appropriate dimension, as well as the vectors c, q_s, b, and h_s, are known. In applications, the number of scenarios S is very large, and the stochastic programming problem in its linear programming form (7.36) may have dimensions exceeding the capabilities of linear programming solvers.

Another way to look at the problem is to consider, for a fixed x, the *second stage problems*

$$\begin{aligned} &\underset{y_s}{\text{minimize}} \ \langle q_s, y_s \rangle \\ &\text{subject to} \ W y_s = h_s - T_s x, \\ &\qquad\qquad\ y_s \geq 0. \end{aligned} \tag{7.37}$$

Their optimal values, denoted by $Q_s(x)$, depend on the first stage decision x, and on the scenario s. For simplicity, we assume that problems (7.37) are always solvable, when x satisfies the constraints $Ax = b$, $x \geq 0$. Much of what we present here can be easily extended to the general case, but with technical complications that would obscure the presentation.

The entire problem can now be compactly written as follows:

$$\begin{aligned} &\underset{x}{\text{minimize}} \ \Big[f(x) \triangleq \langle c, x \rangle + \sum_{s=1}^{S} p_s Q_s(x) \Big] \\ &\text{subject to} \ Ax = b, \ x \geq 0. \end{aligned} \tag{7.38}$$

We know from Section 4.6 that the functions $Q_s(\cdot)$, as optimal values of parametric convex programming problems, are convex. Theorem 4.26 on page 194 provides us with the description of the subdifferential of the optimal value function of problem (7.37) with respect to the vector $r_s = h_s - T_s x$. Any solution $\lambda_s(x)$ of the dual problem

$$\begin{aligned} &\text{maximize} \ \langle h_s - T_s x, \lambda \rangle \\ &\text{subject to} \ W^T \lambda \leq q_s, \end{aligned} \tag{7.39}$$

is a subgradient of the optimal value function,[†] with respect to the vector $r_s = h_s - T_s x$. The chain rule of Lemma 2.83 on page 67 provides us with the subgradients of $Q_s(\cdot)$:

$$\partial Q_s = \big\{ -T_s^T \lambda_s(x) : \lambda_s(x) \text{ is a solution of (7.39)} \big\}.$$

Therefore, a cutting plane for the function $f(x)$ in (7.38) can be obtained as follows. We solve, for $s = 1, \ldots, S$, second stage problems (7.37) and we obtain values of Lagrange multipliers $\lambda_s(x)$ associated with their constraints. Then the vector

$$g(x) = c - \sum_{s=1}^{S} p_s T_s^T \lambda_s(x),$$

is a subgradient of $f(\cdot)$ at x.

Since each $Q_s(\cdot)$ is convex and piecewise linear, the function $f(\cdot)$ is convex and piecewise linear as well, although the number of its pieces may be enormous. By Theorem 7.8, the cutting plane method finds an optimal solution of the problem in

---

[†] Following the conventions of linear programming, we formulate the Lagrangian of problem (7.37) as $L(x, \lambda) = \langle c, x \rangle + \langle \lambda, h_s - T_s x \rangle$, and this results in the opposite sign of subgradients.

finitely many steps. It is convenient to ensure that the multipliers $\lambda_s(x)$ are selected from the set of *basic* optimal solutions of the corresponding dual problems. Then the cutting plane method uses gradients of active pieces as cuts, and deletion of inactive cuts can be employed safely. It is very relevant from the practical point of view, because many cuts need to be generated to solve practical problems.

The cutting plane method also can be specialized to problems when the objective function $f(x)$ is a sum of many components, which are subdifferentiated separately, as in (7.38). It is then possible to construct separate cutting plane approximation for each piece (see Exercise 7.4).

7.2.4 Application to Dual Problems

As we discuss in Section 7.1.3, one of the main applications of nonsmooth optimization methods is the solution of dual problems associated with convex programming problems. To illustrate the way in which a primal solution can be recovered from the solution of the dual problem by a cutting plane method, let us consider again a linearly constrained convex problem:

$$\begin{aligned} \text{minimize} \;\; & f(x) \\ \text{subject to} \;\; & Ax = b, \\ & x \in X_0. \end{aligned} \tag{7.40}$$

We assume that the function $f : \mathbb{R}^n \to \mathbb{R}$ is convex and the set $X_0 \subset \mathbb{R}^n$ is convex and closed. The matrix A of dimension $m \times n$ and the vector $b \in \mathbb{R}^m$ are given. We also assume that the set X_0 is bounded, so that the problem has an optimal solution. We denote by f^* the optimal value of the objective function. Finally, we assume Slater's constraint qualification condition: there exists a point $x_S \in \operatorname{int} X_0$ such that $Ax_S = b$. Then the duality relation holds true and we can apply the dual method.

The dual function has the form

$$L_D(\mu) = \min_{x \in X_0} \big\{ f(x) + \langle \mu, b - Ax \rangle \big\}. \tag{7.41}$$

Identical to the situation analyzed in Section 7.1.3, at every point μ^k generated by the method, a subgradient of the dual function can be calculated as follows: we solve the minimization problem on the right hand side of (7.41) to obtain a point x^k, which yields the subgradient of $L_D(\cdot)$:

$$g^k = b - Ax^k.$$

The dual problem is a maximization problem and thus the cutting plane approximation is concave and piecewise linear. The master problem at iteration

k has the form

$$\begin{aligned} &\text{maximize } \ v \\ &\text{subject to } \ v \le L_D(\mu^j) + \langle b - Ax^j, \mu - \mu^j \rangle, \quad j = 1, \dots, k, \\ &\qquad\qquad -C \le \mu_i \le C, \quad i = 1, \dots, m. \end{aligned} \tag{7.42}$$

We add here the constraints $|\mu_i| \le C$, where C is a sufficiently large constant, to make sure that the cutting plane approximation of the dual problem has a solution. The constant C should be such that the box defined by these artificial constraints contains a solution to the dual problem.

Suppose $(\hat{\mu}, \hat{v})$ is a solution of the master problem, and that $L_D(\hat{\mu}) = \hat{v}$, that is, the stopping test of the cutting plane method is satisfied. Furthermore, let us assume that the artificial constraints $-C \le \mu_i \le C$ are not active at the solution. Then $\hat{\mu}$ is an optimal solution of the dual problem, and $\hat{v}$ is the optimal value of the dual problem, which is equal to the optimal value f^* of problem (7.40).

Let us denote by J^* the set of active cuts in (7.42) and by λ_j, $j \in J^*$, the values of Lagrange multipliers associated with these cuts. The optimality condition with respect to v yields the relation

$$\sum_{j \in J^*} \lambda_j = 1,$$

and, of course, the multipliers λ_j are nonnegative. Consider the point

$$\bar{x} = \sum_{j \in J^*} \lambda_j x^j.$$

We shall show that it is an optimal solution of the primal problem (7.40). It is a convex combination of the points x^j, $j \in J^*$, and therefore belongs to the set X_0. Furthermore, from the optimality conditions for problem (7.42) with respect to μ, we conclude that

$$\sum_{j \in J^*} \lambda_j (b - Ax^j) = 0. \tag{7.43}$$

This can be simply written as

$$b - A\bar{x} = 0,$$

and proves that the point $\bar{x}$ is feasible for problem (7.40).

Let us prove its optimality. The active constraints in (7.42) are satisfied as equations, and thus

$$f^* = \hat{v} = L_D(\mu^j) + \langle b - Ax^j, \hat{\mu} - \mu_j \rangle = f(x^j) + \langle b - Ax^j, \hat{\mu} \rangle, \quad j \in J^*.$$

Multiplying these equations by λ_j and summing up we obtain

$$f^* = \sum_{j \in J^*} \lambda_j f(x^j) + \sum_{j \in J^*} \lambda_j \langle b - Ax^j, \hat{\mu} \rangle.$$

The second sum is zero, by virtue of (7.43). The convexity of $f(\cdot)$ implies that

$$\sum_{j \in J^*} \lambda_j f(x^j) \geq f(\bar{x}).$$

Combining the last two relations we conclude that $f(\bar{x}) \leq f^*$, which proves the optimality of the point $\bar{x}$.

The case of inequality constraints in the primal problem (7.40) is analyzed in an identical way.

7.3 THE PROXIMAL POINT METHOD

7.3.1 Moreau–Yosida Regularization

Consider a convex, proper, extended real-valued function $f : \mathbb{R}^n \to \overline{\mathbb{R}}$. We assume that $f(\cdot)$ is lower semicontinuous, that is, its epigraph is a closed set (Lemma 2.62 on page 47).

For a fixed number $\varrho > 0$, we define the function

$$f_\varrho(w) \triangleq \min_{x \in \mathbb{R}^n} \left\{ \frac{\varrho}{2} \|x - w\|^2 + f(x) \right\}. \tag{7.44}$$

The function $f_\varrho(\cdot)$ is called the *Moreau–Yosida regularization* of $f(\cdot)$.

Lemma 7.10. *For every $\varrho > 0$, the function $f_\varrho(\cdot)$ is real-valued, convex, and continuously differentiable, with $\nabla f_\varrho(w) = \varrho(w - x_\varrho(w))$, where $x_\varrho(w)$ is the solution of* (7.44).

Proof. Since $f(\cdot)$ is proper, there exists a point x_0 such that $f(\cdot)$ is subdifferentiable at x_0 (see the discussion at the beginning of Section 2.6.1). Let $g \in \partial f(x_0)$. Then, for every x we have the inequality

$$f(x) \geq f(x_0) + \langle g, x - x_0 \rangle.$$

Therefore, the function minimized in the right hand side of (7.44) can be bounded below by the function

$$r(x) = \frac{\varrho}{2} \|x - w\|^2 + f(x_0) + \langle g, x - x_0 \rangle.$$

The level sets of the function on the right hand side of (7.44) are included in the level sets of $r(x)$, which are simply balls. Since $f(\cdot)$ is lower semicontinuous, the problem on the right hand side of (7.44) has, for each w, a solution $x_\varrho(w)$. By the strict convexity of the function $x \to \frac{\varrho}{2}\|x - w\|^2 + f(x)$, the solution $x_\varrho(w)$ is unique. We can thus write

$$f_\varrho(w) = \frac{\varrho}{2}\|x_\varrho(w) - w\|^2 + f(x_\varrho(w)).$$

By Theorem 3.5 on page 92, at the minimum $x_\varrho(w)$ there exists a subgradient $g(w) \in \partial f(x_\varrho(w))$ such that

$$\varrho(x_\varrho(w) - w) + g(w) = 0.$$

In other words,

$$\varrho(w - x_\varrho(w)) \in \partial f(x_\varrho(w)). \tag{7.45}$$

Consider a fixed point w_1 and the corresponding solution $x_1 = x_\varrho(w_1)$. From the subgradient inequality and (7.45) at w_1, we deduce that at every point $x \in \mathbb{R}^n$

$$f(x) \geq f(x_1) + \varrho\langle w_1 - x_1, x - x_1\rangle.$$

Therefore for each $w \in \mathbb{R}^n$

$$f_\varrho(w) \geq \min_{x\in\mathbb{R}^n}\left\{\frac{\varrho}{2}\|x - w\|^2 + f(x_1) + \varrho\langle w_1 - x_1, x - x_1\rangle\right\}.$$

The minimum on the right hand side of the last inequality can be calculated analytically, because the function minimized is quadratic. We obtain the solution $x^* = x_1 + w - w_1$. After substituting for x into the last inequality we get

$$\begin{aligned} f_\varrho(w) &\geq \frac{\varrho}{2}\|x_1 - w_1\|^2 + f(x_1) + \varrho\langle w_1 - x_1, w - w_1\rangle \\ &= f_\varrho(w_1) + \varrho\langle w_1 - x_1, w - w_1\rangle, \quad \text{for all} \quad w \in \mathbb{R}^n. \end{aligned} \tag{7.46}$$

It follows that the function $f_\varrho(w_1) + \varrho\langle w_1 - x_1, \cdot - w_1\rangle$ is a linear minorant of $f_\varrho(\cdot)$. This means that $f_\varrho(\cdot)$ is convex, and that the vector $\varrho(w_1 - x_\varrho(w_1))$ is its subgradient at w_1.

By plugging into the right hand side of (7.44) the point $x_1 = x_\varrho(w_1)$, instead of $x_\varrho(w)$, we obtain the inequality

$$\begin{aligned} f_\varrho(w) &\leq \frac{\varrho}{2}\|x_1 - w\|^2 + f(x_1) \\ &= \frac{\varrho}{2}\|(x_1 - w_1) + (w_1 - w)\|^2 + f(x_1) \\ &= f_\varrho(w_1) + \varrho\langle w_1 - x_1, w - w_1\rangle + \frac{\varrho}{2}\|w - w_1\|^2. \end{aligned} \tag{7.47}$$

Both estimates (7.46) and (7.47) prove that the function $f_\varrho(\cdot)$ is in fact differentiable, and that the vector $\varrho(w_1 - x_\varrho(w_1))$ is its gradient at w_1. □

Let us recall that an operator $M : \mathbb{R}^n \to \mathbb{R}^n$ is called *monotone*, if $\langle M(x_2) - M(x_1), x_2 - x_1 \rangle \geq 0$ for all $x_1, x_2 \in \mathbb{R}^n$. The concept of monotonicity plays an important role in convex analysis; we have encountered it already in Exercise 2.14 in connection with the subgradient mapping.

LEMMA 7.11. *The operator $x_\varrho(\cdot)$ is nonexpansive and monotone.*

Proof. Consider two points $w_1, w_2 \in \mathbb{R}^n$ and the corresponding solutions $x_1 = x_\varrho(w_1)$ and $x_2 = x_\varrho(w_2)$. It follows from (7.45) at $w = w_1$ that

$$f(x_2) \geq f(x_1) + \varrho\langle w_1 - x_1, x_2 - x_1 \rangle.$$

Reversing the roles of w_1 and w_2 we also have

$$f(x_1) \geq f(x_2) + \varrho\langle w_2 - x_2, x_1 - x_2 \rangle.$$

Adding the last two inequalities yields

$$0 \geq \langle w_1 - x_1 - w_2 + x_2, x_2 - x_1 \rangle,$$

which can be rewritten as follows:

$$\langle w_2 - w_1, x_2 - x_1 \rangle \geq \|x_2 - x_1\|^2. \tag{7.48}$$

This proves that the operator $x_\varrho(\cdot)$ is monotone. We can also use this relation to write the chain of inequalities

$$\begin{aligned} 0 &\leq \|(w_2 - w_1) - (x_2 - x_1)\|^2 \\ &= \|w_2 - w_1\|^2 - 2\langle w_2 - w_1, x_2 - x_1 \rangle + \|x_2 - x_1\|^2 \\ &\leq \|w_2 - w_1\|^2 - \|x_2 - x_1\|^2. \end{aligned}$$

Consequently, $\|x_2 - x_1\| \leq \|w_2 - w_1\|$, as postulated. □

7.3.2 Application to Convex Optimization

Let us consider the convex optimization problem

$$\underset{x \in \mathbb{R}^n}{\text{minimize}}\ f(x), \tag{7.49}$$

in which $f : \mathbb{R}^n \to \overline{\mathbb{R}}$ is a convex extended real-valued function. We assume that $f(\cdot)$ is proper and lower semicontinuous.

Problem (7.49) is a convenient abstract model of a constrained problem of the form (7.30), where we can simply define the extended real-valued function

$$\bar{f}(x) = \begin{cases} f(x) & \text{if } x \in X, \\ +\infty & \text{otherwise.} \end{cases}$$

and consider the problem of minimizing $\bar{f}(\cdot)$ over $\mathbb{R}^n$.

Using the Moreau–Yosida regularization of $f(\cdot)$, we construct the following iterative process. At iteration k, given the current approximation w^k to the solution of (7.49), we find the point $x^k = x_\varrho(w^k)$, which is the solution of the problem

$$\underset{x\in\mathbb{R}^n}{\text{minimize}} \; \frac{\varrho}{2}\|x - w^k\|^2 + f(x). \tag{7.50}$$

The next approximation is defined according to the formula:

$$w^{k+1} = x_\varrho(w^k), \quad k = 1, 2, \ldots. \tag{7.51}$$

The iterative method (7.51) is called the *proximal point method.* It does not appear to be very practical, because each iteration involves the solution of the optimization problem (7.50), which is comparable in its difficulty to the original problem (7.49). However, the proximal point method is an important theoretical model of various highly efficient methods. We have already encountered it in its pure form in the analysis on the augmented Lagrangian method in Section 6.4.3.

Let us recall that it follows from Lemma 7.10 that problem (7.50) always has a solution, if $f(\cdot)$ is lower semicontinuous. Thus the proximal point method is well defined. Since $f_\varrho(w^k) \le f(w^k)$ by construction, we have $f(w^{k+1}) \le f(w^k)$, $k = 1, 2, \ldots$. Actually, the progress made at each iteration can be estimated very precisely.

Lemma 7.12. *Assume that there exists $\tilde{x} \in \mathbb{R}^n$ such that $f(\tilde{x}) < f(w)$. Then*

$$f_\varrho(w) \le f(w) - \varrho\|\tilde{x} - w\|^2 \varphi\Big(\frac{f(w) - f(\tilde{x})}{\varrho\|\tilde{x} - w\|^2}\Big),$$

where

$$\varphi(\tau) = \begin{cases} 0 & \text{if } \tau < 0, \\ \tau^2 & \text{if } 0 \le \tau \le 1, \\ -1 + 2\tau & \text{if } \tau > 1. \end{cases}$$

Proof. Consider the segment containing points $x = w + t(\tilde{x} - w)$ with $0 \le t \le 1$. Restricting the minimization in (7.50) to these x provides an upper bound for the optimal value:

$$\begin{aligned} f_{\varrho}(w) &\le \min_{0 \le t \le 1} \Big[f((1-t)w + t\tilde{x}) + \frac{\varrho t^2}{2} \|\tilde{x} - w\|^2 \Big] \\ &\le f(w) + \min_{0 \le t \le 1} \Big[t(f(\tilde{x}) - f(w)) + \frac{\varrho t^2}{2} \|\tilde{x} - w\|^2 \Big]. \end{aligned}$$

In the last estimate we also used the convexity of $f(\cdot)$. The value of t that minimizes the above expression is equal to

$$\hat{t} = \min\Big(1, \frac{f(w) - f(\tilde{x})}{\varrho \|\tilde{x} - w\|^2}\Big).$$

Our assertion follows now from a straightforward calculation. □

At the solution $x_{\varrho}(w)$ of problem (7.50) we have

$$f(x_{\varrho}(w)) \le f_{\varrho}(w) \le f(w) - \varrho \|\tilde{x} - w\|^2 \varphi\Big(\frac{f(w) - f(\tilde{x})}{\varrho \|\tilde{x} - w\|^2}\Big).$$

Therefore, if a better point exists, problem (7.50) will find a better point. Consequently, $x = w$ is the minimizer in (7.50) if and only if w is a minimizer of $f(\cdot)$.

In fact, the proximal point method must converge to an optimal solution, if an optimal solution exists.

THEOREM 7.13. *Assume that problem* (7.49) *has an optimal solution. Then the sequence* $\{w^k\}$ *generated by the proximal point method is convergent to an optimal solution of* (7.49).

Proof. Let x^* be an optimal solution. We have the identity

$$\|w^{k+1} - x^*\|^2 = \|w^k - x^*\|^2 + 2\langle w^{k+1} - w^k, w^{k+1} - x^*\rangle - \|w^{k+1} - w^k\|^2. \tag{7.52}$$

Relation (7.45) reads:

$$-\varrho(w^{k+1} - w^k) \in \partial f(w^{k+1}). \tag{7.53}$$

By the definition of the subgradient,

$$f(x^*) \ge f(w^{k+1}) - \varrho\langle w^{k+1} - w^k, x^* - w^{k+1}\rangle.$$

Using this inequality in (7.52) (and skipping the last term) we obtain

$$\|w^{k+1} - x^*\|^2 \le \|w^k - x^*\|^2 - \frac{2}{\varrho}(f(w^{k+1}) - f(x^*)). \tag{7.54}$$

Several conclusions follow from this estimate. First, the sequence $\{w^k\}$ is bounded, because the distance to x^* is nonincreasing. Second, summing up (7.54) from $k = 1$ to ∞, we get

$$\sum_{k=2}^{\infty}(f(w^k) - f(x^*)) \leq \frac{\varrho}{2}\|w^1 - x^*\|^2,$$

so $f(w^k) \to f(x^*)$ as $k \to \infty$. Consequently, for every accumulation point $\tilde{x}$ of $\{w^k\}$ we have $f(\tilde{x}) = f(x^*)$. We choose one such $\tilde{x}$, substitute it for x^* in (7.54), and conclude that the entire sequence $\{w^k\}$ is convergent to $\tilde{x}$. □

For polyhedral $f(\cdot)$ the convergence is finite.

THEOREM 7.14. *Assume that $f(\cdot)$ is a convex polyhedral function and that a minimum of $f(\cdot)$ exists. Then the proximal point method stops after finitely many steps at a minimizer of $f(\cdot)$.*

Proof. Suppose the method does not stop. Therefore, $0 \notin \partial f(w^{k+1})$ for $k = 1, 2, \ldots$. Hence

$$0 \notin \bigcup_{k=1}^{\infty} \partial f(w^{k+1}).$$

As $f(\cdot)$ is polyhedral, only finitely many different subdifferentials $\partial f(w^{k+1})$ exist. Each of them is a convex closed polyhedral set, so the right hand side of the last displayed relation is a union of finitely many closed sets. Thus it is closed. Consequently, there exists $\varepsilon > 0$ such that the ball $B(0, \varepsilon)$ of radius ε about 0 has no common points with this union of subdifferentials. We get

$$B(0, \varepsilon) \cap \partial f(w^{k+1}) = \emptyset, \quad k = 1, 2, \ldots.$$

Because the sequence $\{w^k\}$ is convergent by Theorem 7.13, we have $w^{k+1} - w^k \to 0$. Therefore $\varrho(w^{k+1} - w^k) \in B(0, \varepsilon)$ for large k. This, combined with the last displayed equation, contradicts (7.53). □

The proximal point method can be generalized to a much broader class of problems than convex optimization problems. These are inclusions of the form

$$0 \in M(x), \tag{7.55}$$

where M is a function assigning to points $x \in \mathbb{R}^n$ sets $M(x) \subset \mathbb{R}^n$.

A function M taking arguments in $\mathbb{R}^n$ and assuming as its values sets in $\mathbb{R}^m$ is called a *multivalued operator*, and written $M : \mathbb{R}^n \rightrightarrows \mathbb{R}^m$. For a multivalued operator M we define its *domain* as the set of x for which $M(x)$ is nonempty. Its *graph* is the set $\mathcal{G}(M) \triangleq \{(x, y) : y \in M(x)\}$.

An operator $M : \mathbb{R}^n \rightrightarrows \mathbb{R}^n$ is called *monotone*, if

$$\langle x' - x, y' - y \rangle \geq 0 \quad \text{for all} \quad x, x' \in \operatorname{dom} M, \; y \in M(x), \; y' \in M(x').$$

It is *maximal monotone* if its graph is not properly contained in the graph of any other monotone operator. The subdifferential of a convex lower semicontinuous function is a maximal monotone operator (Exercise 2.14). By virtue of Theorem 3.5, the optimization problem (7.49) is equivalent to the inclusion (7.55) for $M \equiv \partial f$.

However, many other problems which cannot be formulated as optimization problems can be formulated as inclusions (7.55) with maximal monotone operators. Most notably, these are saddle point problems for convex-concave functions, and equilibrium problems in multiperson games.

We can generalize the proximal point method to inclusions of the form (7.55) as follows. At iteration k, given $x^k \in \operatorname{dom} M$ we find x^{k+1} as the solution of the inclusion

$$0 \in M(x) + \varrho(x - x^k).$$

It corresponds to (7.53) for the case of the optimization problem. If the inclusion (7.55) has a solution, the sequence $\{x^k\}$ generated by the proximal point method is convergent to it.

The analysis of inclusions of the form (7.55) and of the proximal point method in the more general setting uses language and techniques which are beyond the scope of this book. The interested reader is referred to the literature discussed at the end of the book.

7.4 THE BUNDLE METHOD

7.4.1 Main Ideas

Let us return to the problem

$$\underset{x \in X}{\text{minimize}} \; f(x), \tag{7.56}$$

in which the function $f : \mathbb{R}^n \to \mathbb{R}$ is convex and the set $X \subset \mathbb{R}^n$ is convex and closed.[†] We do not need to assume the boundedness of the set X, as in the case of the cutting plane method.

[†]We could reformulate the problem as an "unconstrained" problem (7.49), by introducing an extended real-valued function $f_X(x) = f(x)$ for $x \in X$ and $f_X(x) = +\infty$ for $x \notin X$. However, the explicit formulation (7.56) is more convenient here, because we want to use the boundedness of subgradients of $f(\cdot)$ on bounded sets in our analysis.

The principal difficulty associated with the cutting plane method is the growth of the number of cuts that need to be stored in the master problem. Also, there is no easy way to make use of a good starting solution.

To mitigate these difficulties we add a quadratic regularizing term to the polyhedral model used in the master problem (7.32) of the cutting plane method. We obtain the following *regularized master problem*:

$$\underset{x \in X}{\text{minimize}} \; \frac{\varrho}{2}\|x - w^k\|^2 + f^k(x). \tag{7.57}$$

The model $f^k(\cdot)$ is defined similarly to (7.31):

$$f^k(x) \triangleq \max_{j \in J_k} \left[f(x^j) + \langle g^j, x - x^j \rangle \right], \tag{7.58}$$

with $g^j \in \partial f(x^j)$, $j \in J_k$. The set J_k is a subset of $\{1, \dots, k\}$ determined by a procedure for selecting cuts. At this moment we may think of J_k as being equal to $\{1, \dots, k\}$.

In the proximal term $(\varrho/2)\|x - w^k\|^2$, where $\varrho > 0$, the center w^k is updated depending on the relations between the value of $f(x^{k+1})$ at the master's solution, x^{k+1}, and its prediction provided by the current model, $f^k(x^{k+1})$. If these values are equal or close, we set $w^{k+1} := x^{k+1}$ (*descent step*); otherwise $w^{k+1} := w^k$ (*null step*).

In any case, the collection of cuts is updated, and the iteration continues.

If the model $f^k(\cdot)$ were exact, that is $f^k \equiv f$, then problem (7.57) would be identical to the subproblem (7.50) solved at each iteration of the proximal point method and we could just set $w^{k+1} := x^{k+1}$. All steps would be descent steps. However, due to the approximate character of $f^k(\cdot)$, the solution of (7.57) is different than the solution of (7.50). It may not even be better than w^k (in terms of the value of the objective function). This is the reason for introducing conditional rules for updating the center w^k.

On the other hand, the regularized master problem can be equivalently written as a problem with a quadratic objective function and linear constraints:

$$\begin{aligned} &\text{minimize} \;\; v + \frac{\varrho}{2}\|x - w^k\|^2 \\ &\text{subject to} \;\; f(x^j) + \langle g^j, x - x^j \rangle \le v, \quad j \in J_k, \\ &\qquad\qquad\quad x \in X. \end{aligned} \tag{7.59}$$

If the set X is a convex polyhedron, the master problem can be readily solved by specialized techniques, enjoying the finite termination property. This clearly distinguishes it from the "exact" subproblem of the proximal point method, where the original objective function $f(x)$ appears.

Step 0. Set $k := 1$, $J^0 := \emptyset$, $v^1 := -\infty$.

Step 1. Calculate $f(x^k)$ and $g^k \in \partial f(x^k)$.
If $f(x^k) > v^k$ then set $J_k := J_{k-1} \cup \{k\}$; otherwise set $J_k := J_{k-1}$.

Step 2. If $k = 1$ or if

$$f(x^k) \le (1-\gamma) f(w^{k-1}) + \gamma f^{k-1}(x^k),$$

then set $w^k := x^k$; otherwise set $w^k := w^{k-1}$.

Step 3. Solve the master problem (7.59). Denote by (x^{k+1}, v^{k+1}) its solution and set $f^k(x^{k+1}) := v^{k+1}$.

Step 4. If $f^k(x^{k+1}) = f(w^k)$ then stop (w^k is an optimal solution); otherwise continue.

Step 5. Remove from the sets of cuts J_k some (or all) cuts whose Lagrange multipliers λ_j^k at the solution of (7.59) are 0. Increase k by one, and go to Step 1.

Figure 7.3. The bundle method.

Let us observe that problem (7.59) satisfies Slater's constraint qualification condition. Indeed, for every $x_S \in X$ we can choose v_S so large that all constraints are satisfied as strict inequalities. Therefore the optimal solution of the master problem satisfies the necessary and sufficient conditions of optimality of Theorem 3.34 on page 127. We denote by λ_j^k, $j \in J_k$, the Lagrange multipliers associated with the constraints of problem (7.59).

The detailed algorithm is stated in Figure 7.3. The parameter $\gamma \in (0, 1)$ is a fixed constant used to compare the observed improvement in the objective value to the predicted improvement.

7.4.2 Convergence

The main difference between the bundle method and the proximal point method is that the master problem (7.57) uses a model $f^k(\cdot)$ instead of the true function $f(\cdot)$. Its minimizer, x^{k+1}, is no longer guaranteed to be better than w^k. The role of null steps is to correct the model $f^k(\cdot)$, if x^{k+1} is not better than w^k. We shall see that such model improvements ensure that progress will eventually be made whenever any progress is possible.

By the construction of the method, the sequence $\{f(w^k)\}$ is nonincreasing. We first show that the algorithm cannot get stuck at a non-optimal point.

LEMMA 7.15. *Suppose w^k is not an optimal solution of problem* (7.56). *Then there exist $m > k$ such that $f(w^m) < f(w^k)$.*

Proof. Suppose null steps are made at iterations $j = k+1, k+2, \dots$. Thus $w^j = w^k$ and

$$f(x^j) > f^{j-1}(x^j) + (1-\gamma)(f(w^{j-1}) - f^{j-1}(x^j)).$$

Set $\delta_j = f(w^k) - f^{j-1}(x^j)$. Since $w^{j-1} = w^k$ for these j, the last inequality can be rewritten as follows:

$$f(x^j) > f^{j-1}(x^j) + (1-\gamma)\delta_j. \tag{7.60}$$

We shall show that the optimal value of the master problem increases by a quantity related to δ_j, when a new cut is added.

First, the master's optimal value will not change, if we delete inactive cuts. We denote the model without inactive cuts by $\underline{f}^{j-1}(x)$. Subdifferentiating (7.57) at x^j and applying Theorem 3.34 from page 127 we get

$$0 \in \varrho(x^j - w^k) + \partial \underline{f}^{j-1}(x^j) + N_X(x^j).$$

Thus there exists a subgradient $d \in \partial \underline{f}^{j-1}(x^j)$ and a normal vector $h \in N_X(x^j)$ such that

$$d = \varrho(w^k - x^j) - h.$$

This allows us to write for every $x \in X$ the inequality

$$\begin{aligned} f^j(x) \geq \underline{f}^{j-1}(x) &\geq f^{j-1}(x^j) + \langle d, x - x^j \rangle \\ &= f^{j-1}(x^j) + \varrho\langle w^k - x^j, x - x^j \rangle - \langle h, x - x^j \rangle \\ &\geq f^{j-1}(x^j) + \varrho\langle w^k - x^j, x - x^j \rangle. \end{aligned} \tag{7.61}$$

Second, by (7.60), after adding the new cut at x^j we have

$$f^j(x^j) > f^{j-1}(x^j) + (1-\gamma)\delta_j.$$

Therefore, for all $x \in X$,

$$f^j(x) > f^{j-1}(x^j) + (1-\gamma)\delta_j + \langle g, x - x^j \rangle,$$

where $g \in \partial f^j(x^j)$. Combining (7.61) with the last inequality we obtain

$$\begin{aligned} f^j(x) \geq \max\big(&f^{j-1}(x^j) + \varrho\langle w^k - x^j, x - x^j \rangle, \\ &f^{j-1}(x^j) + (1-\gamma)\delta_j + \langle g, x - x^j \rangle\big) \\ \geq\ &f^{j-1}(x^j) + \varrho\langle w^k - x^j, x - x^j \rangle \\ &+ \max\big(0, (1-\gamma)\delta_j + \langle g - \varrho(w^k - x^j), x - x^j \rangle\big). \end{aligned}$$

Consequently, the next value of the master's objective function can be estimated from below as follows:

$$
\begin{aligned}
&f^j(x) + \frac{\varrho}{2}\|x - w^k\|^2 \\
&\quad \geq f^{j-1}(x^j) + \varrho\langle w^k - x^j, x - x^j\rangle + \frac{\varrho}{2}\|x - w^k\|^2 \\
&\qquad + \max\big(0, (1-\gamma)\delta_j + \langle g - \varrho(w^k - x^j), x - x^j\rangle\big) \\
&\quad = f^{j-1}(x^j) + \frac{\varrho}{2}\|x^j - w^k\|^2 + \frac{\varrho}{2}\|x - x^j\|^2 \\
&\qquad + \max\big(0, (1-\gamma)\delta_j + \langle g - \varrho(w^k - x^j), x - x^j\rangle\big).
\end{aligned}
$$

It follows that the master's optimal value,

$$
\theta_j \triangleq f^j(x^{j+1}) + \frac{\varrho}{2}\|x^{j+1} - w^k\|^2,
$$

satisfies the inequality

$$
\begin{aligned}
\theta_j - \theta_{j-1} &\geq \min_{x\in X}\Big[\frac{\varrho}{2}\|x - x^j\|^2 + \max\big(0, (1-\gamma)\delta_j \\
&\qquad\qquad + \langle g - \varrho(w^k - x^j), x - x^j\rangle\big)\Big] \\
&\geq \min_{x\in X}\Big[\frac{\varrho}{2}\|x - x^j\|^2 + \max\big(0, (1-\gamma)\delta_j \\
&\qquad\qquad - \|g - \varrho(w^k - x^j)\|\,\|x - x^j\|\big)\Big].
\end{aligned}
$$

If no descent steps are made after iteration k, then the points x^j are contained in a bounded neighborhood of w^k, and the subgradients of $f(\cdot)$ are bounded in this area. Therefore, we can assume that there exists a constant C such that $\|g\| \leq C$ and $\|\varrho(w^k - x^j)\| \leq C$. Then the last inequality can be simplified into

$$
\theta_j - \theta_{j-1} \geq \min_{\sigma\in\mathbb{R}}\Big[\frac{\varrho\sigma^2}{2} + \max\big(0, (1-\gamma)\delta_j - 2C\sigma\big)\Big],
$$

where σ stands for $\|x - x^j\|$. The right hand side of the above expression can be estimated as follows. If $(1-\gamma)\delta_j \leq 4C^2/\varrho$, we have

$$
\sigma = (1-\gamma)\delta_j/(2C), \qquad \theta_j - \theta_{j-1} \geq \varrho(1-\gamma)^2\delta_j^2/(8C^2);
$$

otherwise,

$$
\sigma = 2C/\varrho, \qquad \theta_j - \theta_{j-1} \geq -2C^2/\varrho + (1-\gamma)\delta_j \geq (1-\gamma)\delta_j/2.
$$

The sequence $\{\theta_j\}$ is increasing and bounded above by $f(w^k)$. Hence $\theta_j - \theta_{j-1} \to 0$. If there are no descent steps after iteration k we conclude that $\delta_j \to 0$. As $f(w^k) \geq \theta_j \geq f(w^k) - \delta_{j+1}$, we have $\theta_j \uparrow f(w^k)$.

On the other hand, the master's objective function value is bounded above by the Moreau–Yosida regularization:

$$\theta_j \leq f_\theta(w^j) = f_\theta(w^k).$$

If w^k is not optimal, Lemma 7.12 yields $f_\theta(w^k) < f(w^k)$ and we obtain a contradiction. □

We are now ready to prove convergence of the bundle method. Our analysis has much in common with the analysis of the proximal point method.

THEOREM 7.16. *Assume that problem* (7.56) *has an optimal solution. Then the bundle method generates a sequence* $\{w^k\}$ *which is convergent to an optimal solution of* (7.56).

Proof. If w^k is optimal for some k, then $w^{j+1} = w^j$ for $j = k, k+1, \ldots$, and the theorem is true. If w^k is not optimal for any k, then, by Lemma 7.15, each series of null steps is finite and is followed by a descent step. Thus, the number of descent steps is infinite. Let us denote by $\mathcal{K}$ the set of iterations at which descent steps occur. If $w^{k+1} = x^{k+1}$ is the optimal solution of the master (7.57), we have the necessary condition of optimality

$$0 \in \partial\Big[f^k(x) + \frac{\varrho}{2}\|x - w^k\|^2\Big] + N_X(x) \quad \text{at} \quad x = w^{k+1}.$$

Hence

$$-\varrho(w^{k+1} - w^k) \in \partial f^k(w^{k+1}) + N_X(w^{k+1}).$$

Let x^* be an optimal solution of (7.56). By the subgradient inequality for $f^k(\cdot)$ we get (for some $h \in N_X(w^{k+1})$) the estimate

$$\begin{aligned} f^k(x^*) &\geq f^k(w^{k+1}) - \varrho\langle w^{k+1} - w^k, x^* - w^{k+1}\rangle - \langle h, x^* - w^{k+1}\rangle \\ &\geq f^k(w^{k+1}) - \varrho\langle w^{k+1} - w^k, x^* - w^{k+1}\rangle. \end{aligned} \tag{7.62}$$

A descent step from w^k to $w^{k+1} = x^{k+1}$ occurs, so the test of Step 2 is satisfied (for $k+1$):

$$f(w^{k+1}) \leq (1-\gamma)f(w^k) + \gamma f^k(w^{k+1}).$$

After elementary manipulations we can rewrite it as

$$f^k(w^{k+1}) \geq f(w^{k+1}) - \frac{1-\gamma}{\gamma}[f(w^k) - f(w^{k+1})]. \tag{7.63}$$

Combining the last inequality with (7.62) and using the relation $f(x^*) \geq f^k(x^*)$ we obtain

$$f(x^*) \geq f(w^{k+1}) + \frac{1-\gamma}{\gamma}[f(w^{k+1}) - f(w^k)] - \varrho\langle w^{k+1} - w^k, x^* - w^{k+1}\rangle.$$

This can be substituted to the identity (7.52) which, after skipping the last term, yields

$$\begin{aligned} \|w^{k+1} - x^*\|^2 \leq \|w^k - x^*\|^2 - \frac{2}{\varrho}[f(w^{k+1}) - f(x^*)] \\ + \frac{2(1-\gamma)}{\gamma\varrho}[f(w^k) - f(w^{k+1})] \quad \text{for all} \quad k \in \mathcal{K}. \end{aligned} \tag{7.64}$$

It is very similar to inequality (7.54) in the proof of Theorem 7.13, and our analysis follows the same line.

The series $\sum_{k=1}^{\infty}[f(w^k) - f(w^{k+1})]$ is convergent, because $\{f(w^k)\}$ is nonincreasing and bounded from below by $f(x^*)$. Therefore we obtain from (7.64) that the distance $\|w^{k+1} - x^*\|$ is uniformly bounded, and $\{w^k\}$ must have accumulation points.

Summing up (7.64) for $k \in \mathcal{K}$ we get

$$\sum_{k\in\mathcal{K}}(f(w^{k+1}) - f(x^*)) \leq \frac{\varrho}{2}\|w^1 - x^*\|^2 + \frac{1-\gamma}{\gamma}[f(w^1) - \lim_{k\to\infty} f(w^k)],$$

so $f(w^{k+1}) \to f(x^*)$, $k \in \mathcal{K}$. Consequently, at every accumulation point $\tilde{x}$ of $\{w^k\}$ one has $f(\tilde{x}) = f(x^*)$. Since $\tilde{x}$ is optimal, we can substitute it for x^* in (7.64). Skipping the negative term we get

$$\|w^{k+1} - \tilde{x}\|^2 \leq \|w^k - \tilde{x}\|^2 + \frac{2(1-\gamma)}{\gamma\varrho}[f(w^k) - f(w^{k+1})].$$

It is true not only for $k \in \mathcal{K}$ but for all k, because at $k \notin \mathcal{K}$ we have a trivial equality here. Summing these inequalities from $k = l$ to $k = m > l$ we get

$$\|w^{m+1} - \tilde{x}\|^2 \leq \|w^l - \tilde{x}\|^2 + \frac{2(1-\gamma)}{\gamma\varrho}[f(w^l) - f(w^{m+1})].$$

Since $\tilde{x}$ is an accumulation point, for every $\varepsilon > 0$ we can find l such that $\|w^l - \tilde{x}\| \leq \varepsilon$. Also, if l is large enough, $f(w^l) - f(w^{m+1}) \leq \varepsilon$ for all $m > l$, because $\{f(w^k)\}$ is convergent. Then $\|w^{m+1} - \tilde{x}\|^2 \leq \varepsilon^2 + 2\varepsilon(1-\gamma)/(\gamma\varrho)$ for all $m > l$, so the entire sequence $\{w^k\}$ is convergent to $\tilde{x}$. □

We have already proved the convergence of the method, but it is instructive to establish its additional technical properties.

LEMMA 7.17. *Under the conditions of Theorem* 7.16,

$$\lim_{k\to\infty} \theta_k = f^*, \tag{7.65}$$

$$\lim_{k\to\infty} f^k(x^{k+1}) = f^*, \tag{7.66}$$

$$\lim_{k\to\infty} (x^{k+1} - w^k) = 0, \tag{7.67}$$

where f^* is the optimal value of our problem.

Proof. We prove at first that

$$f(w^k) - \gamma\,\theta_k \le f(w^{k-1}) - \gamma\,\theta_{k-1}, \quad k = 1, 2, \dots. \tag{7.68}$$

The inequality is true at all null steps, as shown in the proof of Lemma 7.15. If there is a descent step at iteration k, we get from (7.61) that

$$\begin{aligned}
\theta_k &\ge \min_x \Big[f^k(x) + \frac{\varrho}{2}\|x - w^k\|^2 \Big] \\
&\ge \min_x \Big[f^{k-1}(w^k) + \varrho\langle w^{k-1} - w^k, x - w^k\rangle + \frac{\varrho}{2}\|x - w^k\|^2 \Big] \\
&= f^{k-1}(w^k) - \frac{\varrho}{2}\|w^k - w^{k-1}\|^2 \\
&= \theta_{k-1} - \varrho\|w^k - w^{k-1}\|^2.
\end{aligned} \tag{7.69}$$

The test for a descent step is satisfied, so

$$\begin{aligned}
f(w^{k-1}) - f(w^k) &\ge \gamma\,[f(w^{k-1}) - f^{k-1}(w^k)] \\
&= \gamma\,[f^{k-1}(w^{k-1}) - f^{k-1}(w^k)] \ge \gamma\,\varrho\|w^k - w^{k-1}\|^2,
\end{aligned}$$

where in the last transformation we used (7.61) again. Combining the last relation with (7.69) we obtain (7.68), as required.

The optimal value of the master problem satisfies the inequality $\theta_k \le f(w^k)$, so

$$f(w^k) - \gamma\,\theta_k \ge (1-\gamma)f(w^k) \ge (1-\gamma)f^*.$$

It follows from (7.68) that the sequence $\{f(w^k) - \gamma\,\theta_k\}$ is convergent, hence $\{\theta_k\}$ is convergent. If there is a descent step at iteration k, inequality (7.63) implies that

$$f(w^k) \ge \theta_k \ge f^k(w^{k+1}) \ge f(w^{k+1}) - \frac{1-\gamma}{\gamma}[f(w^k) - f(w^{k+1})].$$

Both sides converge to f^* as $k \to \infty$, $k \in \mathcal{K}$, so $\theta_k \to f^*$ at descent steps. But the entire sequence $\{\theta_k\}$ is convergent and (7.65) follows.

The objective of the master problem (7.57) is strictly convex. Therefore its value at w^k can be estimated by using its minimum value, θ_k, and the distance to the minimum, $\|w^k - x^{k+1}\|$, as follows:

$$f(w^k) \geq \theta_k + \frac{\varrho}{2}\|w^k - x^{k+1}\|^2.$$

Therefore,

$$0 \leq \frac{\varrho}{2}\|w^k - x^{k+1}\|^2 \leq f(w^k) - \theta_k.$$

Since the right hand side converges to zero, (7.67) holds. Relation (7.66) follows directly from it. □

If problem (7.56) has feasible solutions but no optimal solution, the bundle method generates a sequence $\{w^k\}$ such that the values $f(w^k)$ converge to the infimum value μ of this problem. To prove this fact, suppose $f(w^k) \geq \mu + \varepsilon$ for some $\varepsilon > 0$. Then we can choose x^* such that $\mu < f(x^*) < \mu + \varepsilon$, and inequality (7.64) remains true. Thus $f(w^k) \to f(x^*)$ as proved in the above theorem. But $f(x^*)$ can be made arbitrarily close to μ by the choice of x^*, and we obtain an absurd.

The efficiency of the bundle method can be improved by dynamically changing the proximal parameter, ϱ. The general principle is clear: if the steps are too long, increase ϱ, if they are too short, decrease ϱ. A good way to decide whether steps are too long is to observe the difference

$$\Delta_k = f(x^k) - f(w^{k-1}).$$

We know that if it is positive (actually, not sufficiently negative) a null step will be made. If Δ_k is large (for example, larger than $f(w^{k-1}) - f^{k-1}(x^k)$), it is advisable to increase ϱ. On the other hand, when $f(x^k) = f^{k-1}(x^k)$, we may conclude that the step is too short, because we do not learn new cuts, so ϱ has to be decreased. Detailed rules are discussed in the literature listed at the end of this chapter.

Another practical question associated with the bundle method is the solution of the master problem (7.57). While for linear master problems, like (7.32), commercially available linear programming solvers may be employed, the regularized master requires a quadratic programming solver. Fortunately, the quadratic regularizing term is just the sum of squares, and there exist many efficient methods for such problems.

7.4.3 Application to Polyhedral Problems

Let us now consider the case when the function $f(\cdot)$ in (7.56) is piecewise linear, that is

$$f(x) = \max_{i \in I} \left\{ \gamma_i + \langle u_i, x \rangle \right\} \quad \text{for all} \quad x \in \mathbb{R}^n, \tag{7.70}$$

where I is a finite index set, and $\gamma_i \in \mathbb{R}$ and $u_i \in \mathbb{R}^n$ are fixed. Furthermore, we assume that X is a convex polyhedron. Problems of this form arise frequently as dual problems to linear or integer programming problems, and in stochastic programming, as illustrated in Examples 4.23 and 7.9.

In the polyhedral case we can prove finite convergence of the method, provided that the method uses as subgradients the gradients of active pieces of (7.34), which we call *basic cuts*, and provided that the deletion rules are slightly refined.

Lemma 7.18. *Assume that problem* (7.56) *has an optimal solution. If the bundle method uses basic cuts and does not stop, then the number of descent steps is infinite and there exists k_0 such that for all $k > k_0$*

$$x^{k+1} = \arg\min_{x \in X} f^k(x), \tag{7.71}$$

and

$$f^k(x^{k+1}) = f^*, \tag{7.72}$$

where f^* denotes the optimal value of $f(\cdot)$.

Proof. By Theorem 7.16, the sequence $\{w^k\}$ is convergent to some optimal solution x^*. If the number of descent steps is finite, then $w^k = x^*$ for all sufficiently large k. We already discussed such a situation prior to the formulation of Lemma 7.15. Since each new cut added at Step 1 cuts the current master's solution off, the optimal value of (7.57) increases after each null step. The number of possible combinations of basic cuts is finite, so the stopping test of Step 4 must activate. Thus, if the method does not stop, the number of descent steps must be infinite.

Let us now look more closely at the master problem (7.57). The necessary condition of optimality for (7.57) implies

$$-\varrho(x^{k+1} - w^k) \in \partial f^k(x^{k+1}) + N_X(x^{k+1}). \tag{7.73}$$

Only finitely many models $f^k(\cdot)$ are possible, and each of them, as a polyhedral function, has finitely many different subdifferentials. Moreover, the set X is a convex polyhedron, so only finitely many different normal cones may occur. Therefore the quantity $\operatorname{dist}\big(0, \partial f^k(x^{k+1}) + N_X(x^{k+1})\big)$ may take only finitely many different values. As the left hand side of (7.73) converges to zero by Lemma 7.17, we must have

$$0 \in \partial f^k(x^{k+1}) + N_X(x^{k+1})$$

for all sufficiently large k, so (7.71) is true.

Since only finitely many different minimum values of the models $f^k(\cdot)$ may occur, Lemma 7.17 implies (7.72) for all sufficiently large k. □

It follows that in the case of infinitely many steps, the descent steps of the bundle method look (for all sufficiently large k) similar to the steps of the cutting plane method. The only role of the regularizing term at these late iterations is to select the solution of the linear master problem that is closest to the current center w^k. We also see that the minimum value of the linear master problem does not change and remains equal to the minimum value of the original problem. We need, therefore, to exclude the possibility of infinitely many such degenerate iterations. To achieve this, we modify the algorithm a little.

The simplest modification is to forbid deletion of cuts at any iteration k at which the value of the linear part of the master's objective function does not change, that is, when

$$f^k(x^{k+1}) = f^{k-1}(x^k).$$

Indeed, by Lemma 7.18, after finitely many steps the bundle method will enter the phase when (7.72) holds. From then on, no deletion will take place. By (7.71) the optimal solution of each master problem is the same as the optimal solution of the master problem (7.32) of the cutting plane method. By Theorem 7.8, the method will stop after finitely many steps.

7.4.4 Aggregation of Cuts

An additional feature of the bundle method is the possibility to drastically reduce the number of cuts in the master problem (7.59). Let us observe that the minimum value of (7.59) and its solution do not change, if all its constraints are replaced by just one aggregate cut:

$$\begin{aligned} &\text{minimize } \ v + \frac{\varrho}{2}\|x - w^k\|^2 \\ &\text{subject to } \ \sum_{j \in J_k} \lambda_j^k \big[f(x^j) + \langle g^j, x - x^j \rangle \big] \le v, \\ &\qquad\qquad\ \ x \in X. \end{aligned} \tag{7.74}$$

Here λ^k is the vector of Lagrange multipliers associated with the cuts in (7.59). Indeed, it follows from the optimality conditions for (7.59) that

$\sum_{j \in J_k} \lambda_j^k = 1$ and $\lambda_j^k \geq 0$, $j \in J_k$. Thus, the constraint in (7.74) is a convex combination of the original cuts, with coefficients λ_j^k. The necessary and sufficient conditions of optimality for (7.74), with the Lagrange multiplier 1 associated with the aggregate cut, are identical to the conditions for the original master problem (7.59). Their solutions are unique, by the strict convexity of the quadratic term. Therefore, at Step 5 of the method, we can collapse all cuts of the master problem to just one aggregate cut:

$$\alpha^k + \langle z^k, x \rangle \leq v,$$

with

$$\alpha_k = \sum_{j \in J_k} \lambda_j^k \big[f(x^j) - \langle g^j, x^j \rangle \big]$$

and

$$z^k = \sum_{j \in J_k} \lambda_j^k g^j.$$

At Step 1 of the next iteration, a new cut will be calculated, and thus the master problem can operate with only two cuts: the new one and the aggregate cut. This can be carried over to all iterations, with the master problem of the form:

$$\begin{aligned} \text{minimize} \;\; & v + \frac{\varrho}{2} \|x - w^k\|^2 \\ \text{subject to} \;\; & \alpha^{k-1} + \langle z^{k-1}, x \rangle \leq v, \\ & f(x^k) + \langle g^k, x - x^k \rangle \leq v, \\ & x \in X. \end{aligned} \tag{7.75}$$

Denote by μ^k and λ^k the Lagrange multipliers associated with constraints of (7.75). It follows from the optimality conditions for this problem that $\mu^k = 1 - \lambda^k$ and that $\lambda^k \in [0, 1]$. The next aggregate cut takes on the form

$$\begin{aligned} \alpha^k &= (1 - \lambda^k)\alpha^{k-1} + \lambda^k \big[f(x^k) - \langle g^k, x^k \rangle \big], \\ z^k &= (1 - \lambda^k) z^{k-1} + \lambda^k g^k. \end{aligned} \tag{7.76}$$

All aggregate cuts are linear minorants of the function $f(\cdot)$. Indeed, convex combinations of linear minorants, as in (7.76), are linear minorants themselves. Therefore the model

$$f^k(x) \triangleq \max \left(\alpha^{k-1} + \langle z^{k-1}, x \rangle, \, f(x^k) + \langle g^k, x - x^k \rangle \right)$$

provides a lower estimate of the function $f(x)$.

The detailed algorithm of the bundle method with aggregation is presented in Figure 7.4. We assume there that at the first iteration there is no aggregate

Step 0. Set $k := 1$, $\alpha^0 := -\infty$, $z^0 := 0$, $v^1 := -\infty$.

Step 1. Calculate $f(x^k)$ and $g^k \in \partial f(x^k)$.

Step 2. If $k = 1$ or if

$$f(x^k) \le (1-\gamma) f(w^{k-1}) + \gamma f^{k-1}(x^k),$$

then set $w^k := x^k$; otherwise set $w^k := w^{k-1}$.

Step 3. Solve the master problem (7.75). Denote by (x^{k+1}, v^{k+1}) its solution and set $f^k(x^{k+1}) := v^{k+1}$.

Step 4. If $f^k(x^{k+1}) = f(w^k)$ then stop (w^k is an optimal solution); otherwise continue.

Step 5. Calculate the new aggregate cut (7.76). Increase k by one, and go to Step 1.

Figure 7.4. The bundle method with aggregation.

cut ($\alpha^0 = -\infty$) and that after solving the first master problem the new aggregate cut is just the first cut obtained: $\alpha^1 = f(x^1) - \langle g^1, x^1 \rangle$, $z^1 = g^1$.

The convergence analysis of the method, as presented in Section 7.4.2, remains almost the same; we just need to use the new definition of the functions $f^k(\cdot)$. However, the property of finite termination for polyhedral problems, as discussed in Section 7.4.3, is no longer valid.

In general, the method with aggregation is less efficient in practical calculations than the full version. The main advantage of the aggregate version is the compact form of the master problem (7.75), which makes it particularly suited for problems having very large dimensions. In particular, in the unconstrained case, when $X = \mathbb{R}^n$, the master problem has a closed-form solution (see Exercise 7.8). This makes the bundle method with aggregation an attractive alternative to the plain subgradient method discussed in Section 7.1.

7.5 THE TRUST REGION METHOD

One of the advantages of the bundle method over the cutting plane method is the ability to control the length of the steps made. It avoids making long shots toward minima of poor approximations and it makes good use of a reasonable initial point. Another way to prevent inefficient long steps is to explicitly limit the step size in the master problem (7.32). We have already used this idea in Section 5.7, where we considered trust region methods for smooth optimization problems. Now we extend it to the nonsmooth case.

The trust region master problem has the form

$$\begin{aligned} &\underset{x \in X}{\text{minimize}} \;\; f^k(x) \\ &\text{subject to} \;\; \|x - w^k\|_\diamond \le \Delta. \end{aligned} \tag{7.77}$$

Here, similar to the bundle method, w^k is the "best" point found so far. The role of the constraint $\|x - w^k\|_\diamond \le \Delta$ is to keep the master's solution, x^{k+1}, in a neighborhood of w^k. The function $f^k(\cdot)$ is the cutting plane approximation of $f(\cdot)$, as defined in (7.58).

From the theoretical point of view, the norm $\|\cdot\|_\diamond$ may be any norm in $\mathbb{R}^n$. However, if we use the Euclidean norm, the master problem (7.77) becomes a quadratically constrained optimization problem. It does not offer much advantage over the master problem of the bundle method (7.57). In fact, if the set X is a convex polyhedron, the master problem of the bundle method is a quadratic programming problem, which is easier to solve than a quadratically constrained problem. Furthermore, problem (7.77) with the Euclidean norm is, in a sense, equivalent to the master problems of the bundle method. Indeed, if λ^k is the Lagrange multiplier associated with the constraint

$$\|x - w^k\|^2 \le \Delta^2,$$

then x^{k+1} is also a solution of the regularized master problem (7.57) with $\varrho = 2\lambda^k$.

For these reasons we consider the trust region method with

$$\|d\|_\diamond \triangleq \|d\|_\infty = \max_{1 \le i \le n} |d_i|.$$

Then the constraint $\|x - w^k\|_\diamond \le \Delta$ can be represented as simple bounds

$$-\Delta \le x_i - w_i^k \le \Delta, \quad i = 1, \dots, n,$$

which do not much complicate the master problem (7.77). Admittedly, the norm $\|\cdot\|_\infty$ does prefer directions with $|d_i| = 1$, $i = 1, \dots, n$, over other directions, but the advantage of having a purely polyhedral master problem outweighs this drawback.

Our presentation of the trust region method is very similar to the description of the bundle method. As before, $\gamma \in (0, 1)$ is a parameter of the method used to judge whether the master's solution, x^{k+1}, is significantly better than w^k. The detailed algorithm is presented in Figure 7.5.

The analysis of convergence of the trust region method is much easier than in the case of the bundle method.

Step 0. Set $k := 1$, $J^0 := \emptyset$, $v^1 := -\infty$.

Step 1. Calculate $f(x^k)$ and $g^k \in \partial f(x^k)$.
If $f(x^k) > v^k$ then set $J_k := J_{k-1} \cup \{k\}$; otherwise set $J_k := J_{k-1}$.

Step 2. If $k = 1$ or if

$$f(x^k) \le (1-\gamma) f(w^{k-1}) + \gamma f^{k-1}(x^k),$$

then set $w^k := x^k$; otherwise set $w^k := w^{k-1}$.

Step 3. Solve the master problem (7.77). Denote by (x^{k+1}, v^{k+1}) its solution and set $f^k(x^{k+1}) := v^{k+1}$.

Step 4. If $f^k(x^{k+1}) = f(w^k)$ then stop (w^k is an optimal solution); otherwise continue.

Step 5. Remove from the sets of cuts J_k some (or all) cuts whose Lagrange multipliers λ_j^k at the solution of (7.77) were 0. Increase k by one, and go to Step 1.

Figure 7.5. The trust region algorithm for nonsmooth optimization.

THEOREM 7.19. *Assume that problem* (7.56) *has an optimal solution. Then the sequence* $\{w^k\}$ *generated by the trust region method has the property that*

$$\lim_{k\to\infty} f(w^k) = f^*,$$

where f^* *is the optimal value of* (7.56).

Proof. Suppose the number of descent steps is finite and let w denote the last point to which a descent step has been made. After the last descent step, the trust region method becomes identical with the cutting plane method, applied to the problem

$$\begin{aligned} &\underset{x\in X}{\text{minimize}} \;\; f(x) \\ &\text{subject to} \;\; \|x - w^k\|_\diamond \le \Delta. \end{aligned} \tag{7.78}$$

By Theorem 7.7, $\{f(x^k)\}$ is convergent to the minimum value of problem (7.78). This minimum value must be equal to $f(w)$, if no descent steps are made after w. Thus $f(w) = f^*$, since otherwise a small step from w toward an optimal solution x^* would guarantee improvement.

Let us now consider the case of infinitely many descent steps. Let x^* be a solution of problem (7.56), and let

$$h_k = \|w^k - x^*\|_\diamond.$$

Suppose a descent step occurs after iteration k, that is, $w^{k+1} = x^{k+1}$. Then (by the rule of Step 2)

$$f(x^{k+1}) - f^* \le (1-\gamma)(f(w^k) - f^*) + \gamma(f^k(x^{k+1}) - f^*). \tag{7.79}$$

If $h_k \le \Delta$, then x^* is feasible for the master problem and

$$f^k(x^{k+1}) \le f^k(x^*) \le f(x^*).$$

This combined with (7.79) yields

$$f(x^{k+1}) - f^* \le (1-\gamma)(f(w^k) - f^*).$$

Suppose now $h_k > \Delta$. Consider the point

$$\tilde{x} = \frac{\Delta}{h_k}x^* + \Big(1 - \frac{\Delta}{h_k}\Big)w^k.$$

By construction, its distance to w^k is Δ. As it is feasible for the master problem,

$$f^k(x^{k+1}) - f^* \le f^k(\tilde{x}) - f^* \le \Big(1 - \frac{\Delta}{h_k}\Big)(f(w^k) - f^*),$$

where we also used the convexity of $f^k(\cdot)$. Combining this inequality with (7.79) we see that

$$f(x^{k+1}) - f^* \le \Big(1 - \frac{\gamma\Delta}{h_k}\Big)(f(w^k) - f^*).$$

In both cases, if there is a descent step after iteration k, we have the inequality

$$f(w^{k+1}) - f^* \le \Big(1 - \frac{\gamma\Delta}{\max(\Delta, h_k)}\Big)(f(w^k) - f^*).$$

Let the index $l = 1, 2, \ldots$ number the descent steps and let us write β_l for the value of $f(w^k) - f^*$ at the lth new center w^k. The last inequality can then be rewritten as

$$\beta_{l+1} \le \Big(1 - \frac{\gamma\Delta}{\max(\Delta, h_{k(l)})}\Big)\beta_l. \quad l = 1, 2 \ldots,$$

where $k(l)$ is the iteration number at which the lth descent step is made. By the triangle inequality for the norm $\|\cdot\|_\diamond$ we have

$$h_{k(l)} \le h_1 + l\Delta.$$

Therefore,

$$\beta_{l+1} \leq \left(1 - \frac{\gamma\,\Delta}{h_1 + l\Delta}\right)\beta_l. \quad l = 1, 2 \ldots, \tag{7.80}$$

The sequence $\{\beta_l\}$ is decreasing and bounded from below by 0. Suppose $\beta_l \geq \varepsilon > 0$ for all l. Then, summing (7.80) from $l = 1$ to m we obtain

$$0 \leq \beta_{m+1} \leq \beta_1 - \varepsilon\gamma\,\Delta \sum_{l=1}^{m} \frac{1}{h_1 + l\Delta},$$

which yields a contradiction as $m \to \infty$, because the series $\sum_{l=1}^{\infty} l^{-1}$ is divergent. Thus $\beta_l \to 0$. □

Similar to the analysis of the bundle method, we can prove that $f(w^k) \to \inf f$, even if the problem has no solution.† Indeed, suppose $f^* > \inf f$ is such that $f(w^k) \geq f^*$ for all k. Then Theorem 7.19 yields $f(w^k) \to f^*$. But f^* can be chosen arbitrarily close to $\inf f$, and the result follows.

In the polyhedral case we can prove finite convergence of the trust region method.

THEOREM 7.20. *Assume that $f(\cdot)$ is a convex polyhedral function, and that X is a convex polyhedron. Then the trust region method finds an optimal solution of problem* (7.56) *after finitely many steps.*

Proof. If the number of descent steps is finite, the result follows from Theorem 7.8. Suppose the number of descent steps is infinite. It follows from Theorem 7.19 that, for sufficiently large k, there exists an optimal solution in the Δ-neighborhood of w^k. Therefore

$$f^k(x^{k+1}) \leq f^*,$$

for all sufficiently large k.

Proceeding as in the proof of Theorem 7.31, we conclude that the number of steps, at which new cuts are added to the master problem, must be finite, because there are finitely many cells of linearity of $f(\cdot)$. Hence

$$f(x^{k+1}) = f^k(x^{k+1})$$

for all sufficiently large k. Combining the last two relations we see that we must have $f(x^{k+1}) = f^*$ for all sufficiently large k. Consequently, only one descent step can be made after that, a contradiction. □

---

†$\inf f$ denotes the infimum of $f(\cdot)$ over the feasible set of (7.56).

As in the case of the cutting plane method, deleting inactive cuts is not easy. If we use basic cuts only, we may afford to delete inactive cuts whenever the optimal value of the master problem increases. For general subgradient cuts, no easy and reliable rule has been found. Things are easier if we use the Euclidean norm for the trust region definition, because the arguments from the analysis of the bundle method apply here. Using Euclidean norms, though, does not provide any significant benefits over the bundle method.

The size of the trust region Δ, similar to the parameter ϱ of the bundle method, can be adjusted in the course of computation. If $f(x^{k+1})$ is significantly larger than $f^k(x^{k+1})$, we may decrease Δ to avoid too long steps. If no new cuts are generated, we may increase Δ to allow longer steps.

7.6 CONSTRAINED PROBLEMS

7.6.1 The Exact Penalty Function

We now consider methods for solving constrained optimization problems of the form

$$\begin{aligned} &\text{minimize} \;\; f(x) \\ &\text{subject to} \;\; g_i(x) \le 0, \quad i = 1, \dots, m, \\ &\qquad\qquad\;\; x \in X_0. \end{aligned} \tag{7.81}$$

We assume that $f : \mathbb{R}^n \to \mathbb{R}$ and $g_i : \mathbb{R}^n \to \mathbb{R}$, $i = 1, \dots, m$, are convex functions, and that X_0 is a convex closed set.

If some of the functions defining the problem are nondifferentiable, we cannot directly apply the methods for constrained optimization discussed in Chapter 6. Specialized nonsmooth optimization methods are needed.

A convenient approach to such a problem is to construct the exact penalty function,

$$P(x) = \sum_{i=1}^{m} \max\big(0, g_i(x)\big), \tag{7.82}$$

and to replace (7.81) with the problem

$$\underset{x \in X_0}{\text{minimize}} \; \big[\Phi_\varrho(x) \triangleq f(x) + \varrho P(x)\big], \tag{7.83}$$

where $\varrho > 0$ is a penalty parameter.[†] We discussed the exact penalty function

---

[†]We may also use different penalty coefficients ϱ_i for different constraints, by considering the function $\Phi_\varrho(x) \triangleq f(x) + \sum_{i=1}^{m} \varrho_i \max\big(0, g_i(x)\big)$. All our considerations easily extend to this case, as they amount to a rescaling of the constraint functions by positive multipliers.

in Section 6.2.3. Now its nonsmoothness is not a crucial obstacle, because the functions $f(\cdot)$ and $g_i(\cdot)$ are assumed to be nonsmooth anyway.

THEOREM 7.21. *Assume that problem* (7.81) *satisfies Slater's constraint qualification condition. Then there exists* $\varrho_0 \geq 0$ *such that for each* $\varrho > \varrho_0$ *a point* $\hat{x}$ *is a solution of problem* (7.81) *if and only if it is a solution of problem* (7.83).

Proof. We apply Theorem 3.34. If $\hat{x}$ is a solution of (7.81) then there exists $\hat{\lambda} \in \mathbb{R}^M_+$ such that

$$0 \in \partial f(\hat{x}) + \sum_{i=1}^{m} \hat{\lambda}_i \partial g_i(\hat{x}) + N_{X_0}(\hat{x}) \tag{7.84}$$

and

$$\hat{\lambda}_i g_i(\hat{x}) = 0, \quad i = 1, \ldots, m. \tag{7.85}$$

Suppose $g_i(\hat{x}) = 0$. It follows from Example 2.80 on page 66 that

$$\partial \max\big(0, g_i(\hat{x})\big) = \bigcup_{0 \leq t \leq 1} t \partial g_i(\hat{x}).$$

If $\varrho \geq \hat{\lambda}_i$ we obtain

$$\hat{\lambda}_i \partial g_i(\hat{x}) \subset \partial\Big[\varrho \max\big(0, g_i(\hat{x})\big)\Big].$$

Suppose $\varrho \geq \hat{\lambda}_i$ for all $i = 1, \ldots, m$. By virtue of Theorem 2.85 on page 68, relations (7.84)–(7.85) and the last displayed inclusion imply

$$\begin{aligned} 0 &\in \partial f(\hat{x}) + \sum_{i=1}^{m} \partial\big[\varrho \max\big(0, g_i(\hat{x})\big)\big] + N_{X_0}(\hat{x}) \\ &= \partial\big[f(\hat{x}) + \varrho P(\hat{x})\big] + N_{X_0}(\hat{x}), \end{aligned} \tag{7.86}$$

which is equivalent to the optimality of $\hat{x}$ in problem (7.83).

Suppose $\varrho > \hat{\lambda}_i$, $i = 1, \ldots, m$, for a dual solution $\hat{\lambda} \in \hat{\Lambda}$. We shall show that every solution $\hat{x}$ of problem (7.83) is a solution of (7.81). If $g_i(\hat{x}) \leq 0$, for $i = 1, \ldots, m$, then $P(x) = 0$, and therefore $f(\hat{x}) = f^*$, where f^* is the optimal value of problem (7.81). Thus $\hat{x}$ is optimal for (7.81). It remains to analyze the case when $g_i(\hat{x}) > 0$ for some i. Since $\varrho > \hat{\lambda}_i$, we conclude that

$$\begin{aligned} f(\hat{x}) + \varrho P(\hat{x}) &> f(\hat{x}) + \sum_{i=1}^{m} \hat{\lambda}_i \max\big(0, g_i(\hat{x})\big) \\ &\geq f(\hat{x}) + \sum_{i=1}^{m} \hat{\lambda}_i g_i(\hat{x}). \end{aligned}$$

On the other hand, for every solution x^* of problem (7.81), we have $P(x^*) = 0$ and thus

$$f^* = f(x^*) + \varrho P(x^*) = f(x^*) + \sum_{i=1}^{m} \hat{\lambda}_i g_i(x^*).$$

As $\hat{\lambda} \in \hat{\Lambda}$, Theorem 4.7 on page 166 implies that the pair $(x^*, \hat{\lambda})$ is a saddle point of the Lagrangian. In particular,

$$f(x^*) + \sum_{i=1}^{m} \hat{\lambda}_i g_i(x^*) \le f(\hat{x}) + \sum_{i=1}^{m} \hat{\lambda}_i g_i(\hat{x}).$$

Putting together the last three relations we get

$$f(x^*) + \varrho P(x^*) < f(\hat{x}) + \varrho P(\hat{x}),$$

and $\hat{x}$ cannot be a solution of problem (7.83). □

An identical result can be obtained for another exact penalty function,

$$P(x) = \max\big(0, g_1(x), \dots, g_m(x)\big),$$

which we have already encountered in Section 7.1.4, in connection with systems of inequalities. Then Theorem 7.21 holds true with

$$\varrho_0 = \sum_{i=1}^{m} \hat{\lambda}_i.$$

More generally, we can consider the constraint violation functions,

$$g_i^+(x) = \max\big(0, g_i(x)\big), \quad i = 1, \dots, m,$$

and the penalty function

$$P(x) = \big\| g^+(x) \big\|_{\diamond}, \tag{7.87}$$

with some norm $\| \cdot \|_{\diamond}$ on $\mathbb{R}^m$. Again, Theorem 7.21 remains valid for this penalty function, provided that

$$\varrho_0 = \|\hat{\lambda}\|_*,$$

where $\| \cdot \|_*$ is the dual norm (Exercise 7.10). This general formulation encompasses all cases discussed before.

The approach via the exact penalty function is very general and reliable, if the original problem has a nonempty set of Lagrange multipliers associated with the constraints. However, when the penalty parameter ϱ is too large, the resulting problem (7.83) may be difficult to solve.

7.6.2 Cutting Plane Methods

When cutting plane methods, such as the basic method of Section 7.2 or the bundle method or the trust region method, are applied to the exact penalty model (7.82)–(7.83), we can further specialize them to facilitate convergence.

Assume that at iterations $j = 1, \dots, k$ we have obtained points x^j, the values of functions appearing in problem (7.81): $f(x^j)$, $g_i(x^j)$, $i = 1, \dots, m$, and the corresponding subgradients:

$$s_f^j \in \partial f(x^j),$$
$$s_{g_i^+}^j \in \partial\big[g_i^+(x^j)\big], \quad i = 1, \dots, m.$$

We use the notation $g_i^+(x) \triangleq \max\big(0, g_i(x)\big)$.

These values and subgradients allow us to calculate a cutting plane for the objective function $\Phi_\varrho(x) \triangleq f(x) + \varrho P(x)$ in problem (7.83):

$$\Phi_\varrho(y) \geq \Phi_\varrho(x^j) + \langle s_\Phi^j, y - x^j \rangle,$$

with

$$s_\Phi^j = s_f^j + \varrho \sum_{i=1}^m s_{g_i^+}^j.$$

The cutting planes can be used in the master problem of the method, exactly as described in the corresponding sections of this chapter. However, we can exploit the specific structure of the function $\Phi_\varrho(\cdot)$ to refine the master problem. The idea is to build separate cutting plane approximations for the components of the function $\Phi_\varrho(\cdot)$, rather than for the sum. We define the piecewise linear functions:

$$f^k(x) \triangleq \max_{1 \leq j \leq k} \big[f(x^j) + \langle s_f^j, x - x^j \rangle\big],$$
$$(g_i^+)^k(x) \triangleq \max_{1 \leq j \leq k} \big[g_i^+(x^j) + \langle s_{g_i^+}^j, x - x^j \rangle\big], \quad i = 1, \dots, m.$$

We know from Example 2.80 on page 66 that subgradients of the functions $g_i^+(\cdot)$ can be calculated as follows: if $g_i(x^j) \leq 0$ then we can set $s_{g_i^+}^j = 0$; if $g_i(x^j) > 0$ then we can choose $s_{g_i^+}^j = s_{g_i}^j \in \partial g_i(x^j)$. We can thus equivalently define the functions $(g_i^+)^k(\cdot)$ in the following way. For each i we define the set $J_i^k \subset \{1, \dots, k\}$ of iterations j at which the point x^j was infeasible with respect to the ith constraint:

$$J_i^k = \big\{j \in \{1, \dots, k\} : g_i(x^j) > 0\big\}.$$

Then we can write the cutting plane approximations of $(g_i^+)(\cdot)$ as follows:

$$(g_i^+)^k(x) \triangleq \max\big(0, g_i^k(x)\big),$$

with

$$g_i^k(x) = \max_{j\in J_i^k}\big[g_i(x^j) + \langle s_{g_i}^j, x - x^j\rangle\big], \quad i = 1,\dots,m.$$

Furthermore, we can define J_0^k as the set of iterations (a subset of $\{1,\dots,k\}$) for which *new* cutting planes were added to the model $f^k(\cdot)$ of the objective function. Hence

$$f^k(x) = \max_{j\in J_0^k}\big[f(x^j) + \langle s_f^j, x - x^j\rangle\big].$$

The cutting plane approximations can be used to define a piecewise linear approximation of $\Phi_\varrho(\cdot)$:

$$\Phi_\varrho^k(x) \triangleq f^k(x) + \varrho\sum_{i=1}^m (g_i^+)^k(x).$$

It is much more accurate than an approximation built by cutting planes evaluated for the whole $\Phi_\varrho(\cdot)$, and therefore it is more suitable for the cutting plane methods. The master problem of the cutting plane method,

$$\operatorname*{minimize}_{x\in X_0} \Phi_\varrho^k(x), \tag{7.88}$$

can also be reformulated as a problem with linear constraints. We introduce auxiliary variables v_i, $i = 0, 1, \dots, m$, and we write the equivalent master problem:

$$\begin{aligned}
\text{minimize}\ \ & v_0 + \varrho\sum_{i=1}^m v_i\\
\text{subject to}\ \ & f(x^j) + \langle s_f^j, x - x^j\rangle \le v_0, \quad j\in J_0^k,\\
& g_i(x^j) + \langle s_{g_i}^j, x - x^j\rangle \le v_i, \quad j\in J_i^k, \quad i = 1,\dots,m,\\
& x\in X_0.
\end{aligned}$$

We leave to the reader the verification that this problem is equivalent to (7.88).

All properties of the cutting plane method presented in Section 7.2 remain the same for the specialized version discussed here. Almost all proofs can be copied verbatim here. Only the property of finite convergence for piecewise linear functions has to be restated by requiring that *all* functions $f(\cdot)$ and $g_i(\cdot)$

Step 0. Set $k := 1$, $J_i^0 := \emptyset$, $i = 0, 1, \dots, m$, $v_0^1 := -\infty$.

Step 1.

(a) Calculate $f(x^k)$ and $s_f^k \in \partial f(x^k)$.
If $f(x^k) > v_0^k$ then set $J_0^k := J_0^{k-1} \cup \{k\}$; otherwise set $J_0^k := J_0^{k-1}$.

(b) For $i = 1, \dots, m$ calculate $g_i(x^k)$ and $s_{g_i}^k \in \partial g_i(x^k)$.
If $g_i(x^k) > 0$ then set $J_i^k := J_i^{k-1} \cup \{k\}$; otherwise set $J_i^k := J_i^{k-1}$.

Step 2. If $f(x^k) = v_0^k$ and $g_i(x^k) \le 0$ for all $i = 1, \dots, m$, then stop (optimal solution found); otherwise continue.

Step 3. Solve the master problem (7.89). If it is infeasible, then stop (the original problem is infeasible); otherwise denote by x^{k+1} its solution, by v_0^{k+1} its objective value, increase k by one, and go to Step 1.

Figure 7.6. The constrained cutting plane method.

are convex piecewise linear, and that X_0 is a convex polyhedron. The proof follows the same line of argument, but is slightly more complicated, because we need to consider families of cells for each cutting plane approximation separately (Exercise 7.7).

In exactly the same way, we can develop specialized versions of bundle and trust region methods for minimizing the function $\Phi_\varrho(x)$ over $x \in X_0$.

If the penalty parameter ϱ is sufficiently large, by virtue of Theorem 7.21, the master problem (7.88) is equivalent to the constrained problem

$$\begin{aligned} &\text{minimize} \;\; f^k(x) \\ &\text{subject to} \;\; g_i^k(x) \le 0, \quad i = 1, \dots, m, \\ &\qquad\qquad\;\; x \in X_0. \end{aligned} \tag{7.89}$$

This problem is a relaxation of the original constrained optimization problem (7.81), and it does not involve any penalty parameter. We can thus view the cutting plane method for the penalty function as a method that solves a piecewise linear approximation (7.89) of the constrained optimization problem, and then updates the approximation by adding new cuts. The cuts for the objective function are called the *objective cuts*, and the cuts for the violated constraints ($g_i(x^k) > 0$) are called the *feasibility cuts*. The operation of the cutting plane method for constrained problems is detailed in Figure 7.6.

THEOREM 7.22. *Assume that the set X_0 is compact. If problem* (7.81) *is infeasible, then the constrained cutting plane method stops at Step 3 after finitely many iterations. If problem* (7.81) *is feasible, the constrained cutting plane method generates a sequence of points $\{x^k\}$, all of whose accumulation points are solutions of problem* (7.81).

Proof. It is clear that the infeasibility test at Step 3 is correct, because the feasible set of the master problem is a relaxation of the feasible set of the original problem.

Suppose the master problem remains feasible for all $k = 1, 2, \ldots$. We shall show that in this case problem (7.81) is feasible, and the method generates a sequence convergent to the set of optimal solutions. We follow the idea of the proof of Theorem 7.7.

Let us focus on a particular constraint function $g_i(\cdot)$. For a fixed $\varepsilon > 0$ we define the set of iterations

$$\mathcal{K}_i(\varepsilon) = \{k : g_i(x^k) > \varepsilon\}.$$

Let $k_1, k_2 \in \mathcal{K}_i(\varepsilon)$ with $k_1 < k_2$. Because $g_i(x^{k_1}) > \varepsilon > 0$ there will be a new feasibility cut generated at x^{k_1}. It will stay in the master problem from k_1 on, so it has to be satisfied at x^{k_2}:

$$g_i(x^{k_1}) + \langle s_{g_i}^{k_1}, x^{k_2} - x^{k_1}\rangle \le 0.$$

On the other hand, by assumption, $g_i(x^{k_2}) > \varepsilon$, which combined with the last inequality yields

$$\varepsilon < g_i(x^{k_2}) - g_i(x^{k_1}) - \langle s_{g_i}^{k_1}, x^{k_2} - x^{k_1}\rangle.$$

The function $g_i(\cdot)$ is subdifferentiable everywhere (because it is finite) and X_0 is compact, so there is a constant C such that $g_i(x_1) - g_i(x_2) \le C\|x_1 - x_2\|$, for all $x_1, x_2 \in X_0$. Subgradients on bounded sets are bounded, and thus we can choose C large enough so that $\|s_{g_i}^k\| \le C$ for all k. It follows that

$$\varepsilon < 2C\|x^{k_1} - x^{k_2}\| \quad \text{for all} \quad k_1, k_2 \in \mathcal{K}_i,\ k_1 \ne k_2.$$

As the set X_0 is compact, there can exist only finitely many points in X_0 having a distance at least $\varepsilon/(2C)$ from each other. Thus the last inequality implies that the set $\mathcal{K}_i$ is finite.

The same applies to all constraint functions and thus the set $\mathcal{K}_1 \cup \cdots \cup \mathcal{K}_m$ is finite. It follows that after finitely many iterations all constraint functions have values at most equal to ε. Since $\varepsilon > 0$ was arbitrary, we conclude that

$$\limsup_{k\to\infty} g_i(x^k) \le 0, \quad i = 1, \ldots, m.$$

Every accumulation point of the sequence $\{x^k\}$ is thus feasible for problem (7.81).

The remaining part of the proof is almost identical to the proof of Theorem 7.7. We already know that if the method does not stop at Step 3, the problem has an optimal solution. We denote by f^* the value of the objective function at an optimal solution. For every $\varepsilon > 0$ we prove, exactly as in Theorem 7.7, that the set of iterations k, at which $f(x^k) > f^* + \varepsilon$, is finite. Consequently,

$$\limsup_{k\to\infty} f(x^k) \le f^*.$$

Combining the last two relations we conclude that every accumulation point of the sequence $\{x^k\}$ is optimal for problem (7.81). □

Application of bundle methods or trust region methods is easier if we work with the penalty problem (7.83), and with the corresponding (regularized or trust region constrained) master problem (7.88). The reason is that in these methods one has to decide about "sufficient progress" for making descent steps. If the point generated by the method is infeasible, it is difficult to define "sufficient progress." Unfortunately, infeasibility of points generated by these methods is typical, if nonlinear constraints are linearized by cutting planes. On the other hand, in the penalty approach we define "sufficient progress" by using values of the function $\Phi_\varrho(x)$. However, if the constraints $g_i(x)$ are polyhedral, we can easily develop bundle and trust region versions of the constrained cutting plane method of Figure 7.6.

Example 7.23. Consider the stochastic programming problem discussed in Example 7.9 on page 362. Recall that the problem can be equivalently formulated as

$$\operatorname*{minimize}_{x\in X_0} \Big[f(x) \triangleq \langle c, x\rangle + \sum_{s=1}^{S} p_s Q_s(x) \Big], \tag{7.90}$$

where

$$X_0 = \{x \in \mathbb{R}^n_+ : Ax = b\},$$

and each $Q_s(x)$ is defined as the optimal vale of the *second stage problem* for scenario s:

$$\begin{aligned} &\operatorname*{minimize}_{y_s} \ \langle q_s, y_s\rangle \\ &\text{subject to } Wy_s = h_s - T_s x, \\ &\qquad\qquad\quad y_s \ge 0. \end{aligned} \tag{7.91}$$

In Example 7.9 we assumed that for each $x \in X_0$ all second stage problems are solvable. Now we relax this assumption, by allowing that $Q_s(x) = +\infty$ for some s and some $x \in X_0$. We still assume, though, that problems (7.91) are bounded from below.

Let us consider, for each scenario s, the auxiliary linear programming problem

$$\begin{aligned} &\underset{y_s, z_s}{\text{minimize}} \ \|z_s\|_1 \\ &\text{subject to } \ W y_s + z_s = h_s - T_s x, \\ &\qquad\qquad\quad y_s \geq 0. \end{aligned} \tag{7.92}$$

Here $\|z\|_1$ denotes the sum of absolute values of the components of the vector z. It is clear that problem (7.92) has an equivalent linear programming formulation. The optimal value of this problem, which always exists and is nonnegative, will be denoted by $g_s(x)$. Now we can write the two-stage stochastic programming problem in a more explicit form:

$$\begin{aligned} &\text{minimize } \left[f(x) \triangleq \langle c, x \rangle + \sum_{s=1}^{S} p_s Q_s(x) \right] \\ &\text{subject to } \ g_s(x) \leq 0, \quad s = 1, \ldots, S, \\ &\qquad\qquad\quad x \in X_0. \end{aligned} \tag{7.93}$$

The constraints $g_s(x) \leq 0$ are called in stochastic programming *induced constraints*. They ensure that, at every x satisfying these constraints, the second stage problems are solvable for all scenarios s. The induced constraints are not given explicitly, but at every point x^k it is possible to calculate the value of $g_s(x^k)$ and a subgradient of $g_s(x^k)$. The latter can be calculated similarly to the subgradient of Q_s in Example 7.9: we obtain values of multipliers μ_s^k associated with the constraints of problem (7.93) and we set the subgradient equal to $-T_s^T \mu_s^k$. Therefore the constrained cutting plane method can be applied to problem (7.93). It is finitely convergent, because all functions involved are polyhedral. The same property holds true for the corresponding versions of bundle and trust region methods.

7.7 COMPOSITE OPTIMIZATION

All our considerations of nonsmooth optimization methods were restricted to *convex* optimization problems, where we could apply our knowledge of subdifferential calculus and of duality theory. While convex nonsmooth problems are frequently encountered in applications, we also have to deal with problems involving *both* nonsmooth and nonconvex functions. The general case of such problems is very difficult to handle numerically. Known methods for nonsmooth and nonconvex optimization are very slow, and their analysis involves advanced techniques, which are beyond the scope of this book. For these reasons we concentrate on a special class of such problems, where nonsmoothness can be associated with convexity, and nonconvexity with smoothness. We considered optimality conditions for such problems in Section 3.6.

As an important case of such problems we consider the *composite optimization problem*

$$\underset{x\in X}{\text{minimize}}\ f(h(x)), \tag{7.94}$$

in which $f : \mathbb{R}^m \to \mathbb{R}$ is a convex and possibly nonsmooth function, while $h : \mathbb{R}^n \to \mathbb{R}^m$ is a continuously differentiable mapping. The set X is assumed to be convex and closed. Example 3.39 on page 135 provides an instance of such a model occurring in approximation theory.

The composition $f(h(\cdot))$ is neither convex nor smooth, but the special form of problem (7.94) makes it possible to develop an efficient computational method. Our first step is to transform (7.94) to an equivalent constrained optimization problem:

$$\begin{aligned} &\text{minimize}\ \ f(y)\\ &\text{subject to}\ \ h(x) - y = 0,\\ &\qquad\qquad\quad x \in X. \end{aligned} \tag{7.95}$$

The objective function of this problem is convex, but the equality constraint defines a nonconvex set in $\mathbb{R}^{n+m}$. Our next idea is thus to linearize the function $h(\cdot)$ at some reference point $w \in X$:

$$h(x) \approx h(w) + h'(w)(x - w).$$

In the formula above, $h'(w)$ denotes the Jacobian of the mapping h.

The linearization is guaranteed to be accurate only in a small neighborhood of w, and therefore we introduce an additional constraint to our approximate problem:

$$\begin{aligned} &\text{minimize}\ \ f(y)\\ &\text{subject to}\ \ h(w) + h'(w)(x - w) - y = 0,\\ &\qquad\qquad\quad x \in X,\\ &\qquad\qquad\quad \|x - w\|_\diamond \le \Delta. \end{aligned} \tag{7.96}$$

The decision variables here are x and y, while w is fixed. The symbol $\|\cdot\|_\diamond$ denotes an arbitrary norm in $\mathbb{R}^n$, and $\Delta > 0$ is the diameter of the neighborhood of w in which we "trust" our approximation. We have already employed this idea in Section 5.7.

We also notice that our approach generalizes the Gauss–Newton method discussed in Section 5.4.3. Indeed, eliminating the variables y from problem (7.96), we obtain a formulation similar to (5.50), when $f(\cdot) = \|\cdot\|^2$.

Let us observe that problem (7.96) is a convex optimization problem, and we can apply to it any of the methods discussed earlier in this chapter. In particular, we can use the cutting plane method, or its more efficient versions: the bundle method or the trust region method.

Our general idea is the following. We solve, for some small Δ, problem (7.96). If the progress is sufficiently good, we move the center w to the solution, and we continue. If no sufficient progress is observed, we decrease Δ. In order to validate this scheme, we need to analyze the relations between problems (7.95) and (7.96).

LEMMA 7.24. *Assume that problem* (7.96) *satisfies Slater's constraint qualification condition. If the pair* $(x, y) = (w, h(w))$ *is an optimal solution of problem* (7.96), *then it satisfies necessary conditions of optimality* (3.57) *for problem* (7.95).

Proof. Because (7.96) satisfies the Slater condition, Theorem 3.34 on page 127 implies that the point $(w, h(w))$ satisfies first order necessary and sufficient conditions of optimality (3.49). They read: there exists a vector $\mu \in \mathbb{R}^m$ such that

$$\begin{aligned} &\mu \in \partial f(y) \quad \text{at} \quad y = h(w), \\ &-\left[h'(w)\right]^T \mu \in N_X(w). \end{aligned} \tag{7.97}$$

They are identical with the necessary conditions (3.57) formulated for problem (7.95) in Theorem 3.38 on page 134. □

It should be stressed that we only know that the point w satisfies the necessary conditions of optimality (is *stationary*), because the original problem is nonconvex. We can also remark that the constraint qualification condition is always satisfied if the constraints of problem (7.96) are convex polyhedral. This is true if the set X is a convex polyhedron and if we use a polyhedral norm, such as $\|\cdot\|_1$ or $\|\cdot\|_\infty$.

Our next result estimates the improvement that can be made in problem (7.94) after finding a solution of problem (7.96). To avoid unnecessarily complicated estimates, we assume that the norm $\|\cdot\|_\diamond$ is the Euclidean norm. For other norms the analysis is almost identical.

LEMMA 7.25. *Assume that the mapping $h'(\cdot)$ is Lipschitz continuous with modulus L and that the function $f(\cdot)$ is Lipschitz continuous with modulus C. If $(x(w, \Delta), y(w, \Delta))$ is an optimal solution of problem* (7.96), *then*

$$f\big(h(x(w, \Delta))\big) \le f\big(y(w, \Delta)\big) + \frac{CL\Delta^2}{2}.$$

Proof. Write, for simplicity, $x = x(w, \Delta)$ and $y = y(w, \Delta)$. By the Newton–Leibniz theorem,

$$h(x) = h(w) + \int_0^1 h'(w + t(x - w))(x - w)\, dt$$
$$= y + \int_0^1 \big[h'(w + t(x - w)) - h'(w)\big](x - w)\, dt.$$

Therefore

$$\|h(x) - y\| \le \int_0^1 \|h'(w + t(x - w)) - h'(w)\| \, \|x - w\|\, dt$$
$$\le L\|x - w\|^2 \int_0^1 t\, dt \le \frac{L\Delta^2}{2}.$$

Using the Lipschitz constant for $f(\cdot)$ we obtain

$$f(h(x)) - f(y) \le C\|h(x) - y\| \le \frac{CL\Delta^2}{2},$$

as required. □

Step 0. Set $k := 1$, $v^1 := -\infty$.

Step 1. Calculate $z^k = h(x^k)$, $h'(x^k)$, $f(z^k)$, and $g^k \in \partial f(z^k)$.

Step 2. If $k = 1$ or if

$$f(z^k) \le (1 - \gamma) f(h(w^{k-1})) + \gamma f(y^k),$$

then set $w^k := x^k$ and $\Delta_k := \Delta_{k-1}$; otherwise set $w^k := w^{k-1}$ and $\Delta_k := \beta\Delta_{k-1}$.

Step 3. Solve the convexified problem (7.96) with $w = w^k$ and $\Delta = \Delta_k$. Denote by (x^{k+1}, y^{k+1}) its solution.

Step 4. If $f(y^{k+1}) = f(w^k)$ then stop (w^k is a stationary point); otherwise increase k by one, and go to Step 1.

Figure 7.7. The trust region method for composite optimization.

It follows from Lemma 7.25 that the error in the objective function value, which is due to the linearization of $h(\cdot)$, is of higher order than the size of the trust region Δ. Thus, we can adjust Δ to achieve convergence, as presented in the detailed algorithm in Figure 7.7. We assume there that the method starts from a feasible point x^1 with some $\Delta_0 > 0$. The method uses a parameter $\gamma \in (0, 1)$ to decide whether sufficient progress has been achieved, and a parameter $\beta \in (0, 1)$ to decrease the trust region if no progress has been made.

The convergence of the method follows now from a simple argument. In the theorem below, as in Figure 7.7, we use the notation $z^k = h(x^k)$.

THEOREM 7.26. *Assume that the set X is compact, and that the assumptions of Lemmas* 7.24 *and* 7.25 *are satisfied at every iteration k. Then:*

(i) *If* $\liminf_{k\to\infty} \Delta_k > 0$ *then every accumulation point of the sequence* $\{(x^k, z^k)\}$*, generated by the trust region method, satisfies the necessary conditions of optimality* (7.97) *for problem* (7.95);

(ii) *If* $\liminf_{k\to\infty} \Delta_k = 0$ *then every accumulation point of the sequence* $\{(x^k, z^k)\}$*, generated by the trust region method at null steps, satisfies the necessary conditions of optimality* (7.97) *for problem* (7.95).

Proof. Since the set X is compact, the sequence $\{w^k\}$ has accumulation points. Consider a convergent subsequence $\{w^k\}_{k\in\mathcal{K}}$ and its limit w^*. We shall show that w^* is a stationary point.

By the construction of the method, the sequence $\{f(h(w^k))\}$ is nondecreasing. As X is compact, this sequence is bounded from below and convergent.

Suppose the trust region radii Δ_k remain bounded from below by some positive constant. This means that after a finite number of steps no decreases of Δ_k occur and $\Delta_k = \Delta > 0$ at all sufficiently large k. As no decrease of Δ_k takes place, only descent steps are made, that is, the inequality

$$f(h(x^{k+1})) \le (1-\gamma) f(h(w^k)) + \gamma f(y^{k+1})$$

is satisfied, and $w^{k+1} = x^{k+1}$. Hence

$$0 \le f(h(w^k)) - f(y^{k+1}) \le \frac{1}{\gamma}\big[f(h(w^k)) - f(h(w^{k+1}))\big].$$

The right hand side is convergent to 0, as $k \to \infty$, and thus

$$\lim_{k\to\infty} \big[f(h(w^k)) - f(y^{k+1})\big] = 0. \tag{7.98}$$

Consider a convergent subsequence $\{w^k\}_{k\in\mathcal{K}}$ and its accumulation point w^*. By choosing a sub-subsequence, if necessary, we can assume that the subsequence $\{(x^{k+1}, y^{k+1})\}_{k\in\mathcal{K}}$ is also convergent to some point (x^*, y^*). We have established in (7.98) that $f(y^*) = f(h(w^*))$. Our next step is to show that (x^*, y^*) is an optimal solution of the limiting linearized problem

$$\begin{aligned} \text{minimize} \quad & f(y) \\ \text{subject to} \quad & h(w^*) + h'(w^*)(x - w^*) - y = 0, \\ & x \in X, \\ & \|x - w^*\|_\diamond \le \Delta. \end{aligned} \tag{7.99}$$

By the continuity of all functions involved in this model, the point (x^*, y^*) is feasible for this problem. To show the optimality of (x^*, y^*) we argue by contradiction. Suppose problem (7.99) has a better feasible solution $(\bar{x}, \bar{y})$, that is, $f(\bar{y}) < f(y^*)$. It is convenient to rewrite the system of constraints of problem (7.99) as follows:

$$\begin{aligned} A^*x - y &= b^*, \\ x - s &= w^*, \\ x &\in X, \\ s &\in B_\Delta, \end{aligned} \tag{7.100}$$

where $A^* = h'(w^*)$, $b^* = h'(w^*)w^* - h(w^*)$, and B_Δ is the ball of radius Δ about 0, associated with the norm $\|\cdot\|_\diamond$. At iterations $k \in \mathcal{K}$ problems (7.96) have systems of constraints of the following form:

$$\begin{aligned} A^kx - y &= b^k, \\ x - s &= w^k, \\ x &\in X, \\ s &\in B_\Delta, \end{aligned} \tag{7.101}$$

where $A^k = h'(w^k)$, $b^k = h'(w^k)w^k - h(w^k)$. We know that $A^k \to A^*$, $b^k \to b^*$ and $w^k \to w^*$, as $k \to \infty$, $k \in \mathcal{K}$. To obtain contradiction, we establish that for every solution $(\bar{x}, \bar{y}, \bar{s})$ of system (7.100) we can find a close to it solution $(x^k_{\text{R}}, y^k_{\text{R}}, s^k_{\text{R}})$ of system (7.101). The distance should be bounded by the difference of the data of both problems. This is a question about the stability of a set-constrained linear system, and the positive answer follows from the metric regularity of system (7.100), by virtue of Theorem A.5 in the Appendix. The regularity condition (A.10) for system (7.101) has the form

$$0 \in \operatorname{int} \begin{bmatrix} b^* \\ w^* \end{bmatrix} - \begin{bmatrix} A^* & -I & 0 \\ I & 0 & -I \end{bmatrix} \begin{bmatrix} X \\ \mathbb{R}^m \\ B_\Delta \end{bmatrix}.$$

The first relation, $0 \in \operatorname{int}(b^* - A^*X + \mathbb{R}^m)$, is obvious. The second relation, $0 \in \operatorname{int}(w^* - X + B_\Delta)$, is true as well, because $w^* \in X$ and because B_Δ has a nonempty interior for $\Delta > 0$. Thus, system (7.101) has a feasible solution (x_R^k, y_R^k, s_R^k) such that

$$\lim_{\substack{k\to\infty\\ k\in\mathcal{K}}} x_R^k = \bar{x},$$

$$\lim_{\substack{k\to\infty\\ k\in\mathcal{K}}} y_R^k = \bar{y}.$$

This means that

$$\lim_{\substack{k\to\infty\\ k\in\mathcal{K}}} f(y_R^k) = f(\bar{y}) < f(y^*) = \lim_{\substack{k\to\infty\\ k\in\mathcal{K}}} f(y^{k+1}).$$

Therefore, for sufficiently large k, the point (x_R^k, y_R^k) is better than (x^{k+1}, y^{k+1}) in problem (7.96), a contradiction. We have thus proved that no feasible solution $(\bar{x}, \bar{y})$ of problem (7.99) can be better than (x^*, y^*). By Lemma 7.24, the point (x^*, y^*), as an optimal solution, satisfies necessary conditions of optimality (7.97) for problem (7.95). The first part of the theorem is proved.

Suppose $\Delta_k \to 0$. Consider the infinite set of iterations $\mathcal{K}$ at which the test of Step 2 fails and Δ_k is decreased. Let w^* be an accumulation point of w^k, $k \in \mathcal{K}$. By choosing a subsequence, if necessary, we may assume that $w^k \to w^*$ when $k \to \infty$, $k \in \mathcal{K}$. Suppose w^* does not satisfy the necessary conditions of optimality (3.57) on page 134 for problem (7.95). This means that the system

$$\mu \in \partial f(h(w^*)),$$

$$-\big[h'(w^*)\big]^T \mu \in N_X(w^*).$$

is inconsistent. In other words,

$$-\big[h'(w^*)\big]^T \partial f(h(w^*)) \cap N_X(w^*) = \emptyset.$$

The argument below reconstructs the derivation of the necessary conditions of optimality, but it is instructive to see its meaning for the problem in question. As both sets intersected above are convex and closed, and the subdifferential $\partial f(h(w^*))$ is compact, we can strictly separate them. By virtue of Theorem 2.17 on page 25, there exist $d \in \mathbb{R}^n$ and $\varepsilon > 0$ such that

$$\langle d, v\rangle \le -\langle d, \big[h'(w^*)\big]^T g\rangle - \varepsilon, \tag{7.102}$$

for all $v \in N_X(w^*)$ and all $g \in \partial f(h(w^*))$. The left hand side of (7.102) is bounded above for all $v \in N_X(w^*)$, which means that $d \in \big[N_X(w^*)\big]^\circ$ (see Lemma 2.26 on page 29). By Definition 2.37 of the normal cone and by Theorem 2.27 on page 29, we conclude that $d \in \overline{\text{cone}}(X - w^*) = T_X(w^*)$

(Lemma 3.13 on page 100). The left hand side of (7.102), as bounded above for v in a cone, is nonpositive, and we obtain

$$\langle d, [h'(w^*)]^T g\rangle \le -\varepsilon \quad \text{for all} \quad g \in \partial f(h(w^*)).$$

This allows us to estimate the directional derivative:

$$f'\big(h(w^*); h'(w^*)d\big) = \sup_{g\in\partial f(h(w^*))} \langle h'(w^*)d, g\rangle \le -\varepsilon. \tag{7.103}$$

The direction $h'(w^*)d$ is a direction of descent for $f(\cdot)$ at the point $h(w^*)$. With no loss of generality we may assume that $\|d\|_\diamond = 1$. Since d is a tangent direction for X at w^*, we can construct a trajectory

$$x(\tau) = w^* + \tau d + r(\tau),$$

which is contained in X for $\tau \in [0, \tau_0]$, with some $\tau_0 > 0$. Here $r(\cdot)$ is infinitely small with respect to τ, that is, $\lim_{\tau\to 0} \|r(\tau)\|/\tau = 0$. Associated with the trajectory $x(\tau)$ is the trajectory

$$y(\tau) = h(w^*) + h'(w^*)(x(\tau) - w^*) = h(w^*) + \tau h'(w^*)d + h'(w)r(\tau).$$

By construction, the pair $(x(\tau), y(\tau))$ is feasible for problem (7.96) at $w = w^*$, provided that $0 \le \tau \le \min(\tau_0, \Delta)$. Since $r(\cdot)$ is negligible with respect to τ for all sufficiently small τ, the trajectory $y(\tau)$ is tangent to the direction $h'(w^*)d$ at $\tau = 0$. The directional derivative (7.103) is negative and we can choose $\tau_0 > 0$ small enough, so that $f(y(\tau)) < f(h(w^*)) - \varepsilon\tau/2$ for all $\tau \in (0, \tau_0)$. Let us, temporarily, fix some $\Delta \in (0, \tau_0)$. We have

$$f(y(\Delta)) < f(h(w^*)) - \varepsilon\Delta/2. \tag{7.104}$$

Consider problems (7.96) at iteration k, but with fixed Δ. They would then be equivalent to (7.101). By the stability considerations in the first part of the proof, for the solution $(x(\Delta), y(\Delta), x(\Delta) - w^*)$ of problem (7.100) we can find a feasible solution $(x_{\text{R}}^k, y_{\text{R}}^k, x_{\text{R}}^k - w^k)$ of problem (7.101), whose distance to $(x(\Delta), y(\Delta), x(\Delta) - w^*)$ is at most proportional to the difference in the data of both problems. Consider the points

$$\bar{x}_{\text{R}}^k(\Delta_k) = w^k + \frac{\Delta_k}{\Delta}\big[x_{\text{R}}^k - w^k\big],$$

$$\bar{y}_{\text{R}}^k(\Delta_k) = h(w^k) + h'(w^k)\big[\bar{x}_{\text{R}}^k(\Delta_k) - w^k\big].$$

They are feasible for problem (7.101) with Δ_k. By the convexity of $f(\cdot)$, we have

$$f(\bar{y}_{\text{R}}^k(\Delta_k)) \le f(h(w^k)) + \frac{\Delta_k}{\Delta}\big[f(y_{\text{R}}^k) - f(w^k)\big].$$

If $k \in \mathcal{K}$ is large enough, then

$$f(y_{\text{R}}^k) \le f(y(\Delta)) + \varepsilon\Delta/8,$$
$$f(h(w^k)) \ge f(h(w^*)) - \varepsilon\Delta/8.$$

After combining the last three estimates we obtain

$$f(\bar{y}_{\text{R}}^k(\Delta_k)) - f(h(w^k)) \le \frac{\Delta_k}{\Delta}\big[f(y(\Delta)) - f(h(w^*))\big] + \frac{\Delta_k\varepsilon}{4}.$$

Substituting (7.104) yields

$$f(\bar{y}_{\text{R}}^k(\Delta_k)) - f(h(w^k)) \le -\frac{\Delta_k\varepsilon}{4}.$$

Since (x^{k+1}, y^{k+1}) is optimal for the convexified problem at iteration k, it is at least as good as $(\bar{x}_{\text{R}}^k(\Delta_k), \bar{y}_{\text{R}}^k(\Delta_k))$:

$$f(y^{k+1}) - f(h(w^k)) \le -\frac{\Delta_k\varepsilon}{4}.$$

By construction of the linearized problem (7.96), we have $y^{k+1} - h(x^{k+1}) = o(\Delta_k)$, and also $f(y^{k+1}) - f(h(x^{k+1})) = o(\Delta_k)$. Hence

$$\begin{aligned} f(h(x^{k+1})) &\le f(h(w^k)) + \big[f(y^{k+1}) - f(h(w^k))\big] + o(\Delta_k) \\ &\le f(h(w^k)) + \gamma\big[f(y^{k+1}) - f(h(w^k))\big] \\ &\quad + (1-\gamma)\big[f(y^{k+1}) - f(h(w^k))\big] + o(\Delta_k) \\ &\le f(h(w^k)) + \gamma\big[f(y^{k+1}) - f(h(w^k))\big] - (1-\gamma)\frac{\Delta_k\varepsilon}{4} + o(\Delta_k). \end{aligned}$$

For all sufficiently large $k \in \mathcal{K}$ we have $o(\Delta_k) \le (1-\gamma)\Delta_k\varepsilon/4$. Consequently, the test at Step 2 will be satisfied, and a descent step will be made at iteration k. This contradicts the definition of $\mathcal{K}$ as the set of iterations when no descent step was made. Consequently, the accumulation point z^* must indeed be a stationary point of problem (7.95). □

A very important feature of the trust region method is that cutting plane approximations of the function $f(\cdot)$, which are constructed in the course of solving the convexified problem (7.96) at iteration k, can be re-used at further iterations, after moving the center w^k or decreasing Δ_k. In fact, it is possible to apply the test of Step 2 *before* reaching the optimal solution of the convexified problem, but rather using the solution of the current cutting plane approximation problem, which provides a lower bound for the optimal value of problem (7.96). Since it involves additional technical complications, and requires modifying the algorithm for convex nonsmooth optimization, we shall not discuss it here. The simple version presented here has the advantage that it uses a convex nonsmooth optimization method as a ready module.

7.8 NONCONVEX CONSTRAINTS

We are now ready to derive numerical methods for a constrained nonsmooth optimization problem

$$\begin{aligned} \text{minimize} \quad & f(x) \\ \text{subject to} \quad & g_i(x) \le 0, \quad i = 1, \dots, m, \\ & h_i(x) = 0, \quad i = 1, \dots, p, \\ & x \in X_0. \end{aligned} \tag{7.105}$$

We assume that $f : \mathbb{R}^n \to \mathbb{R}$ is a convex, possibly nonsmooth function. The functions $g_i : \mathbb{R}^n \to \mathbb{R}$, $i = 1, \dots, m$, are assumed to be either convex, in which case they may be nonsmooth, or continuously differentiable and possibly nonconvex. The functions $h_i : \mathbb{R}^n \to \mathbb{R}$, $i = 1, \dots, p$, are assumed to be continuously differentiable, and the set $X_0 \subset \mathbb{R}^n$ is convex and closed. We analyzed this problem in Section 3.6 and we derived the necessary optimality conditions of Theorem 3.38.

Our approach to the numerical solution of the problem is based on the exact penalty method, outlined in Section 7.6.1. We construct the penalty for violating the constraints

$$P(x) = \sum_{i=1}^{m} \max\big(0, g_i(x)\big) + \sum_{i=1}^{p} |h_i(x)|,$$

and we replace (7.105) with the problem

$$\underset{x \in X_0}{\text{minimize}} \ \big[\Phi_\varrho(x) \triangleq f(x) + \varrho P(x)\big], \tag{7.106}$$

where $\varrho > 0$ is a penalty parameter. Our idea is that for sufficiently large $\varrho > 0$ solutions of problem (7.105) should also be minimal points of the function $\Phi_\varrho(x)$ in X_0, analogously to Theorem 7.21.

Once problem (7.106) has been constructed, we notice that it is a composite optimization problem of the form analyzed in the preceding section.

Assume that the functions $g_i(\cdot)$ are convex. Defining the mapping

$$S(x) \triangleq (x, h(x)),$$

we can equivalently write the function $\Phi_\varrho(x)$ as a composition of a certain convex function $\varphi(\cdot)$ and a smooth function $S(\cdot)$,

$$\Phi_\varrho(x) = \varphi(S(x)),$$

where $\varphi : \mathbb{R}^n \times \mathbb{R}^p \to \mathbb{R}$ has the form

$$\varphi(x, s) \triangleq f(x) + \varrho \sum_{i=1}^{m} \max\big(0, g_i(x)\big) + \varrho \sum_{i=1}^{p} |s_i|.$$

Obviously, $\varphi(\cdot)$ is convex. Thus, problem (7.106) is equivalent to the composite optimization problem

$$\underset{x \in X_0}{\text{minimize}} \ \varphi(S(x)). \tag{7.107}$$

The method for solving composite optimization problems, presented in the preceding section, can be applied to problem (7.106). Many specialized techniques may be added to enhance its operation, such as working with separate cutting plane approximations for the various components of $\Phi_\varrho(\cdot)$, which we discuss in Section 7.6.2. Also, we can use different penalty coefficients for different constraints.

The crucial question is the relation between the original problem (7.105) and the penalty problem (7.106).

Let us develop necessary conditions of optimality for problem (7.105). Assume that $\hat{x} \in X_0$ is a local minimum of problem (7.105). Assume that there exists $x_{\text{S}} \in X_0$ such that $g_i(x_{\text{S}}) < 0$, $i = 1, \dots, m$. Furthermore, let Robinson's constraint qualification (3.56) be satisfied. Application of Theorem 3.38 yields the following necessary conditions of optimality: there exist $\hat{\lambda} \in \mathbb{R}^m_+$ and $\hat{\mu} \in [T_{Y_0}(h(\hat{x}))]^\circ$ such that

$$0 \in \partial f(\hat{x}) + \sum_{i=1}^{m} \hat{\lambda}_i \partial g_i(\hat{x}) + [h'(\hat{x})]^T \hat{\mu} + N_{X_0}(\hat{x}) \tag{7.108}$$

and

$$\hat{\lambda}_i g_i(\hat{x}) = 0, \quad i = 1, \dots, m. \tag{7.109}$$

To formulate the necessary conditions for (7.107) we rewrite the problem as follows:

$$\begin{aligned} &\text{minimize} \ \ \varphi(x, s) \\ &\text{subject to} \ \ h(x) - s = 0, \\ &\qquad\qquad\quad x \in X_0. \end{aligned}$$

The necessary conditions read: there exist multipliers $\mu \in \mathbb{R}^m$ such that

$$\begin{aligned} &0 \in \partial_x \varphi(\hat{x}, \hat{s}) + [h'(\hat{x})]^T \mu + N_{X_0}(\hat{x}), \\ &0 \in \partial_s \varphi(\hat{x}, \hat{s}) - \mu. \end{aligned} \tag{7.110}$$

The first relation can be rewritten as

$$0 \in \partial f(\hat{x}) + \varrho \sum_{i=1}^{m} \alpha_i \partial g_i(\hat{x}) + [h'(\hat{x})]^T \mu + N_{X_0}(\hat{x}),$$

with $\alpha_i = 0$ for $g_i(\hat{x}) < 0$, $\alpha_i = 1$ when $g_i(\hat{x}) > 0$, $\alpha_i \in [0, 1]$ when $g_i(\hat{x}) = 0$.

Consider a local minimum point $\hat{x}$ of problem (7.105) satisfying conditions (7.108)–(7.109). If ϱ is larger than all $\hat{\lambda}_i$, we can set $\alpha_i = \hat{\lambda}_i/\varrho, i = 1, \dots, m$, and $\mu = \hat{\mu}$ to satisfy (7.110). Thus $\hat{x}$ satisfies necessary conditions of optimality for the penalty problem, if ϱ is sufficiently large.

The converse argument is more difficult than in the convex case analyzed in Theorem 7.21. If a local minimum of the penalty problem is feasible for the original problem, Lemma 6.5 immediately implies that the solution of the penalty problem is optimal for the original problem. But even for very large ϱ the feasibility is difficult to ensure theoretically, without restrictive assumptions.

If the functions $g_i(\cdot)$ are nonconvex, but smooth, we can define

$$S(x) \triangleq (x, g(x), h(x)),$$

and

$$\varphi(s_1, s_2, s_3) \triangleq f(s_1) + \varrho \sum_{i=1}^{m} \max\big(0, s_{2i}\big) + \varrho \sum_{i=1}^{p} |s_{3i}|.$$

The convexity of $\varphi(\cdot)$ is evident. Alternatively, by adding slack variables, we can convert smooth inequalities into equations and sign restrictions on the slack variables. If some of the inequality constraints involve convex functions and some smooth nonconvex functions, we combine the approaches. We leave to the reader the derivation of the composite optimization problem in this case, and the comparison of the necessary conditions of optimality.

EXERCISES

7.1. Calculate the direction of steepest descent for the function

$$f(x_1, x_2) = x_1 + x_2 + \max\big(0, (x_1)^2 + (x_2)^2 - 4\big)$$

at the point $x = (0, -2)$.

7.2. Consider the problem of minimizing the function of one variable $f(x) = |x|$.

(a) Carry out several iterations of the subgradient method, starting from $x^1 = 1$, using the step sizes $\tau_k = 1/(k+1)$ and scaling coefficients $\gamma_k = 1$.

(b) Prove that for all $k \geq 4$ we have $|x^k| \leq 1/k$ and thus the method is convergent.

7.3. Develop a specialized version of the subgradient method from Section 7.1.4 for solving systems of linear inequalities $Ax \leq b$, where A is an $m \times n$ matrix, and $b \in \mathbb{R}^m$.

7.4. The function $f : \mathbb{R}^n \to \mathbb{R}$ has the form

$$f(x) = \sum_{l=1}^{L} f_l(x),$$

where each function $f_l(\cdot)$ is convex. Consider the problem

$$\operatorname*{minimize}_{x \in X} f(x),$$

with a convex and compact set X. To solve this problem, we define a cutting plane method using disaggregated cuts. At each point x^k generated by the method, it calculates subgradients $g_l^k \in \partial f_l(x^k)$ for all functions $f_l(\cdot)$, $l = 1, \ldots, L$. They are used in the master problem

$$\begin{aligned} &\text{minimize} \ \sum_{l=1}^{L} v_l \\ &\text{subject to} \ f_l(x^j) + \langle g_l^j, x - x^j \rangle \leq v_l, \quad j = 1, \ldots, k, \quad l = 1, \ldots, L, \\ &\qquad\qquad x \in X. \end{aligned}$$

The x-part of its solution is the next point x^{k+1}. Prove that the master problem is equivalent to minimizing the function

$$f^k(x) \triangleq \sum_{l=1}^{L} f_l^k(x),$$

where each function $f_l^k(\cdot)$ is a cutting plane approximation of $f_l(\cdot)$:

$$f_l^k(x) \triangleq \max_{1 \leq j \leq k} \left\{ f_l(x^j) + \langle g_l^j, x - x^j \rangle \right\}.$$

Using this observation, prove the convergence of the method. Follow Theorem 7.7.

7.5. Develop a specialized cutting plane method for the dual problem to the facility location problem of Example 4.23.

7.6. Derive a specialized cutting plane method for the stochastic programming problem (7.38).

7.7. Prove finite convergence of the specialized cutting plane method for penalty functions (7.88), under the assumption that all functions are convex polyhedral, and the set X_0 is a convex polyhedron. Use the idea of the proof of Theorem 7.8.

7.8. Find the closed-form solution of the master problem (7.75) of the bundle method with aggregation.

7.9. Describe the operation of the bundle method for the eigenvalue optimization problem (4.39):

$$\underset{\mu \in \mathbb{R}^n}{\text{minimize}}\ k\lambda_{\max}\Big(\sum_{i=1}^{m} \mu_i A_i - C\Big) - \langle b, \mu \rangle.$$

where $A_1, \ldots, A_m$ and C are symmetric matrices of dimension n, and $b \in \mathbb{R}^n$.

7.10. Prove Theorem 7.21 for the penalty function (7.87). Show that

$$\varrho_0 = \|\lambda\|_*,$$

where $\|\cdot\|_*$ is the dual norm, is a sufficient minimum value of the penalty parameter.

Appendix A

Stability of Set-Constrained Systems

A.1 LINEAR–CONIC SYSTEMS

Our main interest here is in the analysis of generalized systems of equations

$$\begin{aligned} Ax &= b, \\ x &\in K. \end{aligned} \tag{A.1}$$

In the system above, A is a matrix of dimension $m \times n$, $b \in \mathbb{R}^m$, and K is a convex cone in $\mathbb{R}^n$ containing 0.

Clearly, system (A.1) can be used to model linear equations and inequalities by introducing additional "slack" variables and defining the cone K in a suitable way. Also, systems involving other cones, like the positive semidefinite cone, fall into this framework. Finally, the analysis of (A.1) provides a solid foundation for the analysis of nonlinear inequalities.

One of the fundamental questions associated with such systems is the question of their stability: *Does the system have solutions if the matrix A and vector b are slightly perturbed?* The key to answering this question is the analysis of system (A.1) with perturbations of the right hand side b only.

Let us observe that if (A.1) has a solution for all right hand sides vectors b in a certain ball about 0, say for $\|b\| \le \varepsilon$ with some $\varepsilon > 0$, then it has a solution for all $b \in \mathbb{R}^m$. Indeed, we can solve at first the system with the right hand side equal to $\varepsilon b/\|b\|$. Then we can multiply the solution by $\|b\|/\varepsilon$ to obtain a solution for b.

We do not assume that K is closed, because we want to apply our results to cones of feasible directions, which are not closed, in general. We assume, however, that

$$K \setminus \{0\} + \overline{K} \subset K. \tag{A.2}$$

This condition trivially holds for all closed cones, but it is also true for many important cones which are not closed. We discuss condition (A.2) later in this section.

Lemma A.1. *Assume that system* (A.1) *has a solution for every $b \in \mathbb{R}^m$. Then there exists a constant C such that for every $b \in \mathbb{R}^m$ we can find a* *solution $x(b)$ such that*

$$\|x(b)\| \le C\|b\|.$$

Proof. Define the set

$$Y = \{Ax : x \in K, \ \|x\| \le 1\}.$$

It is convex. By assumption, every $b \in \mathbb{R}^m$ must be an element of kY for some integer k. Hence

$$\mathbb{R}^m = \bigcup_{k \in \mathbb{N}} kY \subset \text{lin}(Y),$$

which implies that $\text{int}\, Y \neq \emptyset$. Suppose $0 \notin \text{int}\, Y$. Then we can separate 0 from the interior: there exists $h \neq 0$ such that

$$\langle h, y \rangle \le 0 \quad \text{for all} \quad y \in \text{int}\, Y.$$

As $\text{int}\, Y \neq \emptyset$ and Y is convex, $Y = \overline{\text{int}\, Y}$. Then the last inequality implies that

$$\langle h, y \rangle \le 0 \quad \text{for all} \quad y \in Y.$$

This can be rewritten as

$$\langle h, Ax \rangle \le 0 \quad \text{for all} \quad x \in K, \ \|x\| \le 1.$$

The left hand side of the last inequality is homogeneous in x and thus

$$\langle h, Ax \rangle \le 0 \quad \text{for all} \quad x \in K.$$

Consider system (A.1) with $b = h$. It has a solution $x \in K$ by assumption, and the last inequality yields $\|h\|^2 \le 0$, a contradiction. Therefore

$$0 \in \text{int}\, Y.$$

It follows that there exists $\varepsilon > 0$ such that for all b with $\|b\| \le \varepsilon$ the system (A.1) has a solution $x(b) \in K$ with $\|x(b)\| \le 1$. For an arbitrary $b \in \mathbb{R}^m$ we can define

$$x(b) = \frac{\|b\|}{\varepsilon} x\Big(\frac{\varepsilon b}{\|b\|}\Big).$$

It is an element of K and $Ax(b) = b$. Moreover,

$$\|x(b)\| \le \frac{1}{\varepsilon}\|b\|,$$

and our assertion holds true with $C = 1/\varepsilon$. □

We shall refer to the constant C defined in this lemma as the *modulus* of the system. As we see, it depends only on A and K.

Let us now consider a perturbed system

$$\begin{aligned} \widetilde{A}x &= \widetilde{b}, \\ x &\in K, \end{aligned} \tag{A.3}$$

with a different matrix $\widetilde{A}$ of dimension $m \times n$, and with a different right hand side vector $\widetilde{b} \in \mathbb{R}^m$.

THEOREM A.2. *Assume that system* (A.1) *has a solution for every* $b \in \mathbb{R}^m$, *and let* C *be its modulus. If* $\|\widetilde{A} - A\| < 1/C$, *then system* (A.3) *has a solution for every* $\widetilde{b} \in \mathbb{R}^m$, *and its modulus is at most*

$$\widetilde{C} = \frac{C}{1 - C\|\widetilde{A} - A\|}.$$

Furthermore, for every solution x *of system* (A.1) *there exists a solution* $\widetilde{x}$ *of system* (A.3) *such that*

$$\|\widetilde{x} - x\| \le \widetilde{C}\Big(\|\widetilde{b} - b\| + \|x\|\|\widetilde{A} - A\|\Big).$$

Proof. We denote by $\hat{x}(r)$ the solution of system (A.1) with the right hand side $b = r$ satisfying $\|\hat{x}(r)\| \le C\|r\|$. It always exists, due to Lemma A.1. Let us select an arbitrary point $x_0 \in K$ and let us define two sequences, $\{x_k\}$ and $\{r_k\}$, as follows:

$$\begin{aligned} r_k &= \widetilde{b} - \widetilde{A}x_k, \\ x_{k+1} &= x_k + \hat{x}(r_k), \quad k = 0, 1, 2, \dots. \end{aligned} \tag{A.4}$$

By construction, $\hat{x}(r_k) \in K$, and thus all points x_k are elements of the cone K.

By direct calculation we obtain

$$\begin{aligned} \widetilde{A}x_{k+1} &= \widetilde{A}x_k + \widetilde{A}\hat{x}(r_k) \\ &= \widetilde{A}x_k + A\hat{x}(r_k) + (\widetilde{A} - A)\hat{x}(r_k) \\ &= \widetilde{A}x_k + r_k + (\widetilde{A} - A)\hat{x}(r_k) \\ &= \widetilde{b} + (\widetilde{A} - A)\hat{x}(r_k). \end{aligned}$$

Therefore

$$\begin{aligned} \|r_{k+1}\| = \|\widetilde{b} - \widetilde{A}x_{k+1}\| &\le \|\widetilde{A} - A\|\|\hat{x}(r_k)\| \\ &\le C\|\widetilde{A} - A\|\|r_k\|, \quad k = 0, 1, 2, \dots. \end{aligned}$$

Defining $\varrho = C\|\widetilde{A} - A\|$ we notice that $\varrho < 1$ and that

$$\|r_k\| \le \varrho^k \|r_0\|, \quad k = 0, 1, 2, \dots .$$

This implies that $r_k \to 0$, as $k \to \infty$. Moreover,

$$\|x_{k+1} - x_k\| \le \|\hat{x}(r_k)\| \le C\|r_k\| \le C\varrho^k \|r_0\|, \quad k = 0, 1, 2, \dots .$$

Consequently, the sequence $\{x_k\}$ is convergent. Denoting by $\widetilde{x}$ its limit and passing to the limit in (A.4) we get $\widetilde{A}\widetilde{x} = \widetilde{b}$.

If $x_0 \in K\setminus\{0\}$ then we have the inclusion

$$\widetilde{x} = x_0 + \sum_{k=0}^{\infty} \hat{x}(r_k) \in K\setminus\{0\} + \overline{K}.$$

Thus, by assumption, $\widetilde{x}$ is an element of K. If $x_0 = 0$ and $r_0 \neq 0$ then $x_1 \in K\setminus\{0\}$ and a similar argument yields $\widetilde{x} \in K$ again. Finally, if $x_0 = 0$ and $r_0 = 0$ then simply $\widetilde{x} = 0$. Therefore, $\widetilde{x}$ is a solution of system (A.3).

Furthermore,

$$\|\widetilde{x} - x_0\| \le \sum_{k=0}^{\infty} \|x_{k+1} - x_k\| \le C \sum_{k=0}^{\infty} \varrho^k \|r_0\| = \frac{C}{1-\varrho}\|r_0\|. \qquad \text{(A.5)}$$

If we start from $x_0 = 0$, the above estimate yields

$$\|\widetilde{x}\| \le \frac{C}{1-\varrho}\|\widetilde{b}\| = \widetilde{C}\|\widetilde{b}\|,$$

as required. If we start from $x_0 \in K$ such that $Ax_0 = b$, then

$$r_0 = \widetilde{b} - \widetilde{A}x_0 = (\widetilde{b} - b) - (\widetilde{A} - A)x_0.$$

Hence estimate (A.5) becomes

$$\|\widetilde{x} - x_0\| \le \frac{C}{1-\varrho}\Big(\|\widetilde{b} - b\| + \|\widetilde{A} - A)\|\|x_0\|\Big),$$

which is our second assertion. □

We can now easily obtain a uniform version of the stability result. Consider two systems

$$\begin{aligned} A_1 x &= b_1, \\ x &\in K, \end{aligned} \qquad \text{(A.6)}$$

and

$$\begin{aligned} A_2 x &= b_2, \\ x &\in K. \end{aligned} \qquad \text{(A.7)}$$

The distance between their solutions can be estimated as follows.

COROLLARY A.3. *Assume that condition* (A.2) *is satisfied and that system* (A.6) *has a solution for every* $b \in \mathbb{R}^n$. *Let* C *be its modulus. Let* A_2 *be such that* $\|A_2 - A_1\| < 1/C$. *Then for each solution* x_1 *of system* (A.6) *there exists a solution* x_2 *of system* (A.7) *such that*

$$\|x_2 - x_1\| \leq \widetilde{C}(\|b_2 - b_1\| + \|x_1\|\|A_2 - A_1\|),$$

with

$$\widetilde{C} = \frac{C}{1 - C\|A_2 - A_1\|}.$$

Proof. Let x_1 be a solution of (A.6). Substituting $d = x - x_1$ we see that system (A.7) is equivalent to

$$A_2 d = h,$$
$$d \in K - \{x_1\},$$

with $h = b_2 - A_2 x_1$. Noting that $K \subset K - \{x_1\}$, we restrict the last system by requiring that $d \in K$. Now we use Theorem A.2 to conclude that this restricted system has a solution $\widetilde{d}$ such that $\|\widetilde{d}\| \leq \widetilde{C}\|h\|$. Since

$$\|h\| = \|b_2 - b_1 + (A_1 - A_2)x_1\| \leq \|b_2 - b_1\| + \|x_1\|\|A_2 - A_1\|,$$

the result follows. □

A.2 SET-CONSTRAINED LINEAR SYSTEMS

Let us now pass to the analysis of more general systems of the form

$$\begin{aligned} Ax &= b, \\ x &\in X_0, \end{aligned} \tag{A.8}$$

where X_0 is a closed convex set in $\mathbb{R}^n$. The other notation is the same as in the preceding section. Assume that x_0 is a solution of the above system. We are interested in the following question: *Is it possible to find solutions of the perturbed system,*

$$\begin{aligned} \widetilde{A}x &= \widetilde{b}, \\ x &\in X_0, \end{aligned} \tag{A.9}$$

which are close to x_0?

The key to the analysis of this question is the following *regularity condition*:

$$0 \in \operatorname{int}\{b - Ax : x \in X_0\}. \tag{A.10}$$

Observe that if X_0 is a cone, this condition is identical with the assumption of Lemma A.1, which requires that system (A.8) has a solution for every b.

While the qualitative conclusions from our analysis follow from the more general nonlinear case to be studied in the next section, it is instructive to see our approach at work in the linear case, which is more transparent.

Our method of analysis is based on transforming system (A.8) to a linear-conic system and using the results of the previous section. At first we present the main tool of this transformation. Recall that D_∞ denotes the recession cone of a set D.

Lemma A.4. *Let D be a closed convex set in $\mathbb{R}^n$ such that $0 \in D$. Then the set*

$$K = \{(d, t) : d \in tD,\ t \geq 0\}$$

is a cone, and

$$\overline{K} \setminus K = \{(\bar{d}, 0) : \bar{d} \in D_\infty,\ \bar{d} \neq 0\}.$$

Proof. If $(d, t) \in K$ and $\alpha \geq 0$, then $\alpha d \in (\alpha t)D$. Thus K is a cone. If $(d_1, t_1) \in K$ and $(d_2, t_2) \in K$ then for all $\alpha \in (0, 1)$ the definition of the sum of sets implies

$$\alpha d^1 + (1 - \alpha)d^2 \in \alpha(t_1 D) + (1 - \alpha)(t_2 D) = (\alpha t_1 + (1 - \alpha)t_2)D.$$

Therefore K is a convex cone.

Suppose $(d^k, t^k) \to (\bar{d}, \bar{t})$, as $k \to \infty$. If $\bar{t} > 0$, then $t_k > 0$ for all sufficiently large k and $d^k/t^k \to \bar{d}/\bar{t}$. AS $d^k/t^k \in D$ and D is closed, we conclude that $\bar{d} \in \bar{t}D$, which means that $(\bar{d}, \bar{t}) \in K$. The case of $(\bar{d}, \bar{t}) \notin K$ may occur only if $\bar{t} = 0$. As $(0, 0)$ is an element of K, $\bar{d}$ is nonzero. This implies that $d^k \neq 0$ for all sufficiently large k. Therefore $t^k > 0$ for all sufficiently large k.

Suppose $\bar{d} \notin D_\infty$. Then there exists $\bar{\alpha} > 0$ such that $\alpha\bar{d} \notin D$ for all $\alpha > \bar{\alpha}$. Let $\alpha > \bar{\alpha}$. We obtain

$$\alpha\bar{d} = (\alpha t^k)\frac{d^k}{t^k} + \alpha(\bar{d} - d^k).$$

As $d^k/t^k \in D$ by construction, and because $\alpha t^k \leq 1$ for all large k, the distance of $\alpha\bar{d}$ to D is no more than $\alpha\|\bar{d} - d^k\|$. The set D is closed. Letting $k \to \infty$ we obtain $\alpha\bar{d} \in D$, a contradiction. $\square$

We can now use the homogenization technique to prove our main result.

THEOREM A.5. *Assume that system* (A.8) *satisfies the regularity condition* (A.10). *Then for every solution x_0 of system* (A.8) *there exists $\varepsilon > 0$ and a constant C_0 such that for all $\widetilde{A}$ and $\widetilde{b}$ with $\|\widetilde{A} - A\| + \|\widetilde{b} - b\| \le \varepsilon$, the perturbed system* (A.9) *has a solution $\widetilde{x}$ with*

$$\|\widetilde{x} - x_0\| \le C_0(\|\widetilde{A} - A\| + \|\widetilde{b} - b\|).$$

Proof. Define the set

$$D = X_0 - \{x_0\}.$$

Our first step is to analyze the perturbed system (A.9) with right hand side perturbations only. Setting $d = x - x_0$ and $h = \widetilde{b} - b$ we can rewrite the perturbed system as follows:

$$\begin{aligned} Ad &= h, \\ d &\in D. \end{aligned} \tag{A.11}$$

We transform this system to a linear–conic system by defining the set K in $\mathbb{R}^n \times \mathbb{R}$,

$$K = \{(d, t) : d \in tD,\ t \ge 0\}.$$

By Lemma A.4 it is a convex cone and all points of $\overline{K} \setminus K$ have the form $(\bar{d}, 0)$, where $\bar{d} \in D_\infty \setminus \{0\}$. Thus K satisfies condition (A.2).

Consider the system

$$\begin{aligned} Ad - 0t &= h, \\ (d, t) &\in K. \end{aligned} \tag{A.12}$$

By assumption, for $h = 0$ the pair $(0, 1)$ is its solution. By virtue of the regularity condition, a solution exists for every h in a certain neighborhood of 0. In view of its homogeneity, system (A.12) has a solution $(d(h), t(h))$ for all $h \in \mathbb{R}^m$. By Lemma A.1, there exists a constant C such that $\|(d(h), t(h))\| \le C\|h\|$.

Now consider (A.12) with a matrix perturbation:

$$\begin{aligned} \widetilde{A}d - ht &= 0, \\ (d, t) &\in K. \end{aligned} \tag{A.13}$$

By Theorem A.2, if $\|\widetilde{A} - A\| + \|h\| \le 1/(2C)$ then system (A.13) has a solution $(\widetilde{d}, \widetilde{t})$ such that

$$\|(\widetilde{d}, \widetilde{t}) - (0, 1)\| \le \widetilde{C}(\|\widetilde{A} - A\| + \|h\|),$$

with

$$\widetilde{C} = \frac{C}{1 - C(\|\widetilde{A} - A\| + \|h\|)} \le 2C.$$

If $\|\widetilde{A} - A\| + \|h\| \le 1/(4C)$, we obtain $|\widetilde{t} - 1| \le 1/2$ and thus $\widetilde{t} \ge 1/2$. Moreover, the point $d = \widetilde{d}/\widetilde{t}$ is a solution of system (A.11) with both matrix and right hand side perturbations:

$$\widetilde{A}d = h,$$
$$d \in D,$$

which is equivalent to (A.9). Setting $\widetilde{x} = x_0 + \widetilde{d}/\widetilde{t}$ we obtain the estimate

$$\|\widetilde{x} - x_0\| = \left\| \frac{\widetilde{d}}{\widetilde{t}} \right\| \le \frac{2C}{\widetilde{t}}(\|\widetilde{A} - A\| + \|h\|) \le 4C(\|\widetilde{A} - A\| + \|h\|).$$

Thus the required estimate holds true with $C_0 = 4C$. □

A.3 SET-CONSTRAINED NONLINEAR SYSTEMS

The methodology developed in the first two sections allows us to analyze stability of nonlinear systems of the form

$$\begin{aligned} h(x, u) &= 0, \\ x &\in X_0. \end{aligned} \tag{A.14}$$

In the system above, $u \in \mathbb{R}^s$ denotes a vector of parameters, $h : \mathbb{R}^n \times \mathbb{R}^s \to \mathbb{R}^m$, and X_0 is a closed convex set in $\mathbb{R}^n$.

Assume that for $u = u_0$ the system has a solution x_0. We are interested in the existence and in the distance to x_0 of solutions of the perturbed system, with u close to u_0.

DEFINITION A.6. System (A.14) is called *stable* at the point (x_0, u_0) if there exist $\varepsilon > 0$ and C such that for all $(\widetilde{x}, \widetilde{u})$ satisfying $\|\widetilde{x} - x_0\| \le \varepsilon$ and $\|\widetilde{u} - u_0\| \le \varepsilon$ we can find $x_{\mathrm{R}} \in X_0$ satisfying the equations

$$h(x_{\mathrm{R}}, \widetilde{u}) = 0,$$

and such that

$$\|x_{\mathrm{R}} - \widetilde{x}\| \le C\Big(\operatorname{dist}(\widetilde{x}, X_0) + \|h(\widetilde{x}, \widetilde{u})\|\Big).$$

We want small perturbations of the parameter and of the initial solution to be compensated by adjustments of the order of the residual in the equations and of the distance to the set X_0. In the special case of the system $h(x) - u = 0$, $x \in X_0$, the concept of stability at the point $(x_0, 0)$ is frequently referred to as *metric regularity*.

To analyze the stability of system (A.14) we make several additional assumptions. First, we assume that $h(\cdot, \cdot)$ is continuous in both arguments, $h(\cdot, u)$ is differentiable for all u, and that its Jacobian with respect to the first argument, denoted by $h'(x, u)$, is continuous with respect to both arguments, x and u.

Second, we assume that the following *regularity condition* holds true:

$$0 \in \text{int}\left\{h'(x_0, u_0)(x - x_0) : x \in X_0\right\}. \tag{A.15}$$

We remark that it is an extension of condition (A.10) to the nonlinear case. It is called *Robinson's condition.*

THEOREM A.7. *Assume that system* (A.14) *for* $u = u_0$ *has a solution* x_0. *If the regularity condition* (A.15) *is satisfied, then system* (A.14) *is stable at* (x_0, u_0).

Proof. With no loss of generality we may assume that $x_0 = 0$. Our idea is to construct a sequence $\{x_k\}$ by solving for each $k = 1, 2, \ldots$ a linearized system with set constraints

$$\begin{aligned} h(x_k, \widetilde{u}) + h'(x_0, u_0)(x - x_k) &= 0, \\ x &\in X_0. \end{aligned} \tag{A.16}$$

We denote by x_{k+1} the solution of this system, and we continue. We shall show that this process is well-defined and that it generates a sequence convergent to a point x_R satisfying our assertion.

Let us at first assume that $\widetilde{x} \in X_0$. In this case we start the process from $x_1 = \widetilde{x}$. Setting $A = h'(x_0, u_0)$ and

$$\widetilde{b}_k = h'(x_0, u_0)x_k - h(x_k, \widetilde{u})$$

we can rewrite system (A.16) as follows:

$$\begin{aligned} Ax &= \widetilde{b}_k, \\ x &\in X_0. \end{aligned} \tag{A.17}$$

Following the technique developed in the preceding section, we transform (A.17) to a linear–conic system. Define

$$K = \{(x, t) : x \in tX_0,\ t \geq 0\}$$

and consider the system

$$\begin{aligned} Ax - 0t &= b, \\ (x, t) &\in K. \end{aligned} \tag{A.18}$$

Owing to the regularity condition, Theorem A.2 implies that system (A.18) has a solution for every $b \in \mathbb{R}^m$. Let C be its modulus. Consider the system with matrix perturbation

$$\begin{aligned} Ax - bt &= 0, \\ (x, t) &\in K. \end{aligned} \tag{A.19}$$

Let $b_k = Ax_k$. Clearly, for $b = b_k$ the above system has a solution $(x_k, 1)$. By Corollary A.3, if $\|\widetilde{b}_k\| < 1/C$ then system (A.19) has a solution (y_{k+1}, t_{k+1}) satisfying the inequality

$$\begin{aligned} \|(y_{k+1}, t_{k+1}) - (x_k, 1)\| &\le \widetilde{C}\|(x_k, 1)\|\|\widetilde{b}_k - b_k\| \\ &= \widetilde{C}\|(x_k, 1)\|\|h(x_k, \widetilde{u})\|, \end{aligned}$$

with

$$\widetilde{C} = \frac{C}{1 - C\|\widetilde{b}_k\|}.$$

Suppose $\|\widetilde{b}_k\| \le 1/(2C)$ and $\|x_k\| \le 1$ (we can guarantee this at this stage for $k = 1$, because $(\widetilde{x}, \widetilde{u})$ may be chosen sufficiently close to (x_0, u_0)). Then we obtain $\widetilde{C} \le 2C$ and the last displayed inequality can be rewritten as follows:

$$\|(y_{k+1}, t_{k+1}) - (x_k, 1)\| \le 4C\|h(x_k, \widetilde{u})\|. \tag{A.20}$$

Let $\varepsilon_1 > 0$ be such that for all (x, u) satisfying $\|x - x_0\| \le \varepsilon_1$ and $\|u - u_0\| \le \varepsilon_1$ we have

$$\|h(x, \widetilde{u})\| \le 1/(8C) \quad \text{and} \quad \|h'(x_0, u_0)x - h(x, \widetilde{u})\| \le 1/(2C).$$

If x_k and $\widetilde{u}$ are in these ε_1-neighborhoods, we obtain from (A.20) the inequality $t_{k+1} \ge 1/2$. Observe that the point

$$x_{k+1} = \frac{y_{k+1}}{t_{k+1}}$$

is a solution of the linearized system (A.16). We obtain

$$\begin{aligned} \|x_{k+1} - x_k\| &\le \frac{1}{t_{k+1}}\|y_{k+1} - x_k\| + \|\frac{x_k}{t_{k+1}} - x_k\| \\ &\le 2\|y_{k+1} - x_k\| + 2|1 - t_{k+1}|\|x_k\| \\ &\le 16C\|h(x_k, \widetilde{u})\|. \end{aligned} \tag{A.21}$$

In the last inequality we employed (A.20).

Let $\lambda > 0$ and $\varepsilon_2 \in (0, \varepsilon_1)$ be such that for all (x, u) satisfying $\|x - x_0\| \le \varepsilon_2$ and $\|u - u_0\| \le \varepsilon_2$ we have

$$\|h'(x, u) - h'(x_0, u_0)\| \le \lambda < 1/(16C).$$

The Newton–Leibniz theorem of calculus yields the representation (with $\bar{x}(\theta) = \theta x_{k+1} + (1 - \theta)x_k$):

$$\begin{aligned} h(x_{k+1}, \widetilde{u}) &= h(x_k, \widetilde{u}) + \int_0^1 h'(\bar{x}(\theta), \widetilde{u})(x_{k+1} - x_k)\, d\theta \\ &= h(x_k, \widetilde{u}) + h'(x_0, u_0)(x_{k+1} - x_k) \\ &\quad + \int_0^1 \big[h'(\bar{x}(\theta), \widetilde{u}) - h'(x_0, u_0)\big](x_{k+1} - x_k)\, d\theta \\ &= \int_0^1 \big[h'(\bar{x}(\theta), \widetilde{u}) - h'(x_0, u_0)\big](x_{k+1} - x_k)\, d\theta. \end{aligned}$$

In the last equality we used the fact that x_{k+1} is a solution to (A.16). If $\|x_k - x_0\| \le \varepsilon_2$, $\|x_{k+1} - x_0\| \le \varepsilon_2$ and $\|\widetilde{u} - u_0\| \le \varepsilon_2$ we obtain

$$\begin{aligned} \|h(x_{k+1}, \widetilde{u})\| &\le \int_0^1 \|h'(\bar{x}(\theta), \widetilde{u}) - h'(x_0, u_0)\| \|x_{k+1} - x_k\|\, d\theta \\ &\le \lambda \|x_{k+1} - x_k\|. \end{aligned} \tag{A.22}$$

Define $\varrho = 16\lambda C < 1$ and let $\varepsilon_0 \in (0, \varepsilon_2/2)$ be such that

$$\|h(x, u)\| \le \frac{\varepsilon_2(1 - \varrho)}{32C},$$

whenever $\|x - x_0\| \le \varepsilon_3$ and $\|u - u_0\| \le \varepsilon_0$. We claim that for all $k = 1, 2, \ldots$ the following inequalities are satisfied:

$$\begin{aligned} \|h(x_k, \widetilde{u})\| &\le \varrho^{k-1} \|h(x_1, \widetilde{u})\|, \\ \|x_{k+1} - x_k\| &\le 16C\varrho^{k-1} \|h(x_1, \widetilde{u})\|. \end{aligned} \tag{A.23}$$

They are true for $k = 1$, as can be seen from (A.21). If they hold for all $j = 1, 2, \ldots, k$ then for all these j we have

$$\|x_{j+1} - x_1\| \le \sum_{i=1}^{j} \|x_{i+1} - x_i\| \le \frac{16C}{1 - \varrho} \|h(x_1, \widetilde{u})\| \le \frac{\varepsilon_2}{2}.$$

Therefore $\|x_{j+1}-x_0\| \le \varepsilon_2$ and and we are allowed to use inequalities (A.21) and (A.22) for k. They immediately yield (A.23) for $k+1$. By induction, relations (A.23) hold true for all k. It follows from these relations that the sequence $\{x_k\}$ has a limit x_{R} such that $h(x_{\mathrm{R}}, \widetilde{u}) = 0$ and

$$\|x_{\mathrm{R}} - \widetilde{x}\| \le \frac{16C}{1-\varrho}\|h(x_1, \widetilde{u})\|. \tag{A.24}$$

The set X_0 is closed, and thus $x_{\mathrm{R}} \in X_0$. As $x_1 = \widetilde{x}$, our assertion is true with $C_0 = 16C/(1-\varrho)$.

Now suppose $\widetilde{x} \notin X_0$. In this case we define $x_1 = \Pi_{X_0}(\widetilde{x})$. Obviously, $\|x_1 - x_0\| \le \|\widetilde{x} - x_0\|$ and we can construct the sequence $\{x_k\}$ exactly as before. The estimate (A.24) remains valid. Since $h(\cdot, \widetilde{u})$ is continuously differentiable, it is Lipschitz continuous in the ε_0-neighborhood of x_0. Thus there exists a constant L such that

$$\|h(x_1, \widetilde{u})\| \le \|h(\widetilde{x}, \widetilde{u})\| + L\,\mathrm{dist}(\widetilde{x}, X_0).$$

Combining this inequality with (A.24) we obtain our assertion with

$$C_0 = \frac{16C(L+1)}{1-\varrho}.$$

□

Consider the special case of right hand side perturbations, when (A.14) becomes

$$\begin{aligned} h(x) &= u, \\ x &\in X_0, \end{aligned} \tag{A.25}$$

with a smooth function $h : \mathbb{R}^n \to \mathbb{R}^m$. As we have mentioned earlier, the stability at $(x_0, 0)$ is called the *metric regularity* of the system. The regularity condition simplifies:

$$0 \in \mathrm{int}\left\{h'(x_0)(x - x_0) : x \in X_0\right\}. \tag{A.26}$$

It turns out that metric regularity is *equivalent* to this condition.

THEOREM A.8. *System* (A.25) *is metrically regular if and only if condition* (A.26) *is satisfied.*

Proof. By Theorem A.7, condition (A.26) implies metric regularity. It remains to prove the reverse implication.

Suppose condition (A.26) is not satisfied:

$$0 \notin \mathrm{int}\left\{h'(x_0)(x - x_0) : x \in X_0\right\}.$$

The set $S = \{h'(x_0)(x - x_0) : x \in X_0\}$ is convex by definition, and so is its interior. Theorem 2.15 implies that we can separate int S and 0: there exists $d \neq 0$ such that $\langle d, s\rangle \leq 0$ for all $s \in S$.

Consider $\widetilde{u} = \tau d$, where $\tau > 0$, and $\widetilde{x} = x_0$. If the system is metrically regular, then for every sufficiently small $\tau >$ we must be able to find $x_{\mathrm{R}}(\tau) \in X_0$ such that

$$\|x_{\mathrm{R}}(\tau) - x_0\| \leq C\tau\|d\|,$$

and

$$h(x_{\mathrm{R}}(\tau)) = \tau d. \tag{A.27}$$

Expanding h in the neighborhood of x_0 we obtain

$$h(x_{\mathrm{R}}(\tau)) = h'(x_0)(x_{\mathrm{R}}(\tau) - x_0) + o(\tau),$$

where $o(\tau)/\tau \to 0$, as $\tau \downarrow 0$. Multiplying by d and using the fact that (by separation)

$$\langle d, h'(x_0)(x_{\mathrm{R}}(\tau) - x_0)\rangle \leq 0,$$

we conclude that

$$\langle d, h(x_{\mathrm{R}}(\tau))\rangle \leq \langle d, o(\tau)\rangle.$$

This combined with (A.27) yields

$$\tau\|d\|^2 \leq \langle d, o(\tau)\rangle.$$

Dividing by τ and letting $\tau \downarrow 0$ we get $\|d\| = 0$, a contradiction. Thus, 0 must be an interior point of S. □

The same ideas can be employed to analyze generalized systems of equations and inequalities of the form

$$\begin{aligned} g(x, u) &\in Y_0, \\ x &\in X_0. \end{aligned} \tag{A.28}$$

Here $g : \mathbb{R}^n \times \mathbb{R}^s \to \mathbb{R}^m$ is continuously differentiable with respect to the first argument, Y_0 is a closed convex set in $\mathbb{R}^m$, and X_0 is a closed convex set in $\mathbb{R}^n$.

One way to deal with conditions of the form (A.28) is to introduce new variables $y \in \mathbb{R}^m$ and to consider the equivalent system

$$\begin{aligned} g(x, u) - y = 0, \\ x \in X_0,\ y \in Y_0, \end{aligned} \tag{A.29}$$

which is a special case of (A.14), with x replaced by (x, y), and X_0 by $X_0 \times Y_0$.

We call (A.28) *stable* at (x_0, u_0) if (A.29) is stable at $((x_0, y_0), u_0)$, where $y_0 = g(x_0, u_0)$. An equivalent characterization of stability of this system can be obtained as follows.

LEMMA A.9. *System* (A.28) *is stable at the point* (x_0, u_0) *if and only if there exist* $\varepsilon > 0$ *and* C *such that for every* $\widetilde{x}$ *and* $\widetilde{u}$ *satisfying* $\|\widetilde{x} - x_0\| \leq \varepsilon$ *and* $\|\widetilde{u} - u_0\| \leq \varepsilon$ *we can find* $x_{\text{R}} \in X_0$ *satisfying the inclusion*

$$g(x_{\text{R}}, \widetilde{u}) \in Y_0,$$

and such that

$$\|x_{\text{R}} - \widetilde{x}\| \leq C\Big(\operatorname{dist}(\widetilde{x}, X_0) + \operatorname{dist}(g(\widetilde{x}, \widetilde{u}), Y_0)\Big). \tag{A.30}$$

Proof. If system (A.29) is stable, then there exist $\varepsilon > 0$ and C such that for every $\widetilde{x}$, $\widetilde{y}$ and $\widetilde{u}$ satisfying $\|\widetilde{x} - x_0\| \leq \varepsilon$, $\|\widetilde{y} - y_0\| \leq \varepsilon$ and $\|\widetilde{u} - u_0\| \leq \varepsilon$ we can find $x_{\text{R}} \in X_0$ and $y_{\text{R}} \in Y_0$ satisfying the equations

$$g(x_{\text{R}}, \widetilde{u}) - y_{\text{R}} = 0,$$

and such that

$$\|x_{\text{R}} - \widetilde{x}\| + \|y_{\text{R}} - \widetilde{y}\| \leq C\Big(\operatorname{dist}(\widetilde{x}, X_0) + \operatorname{dist}(\widetilde{y}, Y_0) + \|g(\widetilde{x}, \widetilde{u}) - \widetilde{y}\|\Big). \tag{A.31}$$

In particular, choosing

$$\widetilde{y} = \Pi_{Y_0}(g(\widetilde{x}, \widetilde{u})),$$

we get $\operatorname{dist}(\widetilde{y}, Y_0) = 0$ and

$$\|g(\widetilde{x}, \widetilde{u}) - \widetilde{y}\| = \operatorname{dist}(g(\widetilde{x}, \widetilde{u}), Y_0).$$

This reduces (A.31) to (A.30).

On the other hand, if (A.30) is true, then the triangle inequality implies

$$\operatorname{dist}(g(\widetilde{x}, \widetilde{u}), Y_0) \leq \operatorname{dist}(\widetilde{y}, Y_0) + \|g(\widetilde{x}, \widetilde{u}) - \widetilde{y}\|,$$

for all $\widetilde{y}$. Therefore (A.30) can be rewritten as follows:

$$\|x_{\text{R}} - \widetilde{x}\| \leq C\Big(\operatorname{dist}(\widetilde{x}, X_0) + \operatorname{dist}(\widetilde{y}, Y_0) + \|g(\widetilde{x}, \widetilde{u}) - \widetilde{y}\|\Big). \tag{A.32}$$

Let us choose

$$y_{\text{R}} = g(x_{\text{R}}, \widetilde{u}).$$

We have

$$\begin{aligned}\|y_{\mathrm{R}} - \widetilde{y}\| &= \|g(x_{\mathrm{R}}, \widetilde{u}) - \widetilde{y}\| \leq \|g(x_{\mathrm{R}}, \widetilde{u}) - g(\widetilde{x}, \widetilde{u})\| + \|g(\widetilde{x}, \widetilde{u}) - \widetilde{y}\| \\ &\leq L\|x_{\mathrm{R}} - \widetilde{x}\| + \|g(\widetilde{x}, \widetilde{u}) - \widetilde{y}\|. \end{aligned} \tag{A.33}$$

In the last inequality we used the fact that $g(\cdot, u)$ is locally Lipschitz continuous around (x_0, u_0) and we denoted by L its Lipschitz constant. Substituting (A.32) into (A.33) and adding the resulting inequality to (A.32) we conclude that

$$\|x_{\mathrm{R}} - \widetilde{x}\| + \|y_{\mathrm{R}} - \widetilde{y}\| \leq (2 + LC)\Big(\operatorname{dist}(\widetilde{x}, X_0) + \operatorname{dist}(\widetilde{y}, Y_0) + \|g(\widetilde{x}, \widetilde{u}) - \widetilde{y}\|\Big).$$

This proves the stability of the system (A.29). □

A sufficient condition for stability of system (A.28) can be derived directly from Theorem A.7.

THEOREM A.10. *If*

$$0 \in \operatorname{int}\big\{g'(x_0, u_0)(x - x_0) - (y - g(x_0)) : x \in X_0,\ y \in Y_0\big\}, \tag{A.34}$$

then system (A.28) *is stable at* (x_0, u_0).

We call (A.34) Robinson's condition. Again, if we restrict the class of systems to

$$\begin{aligned} g(x) - u &\in Y_0, \\ x &\in X_0, \end{aligned}$$

then Robinson's condition is equivalent to their stability (metric regularity in this case). This follows directly from Theorem A.8.

Further Reading

Introduction. Specific topics discussed in the examples are studied in more detail in [3] (image reconstruction), [99] (portfolio optimization), [96] (signal processing), [97, 164] (classification), [60] (optimal control models), and [69, 88] (approximation).

Chapter 1. The first systematic treatment of convex sets, including a separation theorem and extreme points, can be found in [105, 106]. The theorem about the alternative is due to [38]. Most of the modern concepts and results of convex analysis are already included in the lecture notes [40] and in the monograph [140]. The concept of a subgradient was introduced and studied in [110, 139]. Conjugate duality is due to [39, 40, 109]. The analysis of the maximum function is due to [162]. The history of convex analysis is extensively discussed in [146]. For a comprehensive treatment in $\mathbb{R}^n$, see [68, 133, 146]. Convex analysis in infinite dimensional spaces is covered in [5, 33, 71, 83, 92, 122]. Extensions of the separation principle and of the concept of the subgradient to nonconvex sets and functions were proposed and developed in [19, 84, 107]. Applications to the problem of moments are discussed in [83]. Approximation problems are extensively covered in [29, 69, 88]. For conjugate duality of Lorenz curves, see [120], and of risk functions, see [151].

Chapter 2. Optimality conditions for smooth unconstrained problems and the Lagrange multiplier rule for equality constrained problems are classical topics of multivariate calculus. The concept of a tangent cone is due to [13]. For a comprehensive modern treatment, see [6]. Constraint qualification conditions used by us appear in [75, 86, 98, 138, 158]. For recent advances, see [64]. Optimality conditions for inequality constrained problems are credited to [75, 85]. The paper [130] provides an account of earlier contributions along these lines. The approach by cone separation is due to [31, 131]. A comprehensive review is provided in [145]. Nonsmooth problems are discussed in [71, 131, 140]. For extensions to nonconvex nonsmooth problems, see [19, 20, 84, 107]. The monograph [146] covers much of this material and provides a comprehensive literature review. Second order conditions and sensitivity are covered extensively in [7, 12]. First and second order tangent cones in semidefinite programming were analyzed in [12, 155]. The pioneering work on infinite-dimensional constrained problems was [70]. For a comprehensive review and modern contributions, see [12, 19, 20, 71, 95, 107].

Chapter 3. For the origins of the duality theory in the theory of games, see [48, 74, 165]. Linear programming duality and its applications are discussed in [22, 23, 163]. Lagrangian duality for convex problems has been developed in [140, 142]. Optimization under dominance constraints and dual utility functions come from [28]. For further developments, see [33, 57, 146] and the references therein. Duality in semidefinite programming was discussed in [115, 155, 170]. Decomposition methods employing duality are presented in [22, 23, 87]. The idea of convexification by dual problems is due to [37]. For relaxation of integer problems, see [114]. Augmented Lagrangians have been introduced for the purpose of numerical methods in [4, 10, 61, 65, 128]. Duality relations for augmented Lagrangians are explored in [58, 141, 145].

Chapter 4. Numerical methods for unconstrained optimization and their implementation are extensively covered in several monographs: [27, 43, 52, 118, 169]. For the fundamentals of iterative methods and convergence rates, see [21, 121]. The method of steepest descent is very old; it is attributed to [18]. The use of the two-slope test in it goes back to [55]. Local rate of convergence analysis is due to [72]. Newton's method for systems of equations is analyzed in [121]. Rate of convergence analysis is due to [73]. The early influential works on the Gauss–Newton method were [8, 91, 101]. Linear least squares are treated comprehensively in [89]. Early contributions to the conjugate gradient methods are in [45, 67, 124]. Comprehensive treatment of these methods can be found in [66]. First works on quasi-Newton methods were [14, 24, 42, 44, 54, 153]. Their equivalence was established in [30]. For limited memory versions, see [50, 94, 117]. Trust region methods are discussed in [108, 129, 152]. Conjugate directions without derivatives were introduced in [127]. Nongradient methods are thoroughly analyzed in [76]. The truncated Newton method was developed in [25, 26, 113]. For applications to noisy functions, see [77]. Automatic differentiation is discussed in [59].

Chapter 5. For the origins of the projection method, see [9, 56]. The pioneering work on large-scale linearly constrained optimization was [112]. The topic is discussed extensively in [43, 52, 100, 163]. First contributions to exact penalty methods were [34, 123, 173]. Augmented Lagrangians have been introduced for equality constraints in [4, 61, 65, 128]. Extensions to inequality constraints were studied in [16, 141, 166]. Relations to the proximal point method are explored in [143]. Versions with general cone constraints are analyzed in [167], and decomposition in [149, 160]. Penalty and multiplier methods are covered in dedicated monographs [10, 58]. Sequential quadratic programming methods were introduced in [49, 62, 132, 168]. For a comprehensive treatment, see [11, 51, 52, 118]. The modern approach via generalized equations and multivalued Newton's method is discussed in [82, 136, 137]. The early contributions to barrier methods are [17, 47].

The synthesis and basic theory is in [41]. Central path for linear problems was investigated in [159]. Interior point methods for linear and quadratic programming are presented in [11, 147, 163, 172]. For nonlinear extensions, see [1, 46, 115, 116, 154, 171] and the references therein. Specialized factorization schemes are discussed in [2].

Chapter 6. The subgradient method was proposed in [156]. Convergence of versions with diminishing step sizes was established in [35, 125]. Refinements for known objective values are due to [126]. A comprehensive presentation can be found in [157]. Extensions to nonconvex problems are presented in [104, 119]. Stochastic versions are discussed in [36]. The cutting plane method is due to [78]. The bundle methods were developed in [90, 103]. First rigorous convergence analysis and versions with aggregation were provided in [79, 80]. Relations to proximal point methods were exploited in [148], on which we base our presentation. For a comprehensive treatment of bundle and trust region methods, see [11, 68]. Applications to stochastic programming are discussed [150] and to semidefinite programming in [63]. Cutting plane methods using analytic centers are surveyed in [53]. For the background on the proximal point method and its extensions, see [32, 102, 109, 110, 111, 144] and the monograph [146]. Composite optimization is discussed in [15, 93, 161]. Bundle methods for nonconvex problems are analyzed in [81].

Appendix. Our presentation follows [134, 135]. Advanced topics are covered in the monographs [12, 82].

Bibliography

[1] F. Alizadeh, J.-P. A. Haeberly, and M. L. Overton, *Primal-dual interior-point methods for semidefinite programming: Convergence rates, stability and numerical results*, SIAM Journal on Optimization **8** (1998), 746–768.

[2] A. Altman and J. Gondzio, *Regularized symmetric indefinite systems in interior point methods for linear and quadratic optimization*, Optimization Methods and Software **11** (1999), 275–302.

[3] S. R. Arridge and J. C. Hebden, *Optical imaging in medicine. 2. Modelling and reconstruction*, Physics in Medicine and Biology **42** (1997), 841–853.

[4] K. J. Arrow and R. M. Solow, *Gradient methods for constrained maxima with weakened assumptions*, Studies in Linear and Non-Linear Programming (K. J. Arrow, L. Hurwicz, and H. Uzawa, eds.), Stanford University Press, Stanford, California, 1958.

[5] J.-P. Aubin and I. Ekeland, *Applied Nonlinear Analysis*, Wiley, New York, 1984.

[6] J.-P. Aubin and H. Frankowska, *Set-Valued Analysis*, Birkhäuser, Boston, 1990.

[7] B. Bank, J. Guddat, D. Klatte, B. Kummer, and K. Tammer, *Non-Linear Parametric Optimization*, Akademie-Verlag, Berlin, 1982.

[8] A. Ben-Israel, *A Newton-Raphson method for the solution of systems of equations*, Journal of Mathematical Analysis and Applications **15** (1966), 243–252.

[9] D. P. Bertsekas, *On the Goldstein-Levitin-Polyak gradient projection method*, IEEE Transactions of Automatic Control **21** (1976), 174–184.

[10] ———, *Constrained Optimization and Lagrange Multiplier Methods*, Academic Press, New York, 1982.

[11] J.-F. Bonnans, J. C. Gilbert, C. Lemaréchal, and C. Sagastizábal, *Numerical Optimization. Theoretical and Practical Aspects*, Springer-Verlag, Berlin, 2003.

[12] J.-F. Bonnans and A. Shapiro, *Perturbation Analysis of Optimization Problems*, Springer-Verlag, New York, 2000.

[13] G. Bouligand, *Sur les surfaces dépourvues de points hyperlimites*, Annales de la Société Polonaise de Mathématique **9** (1930), 32–41.

[14] C. G. Broyden, *The convergence of a class of of double rank minimization algorithms*, Journal of the Institute for Mathematics and Applications **6** (1970), 222–231.

[15] J. V. Burke and M. C. Ferris, *A Gauss-Newton method for convex composite optimization*, Mathematical Programming **71** (1995), 179–194.

[16] J. D. Buys, *Dual algorithms for constrained optimization*, Ph.D. thesis, Rijksuniversiteit te Leiden, Leiden, The Netherlands, 1972.

[17] C. W. Carroll, *The rerated response surface technique for optimizing nonlinear restrained systems*, Operations Research **9** (1961), 169–184.

[18] A. Cauchy, *Méthode générale pour la résolution des systèmes d'équations simultanées*, Comptes Rendus de l'Académie des Sciences Paris, Série I **25** (1847), 536–538.

[19] F. H. Clarke, *Optimization and Nonsmooth Analysis*, Wiley, New York, 1983.

[20] ———, *Methods of Dynamic and Nonsmooth Optimization*, SIAM Publications, Philadelphia, 1989.

[21] J. W. Daniel, *The Approximate Minimization of Functionals*, Prentice Hall, Englewood Cliffs, New Jersey, 1971.

[22] G. B. Dantzig, *Linear Programming and Extensions*, Princeton University Press, Princeton, New Jersey, 1963.

[23] G. B. Dantzig and M. N. Thapa, *Linear Programming*, Springer-Verlag, New York, 2003.

[24] W. C. Davidon, *Variable metric methods for minimization*, AEC Research and Development Report ANL-5990, Argonne National Laboratory, Argonne, Illinois, 1959.

[25] R. Dembo, R. Eisenstat, and T. Steinhaug, *Inexact Newton methods*, SIAM Journal on Numerical Analysis **19** (1982), 400–408.

[26] R. Dembo and T. Steinhaug, *Truncated Newton methods for large scale optimization*, Mathematical Programming **26** (1983), 190–212.

[27] J. E. Dennis, Jr and R. B. Schnabel, *Numerical Methods for Unconstrained Optimization and Nonlinear Equations*, Prentice Hall, Englewood Cliffs, New Jersey, 1983.

[28] D. Dentcheva and A. Ruszczyński, *Optimization with stochastic dominance constraints*, SIAM Journal on Optimization **14** (2003), 548–566.

[29] F. Deutsch, *Best Approximation in Inner Product Spaces*, Springer-Verlag, New York, 2001.

[30] L. C. W. Dixon, *Quasi-Newton algorithms generate identical points. II. The proofs of four new theorems*, Mathematical Programming **3** (1972), 345–358.

[31] A. Y. Dubovitskii and A. A. Milyutin, *Extremum problems in the presence of restrictions*, U.S.S.R. Computational Mathematics and Mathematical Physics **5** (1965), 1–80 (Russian).

[32] J. Eckstein and D. P. Bertsekas, *On the Douglas-Rachford splitting method and the proximal point algorithm for maximal monotone operators*, Mathematical Programming **55** (1992), 293–318.

[33] I. Ekeland and R. Temam, *Analyse Convexe et Problèmes Variationnelles*, Dunod, Paris, 1971, English transl., North-Holland, Amsterdam, 1976.

[34] I. I. Eremin, *The penalty method in convex programming*, Doklady Akademii Nauk SSSR **8** (1966), 459–462 (Russian).

[35] Yu. M. Ermoliev, *Methods for solving nonlinear extremal problems*, Kibernetika (Kiev) **1** (1966), 1–17 (Russian).

[36] Yu. M. Ermoliev, *Stochastic Programming Methods*, Nauka, Moscow, 1976 (Russian).

[37] J. E. Falk, *Lagrange multipliers and nonconvex programs*, SIAM Journal on Control **7** (1969), 534–545.

[38] J. Farkas, *Theorie der einfachen Ungleichungen*, Zeitschrift für Reine und Angewandte Mathematik **124** (1901), 1–27.

[39] W. Fenchel, *On conjugate convex functions*, Canadian Journal of Mathematics **1** (1949), 73–77.

[40] ——, *Convex Cones, Sets and Functions. Lecture Notes*, Princeton University, Princeton, New Jersey, 1951.

[41] A. V. Fiacco and G. P. McCormick, *Nonlinear Programming: Sequential Unconstrained Minimization Techniques*, Wiley, New York, 1968.

[42] R. Fletcher, *A new approach to variable metric algorithms*, Computer Journal **13** (1970), 317–322.

[43] ——, *Practical Methods of Optimization*, second ed., Wiley, Chichester, 1987.

[44] R. Fletcher and M. J. D. Powell, *A rapidly convergent descent method for minimization*, Computer Journal **6** (1963), 163–168.

[45] R. Fletcher and C. M. Reeves, *Function minimization by conjugate gradients*, Computer Journal **7** (1964), 149–154.

[46] A. Forsgren, P. E. Gill, and M. H. Wright, *Interior methods for nonlinear optimization*, SIAM Review **44** (2002), 525–597.

[47] M. R. Frisch, *La résolution des problèmes de programme linéaire par la méthode du potentiel logarithmique*, Cahiers du Séminaire d'Econometrie **4** (1956), 7–20.

[48] D. Gale, H. W. Kuhn, and A. W. Tucker, *Linear programming and the theory of games*, Activity Analysis of Production and Allocation (T. C. Koopmans, ed.), Wiley, New York, 1951.

[49] U. M. Garcia-Palomares and O. L. Mangasarian, *Superlinearly convergent quasi-Newton methods for nonlinearly constrained optimization problems*, Mathematical Programming **11** (1976), 1–13.

[50] J. Gilbert and C. Lemaréchal, *Some numerical experiments with variable-storage quasi-Newton algorithms*, Mathematical Programming, Series B **45** (1989), 407–435.

[51] P. E. Gill, M. A. Murray, and M. A. Saunders, *SNOPT: An SQP algorithm for large-scale constrained optimization*, SIAM Journal on Optimization **12** (2002), 979–1006.

[52] P. E. Gill, W. Murray, and M. H. Wright, *Practical Optimization*, Academic Press, London, 1981.

[53] J.-L. Goffin and J.-P. Vial, *Convex nondifferentiable optimization: a survey focused on the analytic center cutting plane method*, Optimization Methods and Software **17** (2002), 805–867.

[54] D. Goldfarb, *A family of variable-metric methods derived by variational means*, Mathematics of Computation **24** (1970), 23–26.

[55] A. A. Goldstein, *Cauchy's method of minimization*, Numerische Mathematik **4** (1962), 146–150.

[56] ———, *Convex programming in Hilbert space*, Bulletin of The American Mathematical Society **70** (1964), 709–710.

[57] E. G. Golshtein, *Duality Theory in Mathematical Programming*, Nauka, Moscow, 1971 (Russian).

[58] E. G. Golshtein and N. V. Tretyakov, *Modified Lagrange Functions; Theory and Optimization Methods*, Nauka, Moscow, 1989 (Russian), English transl., Wiley, New York, 1996.

[59] A. Griewank, *Evaluating Derivatives. Principles and Techniques of Algorithmic Differentiation*, SIAM Publications, Philadelphia, 2000.

[60] W. A. Gruver and E. Sachs, *Algorithmic Methods in Optimal Control*, Research Notes in Mathematics, vol. 47, Pitman, Boston and London, 1981.

[61] P. C. Haarhoff and J. D. Buys, *A new method for the minimization of a nonlinear function subject to nonlinear constraints*, Computer Journal **13** (1970), 178–184.

[62] S. P. Han, *Superlinearly convergent variable metric algorithms for general nonlinear programming problems*, Mathematical Programming **11** (1976), 263–282.

[63] C. Helmberg and F. Rendl, *A spectral bundle method for semidefinite programming*, SIAM Journal on Optimization **10** (2000), 673–696.

[64] R. Henrion and J. Outrata, *A subdifferential condition for calmness of multifunctions*, Journal of Mathematical Analysis and Applications **258** (2001), 110–130.

[65] M. R. Hestenes, *Multiplier and gradient methods*, Journal of Optimization Theory and Applications **4** (1969), 303–320.

[66] ——, *Conjugate Direction Methods in Optimization*, Springer-Verlag, Berlin, 1980.

[67] M. R. Hestenes and E. Stiefel, *Methods of conjugate gradients for solving linear systems*, Journal of Research of the National Bureau of Standards **49** (1952), 409–436.

[68] J.-B. Hiriart-Urruty and C. Lemaréchal, *Convex Analysis and Minimization Algorithms*, Springer-Verlag, Berlin, 1993.

[69] R. B. Holmes, *A Course on Optimization and Best Approximation*, Lecture Notes in Mathematics, vol. 257, Springer-Verlag, Berlin, 1972.

[70] L. Hurwicz, *Programming in linear spaces*, Studies in Linear and Non-Linear Programming (K. J. Arrow, L. Hurwicz, and H. Uzawa, eds.), Stanford University Press, Stanford, California, 1958.

[71] A. D. Ioffe and V. M. Tikhomirov, *Theory of Extremal Problems*, Nauka, Moscow, 1974 (Russian), English transl., North-Holland, Amsterdam, 1979.

[72] L. V. Kantorovich, *On an effective method of solution of extremal problems for a quadratic functional*, Doklady Akademii Nauk SSSR **48** (1945), 483–487 (Russian).

[73] ——, *Functional analysis and applied mathematics*, Uspekhi Matematicheskich Nauk **3** (1948), 89–185 (Russian).

[74] S. Karlin, *Mathematical Methods and Theory in Games, Programming and Economics*, Pergamon Press, Oxford, 1959.

[75] W. Karush, *Minima of functions of several variables with inequalities as side conditions*, Master's thesis, University of Chicago, Chicago, Illinois, 1939.

[76] C. T. Kelley, *Iterative Methods for Linear and Nonlinear Equations*, SIAM Publications, Philadelphia, 1995.

[77] C. T. Kelley and E. W. Sachs, *Truncated Newton methods for optimization with inaccurate functions and gradients*, Nonlinear Analysis - Theory, Methods and Applications **24** (1995), 883–894.

[78] J. E. Kelley, *The cutting plane method for solving convex programs*, Journal of SIAM **8** (1960), 703–712.

[79] K. C. Kiwiel, *An aggregate subgradient method for nonsmooth convex minimization*, Mathematical Programming **27** (1983), 320–341.

[80] ______, *Methods of Descent for Nondifferentiable Optimization*, Lecture Notes in Mathematics, vol. 1133, Springer-Verlag, Berlin, 1985.

[81] ______, *Restricted step and Levenberg-Marquardt techniques in proximal bundle methods for nonconvex nondifferentiable optimization*, SIAM Journal on Optimization **6** (1996), 227–249.

[82] D. Klatte and B. Kummer, *Nonsmooth Equations in Optimization*, Kluwer, Dordrecht, 2002.

[83] M. G. Krein and A. A. Nudelman, *The Markov Moment Problem and Extremal Problems*, Nauka, Moscow, 1973 (Russian), English transl., American Mathematical Society, Providence, 1977.

[84] A. Y. Kruger and B. S. Mordukhovich, *Extremal points and the Euler equation in nonsmooth problems of optimization*, Doklady Akademii Nauk BSSR (Bielorussian Academy of Sciences) **24** (1980), 684–687 (Russian).

[85] H. W. Kuhn and A. W. Tucker, *Nonlinear programming*, Proceedings of the Second Berkeley Symposium on Mathematical Stitistics and Probability (J. Neyman, ed.), University of California Press, Berkeley, California, 1951, pp. 481–492.

[86] S. Kurcyusz and J. Zowe, *Regularity and stability for the mathematical programming problem in Banach spaces*, Applied Mathematics and Computation **5** (1979), 49–62.

[87] S. Lasdon, *Optimization Theory for Large Systems*, MacMillan, New York, 1970.

[88] P.-J. Laurent, *Approximation et Optimization*, Hermann, Paris, 1972.

[89] C. L. Lawson and R. J. Hanson, *Solving Least Squares Problems*, Prentice-Hall, Englewood Cliffs, New Jersey, 1974.

[90] C. Lemaréchal, *Nonsmooth optimization and descent methods*, Research Report 78-4, International Institute of Applied Systems Analysis, Laxenburg, Austria, 1978.

[91] K. Levenberg, *A method for the solution of certain nonlinear problems in least squares*, Quarterly of Applied Mathematics **2** (1944), 164–168.

[92] V. L. Levin, *Convex Analysis in Spaces of Measurable Functions and Its Applications in Mathematics and Economics*, Nauka, Moscow, 1985 (Russian).

[93] C. Li and X. Wang, *On convergence of the Gauss-Newton method for convex composite optimization*, Mathematical Programming **91** (2002), 349–356.

[94] D. C. Liu and J. Nocedal, *On the limited memory BFGS method for large scale optimization*, Mathematical Programming **45** (1989), 503–528.

[95] D. G. Luenberger, *Optimization by Vector Space Methods*, Wiley, New York, 1974.

[96] Z. Q. Luo, *Applications of convex optimization in signal processing and digital communication*, Mathematical Programming, Series B **97** (2003), 177–207.

[97] O. L. Mangasarian, *Support vector machine classification via parameterless robust linear programming*, Optimization Methods and Software **20** (2005), 115–125.

[98] O. L. Mangasarian and S. Fromovitz, *The Fritz-John conditions in the presence of equality and inequality constraints*, Journal of Mathematical Analysis and Applications **17** (1967), 73–74.

[99] H. M. Markowitz, *Mean–Variance Analysis in Portfolio Choice and Capital Markets*, Blackwell, Oxford, 1987.

[100] I. Maros, *Computational Techniques of the Simplex Code*, Kluwer, Dordrecht, 2002.

[101] D. W. Marquardt, *An algorithm for least-squares estimation of nonlinear parameters*, Journal of SIAM **11** (1963), 431–441.

[102] B. Martinet, *Régularisation d'inéquations variationnelles par approximations successives*, Revue Française Informatique et Recherche Opérationnelle, Ser. R-3 **4** (1970), 154–158.

[103] R. Mifflin, *A modification and an extension of Lemaréchal's algorithm for nonsmooth minimization*, Nondifferential and Variational Techniques in Optimization (D. C. Sorensen and R. J. B. Wets, eds.), vol. 17, 1982, pp. 77–90.

[104] V. S. Mikhalevich, A. M. Gupal, and V. I. Norkin, *Nonconvex Optimization Methods*, Nauka, Moscow, 1987 (Russian).

[105] H. Minkowski, *Geometrie der Zahlen*, Teubner, Leipzig, 1896.

[106] ———, *Theorie der konvexen Körper, insbesondere Begründung ihres Obenrflächenbegriffs*, Gesammelte Abhandlungen, II, Teubner, Leipzig, 1911.

[107] B. S. Mordukhovich, *Variational Analysis and Generalized Differentiation*, Springer-Verlag, Berlin, 2005.

[108] J. J. Moré, *Recent development in algorithms and software for trust region methods*, Mathematical Programming, the State of the Art (A. Bachem, M. Grötschel, and B. Korte, eds.), Springer-Verlag, Berlin, 1983, pp. 258–287.

[109] J. J. Moreau, *Fonctions convexes duales et points proximaux dans un espace Hilbertien*, Comptes Rendus de l'Académie des Sciences Paris, Série I **255** (1962), 2897–2899.

[110] ———, *Propriétés des applications 'prox'*, Comptes Rendus de l'Académie des Sciences Paris, Série I **256** (1963), 1069–1071.

[111] ———, *Proximité et dualité dans un espace Hilbertien*, Bulletin de la Société Mathématique de France **93** (1965), 273–299.

[112] B. A. Murtagh and M. A. Saunders, *Large-scale linearly constrained optimization*, Mathematical Programming **44** (1978), 41–72.

[113] S. G. Nash, *Preconditioning of truncated Newton methods*, SIAM Journal on Scientific Computing **6** (1985), 599–616.

[114] G. L. Nemhauser and L. A. Wolsey, *Integer and Combinatorial Optimization*, John Wiley & Sons, New York, 1988.

[115] Yu. Nesterov and A. Nemirovskii, *Interior-Point Polynomial Algorithms in Convex Programming*, SIAM Publications, Philadelphia, 1994.

[116] Yu. Nesterov and M. Todd, *Primal-dual interior point methods for self-scaled cones*, SIAM Journal on Optimization **8** (1998), 324–364.

[117] J. Nocedal, *Updating quasi-Newton matrices with limited storage*, Mathematics of Computation **35** (1980), 773–782.

[118] J. Nocedal and S. J. Wright, *Numerical Optimization*, Springer-Verlag, New York, 1997.

[119] V. I. Norkin, *Generalized Gradient Descent for Nonconvex Nonsmooth Optimization, PhD Dissertation*, Institute of Cybernetics of the Ukrainian Academy of Sciences, Kiev, 1983 (Russian).

[120] W. Ogryczak and A. Ruszczyński, *Dual stochastic dominance and related mean–risk models*, SIAM Journal on Optimization **13** (2002), 60–78.

[121] J. M. Ortega and W. C. Rheinboldt, *Iterative Solution of Nonlinear Equations in Several Variables*, Academic Press, London, 1970.

[122] R. R. Phelps, *Convex Functions, Monotone Operators, and Differentiability*, Lecture Notes in in Mathematics, vol. 1364, Springer-Verlag, New York, 1989.

[123] T. Pietrzykowski, *An exact potential method for constrained maxima*, SIAM Journal on Numerical Analysis **6** (1969), 299–304.

[124] E. Polak and G. Ribière, *Note sur la convergence de méthodes de directions conjuguées*, Revue Francaise d'Informatique et de Recherche Opérationnelle **19** (1969), 35–43.

[125] B. T. Polyak, *A general method for solving extremal problems*, Doklady Akademii Nauk SSSR **174** (1967), 33–36 (Russian).

[126] ———, *Minimization of nonsmooth functionals*, U.S.S.R. Computational Mathematics and Mathematical Physics **9** (1969), 509–521.

[127] M. J. D. Powell, *An eficient method for finding the minimum of a function of several variables without calculating derivatives*, Computer Journal **7** (1964), 155–162.

[128] ———, *A method for nonlinear constraints in minimization problems*, Optimization (R. Fletcher, ed.), Academic Press, London, 1969, pp. 283–298.

[129] ____, *Convergence properties of a class of minimization algorithms*, Nonlinear Programming (O. L. Mangasarian, R. R. Meyer, and S. M. Robinson, eds.), Academic Press, New York, 1975, pp. 1–27.

[130] A. Prékopa, *On the development of optimization theory*, The American Mathematical Monthly **87** (1980), 527–542.

[131] B. N. Pshenichnyi, *Necessary Conditions for an Extremum*, Nauka, Moscow, 1969 (Russian), English transl., Marcel Dekker, New York, 1971.

[132] ____, *Algorithms for the general problem of mathematical programming*, Kibernetika (Kiev) **6** (1970), 120–125 (Russian).

[133] ____, *Convex Analysis and Extremal Problems*, Nauka, Moscow, 1980 (Russian).

[134] S. M. Robinson, *Stability theory for systems of inequalities, Part 1: Linear systems*, SIAM Journal on Numerical Analysis **12** (1975), 754–768.

[135] ____, *Stability theory for systems of inequalities, Part 2: Differentiable nonlinear systems*, SIAM Journal on Numerical Analysis **13** (1976), 497–513.

[136] ____, *Generalized equations and their solutions, Part 1: Basic theory*, Mathematical Programming Study **10** (1979), 128–141.

[137] ____, *Generalized equations and their solutions, Part 2: Applications to nonlinear programming*, Mathematical Programming Study **19** (1982), 200–221.

[138] ____, *Local structure of feasible sets in nonlinear programming, Part I: Regularity*, Numerical Methods (V. Pereira and A. Reinoza, eds.), Lecture Notes in in Mathematics, vol. 1005, Springer-Verlag, Berlin, 1983.

[139] R. T. Rockafellar, *Convex Functions and Dual Extremum Problems, PhD Dissertation*, Harvard University, Cambridge, Massachusetts, 1963.

[140] ____, *Convex Analysis*, Princeton University Press, Princeton, New Jersey, 1970.

[141] ____, *Augmented Lagrange multiplier functions and duality in nonconvex programming*, SIAM Journal on Control and Optimization **12** (1974), 268–285.

[142] ——, *Conjugate Duality and Optimization*, SIAM Publications, Philadelphia, 1974.

[143] ——, *Augmented Lagrangians and applications of the proximal point algorithm in convex programming*, Mathematics of Operations Research **1** (1976), 97–116.

[144] ——, *Monotone operators and the proximal point algorithm*, SIAM Journal on Control and Optimization **14** (1976), 877–898.

[145] ——, *Lagrange multipliers and optimality*, SIAM Review **35** (1993), 183–238.

[146] R. T. Rockafellar and R. J.-B. Wets, *Variational Analysis*, Springer-Verlag, Berlin, 1998.

[147] C. Roos, T. Terlaky, and J.-P. Vial, *Theory and Algorithms for Linear Optimization. An Interior Point Approach*, Wiley, Chichester, 1997.

[148] A. Ruszczyński, *A regularized decomposition method for minimizing a sum of polyhedral functions*, Mathematical Programming **35** (1986), 309–333.

[149] ——, *On convergence of an augmented Lagrangian decomposition method for sparse convex optimization*, Mathematics of Operations Research **20** (1995), 643–656.

[150] A. Ruszczyński and A. Shapiro (eds.), *Stochastic Programming*, Elsevier, Amsterdam, 2003.

[151] A. Ruszczyński and A. Shapiro, *Optimization of convex risk functions*, Research report, Optimization On-Line, 2004.

[152] G. A. Schultz, R. B. Schnabel, and R. H. Byrd, *A family of trust-region-based algorithms for unconstrained minimization with strong global convergence properties*, SIAM Journal on Numerical Analysis **22** (1985), 47–67.

[153] D. Shanno, *Conditioning of quasi-Newton methods for function minimization*, Mathematics of Computation **27** (1970), 647–656.

[154] D. F. Shanno and R. J. Vanderbei, *Interior-point methods for nonconvex nonlinear programming: orderings and higher-order methods*, Mathematical Programming, Series B **87** (2000), 303–316.

[155] A. Shapiro, *First and second order analysis of nonlinear semidefinite programs*, Mathematical Programming, Series B **77** (1997), 301–320.

[156] N. Z. Shor, *Application of the method of gradient descent to the solution of the network transportation problem*, Materials of the Scientific Seminar on Theoretical and Applied Questions of Cybernetics and Operations Research, Ukrainian Academy of Sciences, Kiev, 1962, pp. 9–17 (Russian).

[157] ______, *Minimization Methods for Non-Differentiable Functions*, Naukova Dumka, Kiev, 1979 (Russian), English transl., Springer-Verlag, Berlin, 1985.

[158] M. Slater, *Lagrange multipliers revisited: a contribution to non-linear programming*, Discussion paper, Cowles Commission, 1950.

[159] G. Sonnevend, *An analytical centre for polyhedrons and new classes of global algorithms for linear (smooth, convex) programming*, System Modelling and Optimization (Budapest, 1985), Lecture Notes in Control and Information Sciences, vol. 84, Springer-Verlag, Berlin, 1986, pp. 866–879.

[160] G. Stephanopoulos and W. Westerberg, *The use of Hestenes' method of multipliers to resolve dual gaps in engineering system optimization*, Journal of Optimization Theory and Applications **15** (1975), 285–309.

[161] M. Studniarski and V. Jeyakumar, *A generalized mean-value theorem and optimality conditions in composite nonsmooth minimization*, SIAM Journal on Optimization **24** (1995), 883–894.

[162] M. Valadier, *Sous-différentiels d'une borne supérieure et d'une somme continue de fonctions convexes*, Comptes Rendus de l'Académie des Sciences Paris, Série I **268** (1969), 39–42.

[163] R. J. Vanderbei, *Linear Programming: Foundations and Extensions*, Kluwer, Dordrecht, 1997.

[164] V. N. Vapnik, *The Nature of Statistical Learning Theory. Second Edition*, Springer-Verlag, New York, 2000.

[165] J. von Neumann and O. Morgenstern, *Theory of Games and Economic Behavior*, Princeton University Press, Princeton, New Jersey, 1947.

[166] A. P. Wierzbicki, *A penalty function shifting method in constrained static optimization and its convergence properties*, Archiwum Automatyki i Telemechaniki **16** (1971), 395–416.

[167] A. P. Wierzbicki and S. Kurcyusz, *Projection on a cone, penalty functionals and duality theory for problems with inequality constraints in Hilbert space*, SIAM Journal on Control and Optimization **15** (1977), 25–56.

[168] R. B. Wilson, *A simplicial algorithm for concave programming*, Ph.D. thesis, Graduate School of Business Administration, Harvard University, Cambridge, Massachusetts, 1963.

[169] M. A. Wolfe, *Numerical Methods for Unconstrained Optimization*, Van Nostrand Reinhold, New York, 1978.

[170] H. Wolkowicz, R. Saigal, and L. Vandenberghe (eds.), *Handbook of Semidefinite Programming. Theory, Algorithms, and Applications*, Kluwer, Boston, 2000.

[171] S. J. Wright, *Primal-Dual Interior-Point Methods*, SIAM Publications, Philadelphia, 1997.

[172] Y. Ye, *Interior Point Algorithms; Theory and Analysis*, Wiley, New York, 1997.

[173] W. I. Zangwill, *Non-linear programming via penalty functions*, Management Science **13** (1967), 344–358.

Index